AF361957

Recent Progress in Robot Control Systems: Theory and Applications

Recent Progress in Robot Control Systems: Theory and Applications

Editors

Chengxi Zhang
Jin Wu
Chong Li

Basel • Beijing • Wuhan • Barcelona • Belgrade • Novi Sad • Cluj • Manchester

Editors
Chengxi Zhang
School of Internet of Things
Engineering
Jiangnan University
Wuxi
China

Jin Wu
Department of Electronic and
Computer Engineering
Hong Kong University of
Science and Technology
Hong Kong
China

Chong Li
College of Engineering
Ocean University of China
Qingdao
China

Editorial Office
MDPI
St. Alban-Anlage 66
4052 Basel, Switzerland

This is a reprint of articles from the Special Issue published online in the open access journal *Symmetry* (ISSN 2073-8994) (available at: https://www.mdpi.com/journal/symmetry/special_issues/Recent_Progress_Robot_Control_Systems_Theory_Applications).

For citation purposes, cite each article independently as indicated on the article page online and as indicated below:

Lastname, A.A.; Lastname, B.B. Article Title. *Journal Name* **Year**, *Volume Number*, Page Range.

ISBN 978-3-7258-1202-8 (Hbk)
ISBN 978-3-7258-1201-1 (PDF)
doi.org/10.3390/books978-3-7258-1201-1

Contents

About the Editors . . . . **vii**

Preface . . . . **ix**

Chengxi Zhang, Jin Wu and Chong Li
Recent Progress in Robot Control Systems: Theory and Applications
Reprinted from: *Symmetry* **2024**, *16*, 43, https://doi.org/10.3390/sym16010043 . . . . **1**

Xin Nie, Jun Gong, Jintao Cheng, Xiaoyu Tang and Yuanfang Zhang
Two-Step Self-Calibration of LiDAR-GPS/IMU Based on Hand-Eye Method
Reprinted from: *Symmetry* **2023**, *15*, 254, https://doi.org/10.3390/sym15020254 . . . . **7**

Xiaoyu Tang, Sirui Liu, Qiuchi Xiang, Jintao Cheng, Huifang He and Bohuan Xue
Facial Expression Recognition Based on Dual-Channel Fusion with Edge Features
Reprinted from: *Symmetry* **2022**, *14*, 2651, https://doi.org/10.3390/sym14122651 . . . . **22**

Cheng Chen, Zian Wang, Zheng Gong, Pengcheng Cai, Chengxi Zhang and Yi Li
Autonomous Navigation and Obstacle Avoidance for Small VTOL UAV in Unknown
Environments
Reprinted from: *Symmetry* **2022**, *14*, 2608, https://doi.org/10.3390/sym14122608 . . . . **34**

Guirong Wang, Jiahao Chen, Kun Zhou and Zhihui Pang
Industrial Robot Contouring Control Based on Non-Uniform Rational B-Spline Curve
Reprinted from: *Symmetry* **2022**, *14*, 2533, https://doi.org/10.3390/sym14122533 . . . . **79**

Hepeng Ni, Shuai Ji and Yingxin Ye
Redundant Posture Optimization for 6R Robotic Milling Based on
Piecewise-Global-Optimization-Strategy Considering Stiffness, Singularity and Joint-Limit
Reprinted from: *Symmetry* **2022**, *14*, 2066, https://doi.org/10.3390/sym14102066 . . . . **95**

Senlin Wang and Dangmin Nie
Flight Conflict Detection Algorithm Based on Relevance Vector Machine
Reprinted from: *Symmetry* **2022**, *14*, 1992, https://doi.org/10.3390/sym14101992 . . . . **109**

Zheng Gong, Zian Wang, Chengchuan Yang, Zhengxue Li, Mingzhe Dai and Chengxi Zhang
Performance Analysis on the Small-Scale Reusable Launch Vehicle
Reprinted from: *Symmetry* **2022**, *14*, 1862, https://doi.org/10.3390/sym14091862 . . . . **122**

Zan Zhou, Zian Wang, Zheng Gong, Xiong Zheng, Yang Yang and Pengcheng Cai
Design of Thrust Vectoring Vertical/Short Takeoff and Landing Aircraft Stability Augmentation
Controller Based on L1 Adaptive Control Law
Reprinted from: *Symmetry* **2022**, *14*, 1837, https://doi.org/10.3390/sym14091837 . . . . **149**

Shouwan Gao, Jianan He, Honghua Pan and Tao Gong
A Multi-Scale and Lightweight Bearing Fault Diagnosis Model with Small Samples
Reprinted from: *Symmetry* **2022**, *14*, 909, https://doi.org/10.3390/sym14050909 . . . . **163**

Mingjun Liu, Aihua Zhang and Bing Xiao
Velocity-Free State Feedback Fault-Tolerant Control for Satellite with Actuator and Sensor
Faults
Reprinted from: *Symmetry* **2022**, *14*, 157, https://doi.org/10.3390/sym14010157 . . . . **181**

Jianfeng Zheng, Shuren Mao, Zhenyu Wu, Pengcheng Kong and Hao Qiang
Improved Path Planning for Indoor Patrol Robot Based on Deep Reinforcement Learning
Reprinted from: *Symmetry* **2022**, *14*, 132, https://doi.org/10.3390/sym14010132 . . . . **199**

You Li, Haizhao Liang and Lei Xing
Finite-Time Controller for Flexible Satellite Attitude Fast and Large-Angle Maneuver
Reprinted from: *Symmetry* **2022**, *14*, 45, https://doi.org/10.3390/sym14010045 . . . . **211**

Guangying Jin and Guangzhe Jin
Fault-Diagnosis Sensor Selection for Fuel Cell Stack Systems Combining an Analytic Hierarchy
Process with the Technique Order Performance Similarity Ideal Solution Method
Reprinted from: *Symmetry* **2021**, *13*, 2366, https://doi.org/10.3390/sym13122366 . . . . **230**

Kunyi Jiang, Lei Mao, Yumin Su and Yuxin Zheng
Trajectory Tracking Control for Underactuated USV with Prescribed Performance and Input
Quantization
Reprinted from: *Symmetry* **2021**, *13*, 2208, https://doi.org/10.3390/sym13112208 . . . . **247**

Zian Wang, Shengchen Mao, Zheng Gong, Chi Zhang and Jun He
Energy Efficiency Enhanced Landing Strategy for Manned eVTOLs Using L_1 Adaptive Control
Reprinted from: *Symmetry* **2021**, *13*, 2125, https://doi.org/10.3390/sym13112125 . . . . **267**

About the Editors

Chengxi Zhang

Chengxi Zhang was born in Qufu, China. He received his B.S. and M.S. degrees from Harbin Institute of Technology (HIT), China, in 2012 and 2015, and his Ph.D. from Shanghai Jiao Tong University, China, in 2019. He is an associate professor at the School of Internet of Things Engineering, Jiangnan University, Wuxi, China. He is a session chair of the 2022 Jiangsu Annual Conference of Automation (JACA2022) and an invited session chair of the 36th Chinese Control and Decision Conference (CCDC2024, Xi'An, invited by Prof. Hanlin Dong); a program committee member and associate editor of FASTA2024, the Third Conference on Fully Actuated System Theory and Applications, Shenzhen; and the technical program committee for IEEE GEM 2024 (2024 IEEE Gaming, Entertainment, and Media Conference (GEM)). He serves as a guest editor for special issues on robotics and spacecraft control topics for several journals. He is an editorial board member of *IoT (Internet of Things)*, *Applied Math*, *AI and Autonomous Systems*, *Complex Engineering Systems*, *Astrodynamics* (Tsinghua University), and *Aerospace Systems* (Shanghai Jiao Tong University). He is an associate editor of *Frontiers in Aerospace Engineering*. He is listed in the World's Top 2% Scientists in 2022 by John P.A. Ioannidis, Stanford University, and Elsevier (ranked 173th in mainland China in aerospace).

Jin Wu

Jin Wu was born in Zhenjiang, China, in 1994. He received the B.S. degree from the University of Electronic Science and Technology of China, Chengdu, China, in 2016. He is currently pursuing a Ph.D. degree with the Department of Electronic and Computer Engineering at the Hong Kong University of Science and Technology (HKUST), Hong Kong. From 2013 to 2014, he was a visiting student with Groep T, Katholieke Universiteit Leuven (KU Leuven), Leuven, Belgium. He has co-authored over 120 technical papers in representative journals and conference proceedings. Mr. Wu was a committee member of the IEEE CoDIT conference in 2019; the special section chair of the IEEE ICGNC conference in 2021; the special session chair of the 2023 IEEE International Conference on Intelligent Transportation Systems (ITSC); the track chair of the 2024 IEEE International Conference on Consumer Electronics (ICCE); and the chair of the 2024 IEEE CTSoc Gaming, Entertainment, and Media (GEM) Conference. He was selected as the World's Top 2% Scientist by Stanford University and Elsevier in the 2020, 2021, and 2022 year rounds. He was awarded the Outstanding Reviewer Award for IEEE Transactions on Instrumentation and Measurement in 2021. He is currently a review editor for *Frontiers in Aerospace Engineering* and an invited guest editor for five special issues of MDPI. He is also a member of the IEEE Consumer Technology Society (CTSoc) as a committee member and a publication liaison.

Chong Li

Chong Li received the bachelor's and master's degrees from the Ocean University of China, Qingdao, China, in 2009 and 2012, respectively, and the Ph.D. degree from Auburn University, Auburn, AL, USA, in 2016. He was a postdoctoral research fellow with the integrated MEMS laboratory at the Georgia Institute of Technology between 2016 and 2018. He is currently a "Mountain Tai" endowed full professor and director of the microsystems and precision control laboratory at the Ocean University of China. His research interests include micro-systems and precision control, MEMS sensors, control theory and dynamics, and embedded systems.

Preface

In precision engineering, the inherent symmetry of actuators in electronic rotors, aircraft wings, and spacecraft flywheel structures stands as a testament to human ingenuity. Nevertheless, the idyllic symmetry is often marred by machining imperfections, rendering these meticulously controlled objects asymmetric. Moreover, external disturbances further disrupt their harmonious operation. It is imperative, therefore, that we devise algorithms of sophisticated design to uphold the integrity of these systems. The conception and implementation of control systems are pivotal for the assured safety of an array of mechanical constructs, encompassing space-faring vehicles, maritime robotics, and micro-mechanical systems. While contemporary research has laid a substantial foundation, there remains a pressing need for more nuanced, intelligent, and resource-efficient technologies in control system algorithms to parallel the burgeoning growth of the industry.

In "Recent Progress in Robot Control Systems: Theory and Applications," our aspiration is to curate a collection of original research and survey papers that mirror the cutting-edge advancements in both the theoretical underpinnings and methodological innovations that propel the evolution of control system design and its myriad applications.

We wish to express our gratitude to all the scholars and editors who have been involved in the content and construction of this reprint.

Chengxi Zhang, Jin Wu, and Chong Li
Guest Editors

Editorial

Recent Progress in Robot Control Systems: Theory and Applications

Chengxi Zhang [1,*], Jin Wu [2] and Chong Li [3]

1 School of Internet of Things Engineering, Jiangnan University, Wuxi 214082, China
2 Department of Electronic and Computer Engineering, Hong Kong University of Science and Technology, Hong Kong 999077, China; jwucp@connect.ust.hk
3 Department Automation and Measurement, Ocean University of China, Qingdao 266100, China; lichong7332@ouc.edu.cn
* Correspondence: dongfangxy@163.com

1. Introduction

Many engineering systems, such as electronic rotors, aircraft wings, and spacecraft flywheel structures, rely on the symmetry of their actuators. However, symmetry and asymmetry are not absolute in engineering science. Machining defects and external perturbations can introduce asymmetry to these controlled objects, compromising their performance and stability. Therefore, we need to design complex algorithms to preserve the symmetry of these elegant systems. Control system design is critical for the safe operation of various mechanical systems, such as space vehicles, maritime robotics, and micromechanical systems. Although current research has achieved remarkable results, there is still a high demand for more refined, intelligent, and low-resource-consumption technologies for control system algorithms to meet the industry's growth.

In this Special Issue, we collect original research and survey papers that reflect the recent advances in the theory and methodology of control system design and applications.

2. Noteworthy Aspects of the Special Issue

This Special Issue contains fifteen papers that cover the following aspects of recent progress in robot control systems: (1) Robotics Navigation and Control; (2) Aircraft/Spacecraft Systems; and (3) Reliable Designs. The following is a brief summary of the accepted papers. In addition, we will also present some other achievements that exist beyond this Special Issue. They will be included in Section 3.

2.1. Robotic Systems

Multi-line LiDAR and GPS/IMU are essential for autonomous driving and robotics such as SLAM. Multi-sensor fusion requires the calibration of each sensor's extrinsic parameters, which affect the vehicle's positioning control and perception performance. The algorithm obtains accurate extrinsic parameters and their confidence measures as a symmetric covariance matrix. Existing LiDAR-GPS/IMU calibration methods need specific vehicle motion or manual calibration scenes, leading to high costs and low automation. Ref. [1] proposes a new two-step self-calibration method: extrinsic parameter initialization and refinement. The initialization part decouples the rotation and translation parts of the extrinsic parameters and calculates the initial rotation by rotation constraints, and then the initial translation is calculated by a reliable initial rotation, and the LiDAR odometry drift is eliminated by loop closure to build the map. The refinement part obtains LiDAR odometry through scan-to-map registration and couples it tightly with the IMU. The absolute pose constraints in the map refine the extrinsic parameters.

In artificial intelligence, accomplishing emotion recognition in human–computer interaction is a key work. Expressions contain plentiful information about human emotion.

Citation: Zhang, C.; Wu, J.; Li, C. Recent Progress in Robot Control Systems: Theory and Applications. *Symmetry* **2024**, *16*, 43. https://doi.org/10.3390/sym16010043

Received: 5 December 2023
Accepted: 15 December 2023
Published: 28 December 2023

Ref. [2] found that the canny edge detector can help to significantly improve facial expression recognition performance. A canny edge-detector-based dual-channel network using the OI-network and EI-net is proposed, which does not add redundant network layers and training. The method was verified in CK+, Fer2013, and RafDb datasets and achieved a good result.

Ref. [3] presents a novel algorithm for the industrial robot contouring control based on the NURBS (non-uniform rational B-spline) curve. Ref. [4] presents redundant posture optimization for 6R robotic milling based on a piecewise global optimization strategy considering stiffness, singularity, and joint limit.

To solve the problems of the poor exploration ability and convergence speed of traditional deep reinforcement learning in the navigation task of the patrol robot under indoor specified routes, an improved deep reinforcement learning algorithm based on Pan/Tilt/Zoom (PTZ) image information was proposed in ref. [5]. The obtained symmetric image information and target position information are taken as the input of the network, the speed of the robot is taken as the output of the next action, and the circular route with the boundary is taken as the test. The improved reward and punishment function is designed to improve the convergence speed of the algorithm and optimize the path so that the robot can plan a safer path while avoiding obstacles first. Compared with the Deep Q Network (DQN) algorithm, the convergence speed after improvement is shortened by about 40%, and the loss function is more stable.

Ref. [6] is entitled "Trajectory Tracking Control for Underactuated USV with Prescribed Performance and Input Quantization".

2.2. Aircraft/Space Vehicle Systems

Ref. [7] takes autonomous exploration in unknown environments on a small co-axial twin-rotor unmanned aerial vehicle (UAV) platform as the task. The study of fully autonomous positioning in unknown environments and navigation systems without global navigation satellite systems (GNSS) and other auxiliary positioning means is carried out. Algorithms that are based on the machine vision/proximity detection/inertial measurement unit, namely the combined navigation algorithm and the indoor simultaneous location and mapping (SLAM) algorithm, are not only designed theoretically but also realized and verified in real surroundings. Additionally, obstacle detection, the decision-making of avoidance motion, and motion planning methods such as Octree are also proposed, which are characterized by randomness and symmetry.

In response to the problems of slow running speed and high error rates of traditional flight conflict detection algorithms, Ref. [8] proposes a conflict detection algorithm based on the use of a relevance vector machine. A set of symmetrical historical flight data was used as the training set of the model, and they used the SMOTE resampling method to optimize the training set. They obtained relatively symmetrical training data and trained it with the relevance vector machine, improving the kernels through an intelligent algorithm.

According to the symmetrical characteristics of a new type of Reusable Launch Vehicle (RLV) in the recovery phase, [9] studied the basic aerodynamic model data of Starship and the aerodynamic data with rudder deflection, and the causes of its aerodynamic coefficients are expounded. They analyzed its stability and maneuverability. According to the flying quality requirements, the lateral–directional model of Starship in the return phase at a high angle of attack is analyzed. Finally, they analyzed Starship's lateral heading stability and control deviation using the criterion and nonlinear open-loop simulations. The results show that the Starship has pitching and rolling stability, but it only has heading stability in some ranges of angle of attack, and there is no heading stability at a conventional large angle of attack. After the modal analysis and comparison of flight quality, the longitudinal long-period model of the starship degenerates into a real root and it is stable and convergent. The lateral heading roll mode is at level 2 flight quality, the helical mode is at level 1 flight quality, and the Dutch roll mode diverges, which needs to be stabilized and controlled later.

Ref. [10] is entitled "Thrust Vectoring Vertical/Short Takeoff and Landing Aircraft Stability Augmentation Controller Based on L1 Adaptive Control Law". Aiming at the conversion process of thrust vectoring vertical/short takeoff and landing (V/STOL) aircraft with a symmetrical structure in the transition stage of takeoff and landing, there is a problem with the coupling and redundancy of the control quantities. To solve this problem, a corresponding inner loop stabilization controller and control distribution strategy are designed. In this paper, a dynamic system model and a dynamic model are established. Based on the outer loop adopting the conventional nonlinear dynamic inverse control, an L1 adaptive controller is designed based on the model as the inner loop stabilization control to compensate for the mismatch and uncertainty in the system. The key feature of the L1 adaptive control architecture is ensuring robustness in the presence of fast adaptation, to achieve a unified performance boundary in transient and steady-state operations, thus eliminating the need for adaptive rate gain scheduling.

A new landing strategy is presented in ref. [11] for manned electric vertical takeoff and landing (eVTOL) vehicles, using a roll maneuver to obtain a trajectory in the horizontal plane. This strategy rejects the altitude surging in the landing process, which is the fatal drawback of the conventional jumping strategy. The strategy leads to a smoother transition from the wing-borne mode to the thrust-borne mode, and has a higher energy efficiency, meaning a better flight experience and higher economic performance. To employ the strategy, a five-stage maneuver is designed using the lateral maneuver instead of longitudinal climbing. Additionally, a control system based on L1 adaptive control theory is designed to assist manned driving or execute flight missions independently, consisting of the guidance logic, stability augmentation system, and flight management unit. The strategy is verified with the ET120 platform by Monte Carlo simulation for robustness and safety performance, and an experiment was performed to compare the benefits with conventional landing strategies. The results show that the performance of the control system is robust enough to reduce perturbation by at least 20% in all modeling parameters, and ensures consistent dynamic characteristics between different flight modes. Additionally, the strategy successfully avoids climbing during the landing process with a smooth trajectory and reduces the energy consumed for landing by 64%.

2.3. Reliable Designs

Deep-learning-based methods have been widely used in fault diagnosis to improve diagnosis efficiency and intelligence. However, most schemes require a great deal of labeled data and many iterations for training parameters. They suffer from low accuracy and overfitting under the few-shot scenario. In addition, many parameters in the model consume high computing resources, which is far from practical. In ref. [12], a multi-scale and lightweight Siamese network architecture is proposed for fault diagnosis with few samples. The architecture proposed contains two main modules. The first part implements the feature vector extraction of sample pairs. It is composed of two lightweight convolutional networks with symmetrical shared weights. Multi-scale convolutional kernels and dimensionality reduction are used in these two symmetric networks to improve feature extraction and reduce the total number of model parameters. The second part takes charge of calculating the similarity of two feature vectors to achieve fault classification. Multiple datasets with different loads and speeds validate the proposed network. The results show that the model has better accuracy, fewer model parameters, and a scale compared to the baseline approach through our experiments.

Ref. [13] is entitled "Velocity-Free State Feedback Fault-Tolerant Control for Satellites with Actuator and Sensor Faults", Ref. [14] is "Finite-Time Controllers for Flexible Satellite Attitude Fast and Large-Angle Maneuver", and ref. [15] is "Fault-Diagnosis Sensor Selection for Fuel Cell Stack Systems Combining an Analytic Hierarchy Process with the Technique Order Performance Similarity Ideal Solution Method".

3. Exploring Further Advances in Intelligent Robot Control

Furthermore, we are pleased to introduce several noteworthy recent studies for scholarly consideration. While they may not comprehensively showcase all the achievements in robot control, they each possess certain commendable attributes.

3.1. Spacecraft Robot Control

Results in the control of spacecraft and robots: Ref. [16] proposes a control Lyapunov-barrier function-based controller for the stabilization of the spacecraft attitude tracking error system by combining a designed control barrier function and an existing control Lyapunov function. Ref. [17] investigates the distributed predefined-time attitude coordination control problem for multiple rigid spacecraft. Ref. [18] investigates the trajectory tracking control for a new type of cable-driven large space manipulator via sliding mode control techniques.

3.2. Control of Switching Systems

Results in the stability analysis and controller design of switching systems based on matrix equations: Refs. [19,20] propose an explicit iterative algorithm and an implicit iterative algorithm for solving the coupled Lyapunov matrix equation related to the stability analysis of continuous-time Markovian jump linear systems, respectively. Ref. [21] develops some convergence conditions for the multiple tuning parameters iterative algorithm and the single tuning parameter iterative algorithm, which is proposed to solve the discrete periodic Lyapunov matrix equations related to the control design of discrete-time linear periodic systems.

3.3. Self-Learning Control: A Simple and Effective Design

Self-learning control, originally referred to as online learning control, was initially proposed in [22], providing the first method for stability proof. Its most prominent attribute lies in its simple structure and practical applicability. In [22], it solely utilizes a straightforward proportional-derivative (PD) algorithm as the update term, complemented by a learning term with a fixed learning intensity. This approach achieves a remarkably high control precision and response speed while maintaining a simplified form of the control algorithm. In follow-up research [23], the authors further proposed a control algorithm with variable learning intensity to mitigate the controller saturation response. Additionally, in another investigation [24], the authors employed a variable learning intensity approach to enhance a learning observer, significantly reducing the initial transient oscillations during state and parameter estimation. Due to its user-friendly simplicity, the self-learning control method was swiftly applied to gyroscope systems [25], attaining state-of-the-art performance in certain metrics of gyroscope control systems.

3.4. State Estimation of Robotic Systems

Accurately retrieving information about the condition of a control system, particularly within the domain of aircraft and robotics, necessitates the utilization of state estimation as a highly effective approach for minimizing noise influence. In [26,27], the employment of FIR-smoothing methods has resulted in notable enhancements in estimation performance concerning observation data that exhibit time delays. Additionally, the destabilizing effects of faulty signals on controller stability are estimated by leveraging the mean-field theory described in [28,29]. To fortify resistance against disturbances, the methods proposed in [30,31] leverage Bayesian inference to capture the inherent randomness of unknown signals. For the localization of the underground pipe jacking machine, [32] designed a reliable, real-time, and robust INS/OD solution.

3.5. Trends in Control Systems for Micro Gyroscopes

Another notable trend would be focused on micro-electro-mechanical (MEMS) gyroscopes, which are an essential inertial sensor for robotic navigation and attitude control [33]. They may be considered to be black-boxed sensors within the field of robotics, but they are inherently complicated and self-sustained closed loop control systems [25]. By examining the operation principles, it can be found that there are typically more than four control loops in a single working MEMS gyro chip. The control plant would be a pair of orthogonal resonators coupled with the Coriolis effect along with frequency tracking, amplitude adjustment, quadrature nulling, and force-to-rebalance controllers [34]. Improving the fabrication quality of gyro devices is a straightforward way to enhance the sensing metrics, but optimizing control systems is the advanced approach to break the performance barrier [35]. By switching empirical controller designs to model-based force-to-rebalance controllers, the bandwidth can be magnified by $5\times\sim10\times$ [36]. Self-calibration architectures that observe the gyro parametric errors and calibrations made using electrical stimulus are another implementation of advanced control systems [37], whether they are acknowledged by the MEMS designers or not. In any case, advanced control systems will play an essential role in the field of inertial sensors.

Author Contributions: Conceptualization, C.Z.; writing—original draft preparation, C.Z.; writing—review and editing, J.W. and C.L.; funding acquisition, C.Z. All authors have read and agreed to the published version of the manuscript.

Funding: This work is partially supported by the Fundamental Research Funds for the Central Universities (No. JUSRP123063), 111 Project (B23008), National Natural Science Foundation of China (No. 62003112).

Acknowledgments: We would like to thank *Symmetry* and the Editor-in-Chief for the support they have given us.

Conflicts of Interest: The authors declare no conflict of interest.

References

1. Nie, X.; Gong, J.; Cheng, J.; Tang, X.; Zhang, Y. Two-Step Self-Calibration of LiDAR-GPS/IMU Based on Hand-Eye Method. *Symmetry* **2023**, *15*, 254. [CrossRef]
2. Tang, X.; Liu, S.; Xiang, Q.; Cheng, J.; He, H.; Xue, B. Facial Expression Recognition Based on Dual-Channel Fusion with Edge Features. *Symmetry* **2022**, *14*, 2651. [CrossRef]
3. Wang, G.; Chen, J.; Zhou, K.; Pang, Z. Industrial Robot Contouring Control Based on Non-Uniform Rational B-Spline Curve. *Symmetry* **2022**, *14*, 2533. [CrossRef]
4. Ni, H.; Ji, S.; Ye, Y. Redundant Posture Optimization for 6R Robotic Milling Based on Piecewise-Global-Optimization-Strategy Considering Stiffness, Singularity and Joint-Limit. *Symmetry* **2022**, *14*, 2066. [CrossRef]
5. Zheng, J.; Mao, S.; Wu, Z.; Kong, P.; Qiang, H. Improved Path Planning for Indoor Patrol Robot Based on Deep Reinforcement Learning. *Symmetry* **2022**, *14*, 132. [CrossRef]
6. Jiang, K.; Mao, L.; Su, Y.; Zheng, Y. Trajectory Tracking Control for Underactuated USV with Prescribed Performance and Input Quantization. *Symmetry* **2021**, *13*, 2208. [CrossRef]
7. Chen, C.; Wang, Z.; Gong, Z.; Cai, P.; Zhang, C.; Li, Y. Autonomous Navigation and Obstacle Avoidance for Small VTOL UAV in Unknown Environments. *Symmetry* **2022**, *14*, 2608. [CrossRef]
8. Wang, S.; Nie, D. Flight Conflict Detection Algorithm Based on Relevance Vector Machine. *Symmetry* **2022**, *14*, 1992. [CrossRef]
9. Gong, Z.; Wang, Z.; Yang, C.; Li, Z.; Dai, M.; Zhang, C. Performance Analysis on the Small-Scale Reusable Launch Vehicle. *Symmetry* **2022**, *14*, 1862. [CrossRef]
10. Zhou, Z.; Wang, Z.; Gong, Z.; Zheng, X.; Yang, Y.; Cai, P. Design of Thrust Vectoring Vertical/Short Takeoff and Landing Aircraft Stability Augmentation Controller Based on L1 Adaptive Control Law. *Symmetry* **2022**, *14*, 1837. [CrossRef]
11. Wang, Z.; Mao, S.; Gong, Z.; Zhang, C.; He, J. Energy Efficiency Enhanced Landing Strategy for Manned eVTOLs Using L1 Adaptive Control. *Symmetry* **2021**, *13*, 2125. [CrossRef]
12. Gao, S.; He, J.; Pan, H.; Gong, T. A Multi-Scale and Lightweight Bearing Fault Diagnosis Model with Small Samples. *Symmetry* **2022**, *14*, 909. [CrossRef]
13. Liu, M.; Zhang, A.; Xiao, B. Velocity-Free State Feedback Fault-Tolerant Control for Satellite with Actuator and Sensor Faults. *Symmetry* **2022**, *14*, 157. [CrossRef]
14. Li, Y.; Liang, H.; Xing, L. Finite-Time Controller for Flexible Satellite Attitude Fast and Large-Angle Maneuver. *Symmetry* **2022**, *14*, 45. [CrossRef]
15. Jin, G.; Jin, G. Fault-Diagnosis Sensor Selection for Fuel Cell Stack Systems Combining an Analytic Hierarchy Process with the Technique Order Performance Similarity Ideal Solution Method. *Symmetry* **2021**, *13*, 2366. [CrossRef]
16. Wu, Y.-Y.; Sun, H.-J. Attitude Tracking Control with Constraints for Rigid Spacecraft Based on Control-Barrier Lyapunov Functions. *IEEE Trans. Aerosp. Electron. Syst.* **2022**, *58*, 2053–2062. [CrossRef]

17. Sun, H.-J.; Wu, Y.-Y.; Zhang, J. A Distributed Predefined-Time Attitude Coordination Control Scheme for Multiple Rigid Spacecraft. *Aerosp. Sci. Technol.* **2023**, *133*, 108134. [CrossRef]
18. Meng, D.; Xu, H.; Xu, H.; Sun, H.-J.; Liang, B. Trajectory Tracking Control for a Cable-driven Space Manipulator using Time-delay Estimation and Nonsingular Terminal Sliding Mode. *Control. Eng. Pract.* **2023**, *139*, 105649. [CrossRef]
19. Sun, H.-J.; Zhang, Y.; Liu, W. Explicit Iterative Algorithms for Continuous Coupled Lyapunov Matrix Equations. *IEEE Trans. Autom. Control.* **2020**, *65*, 3631–3638. [CrossRef]
20. Wu, A.-G.; Sun, H.-J.; Zhang, Y. An SOR Implicit Iterative Algorithm for Coupled Lyapunov Equations. *Automatica* **2018**, *97*, 38–47. [CrossRef]
21. Zebin, C.; Xuesong, C.; Sun, H.-J. Some Convergence Properties of Two Iterative Algorithms for Discrete Periodic Lyapunov Equations. *IEEE Trans. Autom. Control.* **2023**, *68*, 6751–6757.
22. Zhang, C.; Xiao, B.; Wu, J.; Li, B. On low-complexity control design to spacecraft attitude stabilization: An online-learning approach. *Aerosp. Sci. Technol.* **2021**, *110*, 106441. [CrossRef]
23. Zhang, C.; Ahn, C.K.; Wu, J.; He, W. Online-learning control with weakened saturation response to attitude tracking: A variable learning intensity approach. *Aerosp. Sci. Technol.* **2021**, *117*, 106981. [CrossRef]
24. Zhang, C.; Ahn, C.K.; Wu, J.; He, W.; Jiang, Y.; Liu, M. Robustification of learning observers to uncertainty identification via time-varying learning intensity. *IEEE Trans. Circuits Syst. II Express Briefs.* **2021**, *69*, 1292–1296. [CrossRef]
25. Li, C.; Wang, Y.; Ahn, C.K.; Zhang, C.; Wang, B. Milli-Hertz Frequency Tuning Architecture Toward High Repeatable Micromachined Axi-Symmetry Gyroscopes. *IEEE Trans. Ind. Electron.* **2022**, *70*, 6425–6434. [CrossRef]
26. Zhao, S.; Shmaliy, Y.S.; Liu, F. Batch Optimal FIR Smoothing: Increasing State Informativity in Nonwhite Measurement Noise Environments. *IEEE Trans. Ind. Inform.* **2022**, *19*, 6993–7001. [CrossRef]
27. Zhao, S.; Wang, J.; Shmaliy, Y.S.; Liu, F. Discrete Time q-Lag Maximum Likelihood FIR Smoothing and Iterative Recursive Algorithm. *IEEE Trans. Signal Process.* **2021**, *69*, 6342–6354. [CrossRef]
28. Zhao, S.; Li, K.; Ahn, C.K.; Huang, B.; Liu, F. Tuning-Free Bayesian Estimation Algorithms for Faulty Sensor Signals in State-Space. *IEEE Trans. Ind. Electron.* **2023**, *70*, 921–929. [CrossRef]
29. Zhao, S.; Huang, B.; Zhao, C. Online Probabilistic Estimation of Sensor Faulty Signal in Industrial Processes and Its Applications. *IEEE Trans. Ind. Electron.* **2021**, *68*, 8853–8862. [CrossRef]
30. Zhang, T.; Zhao, S.; Luan, X.; Liu, F. Bayesian Inference for State-Space Models with Student-t Mixture Distributions. *IEEE Trans. Cybern.* **2022**, *53*, 4435–4445. [CrossRef]
31. Zhao, S.; Huang, B. Trial-and-error or avoiding a guess? Initialization of the Kalman filter. *Automatica* **2020**, *121*, 109184. [CrossRef]
32. Zhao, S.; Zhou, Z.; Zhang, C.; Wu, J.; Liu, F.; Shi, G. Localization of underground pipe jacking machinery: A reliable, real-time and robust INS/OD solution. *Control. Eng. Pract.* **2023**, *141*, 105711. [CrossRef]
33. Wang, Y.; Cao, R.; Li, C.; Dean, R.N. Concepts, roadmaps and challenges of ovenized MEMS gyroscopes: A review. *IEEE Sens. J.* **2020**, *21*, 92–119. [CrossRef]
34. Wang, Y.; Hou, J.; Li, C.; Wu, J.; Jiang, Y.; Liu, M.; Hung, J.Y. Ultrafast Mode Reversal Coriolis Gyroscopes. *IEEE/ASME Trans. Mechatron.* **2022**, *27*, 5969–5980. [CrossRef]
35. Qi, Z.; Wu, J.; Li, C.; Zhao, W.; Cheng, Y. Micromachined rate-integrating gyroscopes: Concept, asymmetry error sources and phenomena. *Symmetry* **2020**, *12*, 801. [CrossRef]
36. Li, C.; Wen, H.; Chen, H.; Ayazi, F. A digital force-to-rebalance scheme for high-frequency bulk-acoustic-wave micro-gyroscopes. *Sens. Actuators A Phys.* **2020**, *313*, 112181. [CrossRef]
37. Li, C.; Wen, H.; Wisher, S.; Norouzpour-Shirazi, A.; Lei, J.Y.; Chen, H.; Ayazi, F. An FPGA-based interface system for high-frequency bulk-acoustic-wave microgyroscopes with in-run automatic mode-matching. *IEEE Trans. Instrum. Meas.* **2019**, *69*, 1783–1793. [CrossRef]

Article

Two-Step Self-Calibration of LiDAR-GPS/IMU Based on Hand-Eye Method

Xin Nie [1,*,†], Jun Gong [1,†], Jintao Cheng [2], Xiaoyu Tang [2] and Yuanfang Zhang [3]

[1] State Key Laboratory of Advanced Design and Manufacturing for Vehicle Body, Hunan University, Changsha 410082, China
[2] School of Physics and Telecommunication Engineering, South China Normal University, Guangzhou 510006, China
[3] The Autocity (Shenzhen) Autonomous Driving Co., Ltd., Shenzhen 518000, China
* Correspondence: niexinpiero@163.com
† These authors contributed equally to this work.

Abstract: Multi-line LiDAR and GPS/IMU are widely used in autonomous driving and robotics, such as simultaneous localization and mapping (SLAM). Calibrating the extrinsic parameters of each sensor is a necessary condition for multi-sensor fusion. The calibration of each sensor directly affects the accurate positioning control and perception performance of the vehicle. Through the algorithm, accurate extrinsic parameters and a symmetric covariance matrix of extrinsic parameters can be obtained as a measure of the confidence of the extrinsic parameters. As for the calibration of LiDAR-GPS/IMU, many calibration methods require specific vehicle motion or manual calibration marking scenes to ensure good constraint of the problem, resulting in high costs and a low degree of automation. To solve this problem, we propose a new two-step self-calibration method, which includes extrinsic parameter initialization and refinement. The initialization part decouples the extrinsic parameters from the rotation and translation part, first calculating the reliable initial rotation through the rotation constraints, then calculating the initial translation after obtaining a reliable initial rotation, and eliminating the accumulated drift of LiDAR odometry by loop closure to complete the map construction. In the refinement part, the LiDAR odometry is obtained through scan-to-map registration and is tightly coupled with the IMU. The constraints of the absolute pose in the map refined the extrinsic parameters. Our method is validated in the simulation and real environments, and the results show that the proposed method has high accuracy and robustness.

Keywords: autonomous driving; robot control; LiDAR-GPS/IMU; hand-eye calibration; sensor fusion

1. Introduction

In autonomous driving and robotics, commonly used sensors are LiDAR, camera, and GPS/IMU. The LiDAR market has broad prospects. It is predicted that by 2025, China's LiDAR market will be close to USD 2.175 billion (equal to CNY 15 billion), and the global market will be close to USD 4.35 billion (equal to CYN 30 billion). By 2030, China's LiDAR market will be close to USD 5.075 billion (equal to CYN 35 billion), and the global market will be close to USD 9.425 billion (equal to CYN 65 billion). The annual growth rate of the global market will reach 48.3% (http://www.evinchina.com/newsshow-2094.html, accessed on 2 January 2023). The annual growth rate of vehicle camera modules from 2018 to 2023 is about 15.8%. It is estimated that the annual sales in 2023 will reach USD 5.8 billion (about CYN 39.44 billion), of which robots and cars will account for 75% (https://www.yoojia.com/article/10096726019901689110.html, accessed on 2 January 2023). GPS/IMU is a comprehensive system combining inertial measurement unit (IMU) and global positioning system (GPS), which can provide centimeter-level absolute localization accuracy. A single IMU can only offer a relatively accurate motion estimation in a short time, which depends on the hardware performance of the IMU. The localization error increases rapidly with time.

Citation: Nie, X.; Gong, J.; Cheng, J.; Tang, X.; Zhang, Y. Two-Step Self-Calibration of LiDAR-GPS/IMU Based on Hand-Eye Method. *Symmetry* **2023**, *15*, 254. https://doi.org/10.3390/sym15020254

Academic Editor: Jan Awrejcewicz

Received: 21 December 2022
Revised: 11 January 2023
Accepted: 13 January 2023
Published: 17 January 2023

LiDAR (light detection and ranging) is the most critical ranging sensor, which can estimate the relative motion of vehicles in real-time through point cloud registration. LiDAR's ranging calculation is complex and often needs to be accelerated in a parallel settlement [1], and due to the high cost, FGPA is used as a part of the data processing module to reduce costs [2]. The SLAM based on LiDAR will degenerate where the geometric feature information is sparse. The camera projects the three-dimensional scene onto the two-dimensional image, through which we can obtain visual odometry. However, the visual odometry has scale ambiguity, and the image is also affected by the exposure rate and the light. In order to integrate the advantages of the sensors to achieve better robustness, many tasks need more than one kind of sensor. Multi-sensor fusion has become a trend [3–8]. The study in [3] proposes a pipeline inspection and retrofitting based on LiDAR and camera fusion for an AR system. In [4], for the loss of visual feature tracking of mobile robots, a feature matching method is designed based on a camera and IMU to improve robustness. Efficiently fusing different attributes of the environment captured by the two sensors facilitates a reliable and consistent perception of the environment. In [5], a method for estimating the distance between an autonomous vehicle and other vehicles, objects, and signboards on its path using the accurate fusion of LiDAR and camera is proposed. It can be seen that many tasks tend to use multi-sensor fusion to complete the task. However, since the coordinate systems of the data collected by each sensor are different, we need to unify the coordinate systems of different sensors. This requires different types of data collected by sensors to calibrate the intrinsic parameters of a single sensor and the extrinsic parameters of multiple sensors. For example, Refs. [9–12] are calibration works about multi-LiDAR extrinsic parameters. Refs. [13–15] are calibration works about LiDAR-camera extrinsic parameters. Refs. [16–18] are calibration works about LiDAR-IMU extrinsic parameters. There are few calibration works about LiDAR-GPS/IMU extrinsic parameters. The calibration of LiDAR-GPS/IMU is to complete the calibration of extrinsic parameters of LiDAR and IMU rotation and LiDAR and GPS translation.

Since the intrinsic parameters of IMU and LiDAR are usually provided by the manufacturer at the factory, we mainly solve the calibration of extrinsic parameters between LiDAR and GPS/IMU. At present, there are many challenges in the calibration of extrinsic parameters of LiDAR and GPS/IMU. For example, the movement of the vehicle in automatic driving is not like the movement of the robot arm. The movement of the vehicle is mainly the plane movement of three degrees of freedom, which has fewer constraints on the other three degrees of freedom, so the error of the other three degrees of freedom is relatively large. Due to the large difference in the measurement principle between radar and GPS/IMU, to detect the same object in different sensors for calibrating extrinsic parameters between them, many calibration methods require specific vehicle operation and manual calibration marking scenes, resulting in high cost and low degree of automation.

The motion-based method is the primary method for calibration between LiDAR and GPS/IMU. Geiger et al. [19] proposed to use a hand–eye calibration method to constrain GPS/IMU odometry and obtain LiDAR odometry through point-plane registration. However, it does not have an initial extrinsic parameter calculation, so its accuracy depends on the high-precision registration of the LiDAR odometry. Hand–eye calibration utilizes the hand-eye model proposed in the field of robotics. It is used to solve the rigid transformation between the manipulator and the camera rigidly mounted above, and this problem can be extended to solve the problem of relatively rigid transformation between any two sensors [20,21]. Therefore, we can also use it to solve the rigid transformation between LiDAR and GPS/IMU. Hand–eye calibration mainly solves the equation AX = XB, where A and B are the odometry of two different sensors, respectively, and X is the extrinsic parameters between the two sensors. In [22], the ground is used to solve the Z-axis translation, roll, and pitch transformation, and then the three-dimensional problem is converted into a two-dimensional problem. The remaining three degrees of freedom are estimated by hand–eye calibration. It can be regarded as taking the three degrees of freedom optimized on the ground as the initial value and then performing hand–eye calibration. The initial value calculation is too simple and rough, there are manual operation errors, and the obtained accuracy is relatively low. Baidu's open-source Apollo project divides the calibration process into two steps. First, calculate the initial calibration parameters through the trajectory and then refine the calibration parameters through the feature-based method. Our calibration method shares the same idea, which is to complete the calibration parameters from the coarse to the fine, but the details of the method implementation are different. We adopt more strategies, such as loop closure detection and tight coupling strategies, so the accuracy and robustness are relatively better. In [23], in order to solve the problem of missing constraints caused by the plane motion of vehicle motion, large-range and small-range trajectories are used to solve the extrinsic parameters of rotation and translation, respectively. However, large-range trajectory recording requires a large labor cost and is demanding; therefore, the method is not universal. The advantages and disadvantages of related work are shown in Table 1.

Table 1. The advantages and disadvantages of the related work.

Methods	Advantages	Disadvantages
Geiger et al.	Hand–eye calibration method is proposed	no initial extrinsic parameters
Chen, C et al.	Group calibration	manual operation, the accuracy is low
Baidu Apollo project	Calibrate from coarse to fine	no loop closure detection
Xuan, Y et al.	The large-range trajectories strategy	harsh conditions, not universal

This paper proposes a new two-step self-calibration method, which includes extrinsic parameters initialization and refinement. The initialization part decouples the extrinsic parameters from the rotation and translation part by first calculating the reliable initial rotation through the rotation constraints, then calculating the initial translation after obtaining a reliable initial rotation, and eliminating the accumulated drift of LiDAR odometry by loop closure to complete the map construction. In the refinement part, the LiDAR odometry is obtained through scan-to-map registration and is tightly coupled with the IMU. The constraints of the absolute pose in the map refined the extrinsic parameters. The main contributions of our work can be summarized as follows:

- A new LiDAR and GPS/IMU calibration system can be calibrated daily in an open natural environment and has good universality.
- Like the common localization problem, we divide the calibration problem into two steps and adopt the idea from coarse to fine to make the extrinsic parameters have better robustness and accuracy and effectively eliminate the influence of accumulated drift on the extrinsic parameters.
- The proposed calibration system is verified in both a simulation environment and a real environment.

The rest of the paper is organized as follows: Section 2 is an overview of the system, which describes the core calibration algorithm of the LiDAR-GPS/IMU calibration system, which is divided into two parts to introduce our proposed algorithm and provide a detailed introduction to the algorithm. Section 3 is the experimental results of the simulation and real environment. Section 4 is the summary of this paper and future work. Table 2 is all the abbreviations used in this article.

Table 2. Table of abbreviations.

Abbreviation	Meaning
LiDAR	Light Detection and Ranging
GPS	Global Positioning System
IMU	Inertial Measurement Unit
SVD	Singular Value Decomposition
ICP	Iterative Closest Points
GICP	Generalized ICP
LOAM	Lidar Odometry and Mapping in Real-Time
L-M	Levenberg–Marquard

2. Methods

This section introduces the architecture of the proposed LiDAR-GPS/IMU calibration system, as shown in Figure 1. The system is mainly divided into two parts: parameter initialization and parameter refinement. In the parameter initialization part, the feature-based LiDAR odometry and the interpolated GPS/IMU relative pose were used to construct the hand–eye calibration problem to solve the initial extrinsic parameters and construct the map. In the parameter refinement part, LiDAR and IMU are tightly coupled by the initial extrinsic parameters, and the extrinsic parameters were refined through the constraints of the absolute pose in the map.

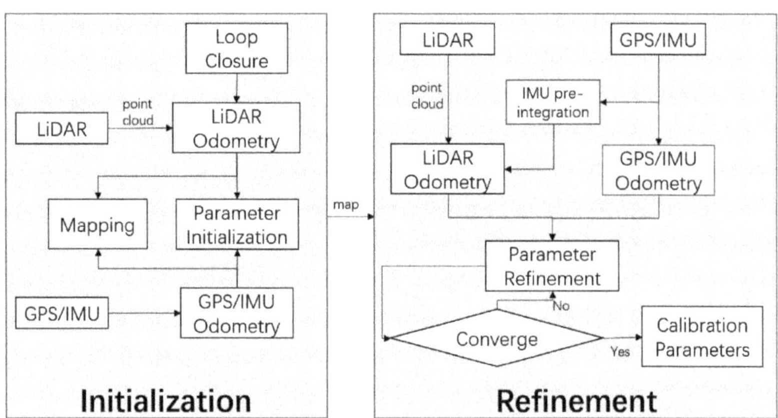

Figure 1. The pipeline of the LiDAR-GPS/IMU calibration system is presented in this paper. In the parameter initialization part, the feature-based LiDAR odometry and the interpolated GPS/IMU relative pose were used to construct the hand–eye calibration problem to solve the initial extrinsic parameters and construct the map. In the parameters refinement part, the initial extrinsic parameters were tightly coupled with the LiDAR and IMU, and the extrinsic parameters were refined through the constraints of the absolute pose in the map. When the relative change in the extrinsic parameters is less than the set threshold during the iterative refinement process, it is considered that the extrinsic parameters are sufficiently convergent, and the refinement ends.

The core of the extrinsic parameter initialization is a hand–eye calibration algorithm. When the vehicle is in motion, Figure 2 shows the relationship between relative and absolute pose during hand–eye calibration of LiDAR and GPS/IMU, where the $\{W\}$ is the first frame of the GPS/IMU, and the $\{L_k\}$ and $\{I_k\}$ represent the kth frame of the LiDAR and the GPS/IMU frame obtained by interpolation at this time, respectively. $\mathbf{T}_{I_k}^{W}$ is the absolute pose of $\{I_k\}$ relative to $\{W\}$. $\mathbf{T}_{I_{k+1}}^{I_k}$ is the relative pose from $\{I_{k+1}\}$ to $\{I_k\}$. $\mathbf{T}_{L_{k+1}}^{L_k}$ is the relative pose from $\{\mathbf{L_{k+1}}\}$ to $\{\mathbf{L_k}\}$. It is easy to see from Figure 2 that we have two ways to obtain the relative pose from $\{\mathbf{L_{k+1}}\}$ to $\{\mathbf{I_k}\}$, so we can obtain the formula as follows:

$$\mathbf{T}_{I_{k+1}}^{I_k}\mathbf{T}_L^I = \mathbf{T}_L^I\mathbf{T}_{L_{k+1}}^{L_k} \tag{1}$$

Equation (1) constitutes the equation AX = XB for hand–eye calibration. The principle of hand–eye calibration will be introduced in detail below.

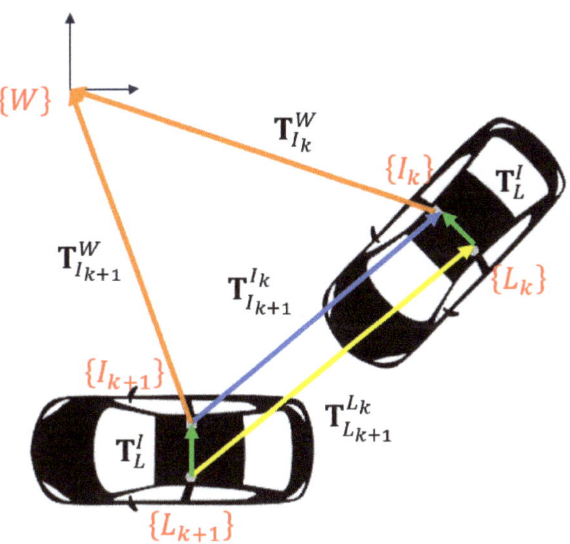

Figure 2. This figure shows the pose relationship of hand–eye calibration. We use $\{\mathbf{W}\}$ as the world coordinate system for mapping. Hand–eye calibration is mainly the relationship between extrinsic parameter $\mathbf{T}_L^I$ and two relative poses $\mathbf{T}_{I_{k+1}}^{I_k}$ and $\mathbf{T}_{L_{k+1}}^{L_k}$, which denote the relative pose from $\{\mathbf{I_{k+1}}\}$ to $\{\mathbf{I_k}\}$ and $\{\mathbf{L_{k+1}}\}$ to $\{\mathbf{L_k}\}$, respectively.

2.1. Hand–Eye Calibration

Equation (1) is usually decoupled into two parts according to [24]: rotation and translation. As the vehicle moves, the following equation holds for any k:

$$\mathbf{R}_{I_{k+1}}^{I_k}\mathbf{R}_L^I = \mathbf{R}_L^I\mathbf{R}_{L_{k+1}}^{L_k} \tag{2}$$

$$\left(\mathbf{R}_{I_{k+1}}^{I_k} - \mathbf{I}_3\right)\mathbf{t}_L^I = \mathbf{R}_L^I\mathbf{t}_{L_{k+1}}^{L_k} - \mathbf{t}_{I_{k+1}}^{I_k} \tag{3}$$

2.1.1. Extrinsic Rotation

Equation (2) is expressed by a quaternion as follows:

$$\begin{aligned}
\mathbf{q}_{I_{k+1}}^{I_k} \otimes \mathbf{q}_L^I &= \mathbf{q}_L^I \otimes \mathbf{q}_{L_{K+1}}^{L_k} \\
\Rightarrow \left(\left[\mathbf{q}_{I_{k+1}}^{I_k}\right]_L - \left[\mathbf{q}_{I_{k+1}}^{I_k}\right]_R\right)\mathbf{q}_L^I &= \mathbf{Q}_{k+1}^k\mathbf{q}_L^I
\end{aligned} \tag{4}$$

where $\otimes$ is the quaternion multiplication operator, and $\left[\mathbf{q}_{I_{k+1}}^{I_k}\right]_L$ and $\left[\mathbf{q}_{I_{k+1}}^{I_k}\right]_R$ are the matrix representation of the left and right quaternion multiplication, respectively [25]. After stacking measurements at different times, we obtain an over-determined equation as follows:

$$
\begin{bmatrix}
w_2^1 \cdot \mathbf{Q}_2^1 \\
\vdots \\
w_{k+1}^k \cdot \mathbf{Q}_{k+1}^k
\end{bmatrix}_{4K\times 4}
\mathbf{q}_L^I = \mathbf{Q}_K \mathbf{q}_L^I = \mathbf{0}_{4K\times 4}
\tag{5}
$$

where K is the number of rotation pairs of the over-determined equation, and w_{k+1}^k is the robust weight to better handle outliers. The angle of the current rotation pair difference in the angular axis is calculated by Equation (4) and taken as the parameters of Huber loss, whose derivative is the weight:

$$
\begin{aligned}
w_{k+1}^k &= \rho'(\theta), \quad \theta = 2\arccos(q_w) \\
\mathbf{q} &= (\mathbf{q}_{I_{k+1}}^{I_k} \otimes \mathbf{q}_L^I)^* \otimes \mathbf{q}_L^I \otimes \mathbf{q}_{L_{K+1}}^{L_k}
\end{aligned}
\tag{6}
$$

where $\rho()$ is Huber loss, q_w is the real part of the quaternion $\mathbf{q}$, and $()^*$ means taking the inverse of the quaternion. We use SVD to solve the over-determined Equation (5), whose closed solution is the right unit singular vector corresponding to the minimum singular value. Meanwhile, according to [26], to ensure sufficient rotation constraints, we need to ensure that the second smallest singular value is large enough, which needs to be larger than the artificial threshold. With the rapid increase of $\mathbf{Q}_{k+1}^k$, the priority queue is used to remove the constraint with the smallest rotation. Until the second smallest singular value exceeds the threshold, we can obtain a reliable initial extrinsic rotation.

2.1.2. Extrinsic Translation

When the extrinsic rotation $\hat{\mathbf{R}}_L^I$ is obtained, translation can be solved by the least square approach according to Equation (3).

$$
\begin{bmatrix}
\mathbf{R}_{I_2}^{I_1} - \mathbf{I}_3 \\
\vdots \\
\mathbf{R}_{I_{k+1}}^{I_k} - \mathbf{I}_3
\end{bmatrix}_{3K\times 3}
\mathbf{t}_L^I =
\begin{bmatrix}
\hat{\mathbf{R}}_L^I \mathbf{t}_{L_2}^{L_1} - \mathbf{t}_{I_2}^{I_1} \\
\vdots \\
\hat{\mathbf{R}}_L^I \mathbf{t}_{L_{k+1}}^{L_k} - \mathbf{t}_{I_{k+1}}^{I_k}
\end{bmatrix}_{3K\times 1}
\tag{7}
$$

However, the vehicle motion is usually a three degrees of freedom plane motion, so the translation of the z axis, t_z, is not reliable. At the same time, since the acceleration of IMU is coupled with gravity, it is related to rotation, so it is also not reliable to calculate the initial translation value through the initial rotation. We can make the plane assumption as [10] and rewrite Equation (7) as follows:

$$
\mathbf{R}_{k+1}
\begin{bmatrix}
t_x \\
t_y
\end{bmatrix}
-
\begin{bmatrix}
\cos(\gamma) & -\sin(\gamma) \\
\sin(\gamma) & \cos(\gamma)
\end{bmatrix}
\mathbf{t}_{k+1}^1 = -\mathbf{t}_{k+1}^2
\tag{8}
$$

Equation (8) is the plane motion constraint generated by the $k+1$th relative pose, where γ is the yaw rotation, t_x and t_y are the translation of the x and y axes, respectively, $\mathbf{R}_{k+1}$ is the 2x2 block matrix in the upper left corner of $\left(\mathbf{R}_{I_{k+1}}^{I_k} - \mathbf{I}_3\right)$, and $\mathbf{t}_{k+1}^1$ and $\mathbf{t}_{k+1}^2$ are the first two elements of the column vectors $\mathbf{R}_L^I \mathbf{t}_{L_{k+1}}^{L_k}$ and $\mathbf{t}_{I_{k+1}}^{I_k}$, respectively.

Equation (8) can be rewritten in the form of the matrix equation AX = b:

$$
\underbrace{\begin{bmatrix} \mathbf{R}_{k+1} & \mathbf{J} \end{bmatrix}}_{\mathbf{A}^{k+1}}
\begin{bmatrix}
t_x \\
t_y \\
-\cos(\gamma) \\
-\sin(\gamma)
\end{bmatrix}
= -\mathbf{t}_{k+1}^2
\tag{9}
$$

where $\mathbf{J} = \begin{bmatrix} \left[\mathbf{t}_{k+1}^1\right]_1 & -\left[\mathbf{t}_{k+1}^1\right]_2 \\ \left[\mathbf{t}_{k+1}^1\right]_2 & \left[\mathbf{t}_{k+1}^1\right]_1 \end{bmatrix}$, $\left[\mathbf{t}_{k+1}^1\right]_i$ denotes the ith element of the column vector $\left[\mathbf{t}_{k+1}^1\right]$. As in Equation (5), by stacking measurements at different times according to

Equation (9), we can obtain the final matrix equation AX = b, which can be solved by the least-squares approach:

$$\underbrace{\begin{bmatrix} \mathbf{A}^2 \\ \vdots \\ \mathbf{A}^{K+1} \end{bmatrix}}_{\mathbf{A}_{2K\times 4}} \underbrace{\begin{bmatrix} t_x \\ t_y \\ -\cos(\gamma) \\ -\sin(\gamma) \end{bmatrix}}_{\mathbf{x}_{4\times 1}} = -\underbrace{\begin{bmatrix} t_2^2 \\ \vdots \\ t_{K+1}^2 \end{bmatrix}}_{\mathbf{b}_{2K\times 1}} \tag{10}$$

The obtained yaw rotation, according to Equation (10), is fused with the original calculated extrinsic rotation $\hat{\mathbf{R}}_L^I$ to obtain the new initial rotation. We can manually measure the value of the extrinsic parameters about the z-axis. Suppose the calculated translation of the extrinsic parameters about the Z-axis differs too much from the measured value. In that case, we can set the translation of the Z-axis to the measured value and refine it through the absolute pose constraint in the refinement part; see Section 2.2.2.

2.2. Methods

This section details the procedure for calculating the initial values of extrinsic parameters by GPS/IMU measurements and LiDAR measurements. First, the GPS/IMU measurements were interpolated against the LiDAR timestamp; then, the LiDAR odometry was estimated as in [27]. The extrinsic parameters were initialized by hand–eye calibration; see Section 2.1. We integrate loop closure into the system by using a factor graph and completing the map construction. Then, we tightly coupled LiDAR and IMU through the initially obtained extrinsic parameters, obtained the LiDAR odometry through scan-to-map feature-based registration, and refined the extrinsic parameters by constructing the constraints of the extrinsic parameters through the obtained absolute pose.

2.2.1. Extrinsic Parameters Initialization

As for LiDAR odometry, there are two methods to calculate the relative transformation of two consecutive frames: 1. based on direct matching, such as ICP [28] and GICP [29], 2. feature-based methods, such as LOAM [27], do not need to calculate all the point clouds but only need to calculate representative points, which not only improves accuracy but also reduces computing efficiency. We use the feature-based method in [30] to obtain the LiDAR odometry.

However, since there are no extrinsic parameters between the LiDAR and IMU, we cannot use the IMU pre-integration as the guess pose of the LiDAR odometry as in [30]. Because offline calibration does not require real-time performance like SLAM, we can add more constraints to make the estimated relative motion more accurate. We first need to compute a predicted pose to prevent the optimization algorithm from converging to a local optimum. With the increase in time, the registration between consecutive frames tends to have a larger drift on the Z-axis than other axes. Therefore, we first use the constant velocity model as the predicted pose and then extract the ground points and calculate the centroid of the ground point to optimize roll, pitch angle, and z-translation, respectively. Since LiDAR is basically installed horizontally, we first extract ground points using geometric features in [31], and then filter the outliers by RANSAC so that the ground points obtained can reach a high accuracy and the guess pose is relatively accurate.

When we obtain the latest scans of the raw LiDAR point cloud, firstly, we de-skew the point cloud to the moment of the first LiDAR point by guess pose and project the skewed point cloud into the range image according to the resolution. We extract the two feature points, edge feature and planar feature points, through the curvature. We distribute the feature points uniformly and remove the unstable feature points as [27]. We denote $\mathbb{F}_k = \left\{ F_k^e, F_k^p \right\}$ as all feature points of kth LiDAR point cloud, and F_k^e and F_k^p are denoted as edge feature points and planar feature points, respectively. We can obtain the distances of point-to-line and point-to-plane through the following:

$$d_{e_k} = \frac{\left| \left(\mathbf{p}^e_{k+1} - \overline{\mathbf{p}}^e_{k,2} \right) \times \left(\mathbf{p}^e_{k+1} - \overline{\mathbf{p}}^e_{k,1} \right) \right|}{\left| \overline{\mathbf{p}}^e_{k,1} - \overline{\mathbf{p}}^e_{k,2} \right|} \tag{11}$$

$$\mathbf{n} = \left(\overline{\mathbf{p}}^p_{k,1} - \overline{\mathbf{p}}^p_{k,2} \right) \times \left(\overline{\mathbf{p}}^p_{k,1} - \overline{\mathbf{p}}^p_{k,3} \right) \tag{12}$$

$$d_{p_k} = \frac{\left| \left(\mathbf{p}^p_{k+1} - \overline{\mathbf{p}}^p_{k,1} \right) \cdot \mathbf{n} \right|}{|\mathbf{n}|} \tag{13}$$

where $\mathbf{p}^e_{k+1} \in {}'\mathbb{F}^e_{k+1}$, $\mathbf{p}^p_{k+1} \in {}'\mathbb{F}^p_{k+1}$. ${}'\mathbb{F}_{k+1} = \left\{ {}'\mathbb{F}^e_{k+1}, {}'\mathbb{F}^p_{k+1} \right\}$ is the feature points de-skewed to the first point of the current point cloud. $\left(\overline{\mathbf{p}}^e_{k,1}, \overline{\mathbf{p}}^e_{k,2} \right)$ are the two edge feature points of $\overline{\mathbb{F}}^e_k$ corresponding to $\mathbf{p}^e_{k+1}$. $\left(\overline{\mathbf{p}}^p_{k,1}, \overline{\mathbf{p}}^p_{k,2}, \overline{\mathbf{p}}^p_{k,3} \right)$ are the three planer feature points of $\overline{\mathbb{F}}^p_k$ corresponding to $\mathbf{p}^p_{k+1}$. $\overline{\mathbb{F}}_k = \left\{ \overline{\mathbb{F}}^e_k, \overline{\mathbb{F}}^p_k \right\}$ is the feature points de-skewed to the last point of the previous point cloud. The moment of the last point of the previous point cloud is equal to the moment of the first point of the current point cloud, so the edge feature points at the same moment are on the same line, and the planner points are on the same surface. Thus, we can obtain the constraint of Equations (11)–(13) and use point-to-line and point-to-surface distances as cost functions. The relative pose can be calculated by minimizing the cost function by using the Levenberg–Marquardt algorithm. The LiDAR odometry algorithm is shown in Algorithm 1.

Algorithm 1 LiDAR Odometry.

Input: $\overline{\mathbb{F}}_k = \left\{ \overline{\mathbb{F}}^e_k, \overline{\mathbb{F}}^p_k \right\}$, $\mathbb{F}_{k+1} = \left\{ \mathbb{F}^e_{k+1}, \mathbb{F}^p_{k+1} \right\}$, $\mathbf{T}^W_{L_k}$ from the last recursion

Output: $\overline{\mathbb{F}}_{k+1} = \left\{ \overline{\mathbb{F}}^e_{k+1}, \overline{\mathbb{F}}^p_{k+1} \right\}$, $\mathbf{T}^W_{L_{k+1}}$

1: Through the constant velocity model, guess pose $\mathbf{T}^W_{L_{k+1}}$ is calculated from $\mathbf{T}^W_{L_k}$
2: **for** each edge point in $\mathbb{F}_{k+1}$ **do**
3:      Extracting ground points through geometric features and RANSAC
4: **end for**
5: Calculate the three degrees of freedom z-translation, roll, and pitch through ground point optimization to obtain a new guess pose $\mathbf{T}^W_{L_{k+1}}$.
6: De-skew the point cloud $\mathbb{F}_{k+1}$ to the moment of the first LiDAR point by guess pose. Detect edge points and planar points in ${}'\mathbb{F}_{k+1}$
7: **for** a number of iterations **do**
8:      **for** each edge point in ${}'\mathbb{F}^e_{k+1}$ **do**
9:          Find an edge line as the correspondence, then compute point to line distance based on (11)
10:      **end for**
11:      **for** each edge point in ${}'\mathbb{F}^e_{k+1}$ **do**
12:          Find a planar patch as the correspondence, then compute point to plane distance based on (13)
13:      **end for**
14:      Using distance as a cost function for nonlinear optimization, update $\mathbf{T}^W_{L_{k+1}}$
15:      **if** the nonlinear optimization converges **then**
16:          Break
17:      **end if**
18: **end for**
19: Reproject $\mathbb{F}_{k+1}$ to the moment of the last LiDAR point according to $\mathbf{T}^W_{L_{k+1}}$ to get $\overline{\mathbb{F}}_{k+1}$
20: **return** $\mathbf{T}^W_{L_{k+1}}$, $\overline{\mathbb{F}}_{k+1}$

After completing the registration between two consecutive frames, we optimize the estimated pose again through scan-to-map point cloud registration. We obtain the keyframe through the relative pose transformation amount greater than the threshold value that is considered to be set and save it. The local point cloud used for the current point cloud registration is obtained through the following two methods: 1. It is formed by superimposing several keyframes adjacent in time. 2. It is formed by superimposing several keyframes adjacent in position. The search for adjacent positions uses a KD-tree and adds the position of each keyframe as a point to the KD-tree, and then the keyframes with adjacent positions are obtained through the nearest neighbor search of the KD-tree. In order to reduce the drift of cumulative errors, we combined the obtained relative pose with z-translation, roll, and pitch angle obtained by ground point optimization to make the relative pose more accurate and robust.

After obtaining the relative pose of LiDAR, the initial calibration parameters can be obtained by hand–eye calibration through the relative pose constraints of the LiDAR odometry and the relative pose constraints of the interpolated GPS/IMU data; see Section 2.1. We denote $\left(\mathbf{q}_{L_{K+1}}^{L_k}, \mathbf{t}_{L_{K+1}}^{L_k}\right)$ and $\left(\mathbf{q}_{I_{K+1}}^{I_k}, \mathbf{t}_{I_{K+1}}^{I_k}\right)$ as the relative pose constraint pair of LiDAR data and GPS/IMU data, respectively. We can construct the following cost function with extrinsic parameters:

$$
\begin{aligned}
\min_{q_I^L, t_I^L} \Bigg\{ &\sum_{k \in N} \left\| (\mathbf{q}_{I_{k+1}}^{I_k} \otimes \mathbf{q}_L^I)^* \otimes \mathbf{q}_L^I \otimes \mathbf{q}_{L_{K+1}}^{L_k} \right\|_{xyz}^2 + \\
&\sum_{k \in N} \left\| \left(\mathbf{R}_{I_{k+1}}^{I_k} - \mathbf{I}_3\right) \mathbf{t}_L^I - \mathbf{R}_L^I \mathbf{t}_{L_{K+1}}^{L_k} + \mathbf{t}_{I_{k+1}}^{I_k} \right\|^2 \Bigg\}
\end{aligned}
\tag{14}
$$

where $\|\ \|_{xyz}$ represents the imaginary part of the quaternion, and there are a total of N relative pose constraint pairs. If we make a planar assumption, the residuals of the translation part can be constructed by Equation (10). After multiple optimizations, the initial value of the calibration parameters can be obtained. The global consistency map is constructed by a loop closure.

2.2.2. Extrinsic Parameters Refinement

In this section, the process of extrinsic parameter refinement is introduced in detail. First, the absolute pose is obtained through the registration of the global map obtained from Section 2.2.1. The LiDAR odometry and IMU and tightly coupled by the initial extrinsic parameters, and the cost function of the calibration parameters was constructed by the absolute pose. After nonlinear optimization, the refined extrinsic parameters can be obtained.

From Section 2.2.1, we can obtain the initial extrinsic parameters, the global map, and the world frame. The global map is built under the world frame. Since we have the initial extrinsic parameters, we can tightly couple the IMU during the construction of the LiDAR odometry. Firstly, the pre-integration is used to de-skew the LiDAR point cloud and serve as the guess pose of the LiDAR odometry. The bias of the IMU is corrected by the optimized LiDAR odometry based on the factor graph.

First, we obtain the local map by filtering the global map and obtain the LiDAR odometry by registration with the local map, $\mathbf{T}_{L_k}^W$. Through Figure 3, we can construct constraints with extrinsic parameters through the following formula:

$$
\mathbf{T}_{L_k}^W = \mathbf{T}_{I_k}^W \widehat{\mathbf{T}}_L^I
\tag{15}
$$

where $\mathbf{T}_{I_k}^W$ is the GPS/IMU pose corresponding to the current LiDAR timestamp.

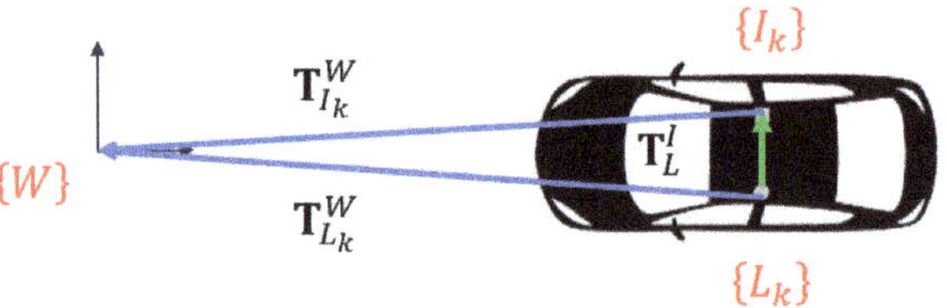

Figure 3. This figure shows the pose relationship during the refinement of extrinsic parameters, where T_{L_k} and $T_{I_k}^{W}$ are the absolute pose of LiDAR and GPS/IMU, respectively, and T_{L}^{I} is extrinsic parameters. The coordinate system of the sensor is indicated in red.

As per Section 2, we can also decouple Equation (15) into the rotation part and the translation part to obtain the following two constraints [24]:

$$
\begin{aligned}
\widehat{\mathbf{q}}_{L_k}^{W} &= \mathbf{q}_{I_k}^{W} \otimes \mathbf{q}_{L}^{I}, \\
\widehat{\mathbf{t}}_{L_k}^{W} &= \mathbf{t}_{I_k}^{W} + \mathbf{R}_{I_k}^{W} \mathbf{t}_{L}^{I}.
\end{aligned}
\tag{16}
$$

The construction of the over-determined equation about rotation and translation is the same as Section 2.2.1, but the assumption of plane motion is not required here, the extrinsic translation parameters can be optimized by the absolute pose, directly.

We can refine the extrinsic parameters by constructing the cost function of the absolute pose and extrinsic parameters according to Equation (16):

$$
\min_{q_{L}^{I}, t_{L}^{I}} \left\{ \sum_{k \in N} \left\| (\mathbf{q}_{L_i}^{W})^{*} \otimes \mathbf{q}_{I_i}^{W} \otimes \mathbf{q}_{L}^{I} \right\|_{xyz}^{2} + \sum_{k \in N} \left\| \mathbf{R}_{I_k}^{W} \mathbf{t}_{L}^{I} + \mathbf{t}_{I_k}^{W} - \mathbf{t}_{L_k}^{W} \right\|^{2} \right\}
\tag{17}
$$

When the cost function is optimized by L-M optimization, the final precise extrinsic parameters can be obtained when the iteration converges.

3. Results and Discussions

3.1. Validation and Results

In this section, we validate the proposed LiDAR-GPS/IMU calibration system in the simulation and real environments, respectively. The simulation environment adopted the Carla simulation platform and was equipped with the required sensors to record multiple data sets under the empty urban road scene. In the real environment, ouster-128 LiDAR and FDI-integrated navigation systems were used to record corresponding data sets in outdoor road scenes, and CAD assembly drawing was used as the ground truth for comparison and verification. In addition to the above hardware configuration, the software configuration in Table 3 is also required. Data communication was performed through ROS in the Ubuntu operating system, the received LiDAR point cloud was processed through PCL, and nonlinear optimization was performed through Ceres Solver. Upon playing the pre-recorded data set, the system receives LiDAR, GPS, and IMU messages through ROS and obtains the global map and initial extrinsic parameters through the method of extrinsic parameters initialization in Section 2.2.1. After the initialization of the extrinsic parameters is completed, the global map and the initial extrinsic parameters are loaded through the method of refining the extrinsic parameters described in Section 2.2.2 to carry out the refinement operation so that the final result of the experiment can be obtained.

Table 3. Software configuration list.

Software	Function
Ubuntu18.04 or 16.04	operating system (OS)
ROS	robot operating system
PCL	Point Cloud Library
Ceres Solver	C++ optimization library

3.1.1. Simulation

As shown in Figure 4, there is the scene diagram built by the Carla simulation platform. Carla [32] is an urban driving simulator and has an ROS interface to support the sensor suite. Carla can be installed through Carla's official website tutorial. Note that installing Carla requires about 130GB of disk space and at least a 6GB GPU. The vehicle is equipped with LiDAR and GPS/IMU, and the ground truth and noise model size are given through the configuration file. After completing the hardware configuration and setting the configuration file, data sets can be recorded in different scenarios. After comparing the data sets in different scenarios, the experimental data were recorded in Table 4. Figure 5 shows the variation trend of the error values of rotation and translation. Both the rotation part and the translation part can achieve high precision. Regarding the rotation part, the error of raw angle and pitch angle is within 0.1 degrees, the error of yaw angle is relatively large, the error of scene 2 and scene 3 is within 0.2 degrees, and the error of scene 1 is about 0.8 degrees. Regarding the translation part, all errors are within 0.1 m.

Figure 4. Outdoor scene diagram of the Carla simulation platform.

Table 4. Carla simulation environment calibration results.

Simulation Data Experiment Result	Rotation (deg)			Translation (m)	
	Row	Pitch	Yaw	x	y
ground truth	0.000000	0.000000	45.00000	1.000000	−0.500000
scene 1	0.064391	0.023962	44.20480	1.042260	−0.556855
scene 2	0.058780	0.050686	44.93890	0.974243	−0.523805
scene 3	0.014632	0.021158	44.83500	0.984068	−0.502618
average error	0.045934	0.031935	0.340433	1.903333×10^{-4}	0.027759

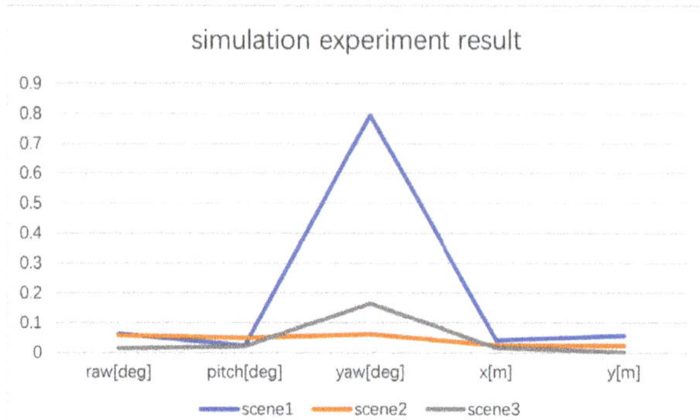

Figure 5. The line chart of the extrinsic parameter error value of the simulation environment. The three broken lines of different colors in the figure represent the error data of extrinsic parameters under three different scenarios. Regarding the rotation part, the error of raw angle and pitch angle is within 0.1 degrees, the error of yaw angle is relatively large, the error of scene 2 and scene 3 is within 0.2 degrees, and the error of scene 1 is about 0.8 degrees. Regarding the translation part, all errors are within 0.1 m.

3.1.2. Real-World Experiments

We assembled the sensors, as shown in Figure 6, to collect data, respectively, in the outdoor scene. The ouster-128 Lidar outputs the point cloud at a frequency of 10 Hz , and the FDI-integrated navigation system outputs the GPS/IMU measurement at a frequency of 100 Hz. Since there is no ground truth of LiDAR-GPS/IMU in the real scene and there is also a lack of an open source LiDAR-GPS/IMU calibration algorithm, we adopted the truth value provided by CAD assembly drawing for verification and checked the repeatability and correctness of calibration results through many experiments. The experimental results are shown in Table 5. Figure 7 shows the variation trend of the error values of rotation and translation. It can be seen that in the real world, the calibration system we proposed can still reach high accuracy. Regarding the rotation part, the errors of the pitch angles of the three scenes are all within 0.2 degrees, the errors of the raw angles are all within 0.45 degrees, and the errors of the yaw angles are all within 0.6 degrees. Regarding the translation part, all errors are within 0.05 m. Comparing the extrinsic parameters errors obtained in the simulation environment and the real environment, the errors of the pitch angle and the translation part are relatively small, and the errors of the yaw angle are relatively large. The difference is that the error of the raw angle in the real environment is larger than that in the simulation environment.

Table 5. Calibration results of our own real data.

Real Data Experiment Result	Rotation (deg)			Translation (m)	
	Row	Pitch	Yaw	x	y
ground truth	90.0000	173.000	−180.000	−0.25000	−0.600000
scene 1	90.3050	173.092	−179.432	−0.25099	−0.585795
scene 2	89.5824	173.054	−179.556	−0.24924	−0.607548
scene 3	90.2324	172.824	−179.724	−0.26456	−0.611476
average error	0.03990	0.25670	−0.42933	0.004930	0.0016063

Figure 6. Illustration of our car equipped with the ouster-128 LiDAR and FDI integrated navigation system.

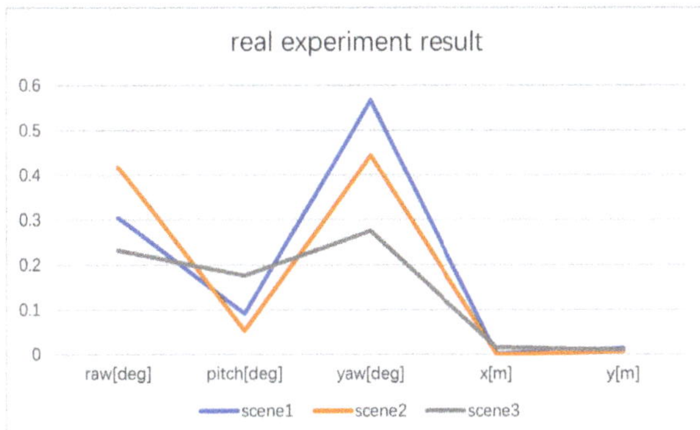

Figure 7. The line chart of the extrinsic parameters error value of the real environment. The three broken lines of different colors in the figure represent the error data of extrinsic parameters under three different scenarios. Regarding the rotation part, the errors of the pitch angles of the three scenes are all within 0.2 degrees, the errors of the raw angles are all within 0.45 degrees, and the errors of the yaw angles are all within 0.6 degrees. Regarding the translation part, all errors are within 0.05 m.

3.2. Discussions

Many existing calibration methods require specific vehicle movement, such as eight-shaped trajectories, or manually marked scenes so that different sensors can measure the same marker for calibration, resulting in low automation and high labor costs. For multi-sensor fusion, it is crucial to calibrate the extrinsic parameters between sensors. This paper proposes a motion-based self-calibration method, which can complete the calibration in the surrounding natural scenes, such as outdoor roads, urban streets, etc., only requiring the completion of data set recording in advance, and then the extrinsic parameters of LiDAR and GPS/IMU can be obtained through two-step offline calibration. However, due to the plane movement of the vehicle, even if the LiDAR odometry and GPS/IMU odometry are accurately estimated from coarse to fine, the z-axis translation of the extrinsic parameters cannot be well estimated. There is still no good solution to this difficulty. It may be due to the large z-axis drift in the registration algorithm of the LiDAR odometry itself. Perhaps this problem can be better solved by studying a new point cloud registration algorithm.

4. Conclusions and Future Work

In this paper, we propose a self-calibration system of LIDAR-GPS/IMU, which can achieve high-precision calibration of extrinsic parameters between LiDAR and GPS/IMU in natural outdoor scenes. The two-step offline self-calibration method was adopted. Firstly, the initial extrinsic parameters were calibrated by hand-eye calibration, and the accumulated drift was removed by loop closure to complete the map construction. Then, the absolute pose, which was obtained by map-based registration, and extrinsic parameters constructed the cost function, and the extrinsic parameters were refined by the optimization algorithm. It has been verified in many experiments in the simulation environment and real environment and has good robustness and accuracy. Since hand–eye calibration is applicable to any calculation of rigid transformation between two sensors, future work involves the calibration of extrinsic parameters between any two sensors between camera, LiDAR, and GPS/IMU.

Author Contributions: Conceptualization, J.G.; methodology, J.G.; software, Y.Z.; validation, J.G., X.T. and J.C.; formal analysis, J.G.; investigation, J.G. and X.N.; resources, Y.Z.; data curation, J.G.; writing—original draft preparation, J.G. and X.T; writing—review and editing, X.N.; visualization, J.C.; supervision, X.N.; project administration, J.G.; funding acquisition, X.N., J.G., J.C., X.T. and Y.Z. All authors have read and agreed to the published version of the manuscript.

Funding: This research was funded by the Hunan Natural Science Foundation of China, grant number 2020JJ4196, and the Liuzhou Science and Technology Foundation of China, grant number 2021CBA0101.

Institutional Review Board Statement: Not applicable.

Informed Consent Statement: Not applicable.

Data Availability Statement: Not applicable.

Conflicts of Interest: The authors declare no conflict of interest.

References

1. Mochurad, L.; Kryvinska, N. Parallelization of Finding the Current Coordinates of the Lidar Based on the Genetic Algorithm and OpenMP Technology. *Symmetry* **2021**, *13*, 666. [CrossRef]
2. Huang, J.; Ran, S.; Wei, W.; Yu, Q. Digital Integration of LiDAR System Implemented in a Low-Cost FPGA. *Symmetry* **2022**, *14*, 1256. [CrossRef]
3. Kumar, G.A.; Patil, A.K.; Kang, T.W.; Chai, Y.H. Sensor Fusion Based Pipeline Inspection for the Augmented Reality System. *Symmetry* **2019**, *11*, 1325. [CrossRef]
4. Zhu, D.; Ji, K.; Wu, D.; Liu, S. A Coupled Visual and Inertial Measurement Units Method for Locating and Mapping in Coal Mine Tunnel. *Sensors* **2022**, *22*, 7437. [CrossRef] [PubMed]
5. Kumar, G.A.; Lee, J.H.; Hwang, J.; Park, J.; Youn, S.H.; Kwon, S. LiDAR and Camera Fusion Approach for Object Distance Estimation in Self-Driving Vehicles. *Symmetry* **2020**, *12*, 324. [CrossRef]
6. Chu, P.M.; Cho, S.; Sim, S.; Kwak, K.; Cho, K. Multimedia System for Real-Time Photorealistic Nonground Modeling of 3D Dynamic Environment for Remote Control System. *Symmetry* **2018**, *10*, 83. [CrossRef]
7. Pan, Y.; Xiao, P.; He, Y.; Shao, Z.; Li, Z. MULLS: Versatile LiDAR SLAM via multi-metric linear least square. In Proceedings of the 2021 IEEE International Conference on Robotics and Automation (ICRA), Xi'an, China, 30 May–5 June 2021; pp. 11633–11640.
8. Le, A.V.; Apuroop, K.G.S.; Konduri, S.; Do, H.; Elara, M.R.; Xi, R.C.C.; Wen, R.Y.W.; Vu, M.B.; Duc, P.V.; Tran, M. Multirobot Formation with Sensor Fusion-Based Localization in Unknown Environment. *Symmetry* **2021**, *13*, 1788. [CrossRef]
9. Lee, H.; Chung, W. Extrinsic Calibration of Multiple 3D LiDAR Sensors by the Use of Planar Objects. *Sensors* **2022**, *22*, 7234. [CrossRef] [PubMed]
10. Jiao, J.; Yu, Y.; Liao, Q.; Ye, H.; Fan, R.; Liu, M. Automatic calibration of multiple 3d lidars in urban environments. In Proceedings of the 2019 IEEE/RSJ International Conference on Intelligent Robots and Systems (IROS), Macau, China, 3–8 November 2019; pp. 15–20.
11. Xue, B.; Jiao, J.; Zhu, Y.; Zhen, L.; Han, D.; Liu, M.; Fan, R. Automatic calibration of dual-LiDARs using two poles stickered with retro-reflective tape. In Proceedings of the 2019 IEEE International Conference on Imaging Systems and Techniques (IST), Abu Dhabi, United Arab Emirates, 9–10 December 2019; pp. 1–6.
12. Zhang, J.; Lyu, Q.; Peng, G.; Wu, Z.; Yan, Q.; Wang, D. LB-L2L-Calib: Accurate and Robust Extrinsic Calibration for Multiple 3D LiDARs with Long Baseline and Large Viewpoint Difference. In Proceedings of the 2022 International Conference on Robotics and Automation (ICRA), Philadelphia, PA, USA, 23–27 May 2022; pp. 926–932.

13. Liu, X.; Yuan, C.; Zhang, F. Targetless Extrinsic Calibration of Multiple Small FoV LiDARs and Cameras using Adaptive Voxelization. *IEEE Trans. Instrum. Meas.* **2022**, *17*, 8502612. [CrossRef]

14. Mishra, S.; Osteen, P.R.; Pandey, G.; Saripalli, S. Experimental evaluation of 3d-lidar camera extrinsic calibration. In Proceedings of the 2020 IEEE/RSJ International Conference on Intelligent Robots and Systems (IROS), Las Vegas, NV, USA, 24 October–24 January 2020; pp. 9020–9026.

15. Yuan, K.; Ding, L.; Abdelfattah, M.; Wang, Z.J. LiCaS3: A Simple LiDAR–Camera Self-Supervised Synchronization Method. *IEEE Trans. Robot.* **2022**, *38*, 3203–3218. [CrossRef]

16. Li, Y.; Yang, S.; Xiu, X.; Miao, Z. A Spatiotemporal Calibration Algorithm for IMU&LiDAR Navigation System Based on Similarity of Motion Trajectories. *Sensors* **2022**, *22*, 7637. [PubMed]

17. Lv, J.; Xu, J.; Hu, K.; Liu, Y.; Zuo, X. Targetless Calibration of LiDAR-IMU System Based on Continuous-time Batch Estimation. In Proceedings of the 2020 IEEE/RSJ International Conference on Intelligent Robots and Systems (IROS), Las Vegas, NV, USA, 24 October–24 January 2020; pp. 9968–9975.

18. Lv, J.; Zuo, X.; Hu, K.; Xu, J.; Huang, G.; Liu, Y. Observability-Aware Intrinsic and Extrinsic Calibration of LiDAR-IMU Systems. *IEEE Trans. Robot.* **2022**, *38*, 3734–3753. [CrossRef]

19. Geiger, A.; Lenz, P.; Stiller, C.; Urtasun, R. Vision meets robotics: The kitti dataset. *Int. J. Robot. Res.* **2013**, *32*, 1231–1237. [CrossRef]

20. Schneider, S.; Luettel, T.; Wuensche, H.J. Odometry-based online extrinsic sensor calibration. In Proceedings of the 2013 IEEE/RSJ International Conference on Intelligent Robots and Systems, Tokyo, Japan, 3–7 November 2013; pp. 1287–1292.

21. Tsai, R.; Lenz, R. A new technique for fully autonomous and efficient 3D robotics hand/eye calibration. *IEEE Trans. Robot. Autom.* **1989**, *5*, 345–358. [CrossRef]

22. Chen, C.; Xiong, G.; Zhang, Z.; Gong, J.; Qi, J.; Wang, C. 3D LiDAR-GPS/IMU Calibration Based on Hand-Eye Calibration Model for Unmanned Vehicle. In Proceedings of the 2020 3rd International Conference on Unmanned Systems (ICUS), Harbin, China, 27–28 November 2020; pp. 337–341.

23. Yuwen, X.; Chen, L.; Yan, F.; Zhang, H.; Tang, J.; Tian, B.; Ai, Y. Improved Vehicle LiDAR Calibration With Trajectory-Based Hand-Eye Method. *IEEE Trans. Intell. Transp. Syst.* **2022**, *23*, 215–224. [CrossRef]

24. Yang, Z.; Shen, S. Monocular visual-inertial fusion with online initialization and camera-IMU calibration. In Proceedings of the 2015 IEEE International Symposium on Safety, Security, and Rescue Robotics (SSRR), West Lafayette, IN, USA, 18–20 October 2015; pp. 1–8.

25. Sola, J. Quaternion kinematics for the error-state Kalman filter. *arXiv* **2017**, arXiv:1711.02508.

26. Jiao, J.; Ye, H.; Zhu, Y.; Liu, M. Robust odometry and mapping for multi-lidar systems with online extrinsic calibration. *IEEE Trans. Robot.* **2021**, *38*, 351–371. [CrossRef]

27. Zhang, J.; Singh, S. Visual-lidar odometry and mapping: Low-drift, robust, and fast. In Proceedings of the 2015 IEEE International Conference on Robotics and Automation (ICRA), Seattle, WA, USA, 26–30 May 2015; pp. 2174–2181.

28. Besl, P.J.; McKay, N.D. Method for registration of 3-D shapes. In Proceedings of the Sensor Fusion IV: Control Paradigms and Data Structures, Boston, MA, USA, 12–15 November 1992; Volume 1611, pp. 586–606.

29. Segal, A.; Haehnel, D.; Thrun, S. Generalized-icp. In Proceedings of the Robotics: Science and Systems, Seattle, WA, USA, 28 June–1 July 2009; Volume 2, p. 435.

30. Shan, T.; Englot, B.; Meyers, D.; Wang, W.; Ratti, C.; Rus, D. Lio-sam: Tightly-coupled lidar inertial odometry via smoothing and mapping. In Proceedings of the 2020 IEEE/RSJ International Conference on Intelligent Robots and Systems (IROS), Las Vegas, NV, USA, 24 October–24 January 2020; pp. 5135–5142.

31. Shan, T.; Englot, B. Lego-loam: Lightweight and ground-optimized lidar odometry and mapping on variable terrain. In Proceedings of the 2018 IEEE/RSJ International Conference on Intelligent Robots and Systems (IROS), Madrid, Spain, 1–5 October 2018; pp. 4758–4765.

32. Dosovitskiy, A.; Ros, G.; Codevilla, F.; Lopez, A.; Koltun, V. CARLA: An Open Urban Driving Simulator. In Proceedings of the Proceedings of the 1st Annual Conference on Robot Learning, Mountain View, CA, USA, 13–15 November 2017; pp. 1–16.

symmetry

Article

Facial Expression Recognition Based on Dual-Channel Fusion with Edge Features

Xiaoyu Tang [1,2,*], Sirui Liu [1], Qiuchi Xiang [2], Jintao Cheng [1], Huifang He [3] and Bohuan Xue [4]

[1] School of Physics and Telecommunication Engineering, South China Normal University, Guangzhou 510006, China
[2] Xingzhi College, South China Normal University, Shanwei 516600, China
[3] Guangdong Engineering Polytechnic, Guangzhou 510520, China
[4] Department of Computer Science and Engineering, The Hong Kong University of Science and Technology, Hong Kong, China
* Correspondence: tangxy@scnu.edu.cn; Tel.: +86-139-2898-6468

Abstract: In the era of artificial intelligence, accomplishing emotion recognition in human–computer interaction is a key work. Expressions contain plentiful information about human emotion. We found that the canny edge detector can significantly help improve facial expression recognition performance. A canny edge detector based dual-channel network using the OI-network and EI-Net is proposed, which does not add an additional redundant network layer and training. We discussed the fusion parameters of α and β using ablation experiments. The method was verified in CK+, Fer2013, and RafDb datasets and achieved a good result.

Keywords: facial expression recognition; channel weighting; feature fusion; edge detection

Citation: Tang, X.; Liu, S.; Xiang, Q.; Cheng, J.; He, H.; Xue, B. Facial Expression Recognition Based on Dual-Channel Fusion with Edge Features. *Symmetry* **2022**, *14*, 2651. https://doi.org/10.3390/sym14122651

Academic Editors: Chengxi Zhang, Jin Wu and Chong Li

Received: 10 November 2022
Accepted: 2 December 2022
Published: 15 December 2022

Publisher's Note: MDPI stays neutral with regard to jurisdictional claims in published maps and institutional affiliations.

1. Introduction

Facial expression recognition (FER) is an important research direction in the affective computing field [1]. The psychologist Mehrabian's research shows that emotional expression = 7% language + 38% voice + 55% facial expressions [2]. This research shows that facial expressions play an important role in human emotional judgment. The accurate recognition of facial expression helps to improve the effect of human–computer interaction. At present, FER has been applied in many fields, such as intelligent teaching, medical facilities, security monitoring, psychological warning, and driver fatigue monitoring.

Feature extraction is a crucial step of FER. Early feature extraction is mainly based on handcrafted methods, such as HOG [3], SIFT [4], and LBP [5]. Among them, HOG and SIFT are calculated by the local gradient of the image. Up to now, these methods have still been common in the FER task, because they can extract the local information of the image in a targeted manner. Both SIFT and HOG have a certain degree of robustness on the impact of illumination, but a common problem with them is a large amount of calculation.

With the development of deep learning, features extracted by deep neural network, an end-to-end method, has become popular. Typical neural network models are AlexNet [6], VGGNet [7], ResNet [8], and GAN [9]. However, there is much redundant information in the extracted features when using the methods of convolutional neural networks. The redundant information is hardly helpful for FER tasks, and some features can be classified as noise. These problems affect the recognition accuracy of FER, which cannot fulfill the current FER needs well.

The gradient information of the image contains much information about the shape of the object, and the edge provides critical information in FER. Although the original images contain the edge information, the deep network trained by original image will lose this information. Aiming at solving the above problem, this paper proposes a dual-channel FER method based on edge feature fusion. The purpose is to effectively focus on the edge information of facial expressions while maintaining high-level semantic features. Adding the network channel for extracting edge features can remove confounding factors in the image. Moreover, the problem domain is simplified, effectively reducing the amount of data and the number of layers required by the deep CNN model. Experiments show that the method proposed in this paper is feasible and can improve the accuracy and robustness of each benchmark dataset. The main contribution of this paper are:

1. This paper proposes a dual-channel FER method based on edge feature fusion to enhance edges, discuss the weight of two channels, and analyze the contribution of edges.

2. This paper proves in the experiment that more than just extracting the edge feature is needed to provide all the information needed for FER. Facts have proved that it can only be used as a supplement to the original image feature, and more information that determines FER is included in the original image.

3. The FER method proposed in this paper performs more robustly on three datasets, including CK+, Fer2013, and RafDb.

2. Related Work

In the area of FER, handcrafted features have been frequently used. Appearance-based features, one of the traditional handcrafted features methods, focus on extracting low-level features, such as edges and corners. Hu et al. [10] proposed a new local feature recognition center-symmetric local octonary pattern (CS-LOP), which improved the LBP algorithm and the CS-LBP algorithm. Meena et al. [11] proposed using graph signal processing (GSP) to solve the problem of HOG high-dimensional feature vectors and computational complexity. In 2021, Shanthi and Nickolas [12] combined LBP features and LNEP features to encode the relationship between pixels, realizing an effective texture representation. These feature extraction methods all focus on low-level features. The handcrafted feature-based methods above have the disadvantages of two points. Firstly, they are useful in datasets with small samples. On the contrary, they are useless in others, such as wild datasets. Secondly, they usually only consider a single feature.

In recent years, convolutional neural networks (CNNs), proposed for image classification tasks, have achieved better recognition performance. Usually, networks with deep layers can extract high-level features but bring more noise and network parameters with excessive redundant information. So, researchers began to find methods to solve these problems. Xie et al. [13] made more targeted improvements to CNNs, mainly including an attention-based salient expression region descriptor (SERD) and a multipath mutation suppression network (MPVS-Net). Minaee et al. [14] proposed an FER method based on an attention-based convolutional network, focusing on critical face parts. Wang, Kai et al. [15] proposed Region Attention Networks (RAN) to solve the obstacles of occlusion and posture changes in FER. They used the attention mechanism and improved CNNs to emphasize the learning of key regions, thereby improving the effectiveness of FER. Actually, low-level features are easily lost. In recent years, some researchers tried to combine handcrafted features with CNNs. G Levi and T Hassner [16] proposed using LBP to preprocess an image and perform deep neural network learning with the original image. H Zhang, B Huang, and G Tian [17] proposed to use the LBP for preprocessing and then weighted fusion with the original image through dual-channel training and added time series, using LSTM to achieve image-sequence-based FER. F. Bougourzi [18] and others proposed the FTDS method, which combined shallow features and in-depth features to identify six basic facial expressions in static images. The paper used HOG, LPQ, and the BSIF to extract low-level features, while using l-PML and VGG-Face networks to extract high-level features. Yu et al. [19] proposed a multitask global–local FER method, using global facial models and part-based models to learn global spatial information features and key dynamic features.

CNNs easily lose low-level features, such as edges. Thought handcrafted feature-based methods can obtain low-level features, these features are sensitive to illumination conditions. However, the edges are stable and contain critical information. This paper proposes a dual-channel FER method to extract features from the original image and edge image, using a shallow network to focus on edge features. The method determined the contribution of the edge in the FER.

3. Proposed Method

This paper uses a dual-channel network model, as shown in Figure 1. We use two channels to extract features from an original image and an edge image, respectively. The two networks are based on VggNet [7]. The channel extracts original image features, called the original image network (OI-Net), which consists of 12 layers of convolution. Another one is called the edge image network (EI-Net), consisting of 8 layers of convolution. The feature fusion of the two channels is performed by the given original image feature parameter α and edge feature parameter β, and Softmax is used for classification.

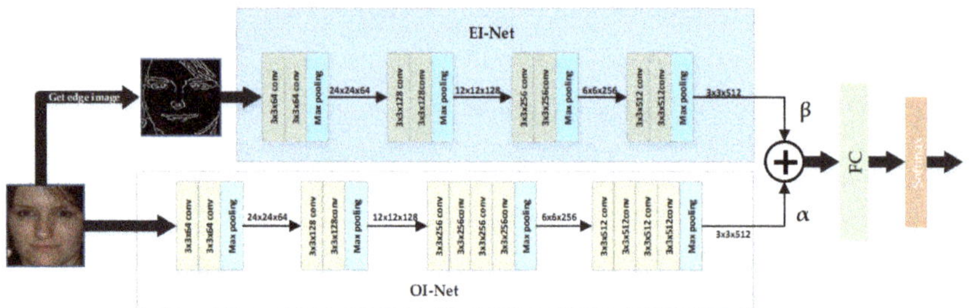

Figure 1. Our proposed network architectur.

3.1. Edge Image Feature Extraction

The edge contains critical information about the face, including three essential senses needed for FER. It contains information such as facial muscle texture and wrinkles corresponding to different expressions, which improve the accuracy of the recognition. Extracting the edge features can effectively reduce redundant information and, meanwhile, distinguish the information that the original image is focused on.

Edge information can be extracted by the shallow network; this low-level feature is important and easily lost in the deeper network. To avoid the lack of edge information, we consider extracting the edge feature as a supplement to the OI-Net and discuss its effect.

When performing edge feature extraction, the gradient information obtained by the solution is very sensitive to noise. Therefore, this paper chooses canny edge detection [20] that can reduce noise interference to extract edge images. Canny edge detection can remove noise while introducing two thresholds, T1 and T2, to better preserve edges.

The specific edge detection steps are as follows. The first is Gaussian filtering. The purpose is to remove noise using formula (1);

$$G(x,y) = \frac{1}{2\pi\sigma^2} e^{\frac{x^2+y^2}{2\sigma^2}} f(x,y) \tag{1}$$

Among them is the gray value of the image for a position, and it is the gray value of the image after Gaussian filtering.

The second step is to calculate the image gradient value and gradient direction; see formulas (2)–(7);

$$G\_x = \begin{bmatrix} -1 & 0 & +1 \\ -2 & 0 & +2 \\ -1 & 0 & +1 \end{bmatrix} \tag{2}$$

$$G\_y = \begin{bmatrix} +1 & +2 & +1 \\ 0 & 0 & 0 \\ -1 & -2 & +1 \end{bmatrix} \tag{3}$$

$$G_x = G(x,y) \times G\_x \tag{4}$$

$$G_y = G(x, y) \times G\_y \tag{5}$$

$$G = \sqrt{(G_x^2 + G_y^2)} \tag{6}$$

$$\theta = arctan\left(\frac{G_y}{G_x}\right) \tag{7}$$

G_x and G_y are the convolution factors needed to calculate the x-direction and the y-direction, respectively. By convolving them with $G(x, y)$ in a plane, the horizontal and vertical brightness difference approximate values G_x and G_y can be obtained. G is the gradient value, and θ is the gradient direction.

The third step is to perform non-maximum suppression on the gradient image. In the process of Gaussian filtering, the edge may be amplified. Use non-maximum suppression to filter non-edge points. The main idea is first to determine the edge, then compare the gradient direction of the edge with the gradient of neighboring points to determine whether to keep or discard.

The fourth step is to use dual thresholds for edge connection. First, a higher threshold is used to detect the edges with a higher degree of certainty, called strong edges, and then a smaller threshold is used to reveal more edges, called weak edges, and choose to keep those edges connected with the strong edges and, finally, form the edges that close the entire image.

In this part, we discuss the image similarity of the same category of facial expression from, respectively, the original image and edge image. Taking the RafDb dataset as the example, the images were cropped to a size of 90*90. The variance indicates the similarity of the image. The smaller the variance, the higher the similarity. We calculated the variance of the pixels and compared the similarity between the original image and the edge image with the same category. Figure 2 shows the average value of pixels in each row of the image. Among the same category, the edge information can show more obvious consistency. When used as a supplement to the features of the original image, an edge can better emphasize the commonality of similar expressions.

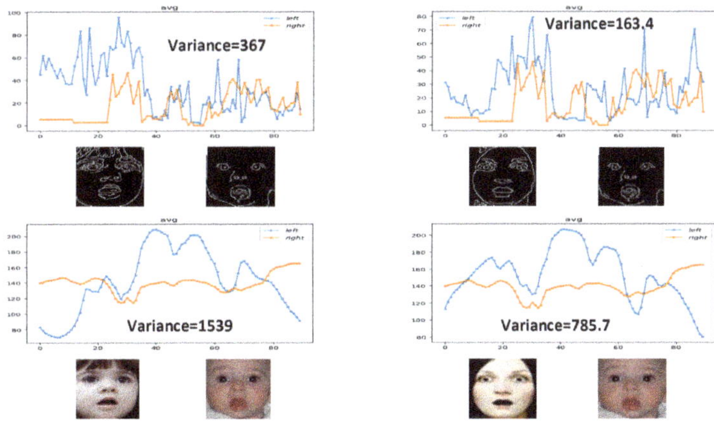

(**a**) Comparison of frightened expressions

Figure 2. *Cont.*

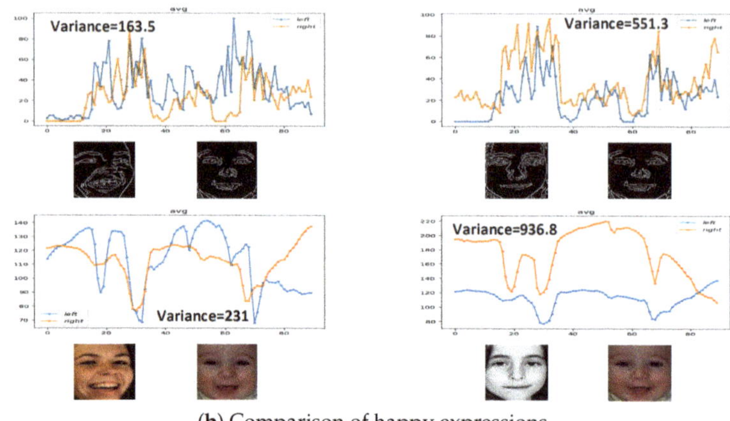

(**b**) Comparison of happy expressions

Figure 2. Image similarity comparison.

3.2. Feature Fusion

The different networks can extract different feature information. The fusion of two dissimilar nets makes the model complementary. To discover and discuss the usefulness of facial edges, we perform a weighted fusion of the features extracted from the two channels.

We use the two parameters of α and β to denote the parameters of OI-Net and EI-Net and calculate the weighted feature of them; see Equations (8) and (9).

$$F_1 = \alpha f_1 \tag{8}$$

$$F_2 = \beta f_2 \tag{9}$$

where f_1 and f_2 are the feature maps of OI-Net and EI-Net, and F_1 and F_2 are the weighted feature maps.

When $\alpha = 1$ and $\beta = 0$, it means that only the feature map from OI-Net is used for softmax classification. Additionally, it is converse when $\alpha = 0$ and $\beta = 1$.

After obtaining the weighted feature map, we use weight fusion to aggregate them.

$$F = Add(F_1, F_2) \tag{10}$$

The size of F_1 and F_2 are 512 dimensions. The size of F is 512 dimensions and is the same. It denotes more supplementary information.

In the next section, we discuss the contribution of F_1 and F_2 with an ablation experiment and explore the important proportion of edges in FER.

4. Experiments

In this section, we conduct a detailed experimental analysis and verify the designed model on three different facial expression datasets, namely CK+ [21], Fer2013 [22], and RafDB [23]. Additionally, the effectiveness of this method is demonstrated through experiments.

4.1. Datasets

- CK+

The CK+ dataset is a relatively extensive laboratory control dataset used for FER. The dataset contains 593 video sequences of 123 subjects, and each sequence contains changes from neutral to peak expressions. According to the facial motion coding system, 327 sequences are labeled with seven basic expression tags (anger, contempt, disgust, fear, happiness, sadness, and surprise).

- Fer2013

This dataset contains 28,709 training images, 3589 verification images, and 3589 test images. Each image has a pixel size of 48*48. It contains seven facial expressions: anger, disgust, fear, happiness, sadness, surprise, and neutral.

- RafDb

The RafDb dataset is a facial expression dataset containing basic expressions or compound expressions annotated by 40 well-trained human annotators on facial expression images. The dataset contains 30,000 facial expression images. In the experiment, we only use 12,271 face images as the training set and 3068 face images as the test set, which contains seven basic expressions: surprised, fear, disgust, happiness, sadness, anger, and neutral expression.

4.2. Experimental Details Settings

The experimental platform configuration is as follows: Ubuntu18.04 system, Intel Xeon Gold 5218 with a CPU frequency of 2.3 GHz, and NVIDIA RTX2080Ti graphics card using Pytorch1.2 learning framework and CUDA framework 10.2.

In the experiment, the same hyperparameters are used for the experiment. Using the SGD optimizer, the weight attenuation coefficient is 5×10^{-4} the momentum is set to 0.9, and the initial learning rate is set to 0.01. The number of iterations on the Fer2013 dataset and RafDB dataset is 150. After 50 iterations, the learning rate is attenuated every five iterations. Each attenuation is 0.9 times the original, and the batch size is set to 128. In order to avoid overfitting, for the Fer2013 dataset, we randomly crop the 48*48 images into the 44*44 size and perform random flips for data enhancement; for the RafDB dataset, we randomly select 100*100 images, cut them into a size of 90*90, and perform random flips for data enhancement. The CK+ dataset uses a 10-fold cross-validation method. The data are randomly divided into ten parts. Each time, nine parts are taken as the training set, and the other part is used as the test set. Then, the accuracy of the ten tests is averaged as the final accurate result of the dataset. The batch size is set to 32 and the number of iterations is set to 40 times; after 15 iterations, the learning rate is attenuated to 0.9 times the original every five iterations, and the 48*48 size image is also randomly cropped to the 44*44 size and randomly flipped for data enhancement.

4.3. Ablation Experiment

4.3.1. Discussion of α and β

In this part, we discuss the parameters of α and β on the RafDb dataset, Fer2013 dataset, and CK+ dataset. In the RafDB dataset, we show the different αs from 0.6 to 1 and the different βs from 0.1 to 0.4. Figure 3a is a broken line chart of the accuracy of the RafDB corresponding to different α and β parameters. We find that when $\alpha = 0.9$ and $\beta = 0.1$, the accuracy rate is as high as 87.58%.

In the Fer2013 dataset, we showed different αs from 0.5 to 1 and different βs from 0.1 to 0.5. Figure 3b is a broken line graph of the Fer2013 accuracy rate when different α and β parameters are selected. We could find that when $\alpha = 1$ and $\beta = 0.2$, the accuracy rate is as high as 73.36%.

In the CK+ dataset, we showed different αs from 0.6 to 1 and βs from 0.1 to 0.4. Figure 3c is a broken line chart of CK+ accuracy when selecting different α and β parameters. When $\alpha = 0.9$ and $\beta = 0.1$, the accuracy rate reaches 98.68%.

From the experiment results between RafDb, Fer2013, and CK+, we found that $\alpha = [0.9, 1]$ and $\beta = [0.1, 0.2]$ can achieve the best recognition effect when performing feature fusion on OI-Net and EI-Net, which shows that features extracted by EI-Net can indeed be used to supplement the features extracted by OI-Net, but not as the primary characterization information. In the FER task, the primary characterization information is still the original image feature; the original image often loses some critical information during feature extraction, and edge features can ameliorate this problem.

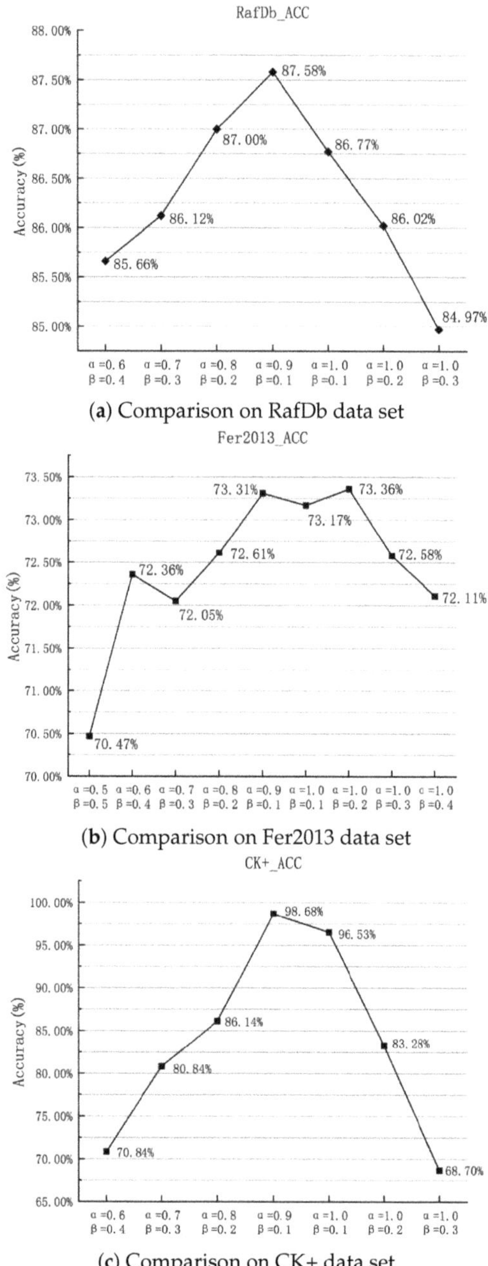

(a) Comparison on RafDb data set

(b) Comparison on Fer2013 data set

(c) Comparison on CK+ data set

Figure 3. Image similarity comparison.

4.3.2. Comparison of Proposed Method with Single Channel

In this part, three methods were compared, respectively, the proposed method (OI-Net+EI-Net), only OI-Net, and only EI-Net. By comparing the recognition rates of the three methods on each expression category, we find that the method proposed in this paper can effectively supplement the facial expression information needed for FER tasks. Figure 4a–c are the comparisons of the three methods on the RafDb, Fer2013, and CK+, respectively.

In RafDb, the recognition rates of fear and disgust are usually low. In Figure 4a, in the absence of edge supplementary features, the recognition rate of fear was 53%, and the recognition rate of disgust was 54%. After adding edge features, the recognition rate of both categories increased by 4%.

Similarly, on the Fer2013 dataset, our method effectively improved the recognition rate of the two categories of anger and disgust. In the absence of edge features, the anger recognition rate is 60%. After adding edge features, the recognition rate reached 63%, an increase of 3%. Similarly, the accuracy rate of disgust also increased by 3%.

On the CK+ dataset, after adding edge features, the recognition rates of the three expression categories of anger, sadness, and contempt were significantly improved, and contempt increased by 15%.

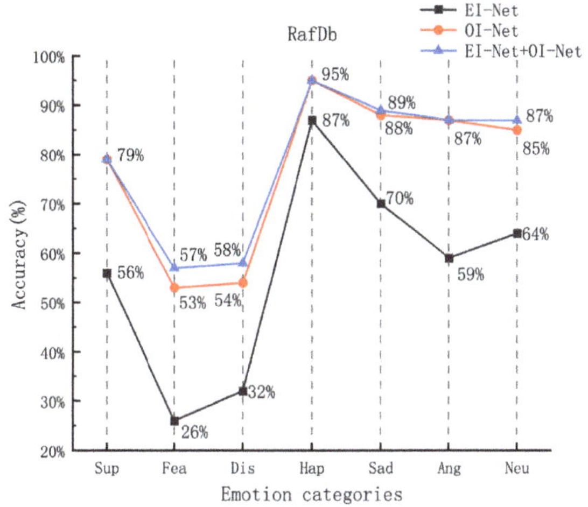

(a) Comparison on RafDb data set

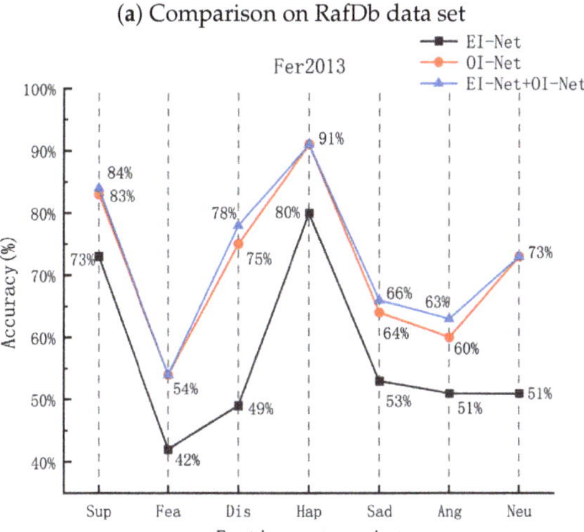

(b) Comparison on Fer2013 data set

Figure 4. *Cont.*

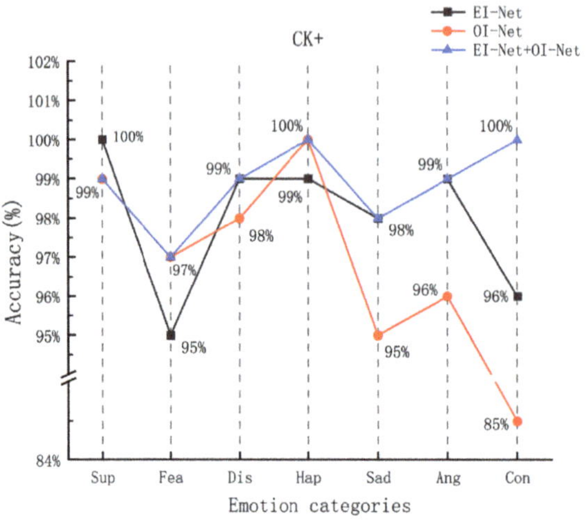

(**c**) Comparison on CK+ data set

Figure 4. Comparison of OI-Net+EI-Net, OI-Net, and EI-Net.

4.4. Confusion Matrices and Comparison with State-of-the-Art Methods

4.4.1. Confusion Matrices

The confusion matrices of the method proposed by the paper on the RafDb, Fer2103, and CK+ datasets can be seen in the Table 1a–c. We achieve outstanding results in both three datasets. However, we should take note of the categories of fear and disgust in RafDb, anger and fear in Fer2013, and sadness in CK+. Additionally, happiness always achieves the best recognition accuracy.

Table 1. Confusion matrices for the CK+, Fer2103, and RafDb datasets.

(a) **RafDb Confusion matrices**						
	Sur	**Fea**	**Dis**	**Hap**	**Sad**	**Ang**	**Neu**
Sur	79%	3%	2%	7%	6%	0	3%
Fea	8%	57%	1%	10%	11%	9%	4%
Dis	5%	0	58%	9%	4%	8%	16%
Hap	0	1%	0	95%	3%	0	1%
Sad	0	1%	0	4%	89%	5%	2%
Ang	0	2%	0	4%	7%	87%	0
Neu	1%	2%	2%	3%	5%	0	87%

(b) **Fer2013 Confusion matrices**						
	Ang	**Dis**	**Fea**	**Hap**	**Sad**	**Sur**	**Neu**
Ang	99%	1%	0	0	0	0	0
Dis	1%	99%	0	0	0	0	0
Fea	0	0	100%	0	0	0	0
Hap	0	0	0	100%	0	0	0
Sad	0	0	0	0	98%	2%	0
Sur	0	0	0	0	0	98%	2%
Con	0	0	0	0	0	0	100%

Table 1. *Cont.*

				(c) CK+ Confusion matrices			
	Ang	**Dis**	**Fea**	**Hap**	**Sad**	**Sur**	**Neu**
Ang	63%	1%	8%	3%	14%	2%	9%
Dis	9%	78%	4%	2%	5%	0	2%
Fea	10%	0	54%	2%	18%	7%	9%
Hap	1%	0	1%	91%	3%	1%	3%
Sad	6%	0	6%	5%	66%	0	17%
Sur	1%	0	7%	3%	2%	84%	3%
Neu	4%	0	4%	4%	14%	1%	73%

4.4.2. Comparison with State-of-the-Art Methods

To verify that the edge features extracted by EI-Net can really supplement the OI-Net, we compare our method with state-of-the-art methods including ReCNN, CNN-SIFT, and so on. Table 2 illustrates the comparison of accuracies between different methods. Our method shows superior performance in RafDb, Fer2013, and CK+. It can be seen that some methods used pretraining, and we have not.

Table 2. Compared with the accuracy of existing methods

Method	Pretraining	RafDb	Fer2013	CK+
[24] Gan et al.	✓	85.69%	-	96.28%
[25] ACNN	✓	85.07%	-	-
[26] SHCNN	-	-	69.10%	-
[27] SCN	✓	87.03%	-	-
[15] Wang et al.	✓	86.90%	-	-
[28] Gao H	-	-	65.2%	-
[14] Minaee et al.	-	-	70.02%	98.0%
[29] MBCC-CNN	-	-	71.52%	98.48%
[30] Multiple CNN	-	-	70.1%	94.9%
[31] Xie et al.	-	-	72.67%	97.11%
[32] CNN+ SIFT	-	-	72.85%	93.46%
[33] DCNN+RLPS	-	72.84%	72.35%	-
[34] ReCNN	✓	87.06%	-	-
[35] LBAN-IL	✓	77.80%	73.11%	-
Ours	-	87.58%	73.36%	98.68%

5. Conclusions

In order to find a simple method to reserve edges and discuss their contribution, we proposed a dual-channel facial expression recognition method to fuse the edge image features and original image features by EI-Net and OI-Net. The weighted fusion method is selected to merge the two network channels, and the fusion parameters are discussed. Through ablation experiments, it is determined that the recognition effect is best when $\alpha = [0.9, 1]$ and $\beta = [0.1, 0.2]$, which also shows that the primary characterization information is still the OI-Net channel.

This paper verifies the proposed method on the three datasets Fer2013, CK+, and RafDb. From the experimental results, the accuracy rate reaches 87.58% on RAFDB, 73.36% on Fer2013, and up to 98.68% on CK+. The experiment demonstrates the effectiveness of the method proposed in this paper.

In the future, we will try to discuss the importance of more low-level features and find a way to achieve feature fusion adaptive parameters.

Author Contributions: Conceptualization, X.T.; methodology, S.L.; software, J.C.; validation J.C. and B.X.; data curation, Q.X. and H.H.; writing—original draft preparation, S.L.; writing—review and editing, Q.X., J.C., H.H. and B.X.; visualization, B.X.; supervision, X.T.; project administration, X.T. All authors have read and agreed to the published version of the manuscript.

Funding: This research was funded by Project of Special Funds for the Cultivation of Guangdong CollegeStudents' Scientific and Technological Innovation ("Climbing Program" Special Funds) grant number pdjh2021a0126.

Data Availability Statement: Not applicable

Acknowledgments: This work was supported by the Project of Special Funds for the Cultivation of Guangdong College Students' Scientific and Technological Innovation ("Climbing Program" Special Funds) (Grant No. pdjh2021a0126), any opinions expressed in this work are those of the authors and do not necessarily represent those of the funding agencies.

Conflicts of Interest: The authors declare that they have no known competing financial interest or personal relationships that could have appeared to influence the work reported in this paper.

References

1. Kumari, J.; Rajesh, R.; Pooja, K. Facial expression recognition: A survey. *Procedia Comput. Sci.* **2015**, *58*, 486–491. [CrossRef]
2. Mehrabian, A.; Russell, J.A. *An Approach to Environmental Psychology*; The MIT Press: Cambridge, MA, USA, 1974.
3. Dalal, N.; Triggs, B. Histograms of oriented gradients for human detection. In Proceedings of the 2005 IEEE Computer Society Conference on Computer Vision and Pattern Recognition (CVPR'05), San Diego, CA, USA, 20–25 June 2005; Volume 1, pp. 886–893.
4. Lowe, D.G. Distinctive image features from scale-invariant keypoints. *Int. J. Comput. Vis.* **2004**, *60*, 91–110. [CrossRef]
5. Ojala, T.; Pietikainen, M.; Maenpaa, T. Multiresolution gray-scale and rotation invariant texture classification with local binary patterns. *IEEE Trans. Pattern Anal. Mach. Intell.* **2002**, *24*, 971–987. [CrossRef]
6. Krizhevsky, A.; Sutskever, I.; Hinton, G.E. Imagenet classification with deep convolutional neural networks. *Commun. ACM* **2017**, *60*, 84–90. [CrossRef]
7. Simonyan, K.; Zisserman, A. Very deep convolutional networks for large-scale image recognition. *arXiv* **2014**, arXiv:1409.1556.
8. He, K.; Zhang, X.; Ren, S.; Sun, J. Deep residual learning for image recognition. In Proceedings of the IEEE Conference on Computer Vision and Pattern Recognition, Las Vegas, NV, USA, 27–30 June 2016; pp. 770–778.
9. Caramihale, T.; Popescu, D.; Ichim, L. Emotion classification using a tensorflow generative adversarial network implementation. *Symmetry* **2018**, *10*, 414. [CrossRef]
10. Hu, M.; Zheng, Y.; Yang, C.; Wang, X.; He, L.; Ren, F. Facial expression recognition using fusion features based on center-symmetric local octonary pattern. *IEEE Access* **2019**, *7*, 29882–29890. [CrossRef]
11. Meena, H.K.; Joshi, S.D.; Sharma, K.K. Facial expression recognition using graph signal processing on HOG. *IETE J. Res.* **2021**, *67*, 667–673. [CrossRef]
12. Shanthi, P.; Nickolas, S. An efficient automatic facial expression recognition using local neighborhood feature fusion. *Multimed. Tools Appl.* **2021**, *80*, 10187–10212. [CrossRef]
13. Xie, S.; Hu, H.; Wu, Y. Deep multi-path convolutional neural network joint with salient region attention for facial expression recognition. *Pattern Recognit.* **2019**, *92*, 177–191. [CrossRef]
14. Minaee, S.; Minaei, M.; Abdolrashidi, A. Deep-emotion: Facial expression recognition using attentional convolutional network. *Sensors* **2021**, *21*, 3046. [CrossRef] [PubMed]
15. Wang, K.; Peng, X.; Yang, J.; Meng, D.; Qiao, Y. Region attention networks for pose and occlusion robust facial expression recognition. *IEEE Trans. Image Process.* **2020**, *29*, 4057–4069. [CrossRef] [PubMed]
16. Levi, G.; Hassner, T. Emotion recognition in the wild via convolutional neural networks and mapped binary patterns. In Proceedings of the 2015 ACM on International Conference on Multimodal Interaction, Seattle, WA, USA, 9–13 November 2015; pp. 503–510.
17. Zhang, H.; Huang, B.; Tian, G. Facial expression recognition based on deep convolution long short-term memory networks of double-channel weighted mixture. *Pattern Recognit. Lett.* **2020**, *131*, 128–134. [CrossRef]
18. Bougourzi, F.; Dornaika, F.; Mokrani, K.; Taleb-Ahmed, A.; Ruichek, Y. Fusing Transformed Deep and Shallow features (FTDS) for image-based facial expression recognition. *Expert Syst. Appl.* **2020**, *156*, 113459. [CrossRef]
19. Yu, M.; Zheng, H.; Peng, Z.; Dong, J.; Du, H. Facial expression recognition based on a multi-task global-local network. *Pattern Recognit. Lett.* **2020**, *131*, 166–171. [CrossRef]
20. Canny, J. A computational approach to edge detection. *IEEE Trans. Pattern Anal. Mach. Intell.* **1986**, 679–698. [CrossRef]
21. Lucey, P.; Cohn, J.F.; Kanade, T.; Saragih, J.; Ambadar, Z.; Matthews, I. The extended cohn-kanade dataset (ck+): A complete dataset for action unit and emotion-specified expression. In Proceedings of the 2010 IEEE Computer Society Conference on Computer Vision and Pattern Recognition-Workshops, San Francisco, CA, USA, 13–18 June 2010; pp. 94–101.
22. Carrier, P.L.; Courville, A.; Goodfellow, I.J.; Mirza, M.; Bengio, Y. *FER-2013 Face Database*; Universit de Montral: Montral, QC, Canada, 2013.

23. Li, S.; Deng, W.; Du, J. Reliable crowdsourcing and deep locality-preserving learning for expression recognition in the wild. In Proceedings of the IEEE Conference on Computer Vision and Pattern Recognition, Honolulu, HI, USA, 21–26 July 2017; pp. 2852–2861.

24. Gan, Y.; Chen, J.; Yang, Z.; Xu, L. Multiple attention network for facial expression recognition. *IEEE Access* **2020**, *8*, 7383–7393. [CrossRef]

25. Li, Y.; Zeng, J.; Shan, S.; Chen, X. Occlusion aware facial expression recognition using CNN with attention mechanism. *IEEE Trans. Image Process.* **2018**, *28*, 2439–2450. [CrossRef]

26. Miao, S.; Xu, H.; Han, Z.; Zhu, Y. Recognizing facial expressions using a shallow convolutional neural network. *IEEE Access* **2019**, *7*, 78000–78011. [CrossRef]

27. Wang, K.; Peng, X.; Yang, J.; Lu, S.; Qiao, Y. Suppressing uncertainties for large-scale facial expression recognition. In Proceedings of the IEEE/CVF Conference on Computer Vision and Pattern Recognition, Seattle, WA, USA, 14–19 June 2020; pp. 6897–6906.

28. Gao, H.; Ma, B. A robust improved network for facial expression recognition. *Front. Signal Process.* **2020**, *4*, 4. [CrossRef]

29. Shi, C.; Tan, C.; Wang, L. A facial expression recognition method based on a multibranch cross-connection convolutional neural network. *IEEE Access* **2021**, *9*, 39255–39274. [CrossRef]

30. Chuanjie, Z.; Changming, Z. Facial Expression Recognition Integrating Multiple CNN Models. In Proceedings of the 2020 IEEE 6th International Conference on Computer and Communications (ICCC), Chengdu, China, 11–14 December 2020; pp. 1410–1414.

31. Xie, W.; Shen, L.; Duan, J. Adaptive weighting of handcrafted feature losses for facial expression recognition. *IEEE Trans. Cybern.* **2019**, *51*, 2787–2800. [CrossRef] [PubMed]

32. Wang, H.; Hou, S. Facial expression recognition based on the fusion of CNN and SIFT features. In Proceedings of the 2020 IEEE 10th International Conference on Electronics Information and Emergency Communication (ICEIEC), Beijing, China, 17–19 July 2020; pp. 190–194.

33. Li, H.; Xu, H. Deep reinforcement learning for robust emotional classification in facial expression recognition. *Knowl.-Based Syst.* **2020**, *204*, 106172. [CrossRef]

34. Xia, Y.; Yu, H.; Wang, X.; Jian, M.; Wang, F.Y. Relation-aware facial expression recognition. *IEEE Trans. Cogn. Dev. Syst.* **2021**, *14*, 1143–1154. [CrossRef]

35. Li, H.; Wang, N.; Yu, Y.; Yang, X.; Gao, X. LBAN-IL: A novel method of high discriminative representation for facial expression recognition. *Neurocomputing* **2021**, *432*, 159–169. [CrossRef]

 *symmetry*

Article

Autonomous Navigation and Obstacle Avoidance for Small VTOL UAV in Unknown Environments

Cheng Chen [1], Zian Wang [2,*], Zheng Gong [3], Pengcheng Cai [3], Chengxi Zhang [4] and Yi Li [5,*]

[1] School of Aerospace Engineering, Shenyang Aerospace University, Shenyang 110000, China
[2] China Academy of Launch Vehicle Technology, Beijing 100076, China
[3] Department of Aerospace Engineering, Nanjing University of Aeronautics and Astronautics, Nanjing 210016, China
[4] Key Laboratory of Advanced Control for Light Industry Processes, Ministry of Education, School of Internet of Things Engineering, Jiangnan University, Wuxi 214122, China
[5] School of Electrical Engineering and Automation, Tianjin University of Technology, Tianjin 300384, China
* Correspondence: wangzian@nuaa.edu.cn (Z.W.); ly@email.tjut.edu.cn (Y.L.)

Abstract: This paper takes autonomous exploration in unknown environments on a small co-axial twin-rotor unmanned aerial vehicle (UAV) platform as the task. The study of the fully autonomous positioning in unknown environments and navigation system without global navigation satellite system (GNSS) and other auxiliary positioning means is carried out. Algorithms that are based on the machine vision/proximity detection/inertial measurement unit, namely the combined navigation algorithm and indoor simultaneous location and mapping (SLAM) algorithm, are not only designed theoretically but also realized and verified in real surroundings. Additionally, obstacle detection, the decision-making of avoidance motion and motion planning methods such as Octree are also proposed, which are characterized by randomness and symmetry. The demonstration of the positioning and navigation system in the unknown environment and the verification of the indoor obstacle-avoidance flight were both completed through building an autonomous navigation and obstacle avoidance simulation system.

Keywords: autonomous navigation; obstacle avoidance; target detection; VI-SLAM

Citation: Chen, C.; Wang, Z.; Gong, Z.; Cai, P.; Zhang, C.; Li, Y. Autonomous Navigation and Obstacle Avoidance for Small VTOL UAV in Unknown Environments. *Symmetry* **2022**, *14*, 2608. https://doi.org/10.3390/sym14122608

Academic Editors: Sergei D. Odintsov and Jan Awrejcewicz

Received: 16 September 2022
Accepted: 17 November 2022
Published: 9 December 2022

Publisher's Note: MDPI stays neutral with regard to jurisdictional claims in published maps and institutional affiliations.

1. Introduction

With the development of UAV technology, UAVs are playing an increasingly essential role in some routine tasks or even under special circumstances in both civil and military applications [1–3]. For example, some UAVs can be used for military reconnaissance, autonomous identification and attack, and they can be also used to explore an unknown region and map it.

The survival capability of drones is a major problem, especially in some complex or even unknown environment; as a result, autonomous navigation is introduced. While external information should be introduced into the navigation system for better effects of flight control, the path planning and obstacle avoidance during autonomous navigation in unknown environments becomes a crucial issue for unmanned surface vehicles (USVs) [4].

A detection and avoidance system was presented for the autonomous navigation of UAVs in urban air mobility (UAM) applications by Enrique Aldao et al. [5]. The principle and navigation method of astronomical spectral velocity measurement, as well as the technical realization of the solar atomic frequency discriminator for autonomous navigation (SAFDAN) based on atomic frequency discrimination velocity measurement were comprehensively introduced by Wei Zhang et al. [6]. A self-trained controller for autonomous navigation in static and dynamic (with moving walls and nets) challenging environments (including trees, nets, windows, and pipe) using deep reinforcement learning, simultaneously trained using multiple rewards was introduced by Ramezani Dooraki Amir [7]. A visual predictive control (VPC) scheme adapted to the autonomous navigation problem among static obstacles was proposed by Durand Petiteville A. [8]. Nowadays the majority of quadrotor drones are manually operated and use global positioning system (GPS) signals for navigation, thus greatly limiting the flight range of drones and consuming a lot of manpower and material resources. To solve the problem, Liu Liwen et al. [9] proposed a method of realizing autonomous flight and conflict avoidance of quadrotor UAVs by using a multi-sensor system and deep learning methods in extreme flight conditions through track prediction. Moreover, in the research of Sina Sajjadi [10], a vision-based target-tracking problem was formulated in the form of a cascaded adaptive nonlinear model predictive control (MPC) strategy. A typical ASV/USV unit with standard radio remote control system to the fully autonomous mode was modernized by Specht C et al. [11]. A method of the obstacle avoidance planning of unmanned surface vehicles based on an improved artificial potential field was proposed by S Xie et al. [12]. Navigation problems of unmanned aerial vehicles (UAVs) flying in a formation in a free and an obstacle-laden environment were investigated in the work of Xiaohua Wang et al. [13]. An unmanned underwater vehicle (UUV) simulator, an extension of the open-source robotics simulator Gazebo to underwater scenarios, was described in the work of Musa Morena Marcusso Manhães et al. [14].

This paper completes the development of an autonomous positioning algorithm and mapping and trajectory planning algorithm. Algorithms that are based on the machine vision/proximity detection/inertial measurement unit, namely the combined navigation algorithm and indoor SLAM algorithm, were designed and realized. Additionally, obstacle detection, the decision-making of avoidance motion and motion planning methods are also proposed. An autonomous navigation and obstacle avoidance simulation system is proposed. A target recognition algorithm was developed and finally the proposed autonomous navigation and obstacle avoidance simulation system was demonstrated and verified through physical experiments. The proposed algorithm and system play an important role in many practical systems and applications, such as sweeping robots, driverless cars, virtual reality technology (VR) and intelligent robots. According to the experiment results, the maximum error along x direction is less than 0.5 m, less than 0.6 m along y direction, and 0.4 m along z direction. The yaw angle error is less than 5°, and absolute error is less than 0.3 m. The calculated closed-loop error is about $0.3/70 = 0.4\%$.

The proposed autonomous navigation and obstacle avoidance system, mainly consisting of three components, namely autonomous positioning, environment mapping and trajectory planning and target detection and recognition, was used to realize autonomous environmental exploration without GPS. Its workflow and output are shown in Figure 1.

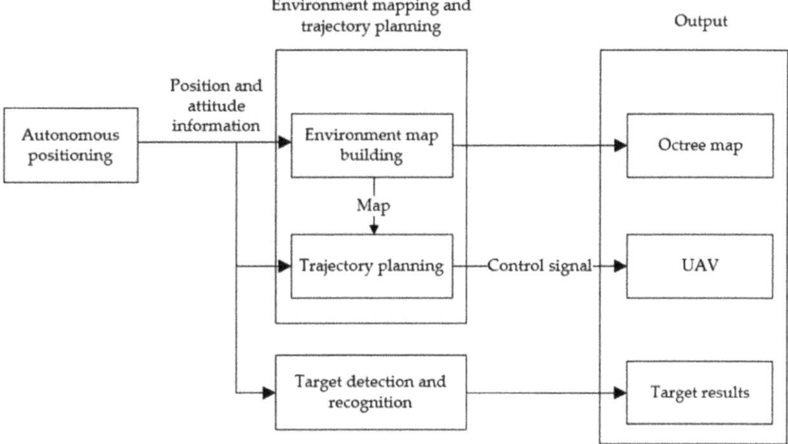

Figure 1. The structural diagram of the autonomous system.

Section 2 introduces the autonomous positioning algorithm and its simulation results. The detailed design and the mechanism of map-building and trajectory-planning algorithm are provided in Section 3. The detailed design of the target detection and recognition algorithm are in Section 4. In Section 5, the validation of the proposed algorithm is verified by a flight test and the test environment; the results of the flight test are also introduced in this part.

1.1. Autonomous Positioning

Visual–inertial simultaneous localization and mapping (VI-SLAM) [15–17] was used to solve the autonomous positioning problem of UAVs without GPS. The system established a global coordinate system by regarding the take-off position as the origin and estimated the relative pose of the UAV by the fusion of the measurement information of the visional and inertial navigation system.

1.2. Map Building and Trajectory Planning

Mapping and path planning were used to solve the motion planning problems of UAVs [18]. By building a raster map and running a path search algorithm, the UAV could be guided to specific targets and avoid known obstacles at the same time.

1.3. Target Detection and Recognition

The target detection and recognition system was used to search, detect and classify the targets in the field of vision, and provide reference information for the subsequent behavior decisions [19].

2. Autonomous Positioning

2.1. The Introduction of the Autonomous Positioning Module

The VI-SLAM algorithm adopts binocular and inertial measurement units (IMU). According to the operation process, the system is divided into four parallel parts: signal preprocessing thread, pose initialization thread, VisualInertial Odometry (VIO) thread, and loopback optimization thread [20,21]. The operation flow of the system is shown in Figure 2:

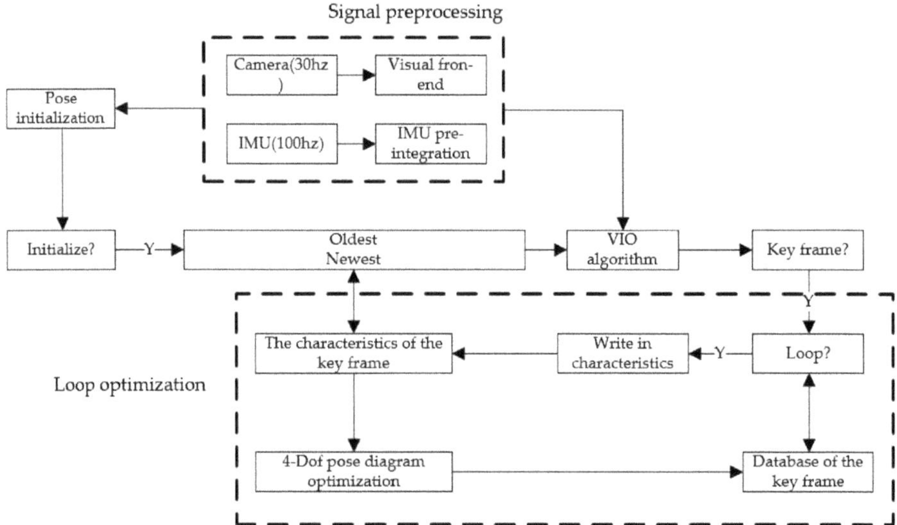

Figure 2. The flow diagram of VI-SLAM.

2.2. The Preprocessing of Signals

This section describes the VIO preprocessing procedures. For visual measurement, we tracked features between successive frames and detected new features in the latest frame. For IMU measurement, pre-integration was adopted between two consecutive frames. Due to the high measurement noise of the low-cost IMU, the offsets of the inertial components were obtained by external calibration during the pre-integration process.

2.2.1. Visual Front-End

The basic task of visual front-end is to extract feature points, which mainly includes two parts: feature tracking and key frame selection.

(1)   Bidirectional Kanade–Lucas–Tomasi (KLT) tracking

For the binocular system, the left and right visual images, R_k and L_k were obtained in each time sequence. Firstly, Harris corner detection was used on the two images, and even distribution of feature was ensured by setting the minimum interval of pixels between two adjacent features. Then a KLT sparse optical flow algorithm and polar line search algorithm were used for feature matching, an RANSAC algorithm of basic matrix model was also used to remove outer points, and the matched binocular feature point pairs were obtained. We then applied the same to R_{k+1} and L_{k+1}.

Next, bidirectional KLT tracking was adopted for the feature points in R_k and R_{k+1}; that is, a KLT matching and RANSAC screening was carried out from R_k to R_{k+1}, and then the remaining matching points were used for a matching and screening from R_{k+1} to R_k to ensure feature stability to the maximum extent.

(2)   The selection strategy of the key frames

At the visual front-end, the key frame selection was performed simultaneously, and there were two selection criteria. The first one was the mean parallax from the previous keyframe. If the mean parallax of the feature points tracked between the current frame and the latest key frame exceeded a certain threshold, the frame would be considered as a new keyframe. Another one was tracking quality. If the number of tracked features was under a certain threshold, we treated this frame as a new key frame, which avoided the complete loss of tracking features.

2.2.2. IMU Pre-Integration

The rotation error of the Euler angle was parameterized by IMU pre-integration. Here, the pre-integration mode proposed by Vins-Mono was used, the covariance transfer function was derived through the IMU error state dynamics under continuous time and the bias correction was introduced to correct the error.

The measurement results of original gyroscope and accelerometer of IMU, $\hat{\omega}$ and $\hat{a}$ are shown as follows:

$$\begin{aligned}
\hat{a}_t &= a_t + b_{a_t} + R^t_w g^w + n_a \\
\hat{\omega}_t &= \omega_t + b_{\omega_t} + n_\omega
\end{aligned} \tag{1}$$

IMU measurement values were measured in the body coordinate system, which is the resultant force that balances the gravity and platform dynamics, and it can be affected by accelerometer offset b_a, gyroscope offset b_ω and additional noise. Under the assumption that the additional noise in the measured value of accelerometer and gyroscope is Gaussian noise, $n_a \sim N(0, \sigma_a^2)$, $n_\omega \sim N(0, \sigma_\omega^2)$. Accelerometer offset and gyroscope offset were modeled as random walks and their derivatives are $n_{b_a} \sim N(0, \sigma_{b_a}^2)$ and $n_{b_\omega} \sim N(0, \sigma_{b_\omega}^2)$, respectively.

$$\begin{aligned}
\dot{b}_{a_t} &= n_{b_a} \\
\dot{b}_{\omega_t} &= n_{b_\omega}
\end{aligned} \tag{2}$$

Given two moments corresponding to the body coordinate system b_k and b_{k+1}, the position, velocity, and direction states can be transmitted by inertial measurement values in the world coordinate system between time intervals $[t_k, t_{k+1}]$:

$$
\begin{aligned}
p_{b_{k+1}}^{\omega} &= p_{b_k}^{\omega} + v_{b_k}^{\omega}\Delta t_k \\
&\quad + \iint_{t\in[t_k,t_{k+1}]} \left(R_t^{\omega}(\hat{a}_t - b_{a_t} - n_a) - g^{\omega}\right)dt^2 \\
v_{b_{k+1}}^{\omega} &= v_{b_k}^{\omega} + \int_{t\in[t_k,t_{k+1}]} \left(R_t^{\omega}(\hat{a}_t - b_{a_t} - n_a) - g^{\omega}\right)dt \\
q_{b_{k+1}}^{\omega} &= q_{b_k}^{\omega} \otimes \int_{t\in[t_k,t_{k+1}]} \tfrac{1}{2}\Omega(\hat{\omega}_t - b_{\omega_t} - n_\omega)q_t^{b_k}dt
\end{aligned}
\tag{3}
$$

where

$$
\Omega(\omega) = \begin{bmatrix} -\lfloor\omega\rfloor_\times & \omega \\ -\omega^T & 0 \end{bmatrix} \cdot \lfloor\omega\rfloor_\times = \begin{bmatrix} 0 & -\omega_z & \omega_y \\ \omega_x & 0 & -\omega_x \\ -\omega_y & \omega_z & 0 \end{bmatrix}
\tag{4}
$$

Δt_k is the span of the interval $[t_k, t_{k+1}]$.

Clearly, the state transmission of IMU requires the rotation, position and velocity of the coordinate system b_k. When these initial states change, we need to retransmit the IMU measurement values. Especially in optimization-based algorithms, IMU measurement values need to be retransmitted between them every time the pose is adjusted, and this transfer strategy is computationally demanding. In order to avoid retransmission, a pre-integration algorithm was introduced.

After changing the reference coordinate system from the world coordinate system to the local coordinate system b_k, pre-integration can be only applied to the relevant part of the linear acceleration $\hat{a}$ and angular velocity $\hat{\omega}$ as follows:

$$
\begin{aligned}
\mathbf{R}_w^{b_k}\mathbf{p}_{b_{k+1}}^w &= \mathbf{R}_w^{b_k}\left(\mathbf{p}_{b_k}^w + \mathbf{v}_{b_k}^w\Delta t_k - \tfrac{1}{2}\mathbf{g}^w\Delta t_k^2\right) + \alpha_{b_{k+1}}^{b_k} \\
\mathbf{R}_w^{b_k}\mathbf{v}_{b_{k+1}}^w &= \mathbf{R}_w^{b_k}\left(\mathbf{v}_{b_k}^w - \mathbf{g}^w\Delta t_k\right) + \beta_{b_{k+1}}^{b_k} \\
\mathbf{q}_w^{b_k} \otimes \mathbf{q}_{b_{k+1}}^w &= \gamma_{b_{k+1}}^{b_k} n
\end{aligned}
\tag{5}
$$

$$
\begin{aligned}
\alpha_{b_{k+1}}^{b_k} &= \iint_{t\in[t_k,t_{k+1}]} \mathbf{R}_t^{b_k}\left(\hat{\mathbf{a}}_t - \mathbf{b}_{a_t} - \mathbf{n}_a\right)dt^2 \\
\beta_{b_{k+1}}^{b_k} &= \int_{t\in[t_k,t_{k+1}]} \mathbf{R}_t^{b_k}\left(\hat{\mathbf{a}}_t - \mathbf{b}_{a_t} - \mathbf{n}_a\right)dt \\
\gamma_{b_{k+1}}^{b_k} &= \int_{t\in[t_k,t_{k+1}]} \tfrac{1}{2}\Omega(\hat{\omega}_t - \mathbf{b}_{\omega_t} - \mathbf{n}_w)\gamma_t^{b_k}dt
\end{aligned}
\tag{6}
$$

Among them, the pre-integration term (6) can be obtained by the IMU measurement value, which regards b_k as the reference frame. $\alpha_{b_{k+1}}^{b_k}$, $\beta_{b_{k+1}}^{b_k}$, $\gamma_{b_{k+1}}^{b_k}$ are only related to the IMU offset in b_k and b_{k+1}, and have nothing to do with other states. When the offset estimation changed, if the change was small, we adjusted $\alpha_{b_{k+1}}^{b_k}$, $\beta_{b_{k+1}}^{b_k}$ and $\gamma_{b_{k+1}}^{b_k}$ according to their first-order approximation to the offset; otherwise, was retransmitted. This strategy saves a lot of computational resources for optimization-based algorithms because the repeated transmission of IMU measurement values is avoided.

Under discrete-time conditions, different numerical integration methods can be used, such as Euler integration, midpoint integration, RK4 integration, etc. The Euler integral is chosen in this section.

At the beginning, $\alpha_{b_k}^{b_k}$ and $\beta_{b_k}^{b_k}$ were 0, and $\gamma_{b_k}^{b_k}$ was a unit quaternion. The average values of α, β and γ in (6) were gradually transmitted as follows. The additional noise

n_a, n_ω were unknown and they were treated as 0 in the actual program. The estimated value of pre-integration is obtained as follows:

$$
\begin{aligned}
\hat{\boldsymbol{\alpha}}_{i+1}^{b_k} &= \hat{\boldsymbol{\alpha}}_i^{b_k} + \hat{\boldsymbol{\beta}}_i^{b_k}\delta t + \tfrac{1}{2}\mathbf{R}\!\left(\hat{\gamma}_i^{b_k}\right)(\hat{\mathbf{a}}_i - \mathbf{b}_{a_i})\delta t^2 \\
\hat{\boldsymbol{\beta}}_{i+1}^{b_k} &= \hat{\boldsymbol{\beta}}_i^{b_k} + \mathbf{R}\!\left(\hat{\gamma}_i^{b_k}\right)(\hat{\mathbf{a}}_i - \mathbf{b}_{a_i})\delta t \\
\hat{\gamma}_{i+1}^{b_k} &= \hat{\gamma}_i^{b_k} \otimes \begin{bmatrix} 1 \\ \tfrac{1}{2}(\hat{\omega}_i - \mathbf{b}_{w_i})\delta t \end{bmatrix}
\end{aligned}
\tag{7}
$$

where i is the discrete moment corresponding to the IMU measurement value during time interval $[t_k, t_{k+1}]$, and δt is the time interval between IMU measurement value i and $i+1$.

We then turned our focus to the covariance transmission problem. Since the four-dimensional rotational quaternion $\gamma_i^{b_k}$ was over-parameterized, we defined its error as the perturbation around its mean value:

$$
\gamma_t^{b_k} \approx \hat{\gamma}_t^{b_k} \otimes \begin{bmatrix} 1 \\ \tfrac{1}{2}\delta\theta_t^{b_k} \end{bmatrix}
\tag{8}
$$

where $\delta\theta_t^{b_k}$ is the three-dimensional small perturbation.

Thus, the linearized equation of the error term under continuous time can be derived as follows:

$$
\begin{aligned}
\begin{bmatrix} \delta\dot{\boldsymbol{\alpha}}_t^{b_k} \\ \delta\dot{\boldsymbol{\beta}}_t^{b_k} \\ \delta\dot{\theta}_t^{b_k} \\ \delta\dot{b}_{a_t} \\ \delta\dot{b}_{w_t} \end{bmatrix} &=
\begin{bmatrix}
0 & I & 0 & 0 & 0 \\
0 & 0 & -R_t^{b_k}\lfloor \hat{a}_t - b_{a_t}\rfloor_\times & -R_t^{b_k} & 0 \\
0 & 0 & -\lfloor \hat{\omega}_t - b_{w_t}\rfloor_\times & 0 & -I \\
0 & 0 & 0 & 0 & 0 \\
0 & 0 & 0 & 0 & 0
\end{bmatrix}
\begin{bmatrix} \delta\boldsymbol{\alpha}_t^{b_k} \\ \delta\boldsymbol{\beta}_t^{b_k} \\ \delta\theta_t^{b_k} \\ \delta b_{a_t} \\ \delta b_{w_t} \end{bmatrix} \\[6pt]
&+ \begin{bmatrix}
0 & 0 & 0 & 0 \\
-R_t^{b_k} & 0 & 0 & 0 \\
0 & -I & 0 & 0 \\
0 & 0 & I & 0 \\
0 & 0 & 0 & I
\end{bmatrix}
\begin{bmatrix} n_a \\ n_\omega \\ n_{b_a} \\ n_{b_w} \end{bmatrix} \\[6pt]
&= F_t \delta z_t^{b_k} + G_t n_t
\end{aligned}
\tag{9}
$$

$\mathbf{P}_{b_{k+1}}^{b_k}$ can be calculated by the recursion and updating of the first-order discrete-time covariance of initial covariance $\mathbf{P}_{b_k}^{b_k} = 0$:

$$
\begin{aligned}
\mathbf{P}_{t+\delta t}^{b_k} &= (\mathbf{I} + \mathbf{F}_t \delta t)\mathbf{P}_t^{b_k}(\mathbf{I} + \mathbf{F}_t \delta t)^T + (\mathbf{G}_t \delta t)\mathbf{Q}(\mathbf{G}_t \delta t)^T \\
& t \in [k, k+1]
\end{aligned}
\tag{10}
$$

where $\mathbf{Q}$ is the diagonal covariance matrix $(\sigma_a^2, \sigma_\omega^2, \sigma_{b_a}^2, \sigma_{b_\omega}^2)$ of the noise.

Meanwhile, the first-order Jacobian matrix $J_{b_{k+1}}$ of $\delta z_{b_{k+1}}^{b_k}$ can also be calculated by the recursion of the initial Jacobian matrix $J_{b_{k+1}} = I$.

$$
\mathbf{J}_{t+\delta t} = (\mathbf{I} + \mathbf{F}_t \delta t)\mathbf{J}_t, \; t \in [k, k+1]
\tag{11}
$$

Using Equation (11), covariance matrix $\mathbf{P}_{b_{k+1}}^{b_k}$ and Jacobian matrix $J_{b_{k+1}}$ were obtained. The first-order approximation of $\alpha_{b_{k+1}}^{b_k}$, $\beta_{b_{k+1}}^{b_k}$, $\gamma_{b_{k+1}}^{b_k}$ relevant to the offset can be expressed as follows:

$$\boldsymbol{\alpha}_{b_{k+1}}^{b_k} \approx \hat{\boldsymbol{\alpha}}_{b_{k+1}}^{b_k} + J_{b_a}^{\alpha}\,\delta\mathbf{b}_{a_k} + J_{b_w}^{\alpha}\,\delta\mathbf{b}_{w_k}$$

$$\boldsymbol{\beta}_{b_{k+1}}^{b_k} \approx \hat{\boldsymbol{\beta}}_{b_{k+1}}^{b_k} + J_{b_a}^{\beta}\,\delta\mathbf{b}_{a_k} + J_{b_w}^{\beta}\,\delta\mathbf{b}_{w_k} \tag{12}$$

$$\boldsymbol{\gamma}_{b_{k+1}}^{b_k} \approx \hat{\boldsymbol{\gamma}}_{b_{k+1}}^{b_k} \otimes \begin{bmatrix} 1 \\ \frac{1}{2}J_{b_w}^{\gamma}\,\delta\mathbf{b}_{w_k} \end{bmatrix}$$

where $J_{b_a}^{\alpha}$ is the subblock matrix of $J_{b_{k+1}}$, and its position corresponds to $\frac{\delta\alpha_{b_{k+1}}^{b_k}}{\delta b_{a_k}}$, which also makes sense for $J_{b_\omega}^{\alpha}$, $J_{b_a}^{\beta}$, $J_{b_\omega}^{\beta}$, $J_{b_\omega}^{\gamma}$.

When the offset estimation changed slightly, we used Equation (12) to approximately correct the results of pre-integration without retransmission.

Hence, the corresponding covariance $\mathbf{P}_{b_{k+1}}^{b_k}$ of the IMU measurement model could be obtained:

$$\begin{bmatrix} \hat{\boldsymbol{\alpha}}_{bk+1}^{b_k} \\ \hat{\boldsymbol{\beta}}_{b_{k+1}}^{b_k} \\ \hat{\boldsymbol{\gamma}}_{b_{k+1}}^{b_k} \\ 0 \\ 0 \end{bmatrix} = \begin{bmatrix} \mathbf{R}_w^{b_k}\left(\mathbf{p}_{b_{k+1}}^w - \mathbf{p}_{b_k}^w + \frac{1}{2}\mathbf{g}^w\Delta t_k^2 - \mathbf{v}_{b_k}^w\Delta t_k\right) \\ \mathbf{R}_w^{b_k}\left(\mathbf{v}_{b_{k+1}}^w + \mathbf{g}^w\Delta t_k - \mathbf{v}_{b_k}^w\right) \\ \mathbf{q}_{b_k}^{w^{-1}} \otimes \mathbf{q}_{b_{k+1}}^w \\ \mathbf{b}_{ab_{k+1}} - \mathbf{b}_{ab_k} \\ \mathbf{b}_{wb_{k+1}} - \mathbf{b}_{wb_k} \end{bmatrix} \tag{13}$$

2.3. Pose Initialization

The pose initialization part is responsible for establishing the coordinate system and maintaining the feature points and the description of UAV in the coordinate system at the early stage of the operation process of system [22,23]. Compared with the monocular, tightly-coupled VIO system, the binocular system can directly recover the depth of feature points to complete initialization under stationary conditions.

2.3.1. The Depth Estimation of Feature Points

Since the binocular camera system was used, the depth of feature points could be calculated directly from the disparity and the relative pose of the camera. The analysis started with ideal conditions: under the assumption that the left and right cameras were in the same plane (the optical axis was parallel) and the camera parameters (focal length f) were identical. Then the depth value could be obtained, as shown in Figure 3:

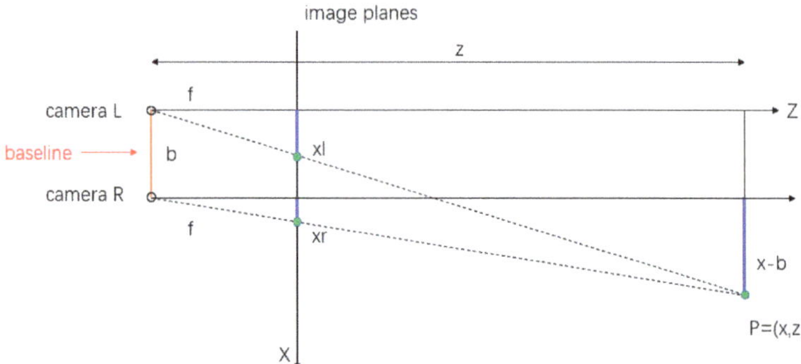

Figure 3. Diagram of the imaging model.

As can be seen from Figure 3 above, an image plane was established with X, Z axes and the distance between point P and the axis of camera R is 'x − b'; the intersection point of the link between camera L and P in X axis is recorded as 'xl' ('l' means 'left') and the same for 'xr'; 'b' is the length of baseline.

According to the triangle similarity:

$$\begin{aligned} \frac{z}{f} &= \frac{x}{xl} = \frac{x-b}{yr} \\ \frac{z}{f} &= \frac{y}{yl} = \frac{y}{yr} \end{aligned} \tag{14}$$

where b is the baseline length, and the optical axes of the two cameras are both located in the XOZ plane.

Then the position of point P can be estimated:

$$\begin{bmatrix} x \\ y \\ z \end{bmatrix} = \begin{bmatrix} xl \cdot z/f \\ yl \cdot z/f \\ f \cdot b/(xl - xr) \end{bmatrix} = \begin{bmatrix} b + xr \cdot z/f \\ yr \cdot z/f \\ f \cdot b/d \end{bmatrix} \tag{15}$$

For non-ideal camera imaging model, the perturbation included optical axis deviation and image distortion. In this case, it was necessary to correct the image and transform it into the ideal situation, as shown in Figure 4.

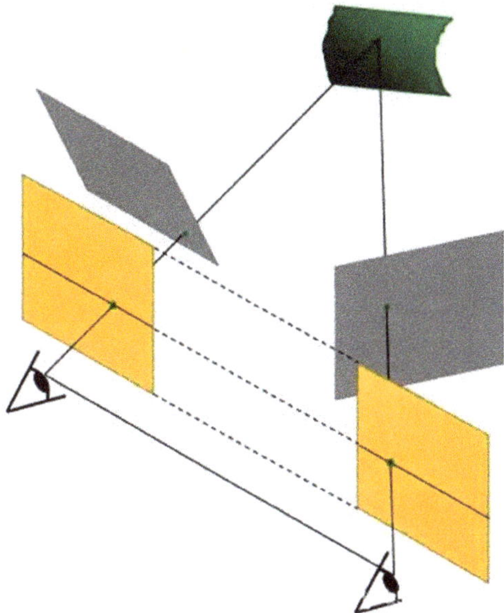

Figure 4. Image correction.

After obtaining the space coordinates of feature points, they needed to be converted into inverse depth to connect them with the SLAM system. Compared with the direct depth expression, the inverse depth error is more consistent with the Gaussian distribution and has better numerical stability. The conversion formula of inverse depth λ is as follows:

$$\begin{aligned} \lambda &= 1/d \\ &= 1/\sqrt{x^2 + y^2 + z^2} \end{aligned} \tag{16}$$

2.3.2. Pose Initialization

When establishing the SLAM coordinate system, the northeast sky coordinate system was established by taking the origin of the camera coordinate system in the first frame as the origin under state of rest. Then the ith landmark feature point m_i in the first frame can be expressed by the inverse depth in the world coordinate system as follows:

$$\begin{bmatrix} \alpha_i \\ \beta_i \\ \lambda_i \end{bmatrix} = \begin{bmatrix} \mathrm{atan}(y/z) \\ \mathrm{atan}(-x/z) \\ 1/\sqrt{x^2+y^2+z^2} \end{bmatrix} \tag{17}$$

In the tracking process of the second frame, the pose of the second frame was obtained by matching the landmark point in the new frame and the counterpart in the first frame and running pose calculation, and the available landmark points were updated for the subsequent pose calculation.

2.4. VIO Algorithm

As the core part of the pose updating of the VI-SLAM algorithm, the VIO algorithm requires accuracy and running speed at the same time. Therefore, the sliding window method based on a nonlinear optimization strategy was selected. The basic idea of the sliding window method is firstly introduced in this section, and then the calculation methods of IMU and visual measurement residual that needed to be updated in the formula are introduced separately.

2.4.1. Sliding Window Method

After the initialization of the estimator, the binocular VIO based on sliding windows was employed for high-precision and robust state estimation. The diagram of the sliding window method is shown in Figure 5:

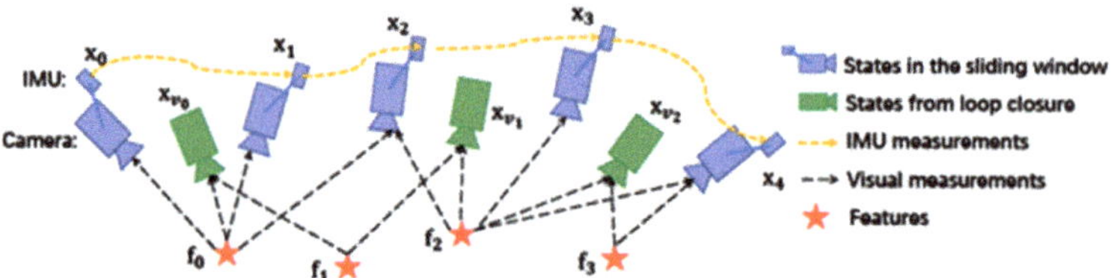

Figure 5. Sliding window method.

The full-state vector in the sliding window is defined as:

$$\begin{aligned}
\mathcal{X} &= \left[x_0, x_1, \cdots x_n, x_c^b, \lambda_0, \lambda_1, \cdots \lambda_m \right] \\
x_k &= \left[p_{b_k}^w, v_{b_k}^w, q_{b_k}^w, b_a, b_g \right], k \in [0, n] \\
x_c^b &= \left[p_c^b, q_c^b \right]
\end{aligned} \tag{18}$$

where x_k is the IMU state when the kth image is captured. It contains the position, velocity and orientation of IMU in the world coordinate system, as well as the accelerometer offset and gyroscope offset in the IMU body coordinate system. n is the total number of keyframes, m is the total number of features in the sliding window, and λ_l is the inverse depth when watching the lth feature the first time.

Visual inertia BA was used here. We minimized the sum of the prior and the Mahalanobis norm [24] of all the measurement residuals to obtain the maximum posterior estimation:

$$\min_{\mathcal{X}} \left\{ \| r_p - H_p \mathcal{X} \|^2 + \sum_{k \in \mathcal{B}} \left\| r_{\mathcal{B}} \left(\hat{z}_{b_{k+1}}^{b_k}, \mathcal{X} \right) \right\|_{P_{b_{k+1}}^{b_k}}^2 + \sum_{(l,j) \in \mathcal{C}} \rho \left(\left\| r_{\mathcal{C}} \left(\hat{z}_l^{c_j}, \mathcal{X} \right) \right\|_{P_l^{c_j}}^2 \right) \right\} \tag{19}$$

where the Huber norm [25] $\rho(s)$ is defined as follows:

$$\rho(s) = \begin{cases} 1 & s \geq 1 \\ 2\sqrt{s} - 1 & s < 1 \end{cases} \tag{20}$$

$\mathbf{r}_{\mathcal{B}}\left(\hat{\mathbf{z}}_{b_{k+1}}^{b_k}, \mathcal{X}\right)$, $\mathbf{r}_{\mathcal{C}}\left(\hat{\mathbf{z}}_l^{c_j}, \mathcal{X}\right)$ are the residuals of IMU and visual measurement, respectively, which are defined in detail in Equations (21) and (22). B is the set of all IMU measurements and C is a set of features observed at least two times in the current sliding window. The ceres nonlinear optimization library was used to solve the algorithm.

2.4.2. The Calculation of IMU Measurement Residual

Taking the IMU measurement between two consecutive frames b_k and b_{k+1} in the sliding window, according to the IMU measurement model defined in (13), the residual of pre-integration IMU measurement can be defined as:

$$\mathbf{r}_{\mathcal{B}}\left(\hat{\mathbf{z}}_{b_{k+1}}^{b_k}, \mathcal{X}\right) = \begin{bmatrix} \delta\boldsymbol{\alpha}_{b_{k+1}}^{b_k} \\ \delta\boldsymbol{\beta}_{b_{k+1}}^{b_k} \\ \delta\boldsymbol{\theta}_{b_{k+1}}^{b_k} \\ \delta b_a \\ \delta b_g \end{bmatrix}$$

$$= \begin{bmatrix} \mathbf{R}_w^{b_k}\left(\mathbf{p}_{b_{k+1}}^w - \mathbf{p}_{b_k}^w + \frac{1}{2}\mathbf{g}^w \Delta t_k^2 - \mathbf{v}_{b_k}^w \Delta t_k\right) - \hat{\boldsymbol{\alpha}}_{b_{k+1}}^{b_k} \\ \mathbf{R}_w^{b_k}\left(\mathbf{v}_{b_{k+1}}^w + \mathbf{g}^w \Delta t_k - \mathbf{v}_{b_k}^w\right) - \hat{\boldsymbol{\beta}}_{b_{k+1}}^{b_k} \\ 2\left[\mathbf{q}_{b_k}^{w^{-1}} \otimes \mathbf{q}_{b_{k+1}}^w \otimes \left(\hat{\boldsymbol{\gamma}}_{b_{k+1}}^{b_k}\right)^{-1}\right]_{xyz} \\ \mathbf{b}_{ab_{k+1}} - \mathbf{b}_{ab_k} \\ \mathbf{b}_{wb_{k+1}} - \mathbf{b}_{wb_k} \end{bmatrix} \tag{21}$$

where $[\cdot]_{xyz}$ is to extract the vector part of quaternion q for error state expression. $\delta\boldsymbol{\theta}_{b_{k+1}}^{b_k}$ is a three-dimensional error state expression of a quaternion. $\left[\hat{\boldsymbol{\alpha}}_{b_{k+1}}^{bk}, \hat{\boldsymbol{\beta}}_{b_{k+1}}^{bk}, \hat{\boldsymbol{\gamma}}_{b_{k+1}}^{bk}\right]^T$ is an IMU measurement term that is obtained through the pre-integration of the measurement values of accelerometer and gyroscope measurements containing only noise during the time interval of two consecutive image frames. Accelerometer and gyroscope offset are also included in the remaining terms of the online correction.

2.4.3. Visual Measurement Residual

In contrast to the traditional pinhole camera models in which the reprojection error is defined on the generalized image plane, the measurement residuals of a camera are defined on the unit sphere. The optics of almost all types of cameras, including wide-angle, fisheye or omnidirectional cameras, can be modeled as unit rays connected to the surface of a unit sphere. Assuming that the lth feature is first observed in the ith image, the residual of the feature observation in the jth image is defined:

$$\mathbf{r}_{\mathcal{C}}\left(\hat{\mathbf{z}}_l^{c_j}, \mathcal{X}\right) = \begin{bmatrix} \mathbf{b}_1 & \mathbf{b}_2 \end{bmatrix}^T \cdot \left(\hat{\mathcal{P}}_l^{c_j} - \frac{\mathcal{P}_l^{c_j}}{\|\mathcal{P}_l^{c_j}\|}\right)$$

$$\hat{\mathcal{P}}_l^{c_j} = \pi_c^{-1}\left(\begin{bmatrix} \hat{u}_l^{c_j} \\ \hat{v}_l^{c_j} \end{bmatrix}\right) \tag{22}$$

$$\mathcal{P}_l^{c_j} = \mathbf{R}_b^c\left(\mathbf{R}_w^{b_j}\left(\mathbf{R}_{b_i}^w\left(\mathbf{R}_c^b \frac{1}{\lambda_l}\pi_c^{-1}\left(\begin{bmatrix} u_l^{c_i} \\ v_l^{c_i} \end{bmatrix}\right) + \mathbf{p}_c^b\right) + \mathbf{p}_{b_i}^w - \mathbf{p}_{b_j}^w\right) - \mathbf{p}_c^b\right)$$

where $\begin{bmatrix} u_l^{c_i} & v_l^{c_i} \end{bmatrix}^T$ is the lth feature which is observed in the ith image the first time. $\begin{bmatrix} \hat{u}_l^{c_j} & \hat{v}_l^{c_j} \end{bmatrix}^T$ is the observation of the same feature in the jth image. π_c^{-1} is a back projection function that converts pixel positions into unit vectors by using internal parameters of camera. Since the degree of freedom of the visual residuals is 2, we project the residual vector onto the tangent plane. As shown below, b_1, b_2 are two randomly chosen orthogonal bases in the tangent plane $\hat{P}_l^{c_j}$, and a group of b_1, b_2 can be easily found. In Equation (22), with fixed length, $P_l^{c_j}$ is the standard covariance in tangent space, as shown in Figure 6.

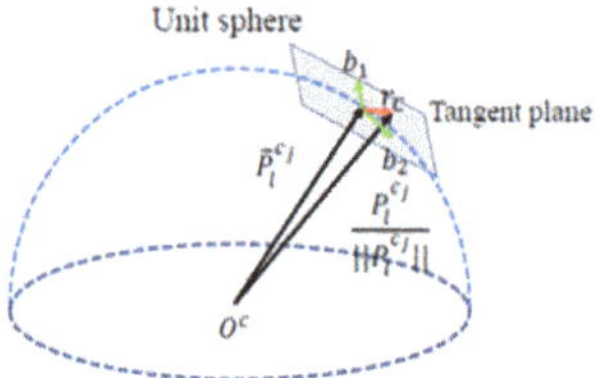

Figure 6. Tangent plane of residual projection.

2.5. Loopback Optimization

Due to measurement and calibration errors, VIO algorithm drifts may cause reduction in positioning accuracy at any time. The loopback optimization method can form additional restraints and suppress the drift problems by estimating the pose changes between some non-adjacent frames. In the loopback optimization part, the DBoW method was first used for loopback detection, then a bidirectional KLT algorithm was used to determine the matching point pairs. The PNP method was used to solve the pose change between two frames, and finally the loopback edge was written into the pose map for overall optimization.

2.5.1. DBoW Loopback Detection

The algorithm, by reference to VINS-MONO, used DBoW2 image similarity evaluation method for loopback detection. The DBoW2 model is the most advanced word bag model, which abstracts images into keyword descriptions for matching. In addition, the pre-stored feature points of the key frame and their descriptors were also used for feature-matching to improve loopback recall. DBoW2 returns loopback detection candidate frames after a temporal and spatial consistency check, as shown in Figure 7.

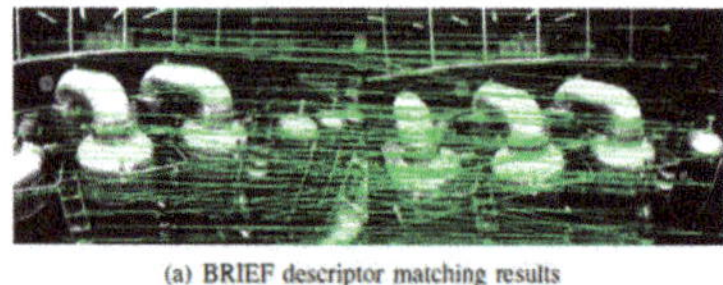
(a) BRIEF descriptor matching results

(b) First step: 2D-2D outlier rejection results

(c) Second step: 3D-2D outlier rejection results.

Figure 7. Loopback detection and exterior point elimination (the same method in VINS-MONO).

2.5.2. Bidirectional KLT Tracking and PNP Relocation

Like in Section 2.2.1 (1), bidirectional KLT tracking was used to obtain matching feature point pairs between two frame feature points with loopback, and then the PNP algorithm was used to obtain the pose changes between two frames.

2.5.3. The Management of 4-Dof Pose Diagram

When creating the pose map, the information $\hat{\phi}, \hat{\theta}$ obtained by IMU estimation was considered as accurate and they were therefore free from optimization. Therefore, the pose map only contained the remaining 4Dof, namely the yaw angle ψ_i and its position information x, y, z, respectively.

Here, the edge residual between frames i and j is defined as:

$$\mathbf{r}_{i,j}\left(\mathbf{p}_i^w, \psi_i, \mathbf{p}_j^w, \psi_j\right) = \begin{bmatrix} \mathbf{R}\left(\hat{\phi}_i, \hat{\theta}_i, \psi_i\right)^{-1}\left(\mathbf{p}_j^w - \mathbf{p}_i^w\right) - \hat{\mathbf{p}}_{ij}^i \\ \psi_j - \psi_i - \hat{\psi}_{ij} \end{bmatrix} \tag{23}$$

Among them, $\hat{\phi}_i, \hat{\theta}_i$ are IMU roll and pitch angle estimations that were directly obtained from monocular VIO.

The whole pose map with sequential edges and loop-back edges is optimized by minimizing the following cost function:

$$\min_{\mathbf{p}, \psi}\left\{\sum_{(i,j)\in\mathcal{S}} \|\mathbf{r}_{i,j}\|^2 + \sum_{(i,j)\in\mathcal{L}} \rho\left(\|\mathbf{r}_{i,j}\|^2\right)\right\} \tag{24}$$

where S is the set of all sequential edges and L is the set of loopback edges. Although tightly coupled relocation was able to reduce false loopbacks, a Huber norm $\rho(\cdot)$ was introduced to further eliminate false loopbacks. In addition, any high-robustness norm was not used between sequential edges, and VIO was considered to have a strong enough elimination mechanism for exterior points.

2.6. Simulation Analysis Test

Before the real flight verification, a physical simulation engine was firstly built in the project, and the ROS Gazebo + Pixhawk scheme was adopted to realize the simulation verification of the algorithm.

2.6.1. Simulation Engine Gazebo

Gazebo is a 3-D dynamic simulator that accurately and effectively simulates robot crowds in complex indoor and outdoor environments, as shown in Figure 8. In the same way that game engines provide high-fidelity visual simulations, Gazebo provides high-fidelity physical simulations as well as a full suite of sensor models, and very user-friendly and programs-friendly interactions.

The typical uses of Gazebo include:

- To test a robot algorithm;
- To design a robot;
- To perform a regression test in actual scenarios.

This engine possesses the following characteristics:

- It contains multiple physics engines;
- It contains a rich library of robot models and environments;
- It contains a variety of sensors;
- The program is convenient to design and has a simple graphical interface.

Gazebo can build a simulation scene for robot tests. It can imitate the real world by adding objects library, garbage bins, ice cream buckets, and even dolls. It can also introduce

2D house design drawings using a building editor and build 3D houses based on the design drawings.

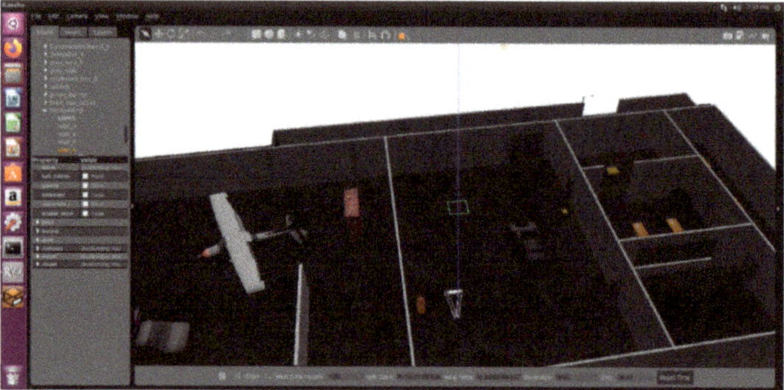

Figure 8. Gazebo.

Gazebo has a very powerful sensor model library, including camera, depth camera, laser, IMU and other sensors that are commonly used by robots. In addition, it has a simulation library, which can be used directly. A new sensor can also be created without any basis and have its specific parameters added. A sensor noise model can even be added to make the sensor more realistic.

2.6.2. Simulation System

The simulation was carried out in the Gazebo engine. A PX4 UAV with a built-in IRIS platform was selected, carrying a RealSense d435i depth camera. The simulation environment was as shown in Figures 9 and 10:

Figure 9. Simulation environment.

The flight path of the UAV is shown below:

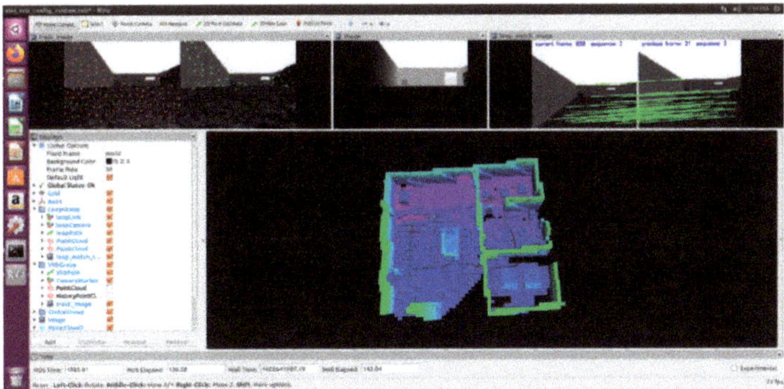

Figure 10. Schematic diagram of the flight path.

Since the final degree of freedom of the pose map was 4Dof, the roll angle and pitch angle directly determined by IMU were ignored in the evaluation process, and the accuracy of the four degrees of freedom of x, y, z and yaw were investigated, as shown in Table 1.

Table 1. Quantitative interpretation and conclusion of Figures 11–15.

Maximum error along X direction	< 0.5 m	Yaw angle error	< 5°
Maximum error along Y direction	< 0.6 m	Absolute error	< 0.3 m
Maximum error along Z direction	< 0.4 m	Calculated closed-loop error	≈ 0.4%
Standard deviation	X		< 0.05 m
	Y		< 0.06 m
	Z		< 0.03 m

In the 12 m × 14 m orbit with a total length of about 70 m, the maximum error along the x direction was less than 0.5 m, less than 0.6 m along the y direction, and 0.4 m along the z direction. The yaw angle error was less than 5°, and absolute error was less than 0.3 m. The yaw error generally stayed at zero, with some small fluctuations when the yaw angle changed abruptly. The calculated closed-loop error was about 0.3/70 = 0.4%.

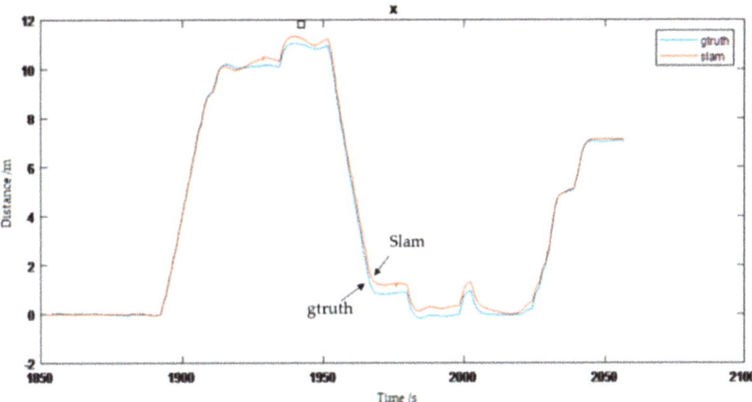

Figure 11. Comparison between SLAM and real value along x axis.

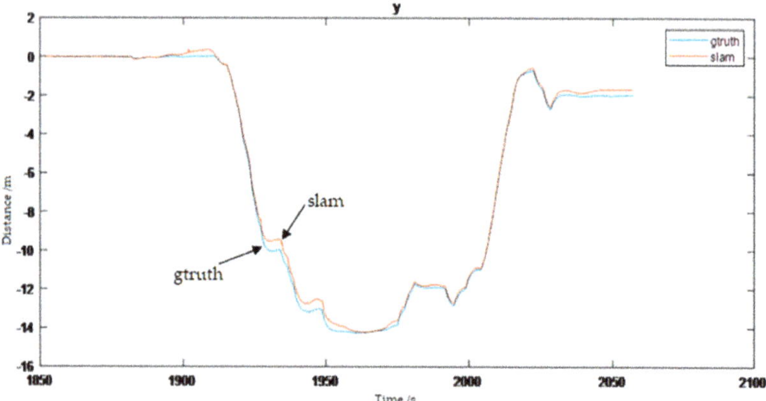

Figure 12. Comparison between SLAM and real value along y axis.

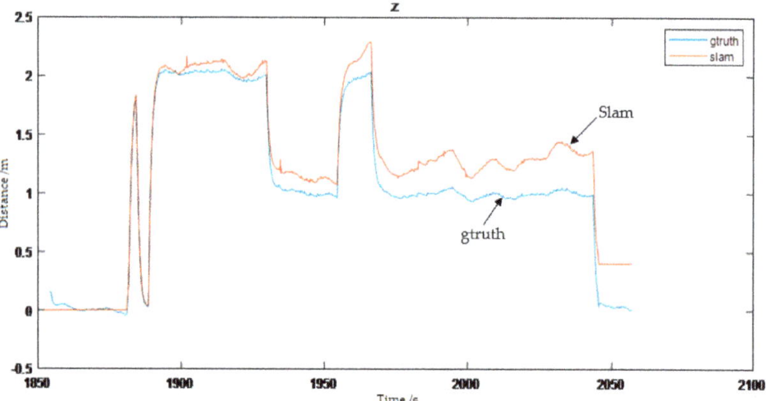

Figure 13. Comparison between SLAM and real value along z axis.

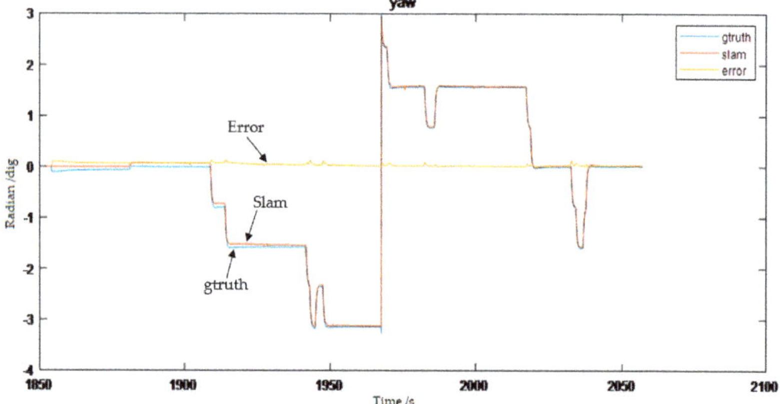

Figure 14. Comparison between SLAM and real value along yaw direction.

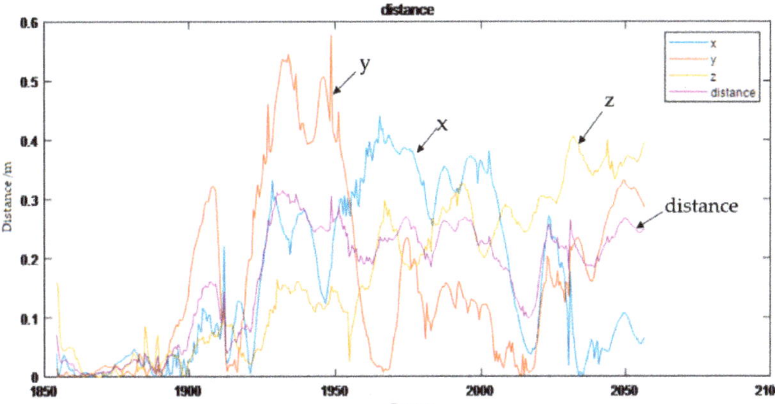

Figure 15. Diagram of position error.

2.7. Section Conclusion

This section introduced the detailed design of the autonomous positioning algorithm and the scene construction and simulation of the algorithm carried out in the Gazebo engine. Through the simulations, under a scene with fixed size, autonomous positioning with a certain extent of accuracy was achieved.

3. Detailed Design of the Map-Building and Trajectory-Planning Algorithm

3.1. The Introduction of the Autonomous Positioning Module

In the mapping and path planning part, the RGB-D camera was selected as a reliable source of in-depth information. An octree map with mature technology was applied to realize the construction of the three-dimensional map. The RRT* algorithm was used to realize obstacle avoidance path planning, and finally the third order spline curve β was used for motion smoothing.

3.2. Octree Map

The point cloud information output by the RGB-D camera can be directly used to construct the point cloud map, but there are several following obvious defects in the application of a point cloud map:

- It has a huge amount of data, and there is serious redundant storage and information redundancy.
- Point cloud maps are stored in continuous space, which means they can't be directly discretized and fast searched.
- This method cannot deal with moving objects and observation errors because we can add objects into the maps but not remove objects from maps.

In order to solve the above problems, the octree map was introduced. This map form has the advantages of flexibility, compressibility, updating and discretization.

3.2.1. The Data Structure of the Octree Map

In a discrete map, it is common to model the 3D space as multiple cubes (voxels). If each facet of the cube is divided into four equally, eight sub-cubes can be obtained until the required precision is reached. If the process of expanding a cube into sub-cubes is regarded as expanding eight sub-nodes from one node, then the process of subdividing the whole space into the smallest sub-space can be regarded as an octo-tree.

The Figure 16 is the octree map structure diagram. The left one shows the process of the cube being split into sub-cubes. If the largest cube is regarded as the root node and the smallest cube as the leaf node, then the octree shown on the right can be formed.

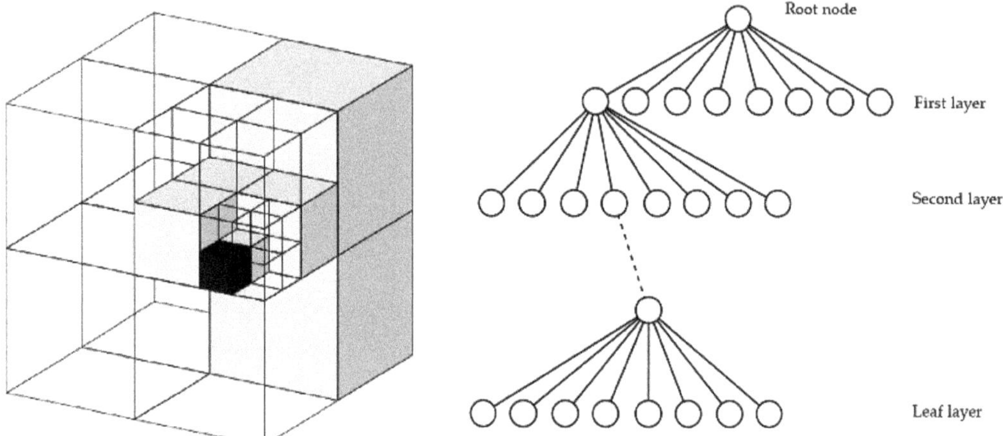

Figure 16. The structure diagram of an octree.

An octree map saves storage space because of its data structure. When all the sub-nodes of a cube are occupied or not occupied, there is no need to continue to expand the node; therefore, only an empty root node is needed when a blank map begins to be established. The actual objects are most closely linked, and it is the same with blank space. Therefore, most octree nodes do not need to be expanded to cotyledon nodes, which can save a lot of storage space.

The occupation information stored in each node of the octree is expressed by the probability: 0 means completely blank and 1 means completely occupied. The initial value is 0.5. If the node is detected to be continuously occupied, the value will increase; otherwise, the value will decrease.

3.2.2. Node Probability Updating

According to the derivation of octree, assuming that when $t = 1,2, \ldots , T$, the observed data is $z_1, \ldots , z_T$, then the information recorded by the nth leaf node is:

$$P(n \mid z_{1:T}) = \left[1 + \frac{1 - P(n \mid z_T)}{P(n \mid z_T)} \frac{1 - P(n \mid z_{1:T-1})}{P(n \mid z_{1:T-1})} \frac{P(n)}{1 - P(n)} \right]^{-1} \tag{25}$$

Since the information expressed directly by probability is too complex to be updated, the algorithm uses log-odds as an alternative description method. Set $y \in \mathbb{R}$ as a probability logarithm, x as the probability value between 0 and 1, then the transformation between them can be described by logit transformation:

$$y = logit(x) = \log\left(\frac{x}{1 - x} \right) \tag{26}$$

And its inverse transformation is shown below:

$$x = logit^{-1}(y) = \frac{\exp(y)}{\exp(y) + 1} \tag{27}$$

When y changes from $-\infty$ to $+\infty$, x correspondingly changes from 0 to 1. When $y = 0$, $x = 0.5$, so we can judge whether a node is occupied or not by storing the value of y. When point clouds are observed continuously in nodes, y increases by a value; when the observed node is empty, y decreases by a certain value. Transfer y to the probability space and utilize the logit inverse transformation when checking the probability.

Set a node as n and the observed data as z. The probability value of this node from the beginning to t is $L(n|z_{1:t})$, and the probability at $t+1$ is as follows:

$$L(n \mid z_{1:t+1}) = L(n \mid z_{1:t-1}) + L(n \mid z_t) \tag{28}$$

With this log probability, the entire octree map can be updated according to RGB-D data. If the depth of a pixel observed in the RGB-D graph is d, it means that an occupied point is observed in the space corresponding to the depth value, and there is no obstacle in the path from the camera optical center to this point.

3.3. Path Planning

Rapidly exploring random tree (RRT) was selected as the path planning algorithm. Traditional path planning algorithms such as the artificial potential field method, the method of fuzzy rules, genetic algorithm, neural network and simulated annealing algorithm, ant colony optimization algorithm, etc., are not suitable for the path planning of multi-degree-of-freedom robots in complex environments because they all require modeling obstacles in a certain space, and the computational complexity has an exponential relationship with the DOF of robots.

RRT effectively solves the problem of path planning under conditions of high-dimensional space and complex constraints because it avoids space modeling by detecting the collision of sampling points in the state space, avoiding the modeling of the space. The characteristic of this method is that it can search the high-dimensional space quickly and effectively and lead detection to blank areas through random sampling points in the state space and then find a planned path from the starting point to the target point, which is suitable for solving the path planning of multi-degree-of-freedom robots in complex and dynamic environments. Note that the RRT algorithm is probabilistically complete and non-optimal, and path planning only finds a feasible path, which may not be optimal.

3.3.1. Basic RRT Algorithm

RRT is an efficient planning method in multi-dimensional space. It takes an initial point as the root node and generates a randomly extended tree by adding leaf nodes through random sampling [26–28]. When the leaf nodes in the random tree contain the target point or enter the target area, a path from the initial point to the target point can be found in the random tree. The workflow of a basic RRT algorithm is as follows:

Initialize the random root node Xinit, which is the starting point of path planning.

A random number P between 0 and 1 is generated. When P < Prob, a sampling point is randomly selected from the state space as Xrand. When P > Prob, the target point is used as Xrand.

Select the nearest point from Xrand in random tree nodes as Xnearest, expand some distance from Xnearest to Xrand to obtain the new node Xnew and the new edge Lnew.

Record the running time: if the run times out, it returns no solution.

If Xnew and Lnew collide with the obstacles in the state space, return to step 2 and repeat it. If there is no collision, then run tree growth, and add Xnew into the random tree as Xnearest's leaf node.

Judge whether Xnew is the target point or not; if it is, then output the current random tree; otherwise, return to step 2 and repeat it.

The basic RRT algorithm process is shown in Figure 17 below:

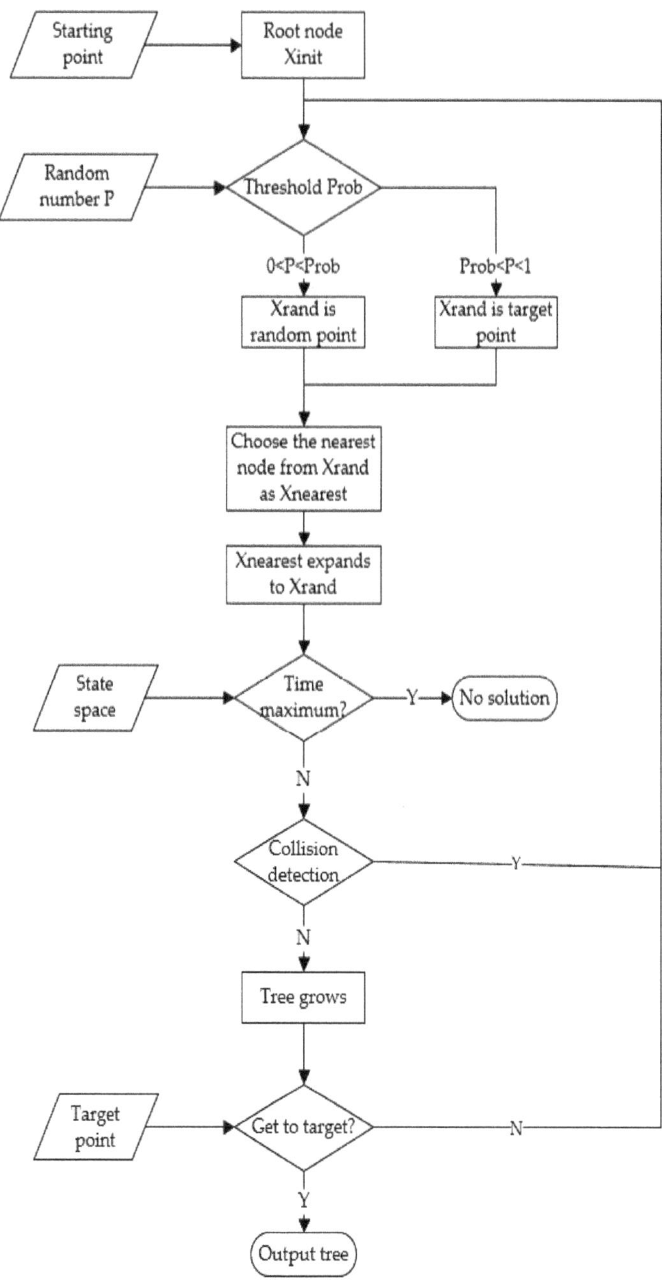

Figure 17. Flow of basic RRT algorithm.

The basic RRT algorithm is not sensitive to the environment and can effectively explore the whole space. However, it also has serious disadvantages in some application conditions:

The basic RRT algorithm is a pure random search algorithm, which degrades the search efficiency significantly when the environment contains many obstacles or narrow channel constraints, as shown in Figure 18.

Figure 18. Performance of the RRT algorithm in a maze.

Because the narrow channel area is small, the probability of being touched is low, and this is why it is difficult to find a path in an environment with narrow passageways, as shown in Figure 19.

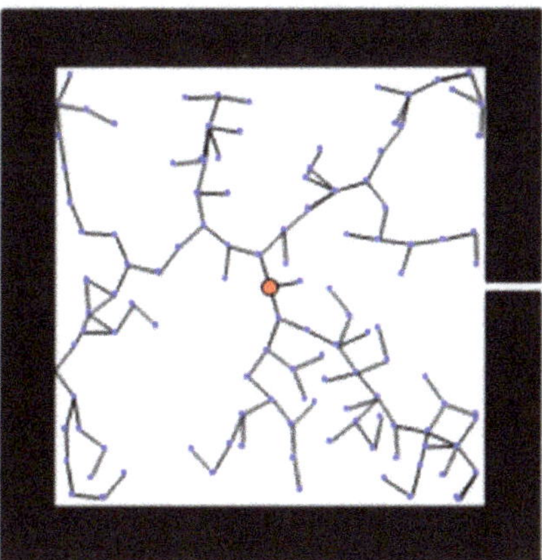

Figure 19. Performance of the RRT algorithm in an environment with narrow passageways.

Because the nodes of the RRT algorithm are completely randomly generated, the path may not be relatively smooth, and it cannot be directly applied for path and motion planning.

3.3.2. RRT* Algorithm

Although RRT is a relatively efficient algorithm that can deal with path planning problems with nonholonomic constraints, and has great advantages in many aspects, the RRT algorithm can't guarantee that the obtained feasible path is relatively optimized. RRT* was improved based on RRT, mainly by reselecting the parent node and rewiring.

In RRT, the nearest point to Xrand is selected as the parent node in the extended node policy, but this choice is not necessarily optimal. The goal of planning is to make this point as close as possible to the starting point. Many improvements have been achieved using RRT* by drawing a small circle around the sampling point after it is added to the path tree and considering whether there are better parent nodes to connect to that point so that the distance from the starting point to the point is shorter (although those nodes may not be the closest points to the sampling point). If a more suitable parent is chosen, then connect them and remove the original wiring (rewiring).

The RRT* algorithm is asymptotically optimized, which means that the resulting path is more and more optimized with the increase of the number of iterations, and it is never possible to obtain the optimal path in limited time. In other words, it takes a certain amount of running time to get a relatively satisfactory and optimized path.

In the rewiring process, the tree structure is optimized by introducing the path length parameter to achieve the optimal path planning. The specific optimization process includes the following 15 main steps. The process and steps of rewiring are introduced as Figures 20–22:

(1) Generate a random point Xrand;
(2) Find the nearest node Xnearest from Xrand on the tree;
(3) Connect Xrand with Xnearest;
(4) With Xrand as the center, search for nodes in the tree with a certain radius and find out the set of potential parent nodes {Xpotential_parent}. The purpose is to update Xrand and observe whether there is a better parent node;
(5) Start with a potential parent, Xpotential_parent;
(6) Calculate the cost of Xnearest being the parent node;
(7) Instead of performing collision detection, connect Xpotential_parent with Xchild (that is, Xrand) and calculate the path cost;
(8) Compare the cost of the new path with that of the initial path. If the cost of the new path is smaller, the collision detection will be carried out; otherwise, the next potential parent node will be replaced;
(9) If collision detection fails, the potential parent node will not act as the new parent node;
(10) Turn to the next potential parent;
(11) Connect the potential parent node to Xchild (that is, Xrand) and calculate the path cost;
(12) Compare the cost of the new path with the cost of the original path. If the cost of the new path is smaller, the collision detection will be carried out; if the cost of the new path is larger, the next potential parent node will be replaced;
(13) The collision detection passes;
(14) Delete the previous edges from the tree;
(15) Add a new edge to the tree, and take the current Xpotential_parent as the parent of Xrand.

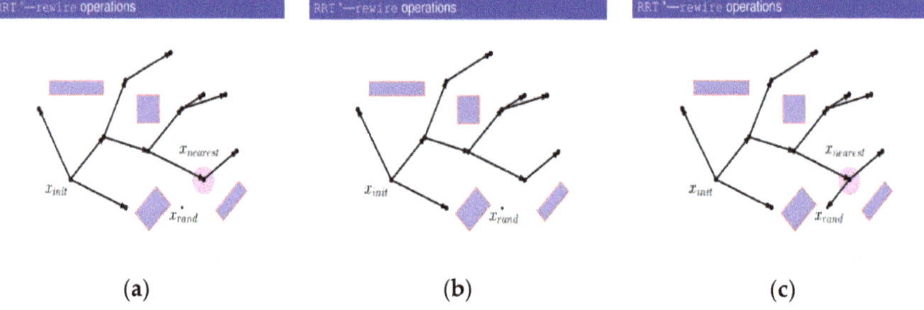

(a) (b) (c)

Figure 20. *Cont.*

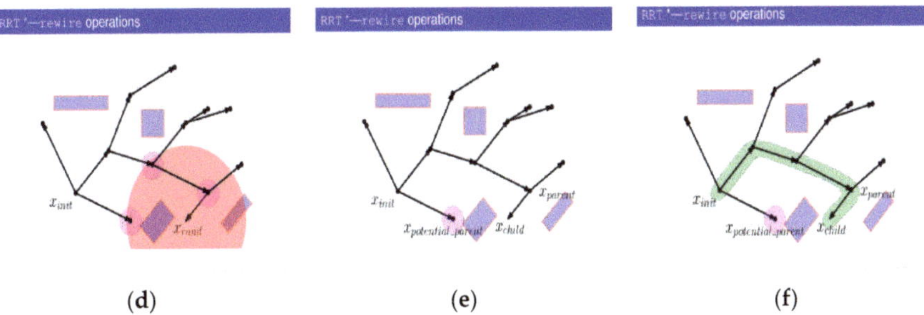

Figure 20. Step (1) to (6). (**a**) Generate random point; (**b**) find the nearest node; (**c**) find initial parent node; (**d**) find potential parent nodes; (**e**) select potential parent node; (**f**) calculate the initial path cost.

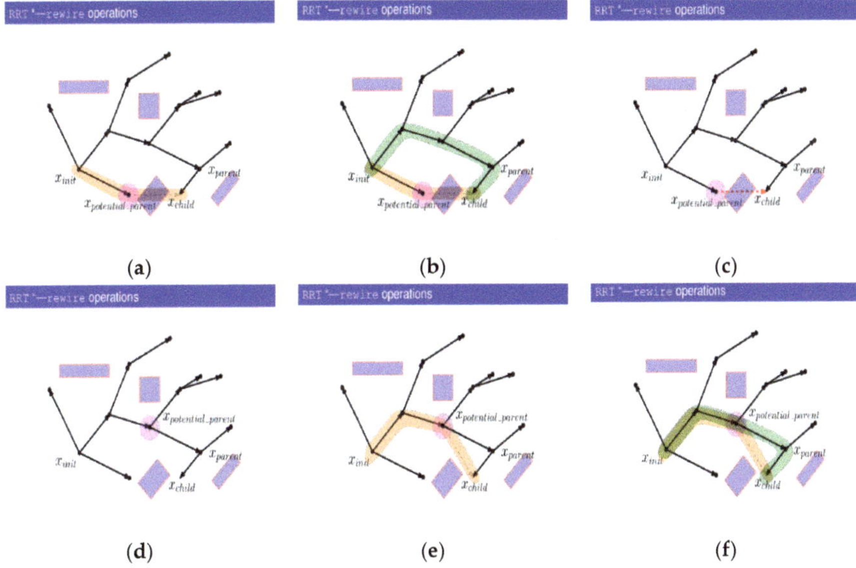

Figure 21. Step (7) to (12). (**a**) Calculate the new path cost; (**b**) compare the cost of new path and initial; (**c**) failure of collision detection; (**d**) select new parent nodes; (**e**) calculate the new path cost; (**f**) the comparison of the cost of new and initial paths.

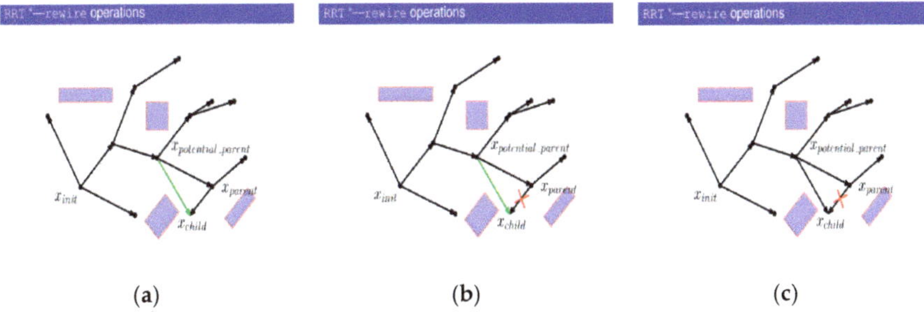

Figure 22. Collision detection passes. (**a**) The collision detection passes; (**b**) delete the previous path edges; (**c**) add new edges.

3.4. Smoothing the Interpolation of Third-Order β Spline

Although the RRT* algorithm improves the optimality and smoothness of the planned trajectory, it still has many sharp points and cannot be directly used for trajectory control. Here, third-order β spline interpolation is used to smoothen the solution of RRT*, which can ensure the continuous acceleration control signal of the motion trajectory.

3.4.1. Node Table

The node table is the key parameter to generating the basic function table, and it is strictly equal to the sum of the number of control points: the number of orders plus one. The parameters of the node table are set artificially. For β spline curve, there are two general ways to set it: sequential method and clamped method. The former is used to make standard β spline open and closed curves, and the latter is used to make a more practical β spline curve.

The order list only needs to be set linearly from 0 to 1, while the clamped list needs to set the nodes of each order plus 1 before and after as 0. Taking the third-order spline curve with six control points as an example, the size of its node table is $6 + 3 + 1 = 10$.

If it is a sequential list, we only need to set it in order:

$$0, \frac{1}{9}, \frac{2}{9}, \frac{3}{9}, \frac{4}{9}, \frac{5}{9}, \frac{6}{9}, \frac{7}{9}, \frac{8}{9}, 1 \tag{29}$$

If it is a clamped list, since it is the third order, the former $3 + 1$ parameters are set as 0, the latter $3 + 1$ parameters are set as 1, and the remaining parameters increase evenly:

$$0, 0, 0, 0, \frac{1}{3}, \frac{2}{3}, 1, 1, 1, 1 \tag{30}$$

3.4.2. Basic Function Tables

The basic function table is essentially a recursive equation, but it is also an intermediate parameter at the same time. The formula is as follows:

$$B_{i,\deg}(t) = \frac{t - knot_i}{knot_{i+\deg} - knot_i} B_{i,\deg-1}(t) + \frac{knot_{i+\deg+1} - t}{knot_{i+\deg+1} - knot_{i+1}} B_{i+1,\deg-1}(t) \tag{31}$$

where, t is the node to be interpolated; $knot_i$ represents the ith element in the node table; $B_{i,\deg}(t)$ is the parameter of the basic function table, whose structure is a two-dimensional array, and its meaning is the value of the ith element of the basic function table at the deg order when the user input is t.

The recursive characteristics of the function table can be seen from (31). The current elements of the deg order need to be calculated by two elements of the deg $-$ 1 order. In addition, β spline curve algorithm requires that when the function table returns to order 0, it can be calculated according to the following formula:

$$B_{i,0} \begin{cases} 1 & knot_i \leq t \leq knot_{i+1} \\ 0 & knot_i > t \text{ or } knot_{i+1} < t \end{cases} \tag{32}$$

3.4.3. Calculation

Assuming that the corresponding position of the t value in the β spline curve is $C(t)$ finally, the calculation formula of the final β spline curve is:

$$C_t = \sum_{i=0}^{n-1} B_{i,\deg}(t) P_i \tag{33}$$

where $B_{i,\deg}(t)$ is the value of the ith deg order basic function table at t, and P_i is the ith interpolation control point.

3.5. Simulation Test and Analysis

The simulation was carried out in the Gazebo engine. A UAV PX4 with a built-in IRIS platform was selected and carried a RealSense D435I depth camera. The simulation environment is shown in Figure 23:

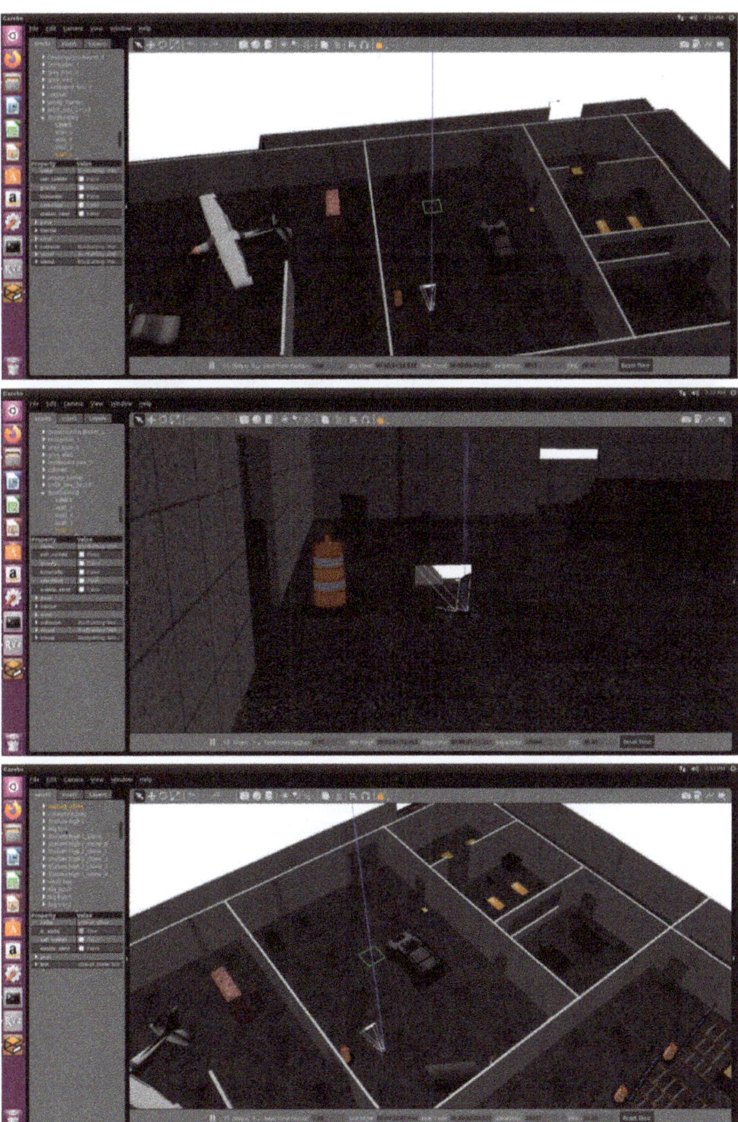

Figure 23. Screenshots of mapping simulation environment.

In the map building test, the UAV control system adopted the default parameters of the simulation system, and the VI-SLAM system constructed in Section 3 was adopted as the positioning system. The results of map building of the simulation environment are shown in Figure 24:

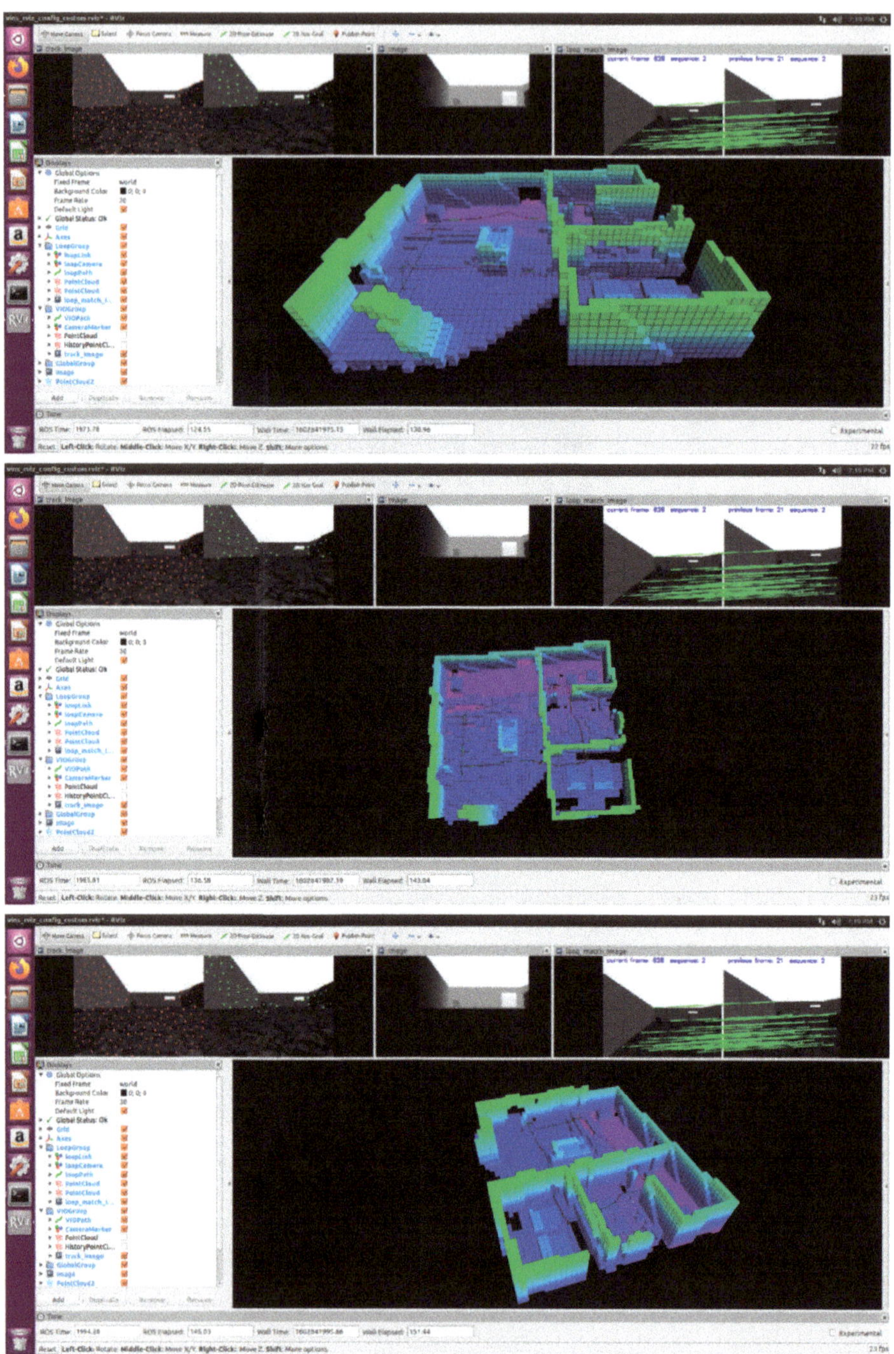

Figure 24. *Cont.*

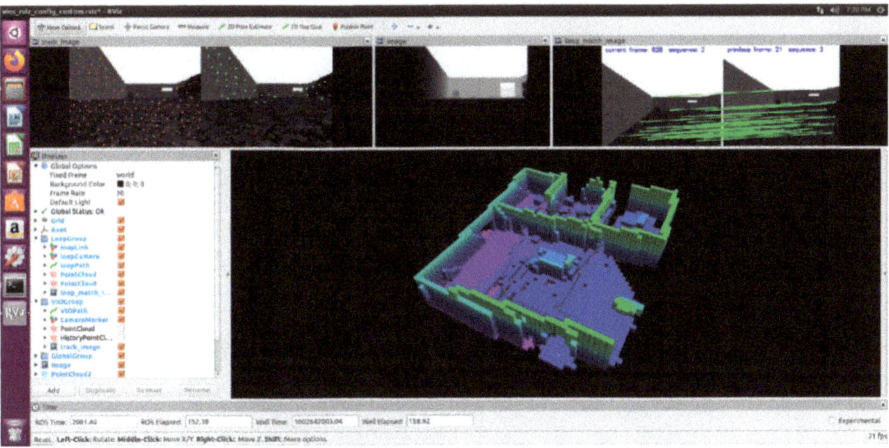

Figure 24. Map building results.

In the path planning test, the UAV control system adopted the default parameters of the simulation system, and the VI-SLAM system constructed in Section 3 was chosen as the positioning system. The path planning environment is shown in Figure 25:

Figure 25. The path planning simulation environment.

Through path planning, the UAV can autonomously avoid obstacles in indoor environments and reach the target location. Parts of the path planning results are shown in Figures 26 and 27:

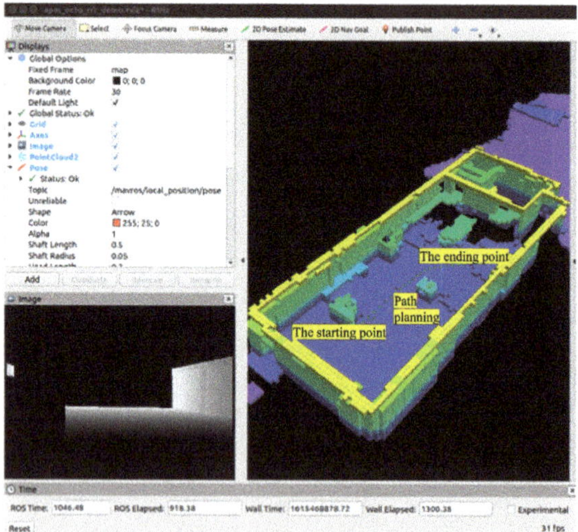

Figure 26. Path planning working condition 1.

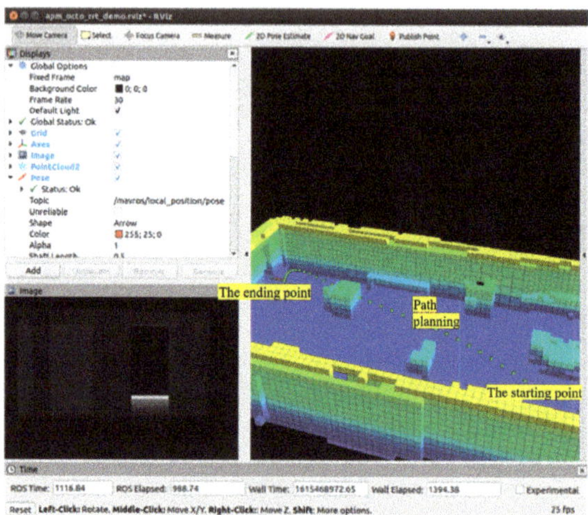

Figure 27. Path planning working condition 2.

3.6. Section Conclusion

In this section, the map building and path planning algorithms were introduced in detail, and the above two algorithms were verified by the Gazebo engine. The results show that the proposed algorithm can fulfill the task requirements well.

4. The Detailed Design of the Target Detection and Recognition Algorithm

4.1. The Introduction of Target Detection and Recognition Module

Since there is no specific cooperation target for detection, the recognition algorithm to be selected needs to be commonly appliable, transferable, and robust. At the same time, the algorithm should be optimized and accelerated under limited performance of the airborne processor to ensure that the high-speed UAV can accurately capture the target [29,30].

Therefore, the Jetson series GPU development board XavierNX was finally selected as the computing platform, the YOLOv3 network was used as the basic detection algorithm, and TensorRT architecture was introduced to achieve GPU inference acceleration.

4.2. Target Detection Network

At present, target detection algorithms can be divided into two categories according to the process. One of them is the region-convolutional neural network (R-CNN) algorithm based on region proposals such as R-CNN, fast R-CNN, etc. These algorithms are two-stage methods, which require the use of heuristic methods or convolutional neural networks to generate candidate regions, and after that the regions need classification and regression. The other category is the one-stage algorithm, e.g., YOLO and SSD, which uses a unified convolutional neural network structure to perform regressive prediction for the location and categorization of targets at the same time. The first kind of method is characterized by high accuracy but slow speed, and the second kind of algorithm runs fast with low accuracy.

Overall, the YOLO algorithm is an end-to-end target recognition network, using a separate full-convolution neural network model, and its workflow is shown in Figure 28: Firstly, the resolution of the input images should be unified as 448×448, then put the images into the convolution neural network, finally these images are processed by the output part, and obtain the location and category information of the target. Compared with R-CNN and other two-stage algorithms, its structure is more concise and unified, with faster processing speed and easier hardware acceleration. At the same time, the YOLO training and processing courses are both end-to-end, and the available network can be directly obtained from the image training set.

Figure 28. The workflow of the YOLO target detection network.

In terms of network structure, YOLO uses a unified convolutional neural network sequence to process images and obtain feature sequences, and then uses a shallow convolutional neural network to perform position regression and category prediction. The specific network structure is shown in Figure 29.

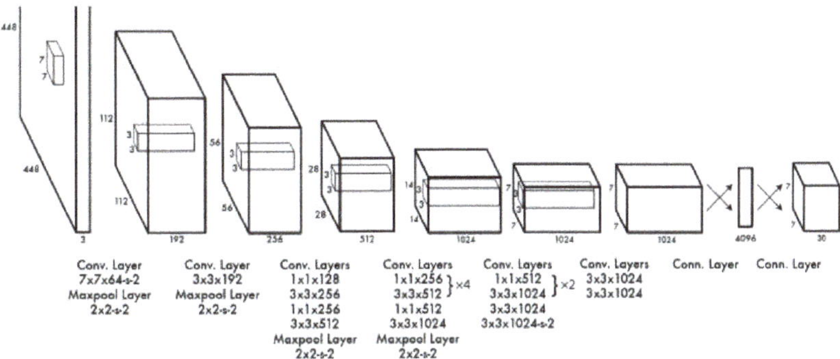

Figure 29. YOLO network structure.

In terms of the internal structure of the network, a 3×3 convolution kernel is mainly used for feature extraction and abstraction, a 1×1 convolution kernel is used for cascade cross-channel parameter pooling, and a LeakyReLU function is adopted as the activation function: $\max(x, 0.1x)$. Note that the activation function at the last layer of the network is replaced with a linear one. The final output of the network is a tensor whose size is $7 \times 7 \times 30$, where $S = 7$ is the number of grids, the first 20 elements in the third dimension represent the degree of confidence of the 20 classifications, elements 21–22 are the degree of confidence of the bounding box $B = 2$, and the last 8 elements are the (x, y, w, h) of the bounding box $B = 2$.

The main features of the YOLO target detector are as follows:

(1)   Features extraction network

Although the YOLOv1 network adopts the structure of the GoogLeNet classification network; it uses 1×1 and 3×3 CNN networks in feature extraction to lower the dimensionality of high-dimensional information and realize the information integration of high and low channels in the network. In the main part of feature extraction, YOLOv2 uses the multi-scale feature fusion method of the single shot multi-box detector (SSD) network and proposes to use the DarkNET-19 network to improve the fine-grained feature extraction in images. Since YOLOv2 only performs feature fusion in the latter layer and produces fixed-size feature maps, this method easily leads to the loss of most fine-grained information in the fusion process of high and low semantics. Thus, YOLOv2 has a poor detection effect for intensive small targets. While maintaining the detection speed, YOLOv3 adopts the simplified residual basic module to replace the 1×1 and 3×3 modules in the original CNN, and a deeper DarkNET-53 network is constructed as the feature extraction backbone network of YOLOv3.

(2)   Residual mechanism

The YOLOv2 feature extraction in DarkNET-19 uses a straight tube network structure such as GoogLeNet or visual geometry group (VGG). Convolution is directly added in DarkNET-19 to deepen the network to realize the purpose of extracting more useful feature information by convolution network. This easily leads to the disappearance or explosion of the loss gradient in the network learning training process. For this reason, YOLOv3 in DarkNET-53, drawing on the concept of ResNet, uses a residual module to achieve the superposition of the output feature map of convolution with the input to solve the contradiction between network depth and gradient disappearance.

(3)   Feature map

In the network, before YOLOv3 outputs the feature map, a method combining the feature pyramid network (FPN) and upsampling is proposed based on the FPN method in SSD, which improves the problem of the loss of fine-grained target feature information in the fusion of multiple high-level information and low-level information in the feature map. The basic idea of this method is: based on the current feature map, the upsampling method is used to concatenate the output features of a convolution layer into a new feature map. This structure can not only improve the feature richness of fine-grained targets, but also help the algorithm to improve the accuracy of target prediction.

4.3. TensorRT Inference Acceleration

NvidiaTensorRT, formerly known as the graphics processing unit (GPU) inference engine (GIE), is a high-performance deep learning inference optimizer that can provide low-latency and high-throughput deployment inference for deep learning applications. TensorRT can be used to accelerate reasoning for exceedingly large-scale data centers, embedded platforms or autonomous driving platforms. TensorRT can now support almost all deep learning frameworks such as TensorFlow, Caffe, Mxnet, Pytorch and so on. Combining TensorRT with NVIDIA GPU, a fast and efficient deployment inference can be realized in almost all frameworks. TensorRT is currently the only programmable inference accelerator that can build and optimize customized network structures in addition to its on-premise network structure, so it can adapt to existing network structures and ones in the near future.

TensorRT has the following optimization methods, the most important of which are the first two kinds of adjustment to the network operation structure:

(1)   Interlayer fusion and tensor fusion

Taking a GoogleNetInception calculation as an example, the left part of Figure 30 below is the calculation chart. There are many layers in this structure. During the deployment of model inference, the calculations of each layer are completed by the GPU, but in the actual computing process, the GPU starts different compute unified device architecture (CUDA) cores to complete the computation. The computing speed of the CUDA core tensor is very fast, and the operation time mainly consists of the slow core startup and read and write processes of the tensor, which cause a large amount of occupation of memory bandwidth and a waste of GPU computing resource. By the transverse and longitudinal merger between layers (the merged structure of convolution, bias and ReLU layers are fused to form a single layer called case-based reasoning (CBR)), the number of layers in the network is greatly reduced while maintaining the original functions. Lateral merging can combine convolution, offset and activation layers into a CBR structure. Vertical merging can combine layers with the same structure but different weights into a wider layer. Both operations can make the optimized structure occupy only one CUDA, and reduce the number of data transfers, memory bandwidth occupation and the time of core start and stop.

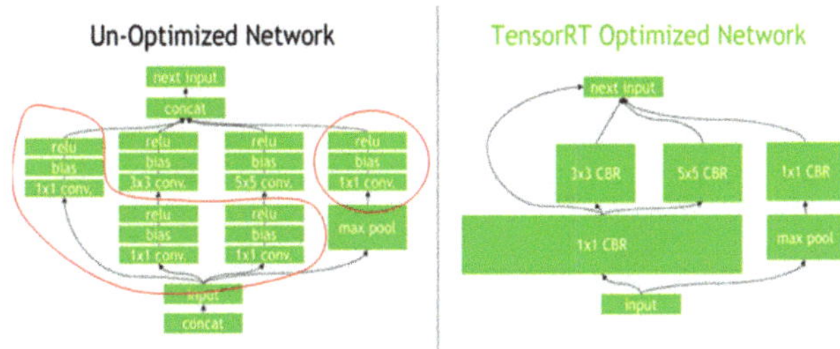

Figure 30. TensorRT optimization model.

The combined calculation is shown on the right side of Figure 30. Fewer computation layers lead to less occupation of CUDA cores, and the entire model structure is more compact and efficient.

(2)   Data accuracy calibration

Most deep learning frameworks train neural networks with tensors at full 32-bit precision (FP32), and once the network is trained, the data length can be reduced to speed up the access and operation because backpropagation is not needed in the process of deploying inference. TensorRT supports FP16 and INT8 data compression acceleration modes. TensorRT provides a fully automatic calibration process to lower FP32 accuracy to INT8 accuracy with best matching performance and minimize performance loss.

(3)   CUDA core optimization

In the inference calculation of the network model, the CUDA core of GPU is called for calculation. TensorRT can adjust CUDA kernel for different algorithms, different network models, and different GPU platforms to ensure the optimal calculation performance of the current model on a specific platform.

(4)   Tensor memory management

Due to the characteristics of the neural network itself, the size of each feature map remains the same during its operation, so the video memory can be allocated in advance. TensorRT assigns certain video memory for every tensor during application, reducing memory usage and improving reuse efficiency.

(5)   GPU multi-stream optimization

For bypass networks that cannot be merged, GPUs generally adopt multi-stream computing and then perform stream synchronization. TensorRT can optimize the flow operation from the hardware aspect to achieve the optimal synchronization effect.

4.4. Analysis Test

The acceleration performance test was performed on a JetsonXavierNX computer. In 15 W working mode, the CPU (no optimization), GPU (CUDA optimization), and GPU (TensorRT optimization) were used for the test. The relative parameters of the device are shown in Table 2.

Table 2. JetsonXavierNX performance parameters.

Ability	10 W Mode	15 W Mode
AI performance	14 TOPS (INT8)	21 TOPS (INT8)
GPU	384-core NVIDIA Volta™ GPU with 48 Tensor cores	
GPU max freq	800 MHz	1100 MHz
CPU	6-core NVIDIA Carmel ARM®v8.2 64-bit CPU 6 MB L2 + 4 MB L3	
CPU max freq	2-core @ 1500 MHz 4-core @ 1200 MHz	2-core @ 1900 MHz 4/6-core @ 1400 Mhz
Memory	8 GB 128-bit LPDDR4x @ 1600 MHz 51.2 GB/s	
Storage	16 GB eMMC 5.1	
Power	10 W ∣ 15 W	

The test used 1280×720 resolution images to identify 1000 groups and average the time. Since CUDA and TensorRT require pre-start of the CUDA core, the time in this section was recorded separately. The results of the speed test are shown in Table 3:

Table 3. Results of processing speed using different computing platforms.

Computing Platforms	Initialization/ms	Average Time/ms
CUP	-	790
GPU(CUDA)	2310	85
GPU(TensorRT)	1103	12

After optimization, the network inference speed was greatly improved, about 66 times as fast as CPU inference, and about 7 times as fast as the CUDA inference, finally reaching about 83FPS. At the same time, due to the simplified network structure, the CUDA core startup process was accelerated by about two times after TensorRT optimization.

Part of the recognition results are shown in Figures 31–33:

Figure 31. Recognition result 1.

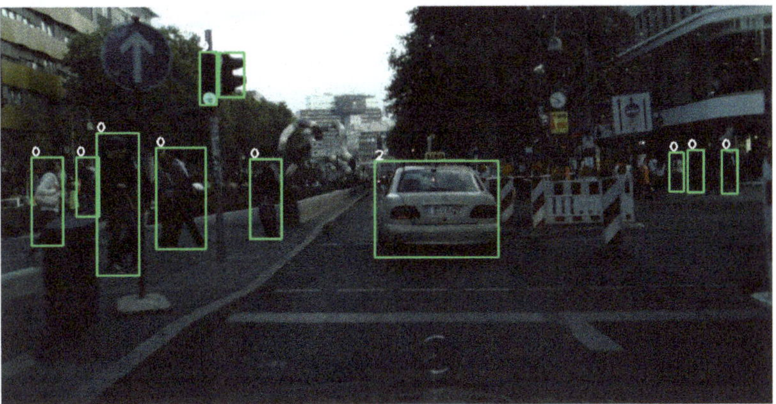

Figure 32. Recognition result 2.

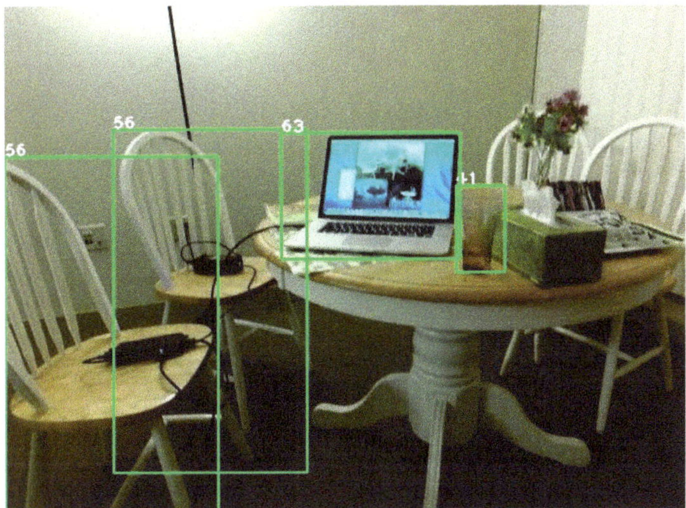

Figure 33. Recognition result 3.

4.5. Section Conclusion

In this section, the target recognition algorithm was introduced in detail, and the algorithm was verified on the corresponding equipment. The results show that YOLO can realize the recognition of targets with high precision and accuracy.

5. Technical Validation

Based on the system mentioned above, an indoor UAV platform that is appliable for indoor environments was built for this paper, which adopted the following plan:

5.1. Introduction of the Platform Plan

The platform consisted of four parts, including the body part, the autonomous navigation system, the ground control system, and the data transfer system. The overall structure is shown in Figure 34:

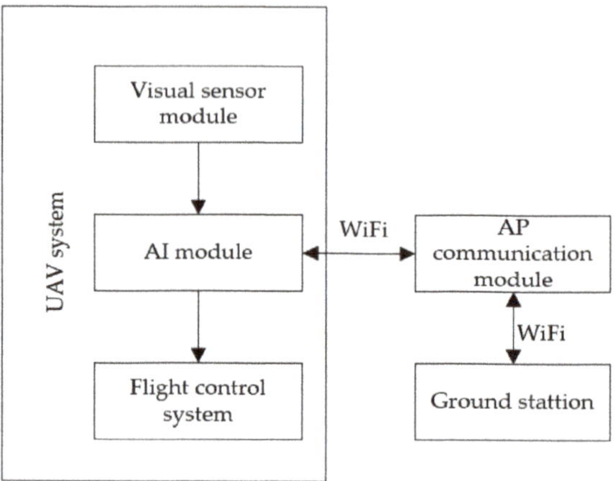

Figure 34. Diagram of overall structure.

According to the content in the figure, the aircraft platform built is shown in Figure 35:

Figure 35. Flight platform.

The subsystems are described as follows.

5.1.1. The Fuselage Part

The body part was composed of power system, frame, and flight control system, which are described as follows.

(1) Power system

The power package was an Air Gear 450, manufactured by Tmotor, as shown in Figure 36:

 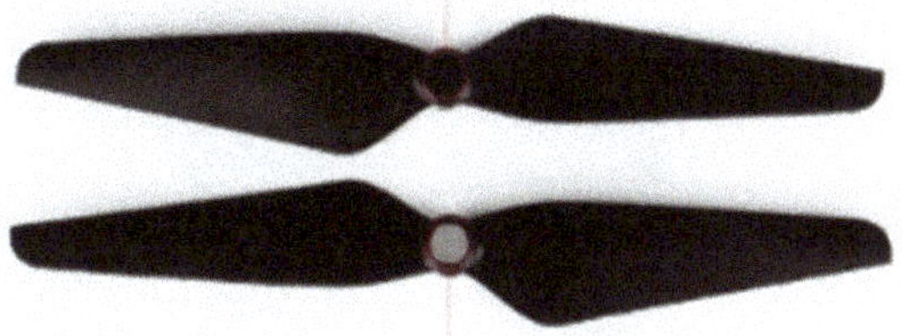

Figure 36. Air Gear 450 power package.

(2)   Frame design

The frame was assembled using carbon fiber with aluminum alloy CNC parts.

(3)   Battery

The Leopard 4S-6000mah was adopted as the battery, and its discharge rate is 60 C, as shown in Figure 37.

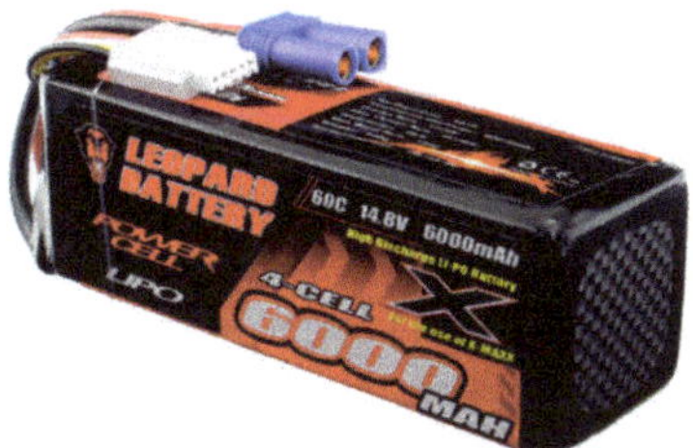

Figure 37. Leopard battery.

(4)   Flight control system

The flight control system adopted the self-developed flight control module, as shown in Figure 38, whose detailed parameters are as follows:
Main control: STM32F103@72MHz frequency;
IMU: ICM20689*2.

Figure 38. Flight control IC.

Since IMU was used as the underlying module, two IMU were installed face-to-face to suppress the gyro drift.
For detailed technical parameters, see the datasheet of IMU and the main control unit.

5.1.2. Autonomous Navigation System

The autonomous navigation system took an NVIDIA module as the processing core and an Intel camera as the sensor.

(1)   Core processing unit

The NVIDIA Jetson Xavier NX was introduced as the processing core unit, as shown in Figure 39.

Figure 39. Jetson XAVIER NX module.

(2)　Sensor

The sensor used an Intel RealSense Camera D435i and T265 as the vision sensing module. D435i was used to provide depth information, and its performance is shown in Figure 40:

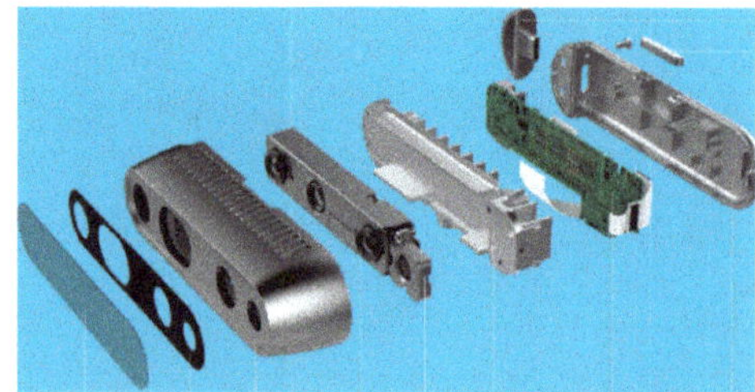

Figure 40. D435i camera.

T265 provides SLAM mapping information, as shown in Figure 41. The T265 contains two fisheye lens sensors, an IMU, and a Movidius Myriad 2 VPU. The camera enjoys low delay and very efficient power consumption. Through extensive performance tests and validation, under expected application conditions, the closed-loop offset was less than 1%. The delay between the pose action and the action reflection was less than 6 milliseconds.

Figure 41. T265 camera.

5.1.3. Data Transfer System

The Huawei AP6750-10T was adopted as data transfer system, as shown in Figure 42:

Figure 42. AP data transfer system.

Its performance indicators are as Table 4.

Table 4. Autonomous navigation system.

Model	AP6750-10T
Type	Distributed wireless router
Wireless standard	IEEE 802.11 a/b/g/ac/ac wave2, support 2×2MIMO
Wireless rate	3000 Mbps
Working frequency range	2.4 GHz, 5 GHz
Support agreement	802.11a/b/g/n/ac/ax
Software parameters	
WPS support	Supports WPS one-click encryption
Safety performance	Support Open System authentication Support WEP authentication, and support 64-bit, 128-bit, 152-bit, and 192-bit encryption bytes Support wap2-psk Support wpa2-802.1x Support wpa3-sae Support wap3-802.1x Support wap-wpa2 Support wap-wpa3
Internet management	Support IEEE 802.3ab standard Support sub-negotiation of rate and duplex mode Compatible with the IEEE 802.1 q Support NAT
Qos support	Based on the WMM, it supports the WMM power saving mode, uplink packet priority mapping, queue mapping, queue mapping, VR/ mobile game application acceleration, and hierarchical HQos scheduling for airports.
Hardware parameters	
Local network interfaces	2×10 GE electrical interface, 1×10 GE SFP+
Other interfaces	one
Type of antenna	Built-in type
Working environment	Temperature: $-10\sim+50$ °C

5.1.4. Ground Control System

The ground control system was coded by Qt software, as shown in Figure 43.

Figure 43. Qt software for coding ground control system.

5.2. Flight Test

5.2.1. Performance Test

(1)   Positioning accuracy of integrated navigation

Four reference points were used for accuracy comparison, which were (0, 0, 0), (0, 0, 0.51), (2, 0, 0.51) and (2, −1, 0.51).

When the UAV was placed at the above four points, the corresponding navigation position was obtained, as shown in Figures 44–47:

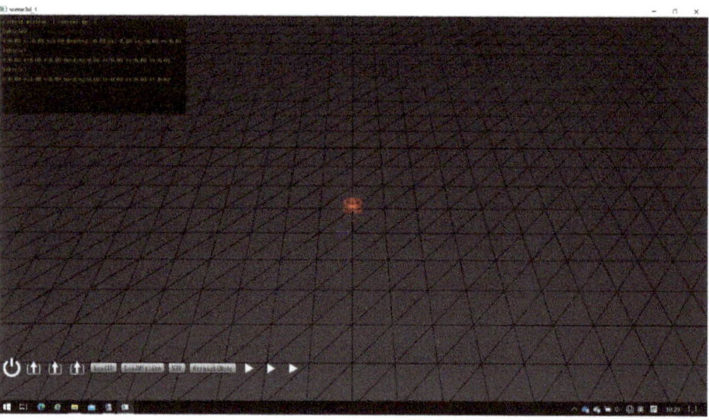

Figure 44. (0, 0, 0) position navigation data map.

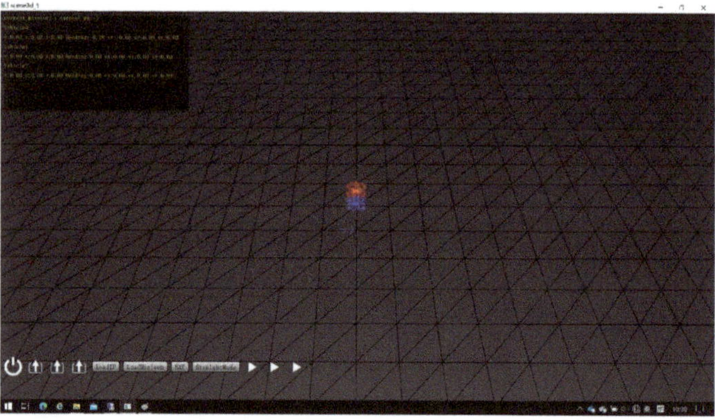

Figure 45. (0, 0, 0.5) position navigation data map.

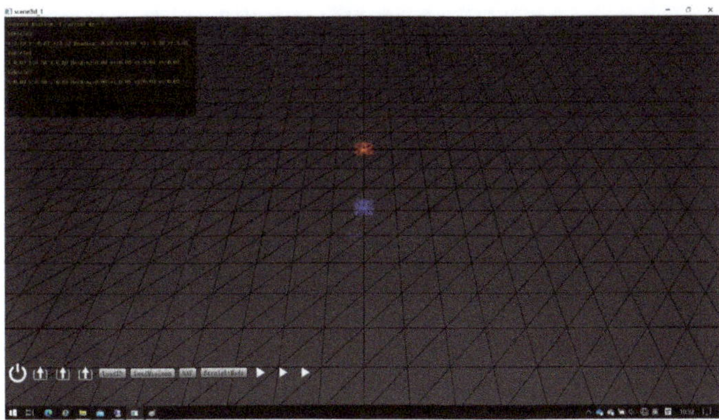

Figure 46. (2, 0, 0.5) position navigation data map.

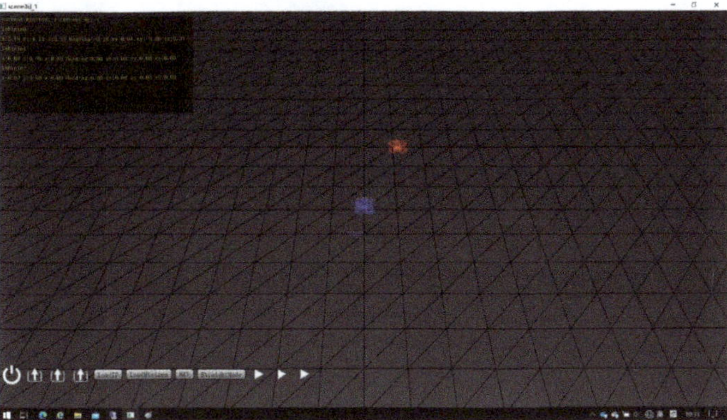

Figure 47. (2, −1, 0.5) position navigation data map.

The actual four-point navigation positions were (0, 0, 0), (0.02, 0,02, 0.54), (2.19, −0.01, 0.52), (2.18, −1.11, 0.52). According to the above results, it can be seen that this meets the requirements of the actual conditions, namely:

- Combined navigation and positioning accuracy (CEP): ≤ 0.2 m;
- Fixed-altitude, fixed-point flight accuracy (CEP): ≤ 0.5 m (RMS).

(2) Obstacles (stools) were directly placed 1 m, 2 m, 3 m, 4 m and 5 m away in front of the UAV for evaluation, respectively. The measurement results are shown in Figures 48–52:

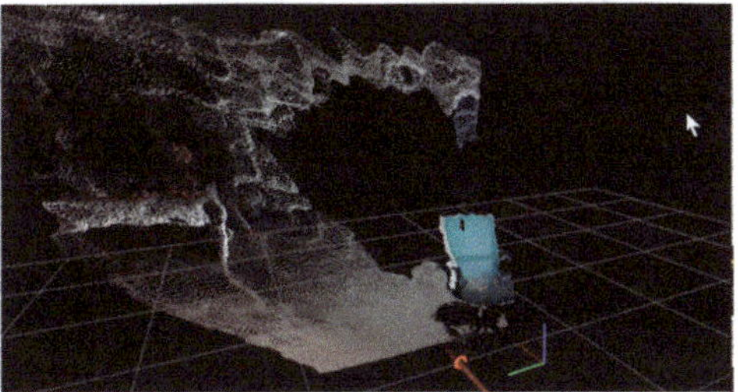

Figure 48. Resultant figure when obstacle was 1 m away.

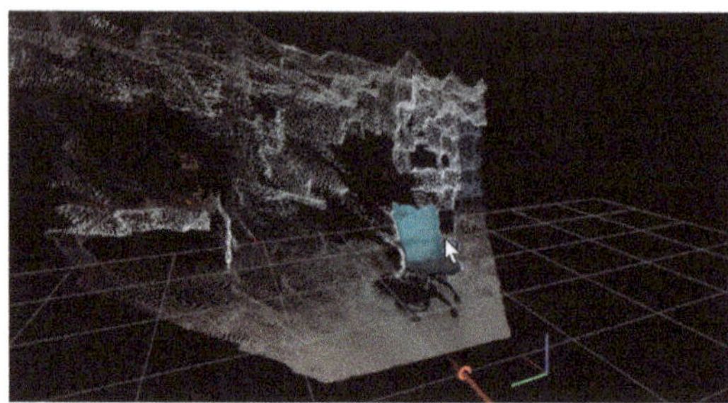

Figure 49. Resultant figure when obstacle was 2 m away.

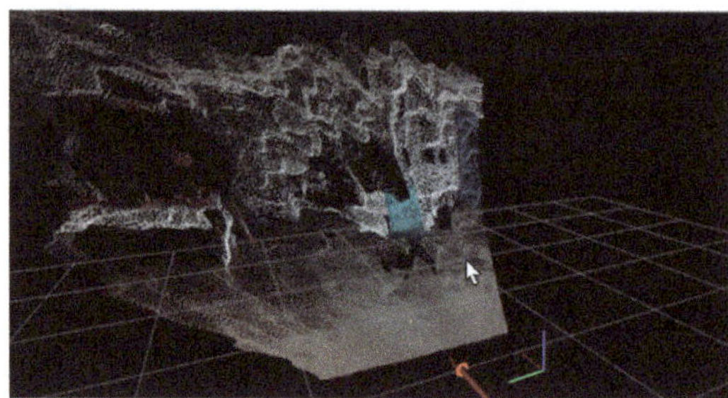

Figure 50. Resultant figure when obstacle was 3 m away.

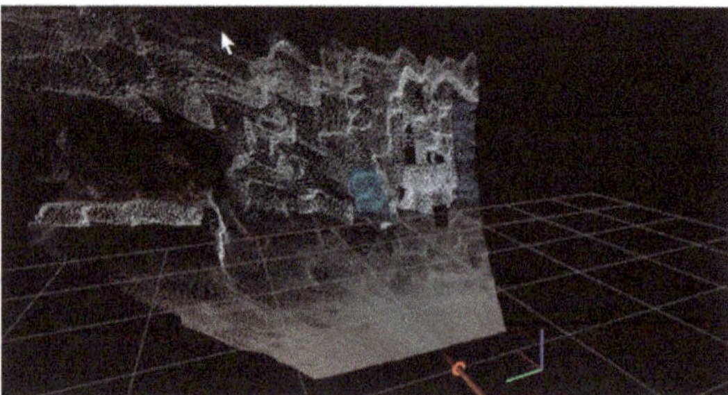

Figure 51. Resultant figure when obstacle was 4 m away.

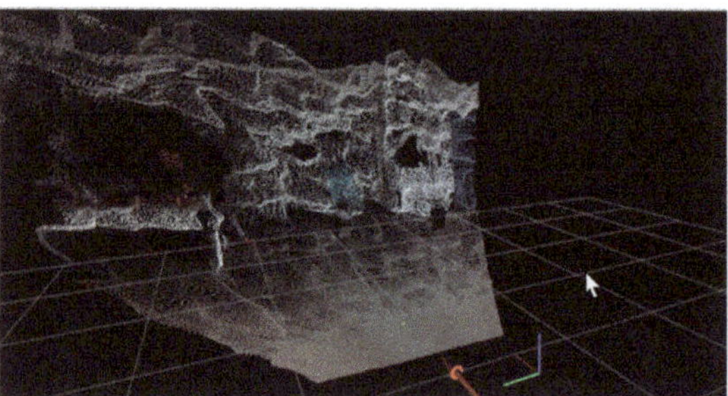

Figure 52. Resultant figure when obstacle was 5 m away.

From the above results, the system meets the requirements of the technical requirements, namely:

Obstacle detection distance: ≥ 5 m.

(3)   Obstacle detection channel and range

Because T265 was used as the obstacle measurement equipment, its field of view (FOV) range was $163 \pm 5°$, and the effect is shown in Figure 53:

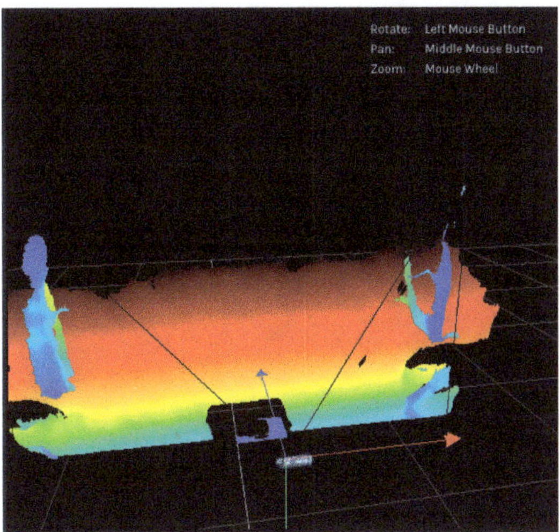

Figure 53. Results of obstacle measurement range.

It can be seen from the results that the system meets the obstacle detection channel and range requirements, namely:

The system possesses detection ability in at least three channels; namely, the front, the top and the bottom. The detection range in each channel ($\geq \pm45°$ horizontally, $\geq \pm45°$ vertically).

(4)   Minimum size of detectable obstacle

The ruler was placed 5 m away from the camera to measure the recognition of obstacle size.

It can be seen from Figure 54 that the system can complete the recognition of obstacles. That is, at 2 m distance, to achieve the recognition of objects with a size of 100×5 mm.

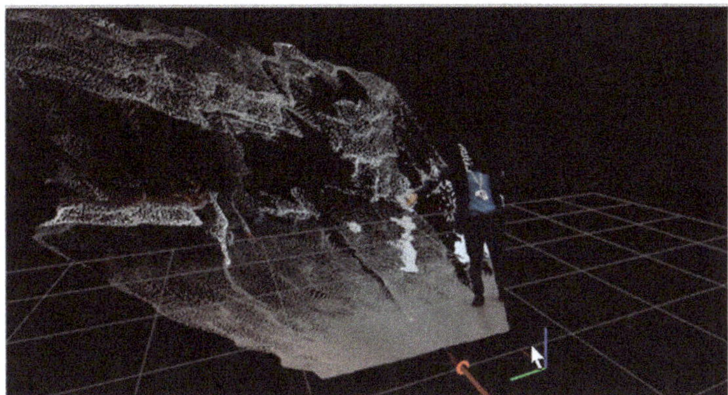

Figure 54. Obstacle size recognition.

(5) Obstacle detection rate

This test was placed in the flight test.

5.2.2. Single-Machine Indoor Autonomous Obstacle Avoidance and Navigation

Scene tests under two conditions were performed in this paper. The maps of the two sites are shown in Figure 55:

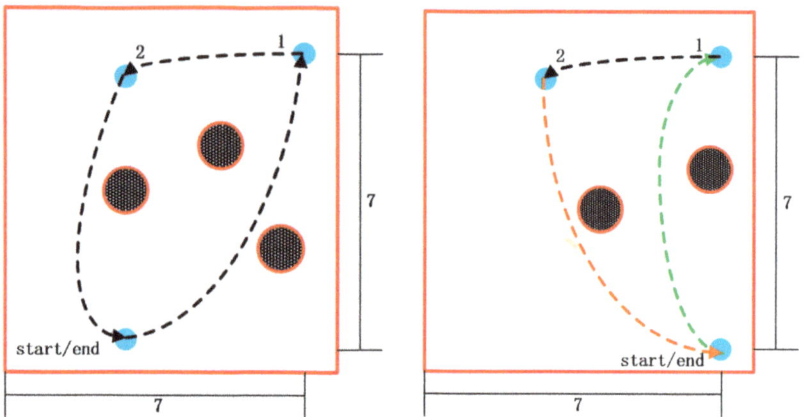

Figure 55. Maps of test sites.

In order to test the indoor autonomous obstacle avoidance and navigation algorithm, a 7 × 7 m field was built in the room, in which two obstacles were placed. The UAV started from the start point in Figure 55, then ran to the first point, the second point, and finally returned to the end point. During this period, the system automatically recognized and avoided the two obstacles in the picture.

(1) Scene experiment with three obstacles

The scene diagram of the three-obstacle experiment is shown in Figure 56:

Figure 56. Three-obstacle scene.

In this scenario, the UAV completed the flight test process from the starting point to the end point well, as shown in Figure 57.

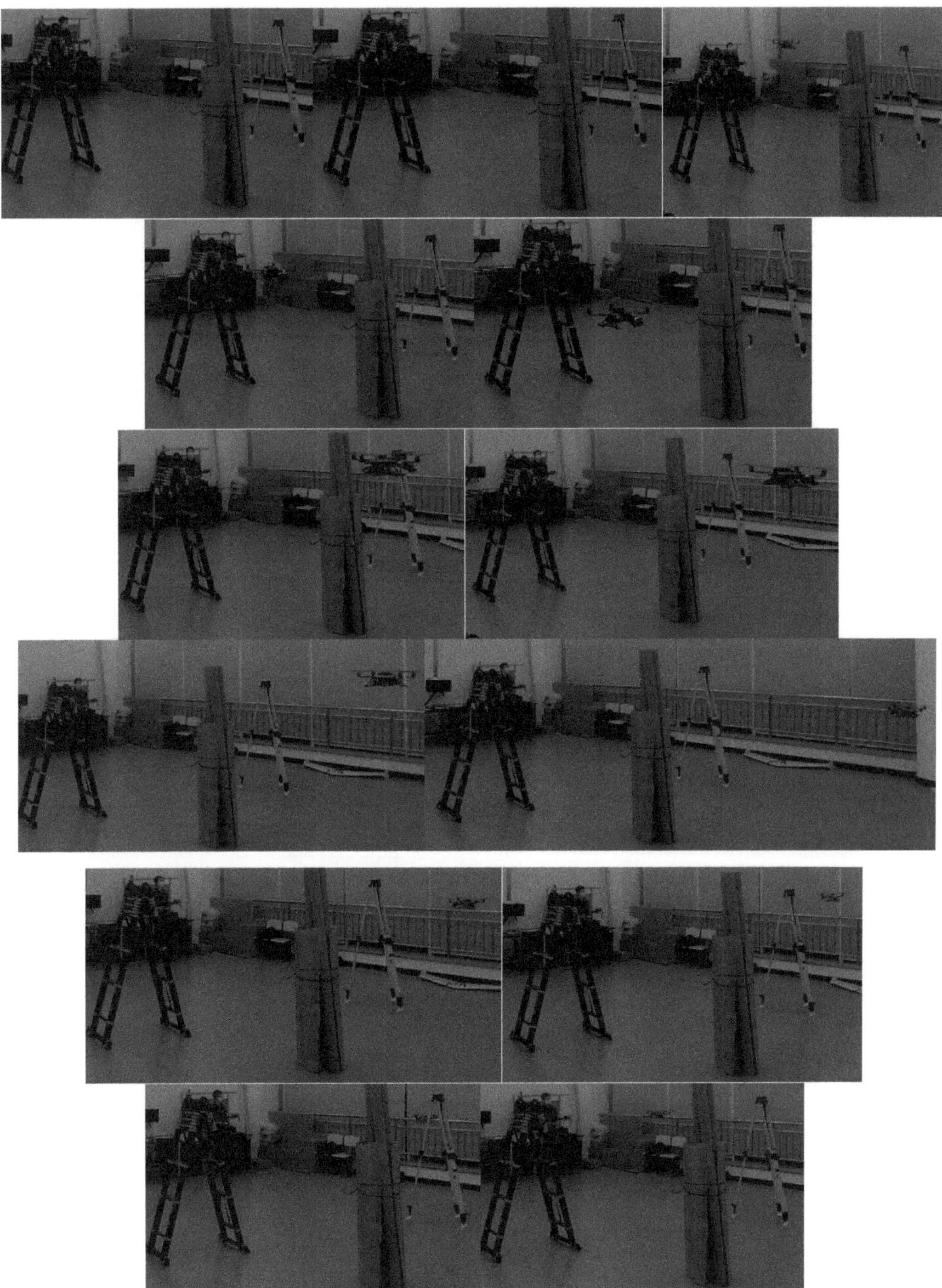

Figure 57. Flight record of three-obstacle scene.

(2) Scene experiment with two obstacles

In the two-obstacle scenario experiments, the flight data of the UAV were recorded as shown in Figure 58:

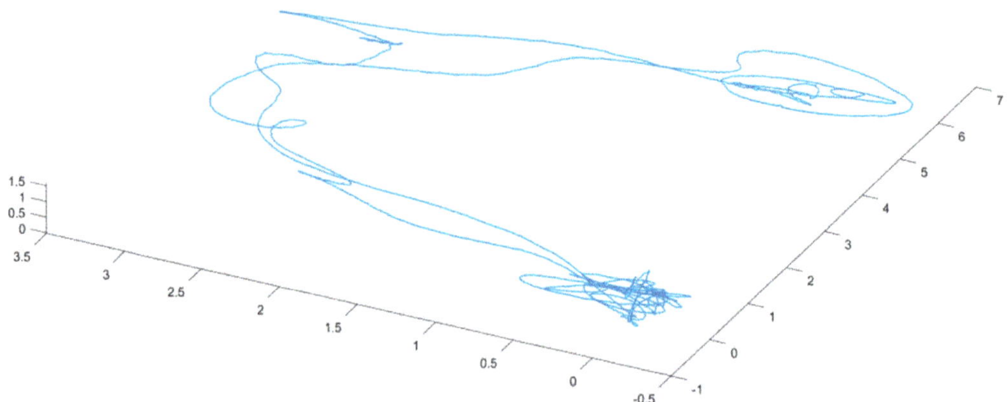

Figure 58. Flight record of two-obstacle scene.

6. Conclusions

This paper completed the following work: Firstly, an integrated navigation algorithm based on machine vision/close-range detection/inertial measurement unit (IMU) was designed and realized. Then, an indoor simultaneous localization and mapping (SLAM) algorithm was designed and realized. Moreover, a method for obstacle detection, obstacle avoidance motion decision and motion planning was designed and realized. At last, an autonomous navigation and obstacle avoidance simulation system was built. In the meantime, the positioning and navigation system in an unknown environment as well as the indoor obstacle avoidance flight was also demonstrated and verified. There are some advantages and weaknesses listed as follows:

➢ Advantages: (1) Uses distributed hardware solution to realize obstacle observation (D435i) and SLAM (T265) functions, which greatly reduces the computational power requirements of the airborne computer; (2) optimizes the YOLO network using TensorRT so it can run in real time on the onboard computer; (3) OCtomAP mapping, RRT* and β spline curve fitting are finished mainly by CPU, target recognition mainly by GPU, making full use of onboard computer resources.

➢ Weaknesses: (1) The basic assumption of a l-SLAM system is that the environment remains static. If the environment moves in whole or part, the localization results will be disturbed; (2) OctomAP requires a cumulative period of stable observations to effectively identify obstacles, which makes the UAV unable to effectively respond to obstacles that suddenly appear; (3) the motion trajectory generated by 3-RRT* and β spline curve only has position command, but no speed and acceleration command, and so cannot guide the UAV to fly at high speed.

There are also some weaknesses that should be noted and explored such as whether the detection of the obstacle and the path computation are influenced by the changing environments and so on.

Author Contributions: Methodology, C.C.; formal analysis, Z.W. and Z.G.; data curation, P.C.; writing—original draft preparation, Z.W. and C.Z.; writing—review and editing, Y.L. All authors have read and agreed to the published version of the manuscript.

Funding: This research received no external funding.

Data Availability Statement: Not applicable.

Conflicts of Interest: The authors declare no conflict of interest.

Nomenclature

UAV	Unmanned aerial vehicle
GNSS	Global navigation satellite system
SLAM	Simultaneous location and mapping
USV	Unmanned surface vehicle
UAM	Urban air mobility
SAFDAN	Solar atomic frequency discriminator for autonomous navigation
VPC	Visual predictive control
MPC	Model predictive control
UUV	Unmanned underwater vehicle
VR	Virtual reality
GPS	Global positioning system
VI-SLAM	Visual–inertial simultaneous localization and mapping
IMU	Inertial measurement unit
VIO	Visual–inertial odometry
KLT tracking	Kanade–Lucas–Tomasi tracking
RRT	Rapidly exploring random trees
CNN	Convolutional neural network
RCNN	Region-convolutional neural network
SSD	Single shot multi-box detector
VGG	Visual geometry group
FPN	Feature pyramid network
GPU	Graphics processing unit
GIE	GPU inference engine
CUDA	Compute unified device architecture
CBR	Case-based reasoning
FP32	Full 32-bit precision
FOV	Field of view

References

1. Li, Z.; Lu, Y.; Shi, Y.; Wang, Z.; Qiao, W.; Liu, Y. A Dyna-Q-based solution for UAV networks against smart jamming attacks. *Symmetry* **2019**, *11*, 617. [CrossRef]
2. Kuriki, Y.; Namerikawa, T. Consensus-based cooperative formation control with collision avoidance for a multi-UAV system. In Proceedings of the 2014 American Control Conference, Portland, OR, USA, 4–6 June 2014; IEEE: New York, NY, USA, 2014; pp. 2077–2082.
3. Kwak, J.; Park, J.H.; Sung, Y. Unmanned aerial vehicle flight point classification algorithm based on symmetric big data. *Symmetry* **2016**, *9*, 1. [CrossRef]
4. Turan, E.; Speretta, S.; Gill, E. Autonomous navigation for deep space small satellites: Scientific and technological advances. *Acta Astronaut.* **2022**, *193*, 56–74. [CrossRef]
5. Kayhani, N.; Zhao, W.; McCabe, B.; Schoellig, A.P. Tag-based visual-inertial localization of unmanned aerial vehicles in indoor construction environments using an on-manifold extended Kalman filter. *Autom. Constr.* **2022**, *135*, 104112. [CrossRef]
6. Li, Z.; Zhang, Y. Constrained ESKF for UAV Positioning in Indoor Corridor Environment Based on IMU and WiFi. *Sensors* **2022**, *22*, 391. [CrossRef]
7. Aldao, E.; González-de Santos, L.M.; González-Jorge, H. LiDAR Based Detect and Avoid System for UAV Navigation in UAM Corridors. *Drones* **2022**, *6*, 185. [CrossRef]
8. Zhang, W.; Yang, Y.; You, W.; Zheng, J.; Ye, H.; Ji, K.; Chen, X.; Lin, X.; Huang, Q.; Cheng, X.; et al. Autonomous navigation method and technology implementation of high-precision solar spectral velocity measurement. *Sci. China Phys. Mech. Astron.* **2022**, *65*, 289606. [CrossRef]
9. Ramezani Dooraki, A.; Lee, D.J. A Multi-Objective Reinforcement Learning Based Controller for Autonomous Navigation in Challenging Environments. *Machines* **2022**, *10*, 500. [CrossRef]
10. Durand Petiteville, A.; Cadenat, V. Advanced Visual Predictive Control Scheme for the Navigation Problem. *J. Intell. Robot. Syst.* **2022**, *105*, 35. [CrossRef]
11. Specht, C.; Świtalski, E.; Specht, M. Application of an autonomous/unmanned survey vessel (ASV/USV) in bathymetric measurements. *Pol. Marit. Res.* **2017**, *nr 3*, 36–44. [CrossRef]

12. Xie, S.; Wu, P.; Peng, Y.; Luo, J.; Qu, D.; Li, Q.; Gu, J. The obstacle avoidance planning of USV based on improved artificial potential field. In Proceedings of the 2014 IEEE International Conference on Information and Automation (ICIA), Hailar, China, 28–30 July 2014; IEEE: New York, NY, USA, 2014; pp. 746–751.

13. Wang, X.; Yadav, V.; Balakrishnan, S.N. Cooperative UAV formation flying with obstacle/collision avoidance. *IEEE Trans. Control. Syst. Technol.* **2007**, *15*, 672–679. [CrossRef]

14. Manhães, M.M.M.; Scherer, S.A.; Voss, M.; Douat, L.R.; Rauschenbach, T. UUV Simulator: A Gazebo-based package for underwater intervention and multi-robot simulation. In Proceedings of the OCEANS 2016 MTS/IEEE Monterey, Monterey, CA, USA, 19–23 September 2016; pp. 1–8.

15. Liu, L.; Wu, Y.; Fu, G.; Zhou, C. An Improved Four-Rotor UAV Autonomous Navigation Multisensor Fusion Depth Learning. *Wirel. Commun. Mob. Comput.* **2022**, *2022*, 2701359. [CrossRef]

16. Sajjadi, S.; Mehrandezh, M.; Janabi-Sharifi, F. A Cascaded and Adaptive Visual Predictive Control Approach for Real-Time Dynamic Visual Servoing. *Drones* **2022**, *6*, 127. [CrossRef]

17. Nascimento, T.; Saska, M. Embedded Fast Nonlinear Model Predictive Control for Micro Aerial Vehicles. *J. Intell. Robot. Syst.* **2021**, *103*, 74. [CrossRef]

18. Ambroziak, L.; Ciężkowski, M.; Wolniakowski, A.; Romaniuk, S.; Bożko, A.; Ołdziej, D.; Kownacki, C. Experimental tests of hybrid VTOL unmanned aerial vehicle designed for surveillance missions and operations in maritime conditions from ship-based helipads. *J. Field Robot.* **2021**, *39*, 203–217. [CrossRef]

19. Hassan, S.A.; Rahim, T.; Shin, S.Y. An Improved Deep Convolutional Neural Network-Based Autonomous Road Inspection Scheme Using Unmanned Aerial Vehicles. *Electronics* **2021**, *10*, 2764. [CrossRef]

20. Zhu, H.; Liu, C.; Li, M.; Shang, B.; Liu, M. Unmanned aerial vehicle passive detection for Internet of Space Things. *Phys. Commun.* **2021**, *49*, 101474. [CrossRef]

21. He, L.; Aouf, N.; Song, B. Explainable Deep Reinforcement Learning for UAV autonomous path planning. *Aerosp. Sci. Technol.* **2021**, *118*, 107052. [CrossRef]

22. Miranda, V.R.; Rezende, A.; Rocha, T.L.; Azpúrua, H.; Pimenta, L.C.; Freitas, G.M. Autonomous Navigation System for a Delivery Drone. *J. Control. Autom. Electr. Syst.* **2022**, *33*, 141–155. [CrossRef]

23. Zhao, X.; Chong, J.; Qi, X.; Yang, Z. Vision Object-Oriented Augmented Sampling-Based Autonomous Navigation for Micro Aerial Vehicles. *Drones* **2021**, *5*, 107. [CrossRef]

24. Ren, Y.; Liu, X.; Liu, W. DBCAMM: A novel density based clustering algorithm via using the Mahalanobis metric. *Appl. Soft Comput.* **2012**, *12*, 1542–1554. [CrossRef]

25. Guitton, A.; Symes, W.W. Robust inversion of seismic data using the Huber norm. *Geophysics* **2003**, *68*, 1310–1319. [CrossRef]

26. Elmokadem, T.; Savkin, A.V. Towards Fully Autonomous UAVs: A Survey. *Sensors* **2021**, *21*, 6223. [CrossRef] [PubMed]

27. Zammit, C.; Van Kampen, E.J. Comparison between A* and RRT algorithms for UAV path planning. In Proceedings of the 2018 AIAA Guidance Navigation, and Control Conference, Kissimmee, FL, USA, 8–12 January 2018; p. 1846.

28. Aguilar, W.G.; Morales, S.; Ruiz, H.; Abad, V. RRT* GL based optimal path planning for real-time navigation of UAVs. In Proceedings of the International Work-Conference on Artificial Neural Networks, Cádiz, Spain, 14–16 June 2017; Springer: Cham, Switzerland, 2017; pp. 585–595.

29. Wang, C.; Meng, M.Q.H. Variant step size RRT: An efficient path planner for UAV in complex environments. In Proceedings of the 2016 IEEE International Conference on Real-Time Computing and Robotics (RCAR), Angkor Wat, Cambodia, 6–10 June 2016; IEEE: New York, NY, USA, 2016; pp. 555–560.

30. Kang, Z.; Zou, W. Improving accuracy of VI-SLAM with fish-eye camera based on biases of map points. *Adv. Robot.* **2020**, *34*, 1272–1278. [CrossRef]

 *symmetry*

Article

Industrial Robot Contouring Control Based on Non-Uniform Rational B-Spline Curve

Guirong Wang *, Jiahao Chen, Kun Zhou and Zhihui Pang

School of Mechanical and Electrical Engineering, China Jiliang University, Hangzhou 310018, China
* Correspondence: lilygrwang@cjlu.edu.cn; Tel.: +86-137-5814-9518

Abstract: This paper presents a novel algorithm about the industrial robot contouring control based on the NURBS (non-uniform rational B-spline) curve. First, aiming at the error between the industrial robot's actual trajectory and the desired trajectory, the contour error is proposed as the trajectory evaluation index, and the estimation algorithm of contour error based on the tangent approximation is proposed. Based on the tangent approximation algorithm, the estimation algorithm of contour error in the local task coordinate frame is proposed to realize the transformation from the Cartesian coordinate frame to the local task coordinate frame. Second, according to the configuration of the industrial robot, a modified cross-coupling control scheme based on the local task coordinate frame is designed. Finally, the Bernoulli's lemniscate curves are constructed by NURBS curve and five-order polynomial curve, respectively, and they are symmetrical. The contrast experiment is designed using the two types of constructed Bernoulli's lemniscate curves as the incentive trajectory. Through the analysis and comparison between the obtained uniaxial tracking error and the contour error curve of the two incentive trajectories, it is concluded that the incentive trajectory constructed by the NURBS curve has better contour control performance than that constructed by the five-order polynomial curve. The results drawn from this paper lay a certain foundation for the future high-precision contouring control of industrial robots.

Keywords: NURBS curve; contour error; local task coordinate frame; cross coupled control; industrial robot

Citation: Wang, G.; Chen, J.; Zhou, K.; Pang, Z. Industrial Robot Contouring Control Based on Non-Uniform Rational B-Spline Curve. *Symmetry* **2022**, *14*, 2533. https://doi.org/10.3390/sym14122533

Academic Editors: Chengxi Zhang, Jin Wu and Chong Li

Received: 26 October 2022
Accepted: 18 November 2022
Published: 30 November 2022

Publisher's Note: MDPI stays neutral with regard to jurisdictional claims in published maps and institutional affiliations.

1. Introduction

In the working process of the industrial robot, the quality of trajectory makes a difference in the overall working process of the robot [1]. Generally speaking, for the industrial robot, given all the critical path points through online-teaching, there exists trajectory error between the actual trajectory and the desired trajectory inevitably because the dynamic response process of the robot always lags behind the reference input [2–4], which are mainly manifested in tracking error and contour error [5,6]. With multi-joint motors of the robot coordinate with each other, the tracking error of single axis motor will superimpose on the operation trajectory, which forms contour errors [7]. Tracking error could be defined as the distance between the desired position and the actual position at a certain moment, while the definition of the contour error is the tangential distance between the actual position at a certain moment and the desired track point. That is to say, contour error could also be defined as the distance between the desired trajectory and the actual trajectory [8]. According to the definition of tracking error and contour error, tracking error describes the distance between two different points, while contour error describes the distance from the actual point to a set of points. In the research of the precision contour error control, the value of contour error is less than or equal to the tracking error's value, and the tracking error could be regarded as the maximum value of e contour error at the very moment. However, the tracking error and contour error does not have a specific relation with each other, and according to the two definitions above, the two kinds of error is completely different [9]. The contour error and tracking error of the robot operation trajectory are shown in Figure 1.

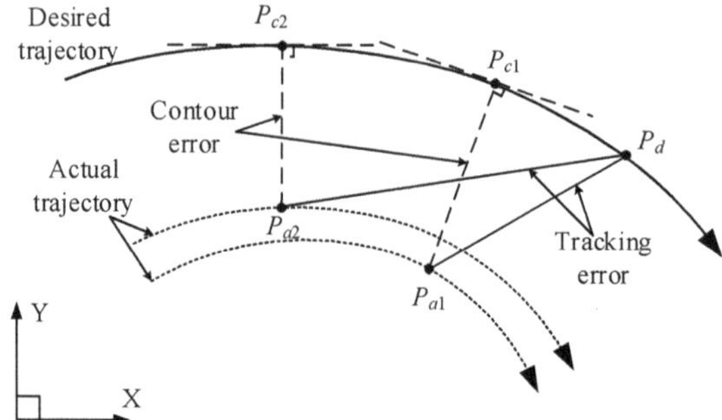

Figure 1. Contour error and tracking error.

In order to explain the difference between the contour error and the tracking error more directly, as shown in Figure 1, P_d is the expected point, while point P_{a1} and P_{a2} are the actual points at different moments. Then, the tracking error corresponding to the actual point P_{a1} and P_{a2} could be expressed as $|\overrightarrow{P_{a1}P_d}|$ and $|\overrightarrow{P_{a2}P_d}|$, meanwhile $|\overrightarrow{P_{a1}P_{c1}}|$ and $|\overrightarrow{P_{a2}P_{c2}}|$ are the contour error corresponding to the actual point P_{a1} and P_{a2}. It could be seen from Figure 1 that the tracking error $|\overrightarrow{P_{a1}P_d}|$ is less than $|\overrightarrow{P_{a2}P_d}|$, but the contour error $|\overrightarrow{P_{a1}P_{c1}}|$ is greater than $|\overrightarrow{P_{a2}P_{c2}}|$. When $|\overrightarrow{P_{a1}P_d}| = 0$, $|\overrightarrow{P_{a1}P_{c1}}| = 0$, the opposite is not true. It is obviously concluded that P_{a2} has more practical significance for the contour control than P_{a1}. According to the above definitions, it is easy to conclude that the contour error only depends on the current actual position and the geometry shape of the desired trajectory, and is irrelevant to the desired point and the tracking error.

The calculation of the contour error depends on the shape of the robot's end-effector trajectory. Contour error could be accurately calculated when the trajectory is a simple curve such as a straight line or arc [10,11]. However, under any common smooth curve, contour error could not be accurately calculated [12], which could be approximated with a variety of approximation algorithms in these cases. Yeh [13] et al. proposed a general curve contour error estimation algorithm based on the tangent approximation of the line contour error, which achieved good estimation effect for curves with small curvature. J. Yang [14] et al. improved the estimation accuracy of the contour error through the osculating circle at any point of the curve as an approximation condition. Y. Zhu [15] et al. proposed new contour error calculation model under task coordinates, and calculated the first order approximation of the contour error, which does not depend on the single axis tracking error. During the calculation process, we just need to know the equation of the desired trajectory, and the coordinates of the actual point, which achieved good results in planar curve contour error calculation. However, the contour error calculation for the spatial curve needs further research [16]. In addition, with the improvement of the processor's computing speed in robot system, F. Huo [17] took the minimum value of the distance between the actual position points and a series of the path points on the desired trajectory as the estimated value of the contour error. When the interpolation period is short, the estimation effect of this algorithm is better for the contour error, but this kind of algorithm needs higher controller performance.

In order to reduce contour error, and improve trajectory performance, L.B [18] et al. applied the single-axis uncoupled control algorithm to control a single axis separately, which reduced the tracking error of each axis and improved track accuracy. Jin.Z [19] established contour error models of the OMPR in straight line, arc, and spiral trajectories are. Then, they established feed-forward combined multi-axis cross-coupled contour control compensation strategy which achieved good control effect. L. Wang [20] et al. designed a cross-coupled controller in which the inputs of the controller were the tracking errors obtained according to the five feed-axes commands and encoder feedbacks. S. Wang [21] et al. designed a self-adaptive fuzzy PID cross-coupled controller which can eliminate the influence of the characteristics mismatching and parameter difference of each axis. N.T. Hu [22] et al. proposed a new structure of cross-coupled position command shaping controller using H_∞ control scheme for the precise tracking in the multiaxial motion control which remarkably reduces contour error. Cross coupling control was applied to solve the multi-axis motion incoordination caused by a large tracking error of single axis which achieved good control effect for the motion platform with orthogonal axes, and multi-axis CNC machining platform [10,19,20,23–25].

However, most of the estimation algorithms mentioned above for the contour error are suitable for the contour error of the planar curve; furthermore, the estimation algorithms for the spatial curve contour error are rarely involved. In recent years, more and more researchers have started to engage in this field. H. Zhao [4] et al. proposed a components-based contouring control structure for a six-degree of freedom robot. A locally iterative robotic contour error estimation approach with high accuracy and high efficiency was designed by compensating the weighted contour error components to the velocity commands in the robotic task space. Z. Wang [26] et al. proposed an Atiken method based on acceleration iterative to achieve higher estimation accuracy of contouring error and reduce the real-time calculation burden, and verified the effectiveness of the proposed method through experiments. The estimation algorithms mentioned above are mostly used in CNC machining platforms with orthogonal axes, and the control algorithm design of the spatial curve contour error is also mostly based on the above platforms [27].

Compared with the CNC machine platform, industrial robot has higher degrees of freedom and more complex actions. Therefore, it is necessary to design a contour error control algorithm which suits the robot contour control [28–30]. According to the configuration of the industrial robot, this paper presents a trajectory planning algorithm in the Cartesian space based on the NURBS curve [31–33]. NURBS curve is often used in robot path generation algorithms. Compared with other algorithms, the NURBS algorithm is more conducive to the generation and processing of multidimensional and irregular curves. G. Wu [34] et al. proposed a path planning method based on the NURBS curve with optimal robot performance. By solving a multi-objective optimization problem, the optimal curve parameters and the execution time distributed along the curve segments can be obtained simultaneously. K. Erwinski [35] et al. presented a NURBS toolpath federate profile generation algorithm for a biaxial linear motor control system, and constructed two-dimensional plane curves of "Bird" and "flower" by using NURBS toolpath with marked control points and polygons. On the basis of existing research results, this paper presents an approximation algorithm to approximate the contour error of the robot, and constructs the incentive trajectory by the NURBS curve. Then, based on the approximation algorithm, a modified cross-coupling controller in the local task coordinate frame is presented, which realizes the contour control of the robot, improves the trajectory performance and quality of industrial robot's working process.

The main contributions of this paper could be summarized as follows:

(1) A contour error approximation algorithm for the spatial curve based on under the local task coordinate frame was proposed, and realizes the transformation from the tracking error of each axis on the desired trajectory to the contour error;

(2) A modified cross-coupling control algorithm is proposed which realizes the error feedback control from the tracking error to the contour error.

(3) The evaluation system of contour error controller with root mean square as evaluation index is established, and Bernoulli's lemniscate curves as the incentive trajectory was constructed by NURBS curve through adjusting its parameters.

This paper is organized as follows. In Section 2, the contour error approximation algorithm of general spatial curve in the local task coordinate frame is proposed. In Section 3, based on the above contour error approximation algorithm, a modified cross-coupling control algorithm is proposed, which is suitable for the industrial robot configuration. In Section 4, based on the improved cross-coupling control algorithm, the NURBS curve trajectory planning algorithm is compared with the five-order polynomial planning algorithm, and illustrates the effectiveness of the NURBS planning and modified cross-coupling controller. In Section 5, this paper arrives at some conclusions.

2. Contour Error Estimation Algorithm

2.1. Contour Error Estimation Based on Tangent Approximation

As shown in Figure 2, the contour error of a line can be accurately calculated. Given the actual point $A(x_a, y_a)$, expected trajectory point $D(x_d, y_d)$, then tracking error could be defined as $\vec{e} = (x_d - x_a, y_d - y_a)$. Given direction vector of the planar line $\vec{s} = (\cos\phi, \sin\phi)$, and β is the included angle between the vector $\vec{e}$ and $\vec{s}$, $\beta + \theta = \pi/2$. Then the contour error of the planar line could be expressed as,

$$\vec{\varepsilon} = \vec{AE} = \vec{e}\cos\theta = \vec{e}\sqrt{1 - \frac{\left(\vec{e}\cdot\vec{s}\right)^2}{\left|\vec{e}\right|^2\cdot\left|\vec{s}\right|^2}} \tag{1}$$

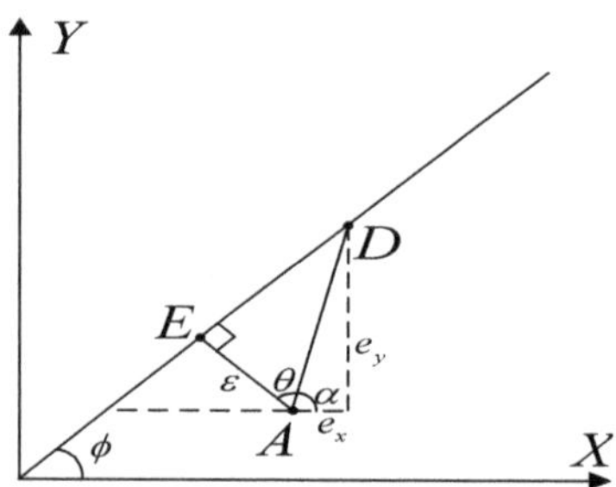

Figure 2. The contour error of a planar line.

Similarly, for a line in the Cartesian space, its parametric equation is shown as,

$$\begin{cases} x = x_d + mt \\ y = y_d + nt \\ z = z_d + pt \end{cases}, \quad \frac{x - xd}{m} = \frac{y - yd}{n} = \frac{z - zd}{p} = t \tag{2}$$

Then, the contour error of a spatial straight line can also be expressed in the form of Equation (1). In addition, a spatial straight line could be regarded as the intersection of two planes, i.e.,

$$\begin{cases} A_1 x + B_1 y + C_1 z + D_1 = 0 \\ A_2 x + B_2 y + C_2 z + D_2 = 0 \end{cases} \tag{3}$$

where $\vec{n}_1 = (A_1, B_1, C_1)$, $\vec{n}_2 = (A_2, B_2, C_2)$ are the normal vectors of the two planes shown in Equation (3). Then the direction vector of the spatial line is,

$$\vec{s} = \vec{n}_1 \times \vec{n}_2 = \begin{vmatrix} \vec{i} & \vec{j} & \vec{k} \\ A_1 & B_1 & C_1 \\ A_2 & B_2 & C_2 \end{vmatrix}$$
$$= \begin{vmatrix} B_1 & C_1 \\ B_2 & C_2 \end{vmatrix}\vec{i} - \begin{vmatrix} A_1 & C_1 \\ A_2 & C_2 \end{vmatrix}\vec{j} + \begin{vmatrix} A_1 & B_1 \\ A_2 & B_2 \end{vmatrix}\vec{k} \tag{4}$$

For common curves, it is difficult to obtain the geometric description of the desired trajectory, thus it is difficult to calculate the contour error of the common curves. However, the calculation of the contour error about a common curve can be simplified as the tangency distance between the actual point and the corresponding desired point, as shown in Figure 3.

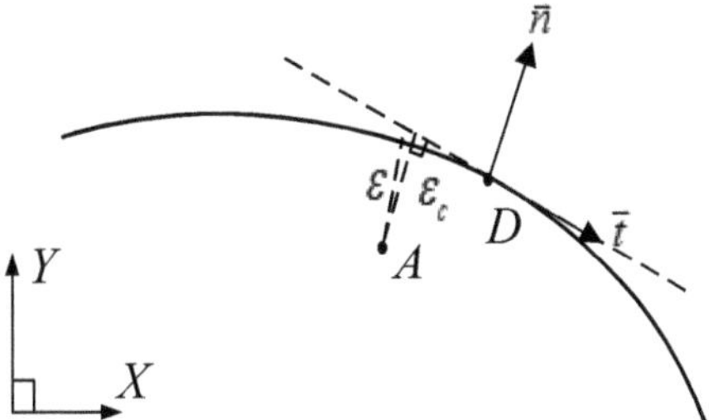

Figure 3. Contour error estimation based on tangential approximation.

In Figure 3, contour error ε could be approximated by ε_c which is the distance of the tangent line from point A to point D, and $\vec{t}$ is the unit tangent vector at point D, $\vec{n}$ is the unit normal vector, then,

$$\varepsilon_c = \vec{AD} \cdot \vec{n} \tag{5}$$

The parametric equation of the spatial curves could be expressed as,

$$\Gamma: \quad r(t) = \begin{cases} x = x(t) \\ y = y(t) \\ z = z(t) \end{cases} \tag{6}$$

Tangent vector at any point of the spatial curve Γ is $\vec{s} = (x'(t), y'(t), z'(t))$. Thus,

$$\vec{\varepsilon_c} = \vec{e} \cos\theta = \vec{e} \sqrt{1 - \frac{\left(\vec{e} \cdot \vec{s}\right)^2}{\left|\vec{e}\right|^2 \cdot \left|\vec{s}\right|^2}} \tag{7}$$

In addition, a spatial curve can be thought as the intersection of surfaces, i.e.,

$$\begin{cases} F(x, y, z) = 0 \\ G(x, y, z) = 0 \end{cases} \tag{8}$$

where $\vec{n}_1 = (F_x, F_y, F_z)$, $\vec{n}_2 = (G_x, G_y, G_z)$ are the normal vectors of the two surfaces shown in Equation (8). Then the tangent vector of the common spatial curve is,

$$\vec{s} = \vec{n}_1 \times \vec{n}_2 = \begin{vmatrix} F_y & F_z \\ G_y & G_z \end{vmatrix} \vec{i} - \begin{vmatrix} F_x & F_z \\ G_x & G_z \end{vmatrix} \vec{j} + \begin{vmatrix} F_x & F_y \\ G_x & G_y \end{vmatrix} \vec{k} \tag{9}$$

2.2. Contour Error in Local Task Coordinate Frame

In Figure 3, $\vec{t}$ and $\vec{n}$ are orthogonal to each other at point D on the desired trajectory, where $\vec{t}$ is the unit tangent vector, and $\vec{n}$ is the unit normal vector, which constitute the local task coordinate frame at point D. From Cartesian coordinates to the local task coordinates, we have the following transformation of the coordinates,

$$\varepsilon = Te, \begin{bmatrix} \varepsilon_n \\ \varepsilon_t \end{bmatrix} = \begin{bmatrix} -\sin\phi & \cos\phi \\ \cos\phi & \sin\phi \end{bmatrix} \begin{bmatrix} e_x \\ e_y \end{bmatrix} \tag{10}$$

where T is the transformation matrix from the Cartesian coordinate frame to the local task coordinate frame, and,

$$T^{-1} = T^T = T \tag{11}$$

In addition, Equation (10) is equivalent to the Equation (1). For the common curves, the approximate value of the tangent approximation could be used to estimate the actual contour error.

For common planar curve whose parametric equation is,

$$\begin{cases} x = x(t) \\ y = y(t) \end{cases} \tag{12}$$

and the unit tangent vector is,

$$\vec{t} = \left(-\frac{x'}{\sqrt{x'^2 + y'^2}}, \frac{y'}{\sqrt{x'^2 + y'^2}} \right) \tag{13}$$

the unit normal vector is,

$$\vec{n} = \left(-\frac{y'}{\sqrt{x'^2 + y'^2}}, \frac{x'}{\sqrt{x'^2 + y'^2}} \right) \tag{14}$$

then,

$$\begin{bmatrix} \varepsilon_n \\ \varepsilon_t \end{bmatrix} = \begin{bmatrix} -\frac{y'}{\sqrt{x'^2+y'^2}} & \frac{x'}{\sqrt{x'^2+y'^2}} \\ \frac{x'}{\sqrt{x'^2+y'^2}} & \frac{y'}{\sqrt{x'^2+y'^2}} \end{bmatrix} \begin{bmatrix} e_x \\ e_y \end{bmatrix} \tag{15}$$

where Equation (10) is equivalent to Equation (15).

The above is the case of a planar curve. In the case of the spatial curve, the Frenet local task coordinate frame as shown in Figure 4 can be established at point D,

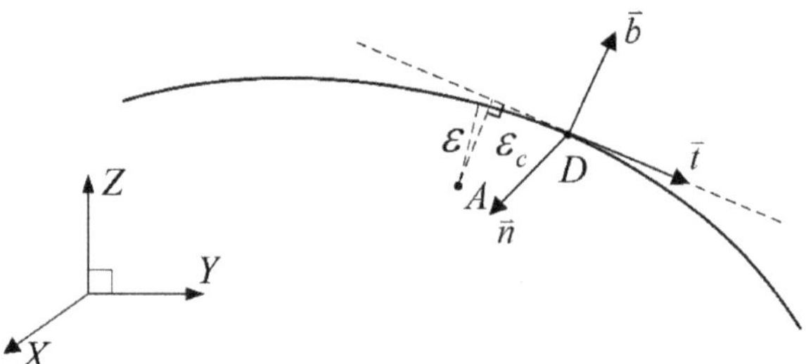

Figure 4. Frenet local task coordinate frame of spatial curve.

Similar with the case of a planar curve, $\vec{t}$ is the unit tangent vector, $\vec{n}$ is the unit principal normal vector, $\vec{b}$ is the unit binormal vector, furthermore, $\vec{t}, \vec{n}, \vec{b}$ are orthogonal to each other, and satisfy the right-hand coordinate frame. Vector $\vec{t}$ and $\vec{n}$ form the osculating plane at point D, $\vec{t}$ and $\vec{b}$ form the rectifying plane, $\vec{n}$ and $\vec{b}$ form the normal plane.

Referring to Equation (6) for parametric equation of spatial curve, then,

$$\vec{t} = \frac{r'(t)}{|r'(t)|} = \frac{(x', y', z')}{\sqrt{x'^2 + y'^2 + z'^2}} \tag{16}$$

$$\vec{b} = \frac{r'(t) \times r''(t)}{|r'(t) \times r''(t)|} = \frac{\left(\begin{vmatrix} y' & z' \\ y'' & z'' \end{vmatrix}, -\begin{vmatrix} x' & z' \\ x'' & z'' \end{vmatrix}, \begin{vmatrix} x' & y' \\ x'' & y'' \end{vmatrix} \right)}{\sqrt{\begin{vmatrix} y' & z' \\ y'' & z'' \end{vmatrix}^2 + \begin{vmatrix} x' & z' \\ x'' & z'' \end{vmatrix}^2 + \begin{vmatrix} x' & y' \\ x'' & y'' \end{vmatrix}^2}} \tag{17}$$

and,

$$\vec{n} = \vec{b} \times \vec{t} = \frac{\begin{vmatrix} \vec{i} & \vec{j} & \vec{k} \\ \begin{vmatrix} y' & z' \\ y'' & z'' \end{vmatrix} - \begin{vmatrix} x' & z' \\ x'' & z'' \end{vmatrix} & \begin{vmatrix} x' & y' \\ x'' & y'' \end{vmatrix} \\ x' & y' & z' \end{vmatrix}}{\sqrt{x'^2 + y'^2 + z'^2}\sqrt{\begin{vmatrix} y' & z' \\ y'' & z'' \end{vmatrix}^2 + \begin{vmatrix} x' & z' \\ x'' & z'' \end{vmatrix}^2 + \begin{vmatrix} x' & y' \\ x'' & y'' \end{vmatrix}^2}} \tag{18}$$

The equations of normal plane Π_1, osculating plane Π_2, and rectifying plane Π_3 at point D are:

$$\begin{cases} \Pi_1 : (X_d - r(t)) \cdot \vec{t} = 0 \\ \Pi_2 : (X_d - r(t)) \cdot \vec{n} = 0 \\ \Pi_3 : (X_d - r(t)) \cdot \vec{b} = 0 \end{cases} \tag{19}$$

The estimated value ε_c of the contour error at point A can be decomposed into the distance ε_b from point A to the osculating plane Π_2, and the distance ε_n from point A to the rectifying plane Π_3, which could be expressed in the form of:

$$\varepsilon = Te \tag{20}$$

i.e.,

$$\begin{bmatrix} \varepsilon_b \\ \varepsilon_n \\ \varepsilon_t \end{bmatrix} = \begin{bmatrix} T_{11} & T_{12} & T_{13} \\ T_{21} & T_{22} & T_{23} \\ T_{31} & T_{32} & T_{33} \end{bmatrix} \begin{bmatrix} e_x \\ e_y \\ e_z \end{bmatrix} \tag{21}$$

where the first, second, and third lines of T are the X, Y, and Z components of $\vec{b}$, $\vec{n}$, and $\vec{t}$ respectively, referring to Equations (16)–(18). Obviously, Equation (20) is equivalent to Equation (21).

Similar with the case of planar curve, $\vec{t}$, $\vec{n}$ and $\vec{b}$ are orthogonal to each other, then we can get the conclusion shown in Equation (11).

The approximation algorithm presented in Section 2.1 can obtain the effective estimation value of the contour error for general spatial curve, whereas, the estimation value is expressed as a scalar, which is not suitable for the coordinate transformation from the tracking error to the contour error. Therefore, it is necessary to establish contour error approximation model in the local task coordinate frame as shown in Section 2.2.

3. Modified Cross Coupling Control in the Local Task Coordinate Frame

According to the first section of this paper, traditional cross-coupling control algorithm has better performance for the contour control of the experimental platform with fewer and orthogonal axes. However, industrial robots are different from orthogonal platform in configuration. They have higher degrees of freedom. The motion among adjacent axis motors is transformed through the robot's link coordinate frame, and the coordinate transformation from the base coordinate frame to the end-effector coordinate frame is more complicated. This paper takes SR4C robot as the experimental platform, and its DH parameter is shown in Table 1.

Table 1. The DH parameter of the SR4C robot.

Link	a_i	α_i	d_i	θ_i	Limit (Deg)
1	40	90	330	θ_1	$-180\sim180$
2	315	0	0	θ_2	$-130\sim80$
3	70	90	0	θ_3	$-70\sim160$
4	0	-90	310	θ_4	$-240\sim240$
5	0	90	0	θ_5	$-30\sim200$
6	0	0	70	θ_6	$-360\sim360$

The link coordinate frame of the SR4C industrial robot is shown in Figure 5.

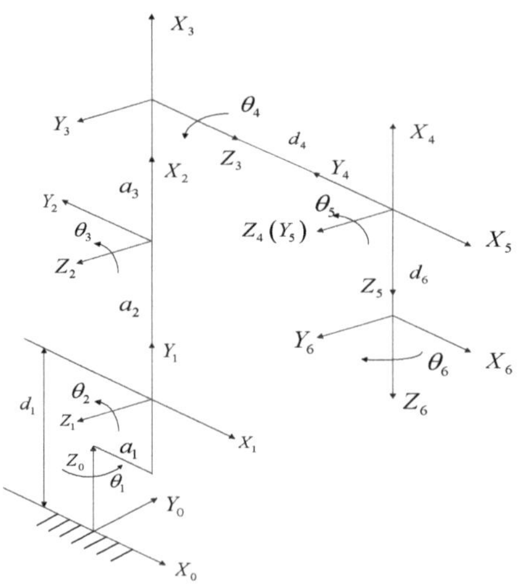

Figure 5. The link coordinate frame of the SR4C robot.

The transformation matrix from base coordinate frame to the end-effector coordinate frame is Equations (22) and (23).

$$T = T_1 T_2 T_3 T_4 T_5 T_6 = \begin{bmatrix} n_x & o_x & a_x & p_x \\ n_y & o_y & a_y & p_y \\ n_z & o_z & a_z & p_z \\ 0 & 0 & 0 & 1 \end{bmatrix} \tag{22}$$

$$T_{i(i=1\sim6)} = \begin{bmatrix} \cos\theta_i & -\cos\alpha_i\sin\theta_i & \sin\alpha_i\sin\theta_i & \alpha_i\cos\theta_i \\ \sin\theta_i & \cos\alpha_i\cos\theta_i & -\sin\alpha_i\cos\theta_i & \alpha_i\sin\theta_i \\ 0 & \sin\alpha_i & \cos\alpha_i & d_i \\ 0 & 0 & 0 & 1 \end{bmatrix} \tag{23}$$

In traditional cross coupling control application, such as linear motor platforms, the transformation matrix T from tracking error to the contour error only involves the transformation of position, and the calculation is relatively simple, but traditional cross coupling control is not suitable for the robot contour control. Based on the traditional cross coupling control, an improved cross coupling controller is proposed in the local task coordinate frame. It is shown in Figure 6.

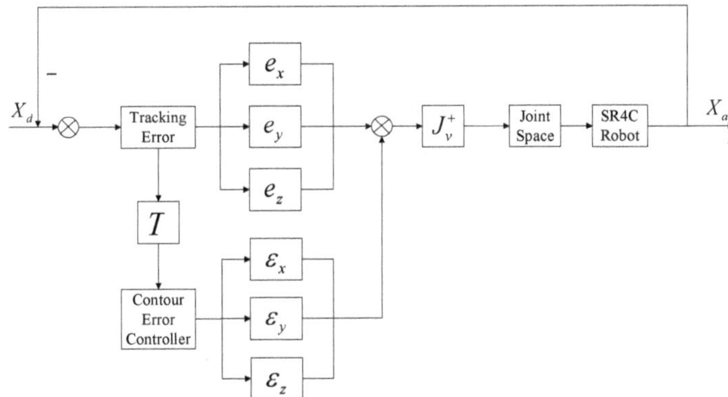

Figure 6. Cross coupled control based on the local task coordinate frame.

The contour error control flow chart based on the control block diagram of Figure 6 is showed in Figure 7.

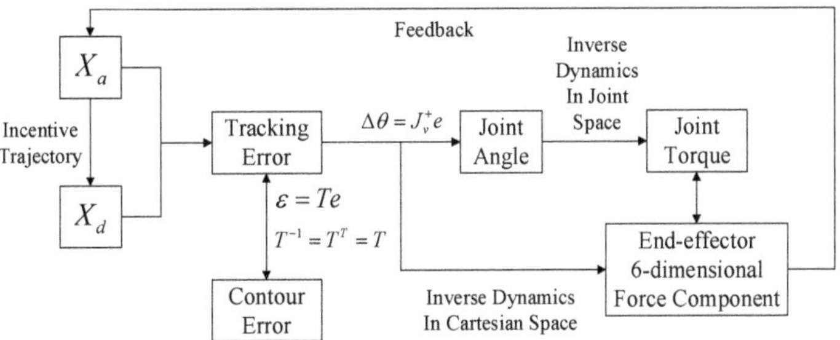

Figure 7. Control flow chart of the contour error.

The controller proposed in the literature [23] is shown in Figure 8.

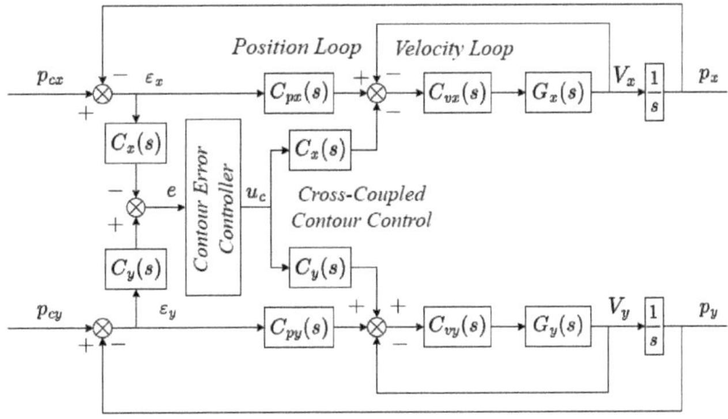

Figure 8. Block diagram of the typical cross-coupled control system.

In Figure 7, first, we get the corresponding point D on the desired trajectory according to the parameter value u of the NURBS curve corresponding to point A on the actual trajectory, and the tracking error from point A to point D could be calculated. The contour error at point D is obtained from Equation (21). Then, the end-effector coordinate frame where the tracking error locates is transformed to the joint space where each axis is located through the inverse kinematics equation of the industrial robot in Equation (22). Next, regarding the joint angle solved from Equation (22) as the input of the inverse dynamics equation in the robot joint space, and the solution of robot inverse dynamics, that is to say, joint torque as the output. At the same time, the inverse dynamics equation in joint space can be transformed to the Cartesian space, that is to say, the tracking error is taken as the input of the inverse dynamics equation. Comparing with the controller in Figure 8 which is proposed in the literature [23], controller in Figure 6, the six-dimensional force component at the end of the output is fed back to the actual trajectory, so that we could realize the closed-loop control of the contour error. Thus, the control accuracy of contour error can be improved.

The inverse dynamics equations in joint space and Cartesian space are shown in Equations (24) and (25).

$$M(\theta)\ddot{\theta} + C\left(\theta,\dot{\theta}\right)\dot{\theta} + G(\theta) = \tau \tag{24}$$

$$M_X(\theta)\ddot{X} + C_X\left(\theta,\dot{\theta}\right)\dot{X} + G_X(\theta) = F \tag{25}$$

where,

$$\begin{cases} F = J^{-T}\tau \\ \dot{X} = J\dot{\theta} \\ \ddot{X} = J\ddot{\theta} + \dot{J}\dot{\theta} \\ \ddot{\theta} = J^{-1}\ddot{X} - J^{-1}\dot{J}\dot{\theta} \end{cases} \tag{26}$$

J is the Jacobi matrix of the robot, see Equation (27),

$$J = \begin{bmatrix} J_v \\ J_\omega \end{bmatrix}, J_v^+ = J_v^T\left(J_v J_v^T\right)^{-1} \tag{27}$$

Combining Equation (19) with (20) and (26), it can get Equation (28),

$$\begin{cases} M_X(\theta) = J^{-T}M(\theta)J^{-1} \\ C_X(\theta,\dot{\theta}) = J^{-T}\left[C(\theta,\dot{\theta}) - M(\theta)J^{-1}\dot{J}\right]J^{-1} \\ G_X(\theta) = J^{-T}G(\theta) \end{cases} \tag{28}$$

For the contour error control, $X = [x,y,z]^T$. Suppose that the coordinate of the desired point is $X_d = [x_d, y_d, z_d]^T$, and the coordinate of the actual point is $X = [x,y,z]^T$, then the tracking error is $e = X - X_d$. When D is a constant point, $X_d = [x_d, y_d, z_d]^T$ also becomes constant, and

$$\dot{X}_d = \ddot{X}_d = 0 \tag{29}$$

Then the inverse dynamics equation in Equation (25) could be written as,

$$M_X(\theta)\ddot{e} + C_X\left(\theta,\dot{\theta}\right)\dot{e} + G_X(\theta) = F \tag{30}$$

Combining Equation (10) with (11), it can get,

$$\dot{e} = T\dot{\varepsilon} + \dot{T}\varepsilon, \quad \ddot{e} = T\ddot{\varepsilon} + 2\dot{T}\dot{\varepsilon} + \ddot{T}\varepsilon \tag{31}$$

Substitute Equation (31) into Equation (30), we could get the dynamics equation of the robot in the local task coordinate frame, which could be expressed as,

$$M_T\ddot{\varepsilon} + C_T\dot{\varepsilon} + D_T\varepsilon + G_T = F_T \tag{32}$$

where,

$$\begin{aligned} M_T = TM_X T, \ C_T = 2TM_X\dot{T} + TC_X T \\ D_T = TM_X\ddot{T} + TC_X\dot{T}, \ G_T = TG_X, F_T = TF \end{aligned} \tag{33}$$

4. Analysis of Simulation Experiment Results

In this paper, SR4C robot is taken as the experimental object, and the dynamic parameters are shown in Table 2.

Table 2. The dynamical parameter of the SR4C robot.

Link	Mass (kg)	Center of Mass (m)			Interia Matrix (kg·m^2)					
		x	y	z	I_{xx}	I_{yy}	I_{zz}	I_{xy}	I_{xz}	I_{yz}
1	1.2228	0.0729	−0.0113	−0.0053	0.0076	0.0130	0.0144	0.0011	0.0013	0.0001
2	1.5967	0.1211	−0.0124	0.0068	0.0071	0.0464	0.047	0.0014	−0.0016	0.00009
3	0.8378	0.0367	−0.0024	−0.021	0.0025	0.0049	0.0043	0.00008	0.0012	0.00002
4	0.5312	−0.0005	0.0008	−0.1167	0.011	0.0111	0.001	0	0	0.00006
5	0.1376	0.00015	0	−0.0111	0.00015	0.00017	0.0001	0	0	0
6	0.0817	0	0	−0.0128	0.00004	0.00004	0.00005	0	0	0

According to the kinetic parameters in Table 2, the kinetic equation described in Equation (24) could be calculated.

Generally speaking, a k-order NURBS curve can be expressed in the following Equation (34),

$$P(u) = \begin{bmatrix} x(u) \\ y(u) \\ z(u) \end{bmatrix} = \frac{\sum\limits_{i=0}^{n} \omega_i d_i N_{i,k}(u)}{\sum\limits_{i=0}^{n} \omega_i N_{i,k}(u)} \tag{34}$$

where d_i stands for $n+1$ control points, $i = 0, 1, \ldots, n$; ω_i is the weight factor corresponding to the control point, $\omega_0 > 0$, $\omega_n > 0$, the rest $\omega_i \geq 0$; $U = [u_0, u_1, u_{n+k+1}]$ is the node vector, and all the u_i is not decrease; $0 \leq u \leq 1$ is the normalization factor, $u_1 = u_2 = \cdots = u_k = 0$, $u_{n+1} = u_{n+2} = \cdots = u_{n+k+1} = 1$; the step size of the rest u_i in the middle is $1/(n+1-k)$, i.e., $u_{k+1} = 1/(n+1-k)$, $u_{k+2} = 2/(n+1-k)$, $\ldots$ $u_n = (n-k)/(n+1-k)$; $N_{i,k}(u)$ is the k-order B-spline basis function, which is defined by the recursion Equation of Cox-de Boor as,

$$\begin{cases} N_{i,0}(u) = \begin{cases} 1, u_i \leq u \leq u_{i+1} \\ 0, else \end{cases} \\ N_{i,k}(u) = \frac{u-u_i}{u_{i+k}-u_i} N_{i,k-1}(u) + \frac{u_{i+k+1}-u}{u_{i+k+1}-u_{i+1}} N_{i+1,k-1}(u) \\ define \frac{0}{0} = 0 \end{cases} \tag{35}$$

The parameters of the NURBS curve are as follows: the order of the NURBS curve $k = 3$; the weight factor vector is $\omega = [1, 1, \ldots, 1]$; the node vector is $U = [0, 0, 0, 0, 1/313, 2/313, \ldots, 312/313, 1, 1, 1, 1]$; and the control points could be determined by the following Equation (36),

$$\begin{cases} x_i = 420 \\ y_i = 100 \cos \theta_i \sqrt{\cos 2\theta_i} \\ z_i = 715 + 100 \sin \theta_i \sqrt{\cos 2\theta_i} \end{cases} \tag{36}$$

where, $i = 0, 1, \ldots, 315$, the number of control points is $n+1 = 316$; and the step size of θ_i in Equation (36) is 0.01 rad.

Bernoulli's lemniscate is commonly used on the robots, which is selected as the incentive trajectory, as shown in Figure 9.

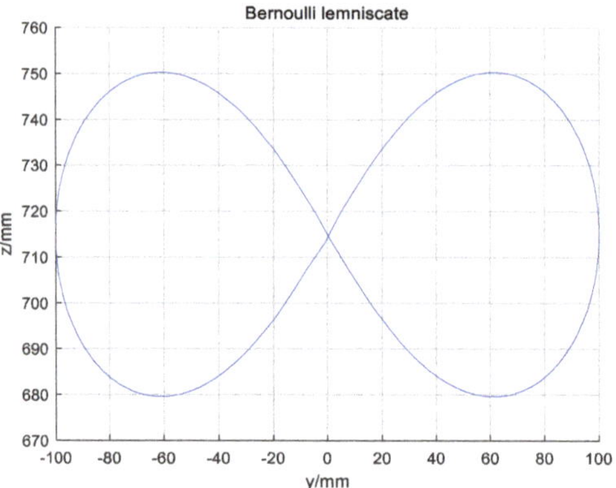

Figure 9. Incentive trajectory.

According to Equation (36), Bernoulli's lemniscate curve is constructed by using NURBS curve. At the same time, Bernoulli's lemniscate curve is also constructed by using the five-order polynomial curve.

The partial enlargement of the above incentive trajectory is shown in Figure 10.

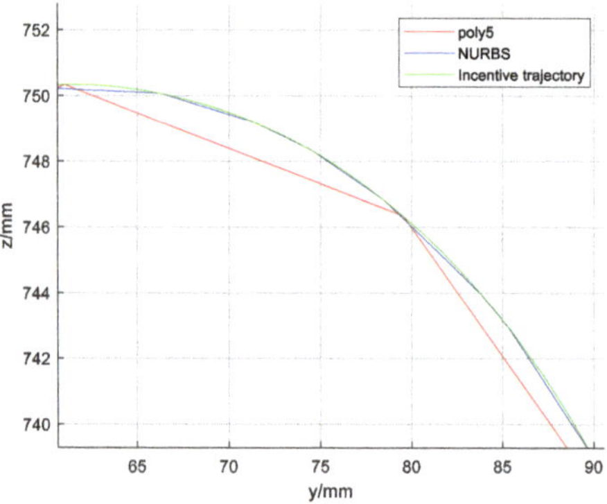

Figure 10. Partial enlarged view of the incentive trajectory.

To evaluate the performance of the controller, the following indicators are used:

$$|\varepsilon_c|_{rms} = \sqrt{\frac{1}{T}\int_0^T |\varepsilon_c|^2 dt} \tag{37}$$

Equation (37) is the root mean square of the contour error, where T is the total planning duration, which could be used for measuring the average contour error control performance.

$\max(|\varepsilon_c|)$ is the maximum absolute value of the contour error, which measures the instant performance.

Figure 11 shows the uniaxial tracking error chart of the NURBS curve, Figure 12 shows the uniaxial tracking error chart of the five-order polynomial curve, and Figure 13 shows the contour error comparison chart of the NURBS curve and the five-order polynomial curve:

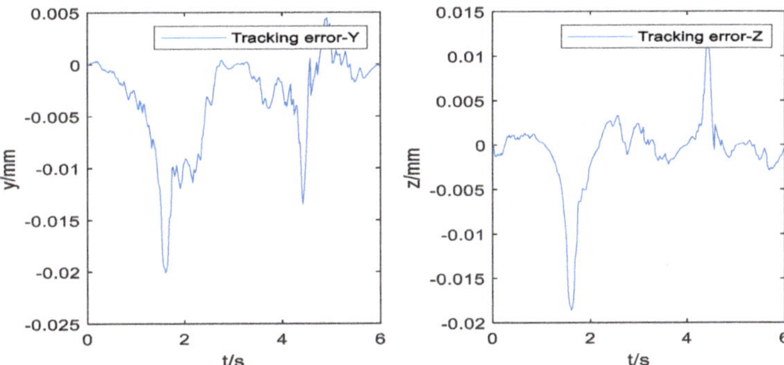

Figure 11. Tracking error of the NURBS curve.

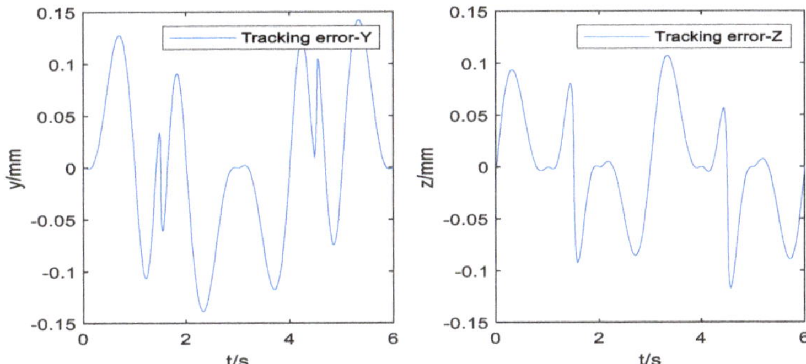

Figure 12. Tracking error of the five-order polynomial curve.

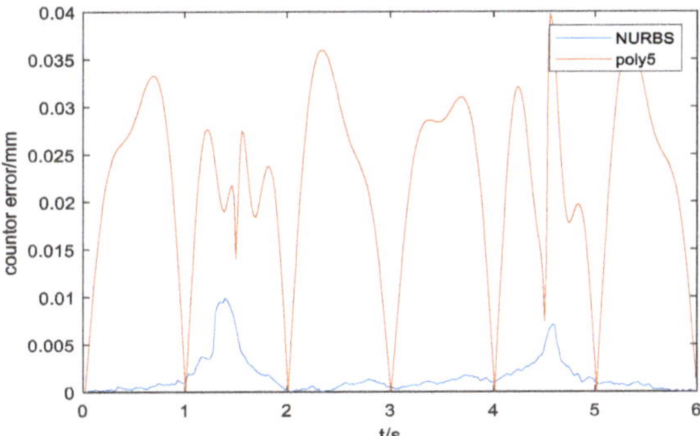

Figure 13. Contour error comparison between the NURBS curve and the five-order polynomial curve.

The tracking error and the contour error of the NURBS curve and five-order polynomial curve can be seen in Tables 3 and 4.

Table 3. Error value of the NURBS curve.

Error Value	NURBS Curve	
	$\max(|\varepsilon_c|)/\mu m$	$|\varepsilon_c|_{rms}/\mu m$
Tracking error-Y	27.792	4.721
Tracking error-Z	24.231	3.134
Contour error	9.922	1.536

Table 4. Error value of the five-order polynomial curve.

Error Value	Five-Order Polynomials	
	$\max(\lvert \varepsilon_c \rvert)/\mu m$	$\lvert \varepsilon_c \rvert_{rms}/\mu m$
Tracking error-Y	142.72	76.03
Tracking error-Z	117.13	53.14
Contour error	39.72	24.06

It could be concluded from Figure 9 that the curve constructed by the NURBS curve trajectory planning is closer to the desired trajectory than the curve constructed by the five-order polynomials. Furthermore, the improved cross-coupling controller in the local task coordinate frame designed in this paper is used to control the profiles of the above two curves. Then, the single axis tracking error diagram, and the contour error result diagram of the two kinds of the two kinds of curves are obtained. Comparing Figures 10 and 11, single axis tracking error curve, it is obvious that the single axis tracking error precision of the NURBS curve planning is significantly higher than the single axis tracking error precision of the five-order polynomial curve, and the single axis tracking error of the NURBS curve planning fluctuates far less than the single axis tracking error of the five-order polynomial curve.

In the comparison of the contour error curves in Figure 12, the contour error of the NURBS curve planning is significantly lower than that of the five-order polynomial curve, and the amplitude fluctuation of the NURBS curve contour error is smaller, and the fluctuation is less. Combining with the contour control data of the two curves in Tables 3 and 4, it could be concluded that the curve constructed by the NURBS curve planning controls the contour error at the level of 10 μm, and the root mean square value of the contour error decreases from 39.72 μm of the five-order polynomial curve to 9.922 μm of the NURBS curve.

5. Conclusions

In this paper, the difference between tracking error and contour error is described, and the significance of contour error for trajectory evaluation obtained from trajectory planning is discussed. Then, the estimation algorithm of the contour error based on the tangent approximation is proposed. Next, the estimation algorithm of the contour error in the local task coordinate frame is proposed. Then, according to the configuration characteristics of the industrial robot, an improved cross-coupling control scheme based on the local task coordinate frame is designed to control the profiles of the two incentive trajectories which are constructed by the NURBS curve and the five-order polynomial curve. The obtained uniaxial tracking error and contour error curve were compared and analyzed. Through the analysis of simulation experiment results, it is concluded that the NURBS curve has better contour control performance than the five-order polynomial curve. The research results of this manuscript provide practical application value for the high precision contouring control for industrial robots.

Author Contributions: Conceptualization, G.W. and Z.P.; methodology, G.W.; software, G.W. and J.C.; validation, G.W. and K.Z.; formal analysis, J.C.; investigation, Z.P.; resources, G.W.; data curation, K.Z.; writing—original draft preparation, G.W. and Z.P.; writing—review and editing, G.W.; visualization, J.C.; supervision, Z.P.; project administration, G.W.; funding acquisition, G.W. All authors have read and agreed to the published version of the manuscript.

Funding: This research was funded by Natural Science Foundation of Zhejiang Province, grant number LGG22F030001.

Data Availability Statement: Not applicable.

Conflicts of Interest: The authors declare no conflict of interest.

References

1. Sun, J.; Han, X.; Zuo, Y.; Tian, S.; Song, J.; Li, S. Trajectory Planning in Joint Space for a Pointing Mechanism Based on a Novel Hybrid Interpolation Algorithm and NSGA-II Algorithm. *IEEE Access* **2020**, *8*, 228628–228638. [CrossRef]
2. Lin, J.; Ye, C.; Yang, J.; Zhao, H.; Ding, H.; Luo, M. Contour error-based optimization of the end-effector pose of a 6 degree-of-freedom serial robot in milling operation. *Robot. Comput. Manuf.* **2022**, *73*, 102257. [CrossRef]
3. Dachang, Z.; Baolin, D.; Aodong, C.; Puchen, Z. Adaptive Non-singular Terminal Sliding Mode Fault-tolerant Control of Robotic Manipulators Based on Contour Error Compensation. *arXiv* **2021**, arXiv:2110.08705.
4. Zhao, H.; Li, X.; Ge, K.; Ding, H. A contour error definition, estimation approach and control structure for six-dimensional robotic machining tasks. *Robot. Comput. Manuf.* **2022**, *73*, 102235. [CrossRef]
5. Deng, K.; Gao, D.; Ma, S.; Lu, Y. Contouring Errors and Feedrate Fluctuation of Serial Industrial Robot in Complex Toolpath with Different Controller. *Intell. Robot. Appl.* **2021**, *13013*, 100–108.
6. Li, K.; Boonto, S.; Nuchkrua, T. Online Self Tuning of Contouring Control for High Accuracy Robot Manipulators under Various Operations. *Int. J. Control Autom. Syst.* **2020**, *18*, 1818–1828. [CrossRef]
7. Chen, R.; Li, K.; Boonto, S.; Nuchkrua, T. Contouring Control Consensus for Robot Manipulators. In Proceedings of the 2019 58th Annual Conference of the Society of Instrument and Control Engineers of Japan (SICE), Hiroshima, Japan, 10–13 September 2019.
8. Kim, N.; Shim, J.; Oh, D.; Kim, H.; Lee, W. Pose optimization of robot machining system for improving position accuracy. In Proceedings of the 19th International Conference of the European Society for Precision Engineering and Nanotechnology, Bilbao, Spain, 3–7 June 2019.
9. Ouyang, P.R.; Pano, V.; Acob, J. Contour Tracking Control for Multi-DOF Robotic Manipulators. In Proceedings of the 2013 10th IEEE International Conference on Control and Automation, Hangzhou, China, 12–14 June 2013.
10. Li, X.; Zhao, H.; Yang, J.; Ding, H. A High Accuracy On-Line Contour Error Estimation Method of Five-axis Machine Tools. In Proceedings of the International Conference on Intelligent Robotics and Applications, ICIRA 2015, Portsmouth, UK, 24–27 August 2015; Lecture Notes in Computer Science. Springer: Cham, Swizerland, 2015.
11. Su, K.-H.; Chen, H.-R.; Cheng, M.-Y. Free-form Curves Contour Error Estimation Using the Backward Arc Length Approach. In Proceedings of the 2014 IEEE/SICE International Symposium on System Integration, Tokyo, Japan, 13–15 December 2014.
12. Cheng, M.-Y.; Lee, C.-C. Motion Controller Design for Contour-Following Tasks Based on Real-Time Contour Error Estimation. *IEEE Ind. Electron. Mag.* **2007**, *54*, 1686–1695. [CrossRef]
13. Yeh, S.-S.; Hsu, P.-L. Estimation of the contouring error vector for the cross-coupled control design. *IEEE/ASME Trans. Mechatron.* **2002**, *7*, 44–51. [CrossRef]
14. Yang, J.; Li, Z. A Novel Contour Error Estimation for Position Loop-Based Cross-Coupled Control. *IEEE/ASME Trans. Mechatron.* **2011**, *16*, 643–655. [CrossRef]
15. Hu, C.; Hu, Z.; Zhu, Y.; Wang, Z. Advanced GTCF-LARC Contouring Motion Controller Design for an Industrial X–Y Linear Motor Stage With Experimental Investigation. *IEEE Trans. Ind. Electron.* **2017**, *64*, 3308–3318. [CrossRef]
16. Zheng, M.; Zhang, F.; Zhu, J.; Zuo, Z. A fast and accurate bundle adjustment method for very large-scale data. *Comput. Geosci.* **2020**, *142*, 104539. [CrossRef]
17. Huo, F.; Poo, A.N. Free-form Two-dimensional Contour Error Estimation Based on NURBS Interpolation. In Proceedings of the Applied Mechanics and Materials. *Trans. Tech. Publ.* **2012**, *157*, 236–240.
18. Biagiotti, L.; Califano, F.; Melchiorri, C. Repetitive Control Meets Continuous Zero Phase Error Tracking Controller for Precise Tracking of B-Spline Trajectories. *IEEE Trans. Ind. Electron.* **2020**, *67*, 7808–7818. [CrossRef]
19. Jin, Z.; Cheng, G.; Meng, Y. Feedforward Combined Multi-axis Cross-coupling Contour Control Compensation Strategy of Optical Mirror Processing Robot. *Robotica* **2022**, *40*, 3476–3498. [CrossRef]
20. Wang, L.; Kong, X.; Yu, G.; Li, W.; Li, M.; Jiang, A. Error Estimation and Cross-coupled Control Based on a Novel Tool PoseRepresentation Method of a Five-axis Hybrid Machine Tool. *Int. J. Mach. Tools Manuf.* **2022**, *182*, 103955. [CrossRef]
21. Wang, S.; Chen, Y.; Zhang, G. Adaptive fuzzy PID cross coupled control for multi-axis motion system based on sliding mode disturbance observation. *Sci. Prog.* **2021**, *104*, 1–19. [CrossRef]
22. Hu, N.T.; Chen, L.Y.; Chen, C.S. Novel Cross-coupling Position Command Shaping Controller using in Multi-axis Motion Systems. *IEEE Trans. Ind. Electron.* **2022**, *69*, 13099–13110. [CrossRef]
23. Li, B.; Wang, T.; Wang, P. Cross-coupled Control Based on Real-time Double Circle Contour Error Estimation for Biaxial Motion System. *Meas. Control* **2021**, *54*, 324–335.
24. Ji, W.; Cui, X.; Xu, B.; Ding, S.; Ding, Y.; Peng, J. Cross-coupled Control for Contour Tracking Error of Free-form Curve Based on Fuzzy PID Optimized by Im-proved PSO Algorithm. *Meas. Control* **2022**, *55*, 807–820. [CrossRef]
25. Jiang, Y.; Chen, J.; Zhou, H.; Yang, J.; Hu, P.; Wang, J. Contour error modeling and compensation of CNC machining based on deep learning and reinforcement learning. *Int. J. Adv. Manuf. Technol.* **2022**, *118*, 551–570. [CrossRef]
26. Wang, Z.; Hu, C.; Zhu, Y. Accelerated Iteration Algorithm Based Contouring Error Estimation for Multiaxis Motion Control. *IEEE/ASME Trans. Mechatron.* **2022**, *27*, 452–462. [CrossRef]
27. Wu, J.; Xiong, Z.; Ding, H. Integral design of contour error model and control for biaxial system. *Int. J. Mach. Tools Manuf.* **2015**, *89*, 159–169. [CrossRef]
28. Alandoli, E.A.; Lee, T.; Vijayakumar, V.; Lin, Y.; Mohammed, M.Q. Dynamic model and integrated optimal controller of hybrid arms robot for laser contour machining. *J. Vib. Control* **2022**, 1–19. [CrossRef]

29. Zhang, X.; Lu, W.; Su, M.; Xu, W. Research on Synchronous Control Strategy of Robot Arm Based on Cross-Coupling Control. *Int. J. Innov. Comput. Inf. Control* **2020**, *16*, 1987–2005.
30. Nuchkrua, T.; Kornmaneesang, W.; Chen, S.L.; Boonto, S. Precision Contouring Control of 5 DOF Dual-arm Robot Manipulators with Holonomic Constraints. In Proceedings of the 2017 11th Asian Control Conference (ASCC), Gold Coast, Australia, 17–20 December 2017.
31. Xu, G.; Zhang, H.; Meng, Z.; Sun, Y. Automatic interpolation algorithm for NURBS trajectory of shoe sole spraying based on 7-DOF robot. *Int. J. Cloth. Sci. Technol.* **2022**, *34*, 434–450. [CrossRef]
32. Zhang, H.; Ni, X.; Hu, P.; Yuan, Y. Real-Time Contour Error Estimation with NURBS Interpolator. In Proceedings of the 2018 2nd International Conference on Robotics and Automation Sciences (ICRAS), Beijing, China, 23–25 June 2018.
33. Erwinski, K.; Paprocki, M.; Wawrzak, A.; Grzesiak, L.M. Pso Based Feedrate Optimization with Contour Error Constraints for NURBS Toolpaths. In Proceedings of the 2016 21st International Conference on Methods and Models in Automation and Robotics (MMAR), Miedzyzdroje, Poland, 29 August–1 September 2016.
34. Wu, G.; Zhao, W.; Zhang, X. Optimum time-energy-jerk trajectory planning for serial robotic manipulators by reparameterized quintic NURBS curves. *Proc. Inst. Mech. Eng. Part C J. Mech. Eng. Sci.* **2021**, *235*, 4382–4393. [CrossRef]
35. Erwinski, K.; Wawrzak, A.; Paprocki, M. Real-Time Jerk Limited Feedrate Profiling and Interpolation for Linear Motor Multiaxis Machines Using NURBS Toolpaths. *IEEE Trans. Ind. Inform.* **2022**, *18*, 7560–7571. [CrossRef]

Article

Redundant Posture Optimization for 6R Robotic Milling Based on Piecewise-Global-Optimization-Strategy Considering Stiffness, Singularity and Joint-Limit

Hepeng Ni [1], Shuai Ji [2,*] and Yingxin Ye [1]

[1] School of Mechanical and Electronic Engineering, Shandong Jianzhu University, Jinan 250101, China
[2] School of Mechanical Engineering, Shandong University, Jinan 250061, China
*   Correspondence: jishuai@sdu.edu.cn

Abstract: Robotic machining has obtained growing attention recently because of the low cost, high flexibility and large workspace of industrial robots (IRs). Multiple degrees of freedom of IRs improve the dexterity of machining while causing the problem of redundancy. Meanwhile, the performance of IRs, such as their stiffness and dexterity, is affected by their position and posture obviously. Therefore, a redundant posture optimization method for robotic milling is proposed to improve the machining performance of the robot. The multiple characteristics of the robot are considered, including the joint-limit, singularity and stiffness, which have symmetry in its workspace. Firstly, the joint-limit is regarded as the constraint. And a symmetrical and effective constraint method is proposed to simply guarantee that all the interpolation points can avoid joint interference. Then, the performance indices of singularity and stiffness are designed as the optimization target. On this basis, the piecewise-global-optimization-strategy (PGOS) is proposed for redundant optimization. Owning to the PGOS, all the given planned tool points in their corresponding segment are considered simultaneously to avoid the gradual deterioration in traditional methods, which is especially suitable for the machining process with a continuous path. Moreover, the computational load of the optimization solution is considered and limited by the designed segmentation strategy. Finally, a series of comparative simulations are conducted to validate the good performance of the proposed method.

Keywords: robotic milling; redundant posture optimization; joint-limit avoidance; stiffness; singularity

Citation: Ni, H.; Ji, S.; Ye, Y. Redundant Posture Optimization for 6R Robotic Milling Based on Piecewise-Global-Optimization-Strategy Considering Stiffness, Singularity and Joint-Limit. *Symmetry* **2022**, *14*, 2066. https://doi.org/10.3390/sym14102066

Academic Editors: Chengxi Zhang, Jin Wu and Chong Li

Received: 29 August 2022
Accepted: 24 September 2022
Published: 4 October 2022

Publisher's Note: MDPI stays neutral with regard to jurisdictional claims in published maps and institutional affiliations.

1. Introduction

Presently, CNC machine tools are the mean equipment for metal cutting, which is suitable for production with high precision and large quantity [1]. However, conventional CNC machine tools suffer from several limitations in the production of large size and small batches, such as high cost and low flexibility [2,3]. In recent years, robotic machining by industrial robots (IRs), especially robotic milling, has attracted growing attention owing to the low cost, high flexibility and large workspace of IRs [4].

As shown in Figure 1, a six revolute (6R) serial robot is usually employed to construct a robotic milling system where the spindle is mounted as the end-effector (EE). The computer-aided manufacturing (CAM) system designed for five-axis CNC milling, such as the Mastercam and Siemens NX, is usually used to generate the target milling path with a series of tool points [5,6]. During the milling process, the motion control of the 6R robot needs six coordinates at each milling point, including three position coordinates to locate the tool center point (TCP) and three posture coordinates to orient the EE [7]. However, the typical CAM system can only provide five coordinates without the rotational degree of freedom (DoF) around the tool axis [8], which can be represented by a γ coordinate in the Euler frame as shown in Figure 1. Hence, the absent γ needs to be determined for the robot controller for the following posture tracking [9,10], which can be summed up as a planning problem of redundant DoF.

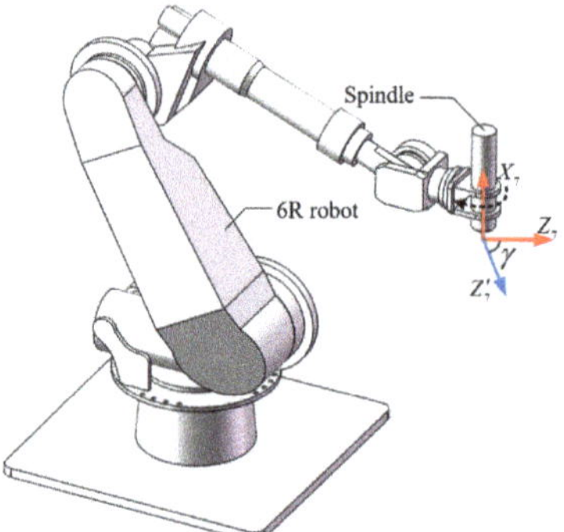

Figure 1. Layout of the robotic milling system and the redundant γ coordinate.

The γ coordinate can be directly selected as a fixed value, which solves the redundancy problem reluctantly. However, the dexterity of the 6R robot is lost. In addition, the performance, such as the singularity and stiffness of EE, changes with the position and posture of IRs, which affects the machining quality directly [11]. Therefore, optimization planning considering the performance of the robot is the most reasonable and valuable mode for the redundancy problem [12]. For 6R robotic milling, three main performances are widely considered, including joint-limit, singularity and stiffness of EE, which are introduced in detail as follows:

- Joint-limit performance

Joint-limit performance is always regarded as one of the optimization targets to avoid joint interference. Zhu et al. [13] define a joint-limit index to keep each joint angle away from the limit boundary and its value range is $[1, +\infty]$. A similar joint-limit index is designed and applied in [14], which obtains a similar optimization result.

In general, the purpose of these performance indices and their optimization is to maintain each joint angle close to the middle of the limit range. In fact, this is not necessary because the joint-limit is to avoid joint interference of mechanical structure but has no effect on the motion performance when approaching the boundary. Therefore, the joint-limit does not need to be the optimization target but should be the constraint for judgment.

In this regard, some researchers use the joint-limit as the constraint to judge whether each joint angle of the planned tool points is within the limit range [15,16]. However, these methods can only ensure the joint-limit avoidance of the planned tool points. But the middle interpolation points between the planned points are ignored, which might result in the risk of joint interference due to the complex nonlinear mapping between Cartesian space and the joint space of IRs.

- Singularity performance

The singularity of IRs affects their motion performance obviously and the singular configuration should be kept away during the milling process. Hence, singularity performance should be regarded as one of the optimization targets.

Several singularity performance indices, such as the manipulability index [17] and the condition number of the Jacobian matrix [18], are designed as the distance metrics to avoid the singularity. The condition number of Jacobian has many different forms and reflects the transfer relationship between force and motion [13]. Among them, the condition number defined in the Frobenius norm [19] is the most widely used owing to the low computational load. These singularity performance indices have different value ranges and can be selected according to the application requirements.

- Stiffness performance

The stiffness performance affects the machining accuracy and surface quality directly and is the most important factor limiting the application of robot machining compared with CNC machine tools [20]. The greater stiffness of robot EE can obtain better machine quality [21]. Therefore, stiffness performance should be regarded as one of the optimization targets.

The stiffness matrix of robot EE K_{car} in Cartesian space can reflect the operational stiffness. However, K_{car} is the tensor index with various parameters and is difficult to use in optimization solutions directly. Hence, the scalar index needs to be constructed. Assuming the robot link is rigid and the joint is elastic, the stiffness distribution of robot EE is an ellipsoid, which is called a stiffness or compliance ellipsoid. Several scalar stiffness indices are designed based on stiffness ellipsoids for different occasions. To improve the accuracy and efficiency of robotic drilling, Jiao et al. [14] and Chen et al. [22] select maximizing the stiffness in the normal direction of the workpiece as the optimization target. Correspondingly, the stiffness indices are designed to describe the deviation of the long axis of the stiffness ellipsoid from the normal direction. Xiong et al. [23] proposes a feed direction stiffness index where the stiffness along the feed direction is maximized to obtain a high feed rate. Guo et al. [7] selects the volume of the stiffness ellipsoid as a performance index to improve the overall machining quality. Similarly, several overall indices are defined in [22,24,25]. These indices realize the scalar metrics of stiffness performance in different aspects and can be selected for specific applications.

Based on the above constraint and performance indices, various redundant posture optimization methods are developed. The stiffness and singularity performance are considered respectively in [8,22]. For better comprehensive performance, the joint-limit and singularity indices are combined as the optimization target in [13,23]. However, as mentioned before, taking a joint-limit as an optimization objective is unnecessary and might limit the performance of IRs. Jiao et al. [14] and Xiong et al. [24] consider three factors simultaneously, where the joint-limit and singularity are used as judgmental constraints. Nevertheless, the threshold of the singularity index is difficult to determine due to the lack of clear physical meaning. Lin et al. [16] takes the stiffness and singularity indices as the optimization targets, respectively. Meanwhile, the constraint for the variation range of γ between two adjacent planned tool points, which is called the displacement constraint of γ, is considered to ensure the machining efficiency and quality. And the corresponding constraint strategy is proposed. However, the velocity planning of the other five DoFs is needed previously, which is complex and has low accuracy. Most importantly, the above optimization methods all employ the sequential-single-point-optimization-strategy (SS-POS), where only one point is considered in each optimization process. Thus, SSPOS is easy to lead to a gradual deterioration in subsequent optimization. Moreover, due to the variation range constraint of γ, the optimization process might even fall into the bad region and cannot jump out.

To overcome these problems, a novel redundant posture optimization method considering joint-limit, singularity and stiffness is proposed in this paper. The main contributions can be described as follows:

- A symmetrical judgment method for joint-limit avoidance is proposed to guarantee that both the planned tool points and their middle points can satisfy the joint-limit constraint, which is effective and simple to apply;
- A new stiffness index based on the stiffness ellipsoid and its symmetry is designed to balance the effects of stiffness and singularity indices in a weighted combination, which can prevent stiffness from being submerged by the singularity index in value;
- Corresponding to SSPOS, the piecewise-global-optimization-strategy (PGOS) and its redundant optimization method are proposed, which can comprehensively consider all the given tool points and the computational load. Meanwhile, a simple displacement constraint method of γ is designed.

The remainder of this paper is organized as follows. Section 2 presents the performance indices and their combination method. In Section 3, the proposed posture optimization method is introduced in detail. The simulation and experiment results are analyzed and compared with previous research work in Section 4. And the conclusions are given in Section 5.

2. Performance Indices and Their Combination Method

In this section, the join-limit is discussed as the judgmental constraint and the corresponding judgment method is proposed. Meanwhile, the singularity and stiffness performance indices and their combination are illustrated.

2.1. Joint-Limit Constraint

Before the trajectory planning of robot milling, a series of tool points on the target milling path is obtained by the CAM system. For the posture optimization of these tool points, the traditional joint-limit judgment method is to directly compare the current joint angle with the corresponding limit range as follows:

$$\theta^j_{\min} \leq \theta^j_i \leq \theta^j_{\max} \tag{1}$$

where θ^j_i is the angle of j-th joint in i-th tool point, $[\theta^j_{\min}, \theta^j_{\max}]$ is the value range of i-th joint. However, Equation (1) only ensures that the optimized points can satisfy the joint-limit constraints, which is suitable for the machining process without contour motion, such as robot drilling. But for the milling process, joint interference might occur at the middle points between adjacent planned points.

The distance between the given tool points is generally close because of the accuracy constraint. Hence, the change range of each joint angle between two given points is small when they are away from the singular position. Therefore, a simple judgment method of joint-limit is designed as follows:

$$\theta^j_{\min} + \Delta\theta^j \leq \theta^j_i \leq \theta^j_{\max} - \Delta\theta^j \tag{2}$$

where $\Delta\theta^j > 0$ is the designed allowance, which acts on the initial value boundary symmetrically. The value of $\Delta\theta^j$ can be set according to the distance between the two planned points. The farther the distance, the larger the value of $\Delta\theta^j$. Equation (2) provides a simple and effective method to guarantee that both the optimized points and the corresponding middle points can be limited to avoid joint interference.

2.2. Singularity Performance Index

There are two main, widely used singularity performance indices that are manipulability and condition number of the Jacobian matrix. The manipulability index K_{mani} can be defined as follows:

$$K_{mani} = \sqrt{\det[J(\theta)J^T(\theta)]} \tag{3}$$

where $J(\theta)$ is the Jacobian matrix and $K_{mani} \in [0, +\infty)$. The larger the value of K_{mani}, the better the manipulability.

In general, the condition number index of the Jacobian matrix can be defined as follows:

$$K_{cond} = ||J(\theta)||||J^{-1}(\theta)|| \tag{4}$$

where $||\cdot||$ denotes the condition number of matrix. In particular, the condition number defined in the Frobenius norm form is the analytic function of $J(\theta)$ and does not need to calculate the singular value. Therefore, the singularity index $K_{\sin}$ in the Frobenius norm form is employed in this paper as follows:

$$K_{\sin} = \frac{1}{m}\sqrt{tr(HH^T)tr[(HH^T)^{-1}]} \tag{5}$$

where $K_{sin} \in [1, +\infty)$, $H = \begin{bmatrix} \frac{1}{L}I_{3 \times 3} & 0_{3 \times 3} \\ 0_{3 \times 3} & I_{3 \times 3} \end{bmatrix} J(\theta)$, m is the number of rows of $J(\theta)$, L is the characteristic length of the robot. The closer the distance to the singular position, the greater the value of K_{sin}.

2.3. Stiffness Performance Index

Under the assumption of flexible joints and rigid links, the compliance matrix of robot EE in the Cartesian space can be obtained as follows:

$$C(\theta) = J(\theta)K_\theta J^T(\theta) \tag{6}$$

where K_θ is the diagonal matrix of joint stiffness. For 6R IRs shown in Figure 1, $C(\theta)$ is a 6 $\times$ 6 matrix and can be partitioned as follows:

$$C(\theta) = \begin{bmatrix} C_{fd}(\theta) & C_{f\delta}(\theta) \\ C_{td}(\theta) & C_{t\delta}(\theta) \end{bmatrix} \tag{7}$$

where $C_{fd}(\theta)$, $C_{f\delta}(\theta)$, $C_{td}(\theta)$ and $C_{t\delta}(\theta)$ are 3 $\times$ 3 compliance submatrices and reflect force-linear displacement, force-angular displacement, torque-linear displacement and torque-angular displacement, respectively.

During the robotic milling process, the cutting force is small, owing to the shallow cutting depth and high spindle speed. Thus, to reduce the complexity and computational load, only the force-linear displacement is considered. The force-linear displacement can be described as follows:

$$d = \begin{bmatrix} d_x \\ d_y \\ d_z \end{bmatrix} = C_{fd}(\theta) \begin{bmatrix} f_x \\ f_y \\ f_z \end{bmatrix} = C_{fd}(\theta)f \tag{8}$$

where f is the cutting force and d is the corresponding displacement vector with three elements. Assuming that unit deformation occurs at robot EE, it can be written as follows:

$$\|d\| = d^T d = 1 \tag{9}$$

Based on Equations (8) and (9), the following relationship can be obtained:

$$f_{uni}^T C_{fd}^T(\theta)C_{fd}(\theta)f_{uni} = 1 \tag{10}$$

where f_{uni} is the fore vector causing unit deformation. As shown in Figure 2, the distribution of f_{uni} in Cartesian space can be described as an ellipsoidal surface called force-linear stiffness ellipsoid, which is symmetrical in space. The volume, shape and posture of the stiffness ellipsoid reflect the distribution of the end stiffness in space. The values and directions of the short and long semi-axes are the magnitude and directions of the minimum and maximum stiffness, respectively.

Based on the stiffness ellipsoid and its symmetry, various stiffness indices are designed. In this paper, the omnidirectional index is employed based on the volume of stiffness ellipsoid as follows:

$$K_{sti} = \frac{1}{V_{se}} \tag{11}$$

where V_{se} is the volume of stiffness ellipsoid and can be calculated as follows:

$$V_{se} = \frac{4}{3}\pi \lambda_1 \lambda_2 \lambda_3 \tag{12}$$

where λ_1, λ_2 and λ_3 are the eigenvalues of $C_{fd}^T(\theta)C_{fd}(\theta)$, respectively. $K_{sti} \in (0, +\infty)$ and the smaller the K_{sti}, the better the stiffness performance of robot EE.

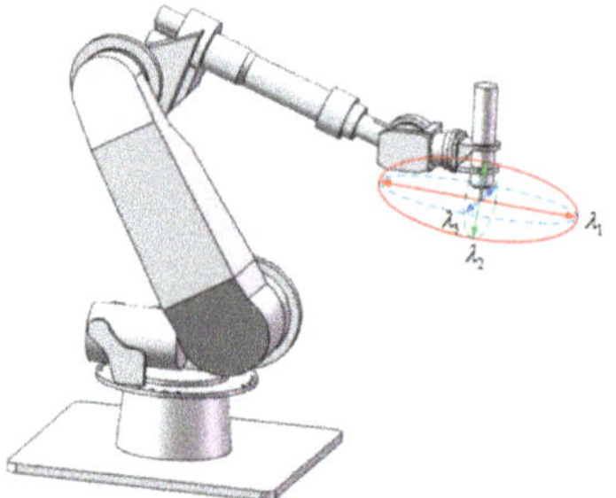

Figure 2. Force-linear stiffness ellipsoid.

2.4. Combination of Singularity and Stiffness Indices

The singularity and stiffness indices need to be combined as one scalar index for the following optimal process, where the weighted combination is the most wildly used method. As can be seen from Equations (12) and (13), the value range of K_{sti} is $(0, +\infty)$. However, the stiffness of each joint is finite for a real IR. Meanwhile, the volume of the stiffness ellipsoid changes gently at the same tool point with different postures. Therefore, there is a big gap between K_{sin} and K_{sti} in terms of value boundary, change amplitude and order of magnitude. In some tool points, the effect of K_{sti} might be submerged by K_{sin} in a directly weighted combination. Hence, a new stiffness index is proposed in this paper based on K_{sti} as follows:

$$K_{sti}^{new} = \sqrt{\frac{(K_{stimax} - K_{stimin})^2}{(K_{stimax} - K_{sti})^2}} \tag{13}$$

where K_{stimax} and K_{stimin} are the maximum and minimum value according to Equation (12) at the same point with different γ posture. $K_{sti}^{new} \in [1, +\infty)$ and the smaller the K_{sti}^{new}, the batter the overall stiffness performance of the robot EE.

According to Equations (5) and (13), K_{sti}^{new} and K_{sin} have same value range and order of magnitude and can be combined with the following form:

$$K_{com} = \omega_1 K_{sin} + \omega_2 K_{sti}^{new} \tag{14}$$

where ω_1 and ω_2 are the weight factors and belong to $(0, 1]$. At present, K_{com} can be selected as the optimization target for the planning of γ. The smaller the K_{com}, the batter the performance of robotic milling.

3. Redundant Posture Optimization Based on Piecewise Global Optimization Strategy

3.1. Fundamental of the Piecewise Global Optimization Strategy

The traditional SSPOS is the most commonly used method but is easy to lead to poor optimization effects on subsequent points since only one point is considered in each optimization process. Moreover, due to the displacement constraint of γ between two adjacent tool points, the optimization process might even fall into a bad region and cannot jump out. Hence, SSPOS is only suitable for the machining process with a single-point operation such as robotic drilling.

The global optimization strategy, which can comprehensively consider all the given tool points, is an effective method to avoid the problems caused by SSPOS. Nevertheless, the quantity of tool points is usually large, which leads to a high computational load for global optimization. Therefore, a piecewise global optimization strategy (PGOS) is proposed, and its flowchart is shown in Figure 3. Firstly, the segmentation is conducted for the given path on the principle of fixed length or fixed-point quantity. Then, globe optimization is employed in each segment until all tool points are optimized, which is introduced in detail in Section 3.2. In particular, the last tool point of the previous segment provides the displacement constraint of γ for the first point of the next segment.

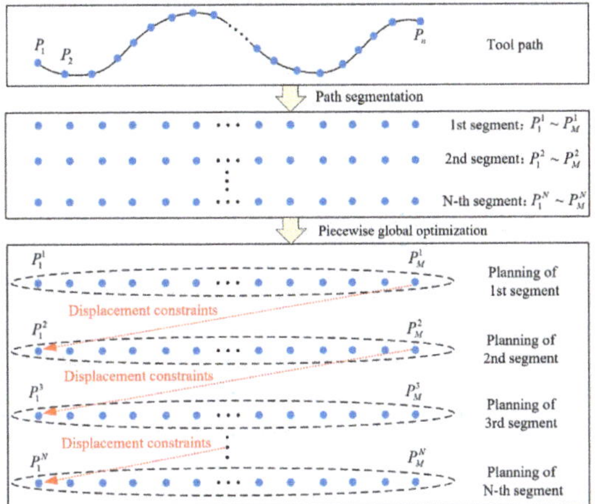

Figure 3. Procedures of the proposed PGOS.

3.2. Redundant Posture Optimization Method Based on PGOS

Based on PGOS given in Section 3.1, the proposed optimization method can be described as follows with the flowchart shown in Figure 4.

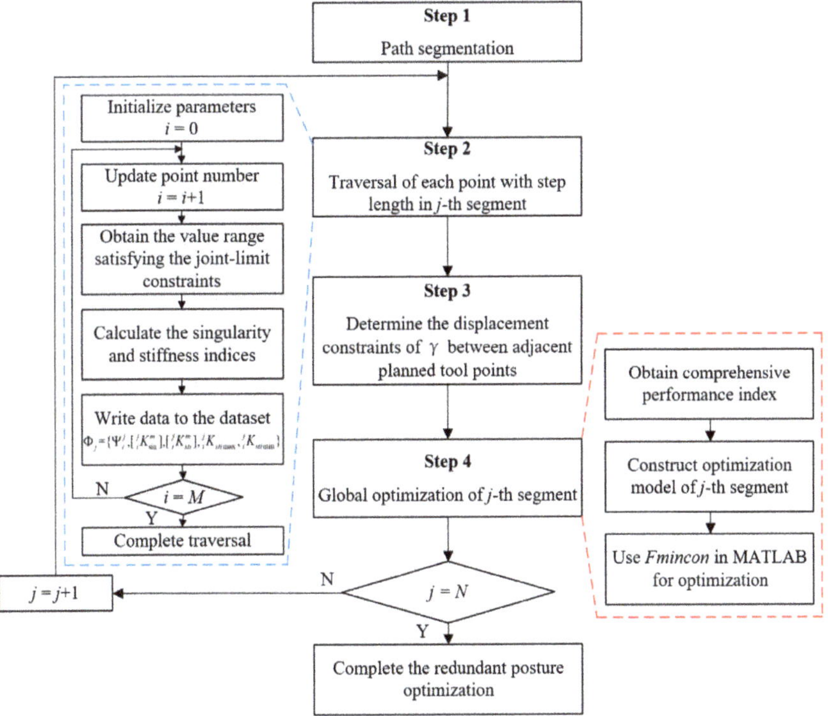

Figure 4. Redundant posture optimization method based on PGOS.

- Step 1: Path segmentation

The fix point quantity principle is employed to divide the target path into several segments. It is assumed that N segments are generated with M tool points in each segment. Set $j = 1$ and go to Step 2.

- Step 2: Traversal of each point with step length $\Delta\gamma$ in j-th segment.

For i-th point, the initial value range of γ_i^j is $[-\beta, \beta]$. In this range, $\Delta\gamma$ is set to be the step length for traversal to obtain the critical information. Firstly, the value range of γ_i^j satisfying the joint-limit constraints can be obtained according to Equation (2) and described as the data range Ψ_i^j.

Secondly, the singularity and stiffness indices of each step in Ψ_i^j can be calculated according to Equations (5) and (11) and stored into the data sequence $[_i^j K_{\sin}^m]$ and $[_i^j K_{sti}^m]$ respectively, where $m = 1, 2, \ldots, n$ denotes the step number. After the traversal of i-th point in j-th segment, the maximum and minimum stiffness indices $_i^j K_{stimax}$ and $_i^j K_{stimin}$ in Equation (13) can be obtained from $[_i^j K_{sti}^m]$.

When all M points are traversed in j-th segment, the critical data obtained above is stored to the dataset $\Phi_j = \{\Psi_i^j, [_i^j K_{\sin}^m], [_i^j K_{sti}^m], _i^j K_{stimax}, _i^j K_{stimin}\}$ $(i = 1, 2, .., M)$. Then, go to Step 3.

- Step 3: Determination of the displacement constraints of γ between adjacent planned tool points

The displacement constraint of γ should be given to limit its variation range between two adjacent planned points. In traditional method, the velocity planning of other five DoFs needs to be executed repeatedly, which is unnecessary and has high computational load. Therefore, a simple method according to the maximum allowable velocity is designed as follows:

$$\gamma_i^j \in [\gamma_{i-1}^j - \Delta\gamma_i^j, \gamma_{i-1}^j + \Delta\gamma_i^j]$$
$$\Delta\gamma_i^j = \max\left(\frac{\left|P_i^j - P_{i-1}^j\right|}{v_{pmax}}, \frac{\left|\alpha_i^j - \alpha_{i-1}^j\right|}{\omega_{\alpha max}}, \frac{\left|\beta_i^j - \beta_{i-1}^j\right|}{\omega_{\beta max}}\right)\omega_{\gamma max} \tag{15}$$

where $P_i^j = [p_{ix}^j, p_{iy}^j, p_{iz}^j]$ is the position vector, α_i^j and β_i^j are the posture coordinate, v_{pmax}, $\omega_{\alpha max}$, $\omega_{\beta max}$ and $\omega_{\gamma max}$ are the given maximum allowable velocities, respectively. The displacement constraint given in Equation (15) is effective and efficiency without prior velocity planning. In particular, the displacement constraint of γ_1^j $(j > 1)$ should be determined by γ_M^{j-1} as shown in Figure 3. Then, go to Step 4.

- Step 4: Globe optimization of j-th segment

For j-th segment, the globe optimization model of M points can be constructed as follows:

$$\begin{cases} \min \sum\limits_{i=1}^{M} K_{com}(\gamma_i^j) \\ s.t.\, \gamma_i^j \in \Psi_j \qquad \text{C1} \\ \quad \gamma_{i-1}^j - \Delta\gamma_i^j \leq \gamma_i^j \leq \gamma_{i-1}^j + \Delta\gamma_i^j \ \ \text{C2} \end{cases} \tag{16}$$

where condition C1 can be obtained from dataset Φ_j, C2 can be determined by Equation (15). During the calculation of $K_{com}(\gamma_i^j)$ according to Equation (14), the performance indices $K_{\sin}(\gamma_i^j)$ and $K_{sti}^{new}(\gamma_i^j)$ can be obtained by linear interpolation based on Φ_j and without complex calculation process, such as the inverse kinematics of robot.

The *Fmincon* function in MATLAB is employed to conduct the optimization solution where condition C2 is used as the linear inequality constraint. After the optimization of j-th segment, make the following judgment:

If $j = N$, the redundant posture optimization of the total path is finished;

If $j < N$, then empty the dataset Φ_j, $j = j + 1$ and back to Step 2 to conduct the optimization of next segment.

4. Simulation and Validation

In this section, the posture optimization simulation of a complex spatial path is performed to evaluate the good performance of the proposed method. Analysis and comparisons are conducted with the representative method.

4.1. Environment Setup

As shown in Figure 5a, the YASKAWA ES165D serial robot assembled with a high-speed spindle is employed in the simulation and experiments. The Denavit–Hartenberg (D-H) parameters under the modified D-H method are shown in Figure 5b and Table 1. The joint limit parameters are given in Table 2.

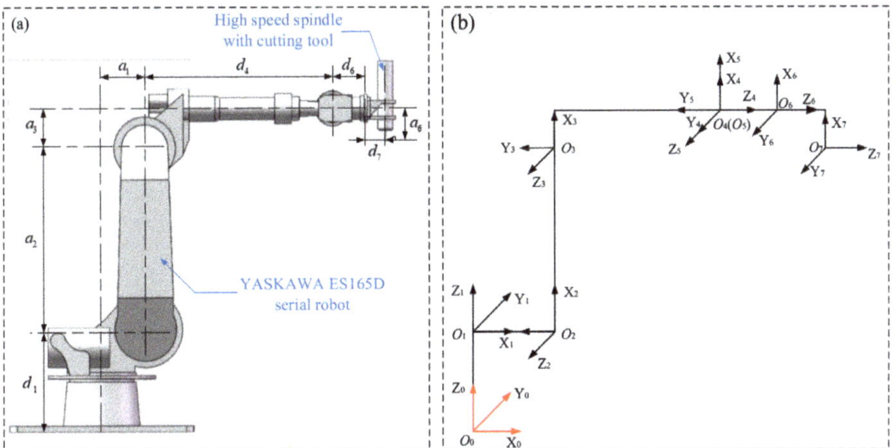

Figure 5. Robotic milling system. (**a**) YASKAWA ES165D serial robot with a high-speed spindle. (**b**) D-H model of robotic milling system.

Table 1. D-H parameters of ES165D robot.

Coordinate System i	a_{i-1} (mm)	α_{i-1} (Degree)	d_i (mm)	θ_i
1	0	0	650	θ_1
2	285	90	0	θ_2
3	1150	0	0	θ_3
4	250	90	1225	θ_4
5	0	−90	0	θ_5
6	0	90	225	θ_6
7	250	0	123	0

Table 2. Joint-limit parameters of ES165D robot.

Joint i	Positive Limit (Degree)	Negative Limit (Degree)
1	180	−180
2	166	30
3	120	−80
4	360	−360
5	130	−130
6	360	−360

The joint stiffness of the ES165D robot can be measured and identified by the loading and measuring experiments. The force loaded on the robot EE is measured by a six-dimensional force sensor and the corresponding translational deformation is obtained by the RADIAN Core (API) laser tracker. The stiffness of each joint is illustrated in Table 3.

Table 3. Joint stiffness of ES165D robot (Nmm/rad).

$k_{\theta1}$	$k_{\theta2}$	$k_{\theta3}$	$k_{\theta4}$	$k_{\theta5}$	$k_{\theta6}$
1.187×10^9	2.578×10^9	3.301×10^9	3.401×10^8	2.608×10^8	3.158×10^7

4.2. Simulation Results of Posture Optimization

As shown in Figure 6, the intersecting line constructed by two orthogonal cylindrical surfaces is employed as the test path. And the position and posture relationship between the test intersecting line and the milling robot is shown in Figure 7 where $O_W X_W Y_W Z_W$ is the workpiece coordinate system. The mathematical expression of intersecting line in $O_W X_W Y_W Z_W$ is defined as follows:

$$\begin{cases} x = r\cos\theta \\ y = r\sin\theta \\ z = R^2 - r^2\sin^2\theta \end{cases} \tag{17}$$

where $\theta \in [0, 2\beta]$, $R = 500$ and $r = 300$ are the radius of two cylindrical surfaces, respectively. The blue line containing 100 points is the target path of the tool center point (TCP). And the red arrow is the posture of the milling tool. Hence, the γ coordinate rotating around the red arrow is redundant and needs to be optimized.

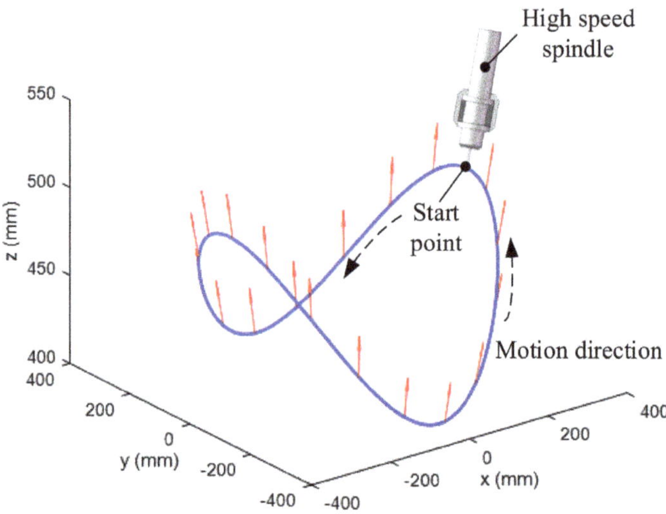

Figure 6. Test intersecting line.

Based on the performance indices given in Section 2, the comprehensive performance of the robot in each tool point with different postures can be illustrated in Figure 8. The white areas indicate the region of joint-limit, where the corresponding γ cannot be chosen. In other areas, according to the color bar on the right, the darker the color, the better the comprehensive performance. Therefore, the redundant posture optimization is to obtain a continuous path in Figure 8, which can go through dark areas.

As shown in Figure 9a, the optimization results obtained by the traditional SSPOS method are illustrated by the red curve. As can be seen, due to the unreasonable joint-limit judge method, the middle points between the adjacent given tool points might not be able to satisfy the joint-limit constraint, such as the B area, especially when the given tool points have a long distance. Meanwhile, the SSPOS can easily to lead to the gradual deterioration of the optimization process. As can be seen from the A area, the γ optimization enters the bad area and cannot jump out because of the displacement constraint of γ between adjacent points. Therefore, the SSPOS has good performance for single-point optimization, such as robotic drilling, but is not suitable for robotic milling with a continuous path.

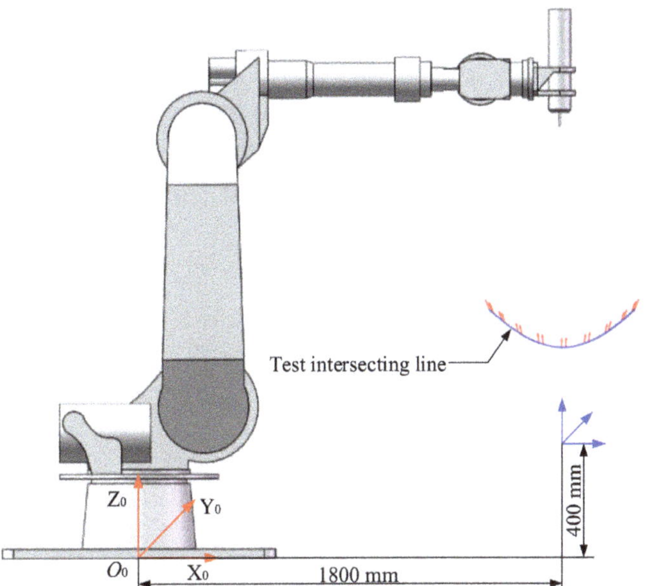

Figure 7. Pose and posture relationship between the test intersecting line and milling robot.

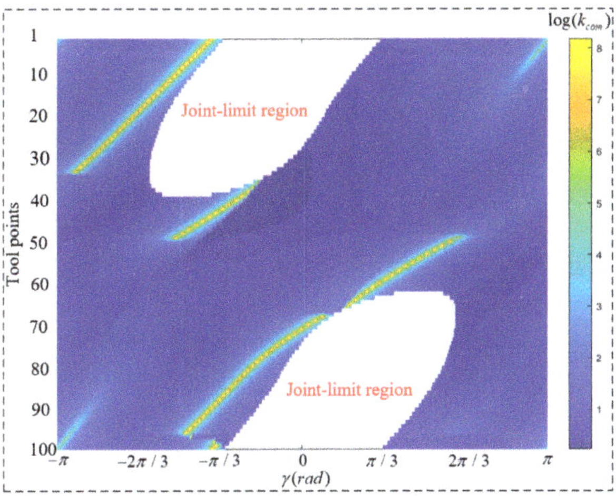

Figure 8. The comprehensive performance of robot in each tool point with different posture.

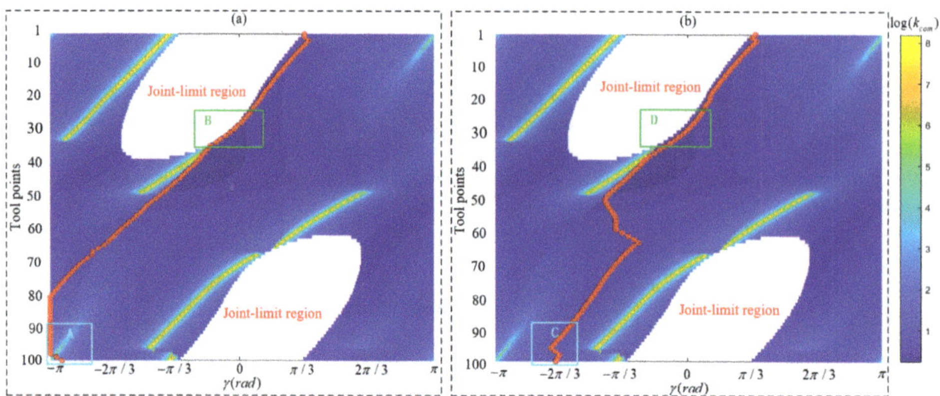

Figure 9. Optimization results of redundant posture. (**a**) Result by SSPOS. (**b**) Result by PGOS.

The optimization results obtained by PGOS are shown in Figure 9b, and the corresponding joint angles are shown in Figure 10. For the first 50 points, the optimization results are similar to SSPOS. The difference is that both the optimized tool points and their middle points in the D area can be guaranteed to satisfy the joint-limit constraints where $\Delta\theta_i = 5\,^\circ\ (i = 1, 2, .., 6)$ in Equation (2). It means that the joint-limit constraint strategy given in Equation (2) is a simple and effective scheme, especially for the areas near the joint-limit boundary. For the subsequent 50 points, the PGOS finds the optimal path, which is away from the A area in Figure 9a. Overall, the robot always has a good comprehensive performance by the proposed PGOS.

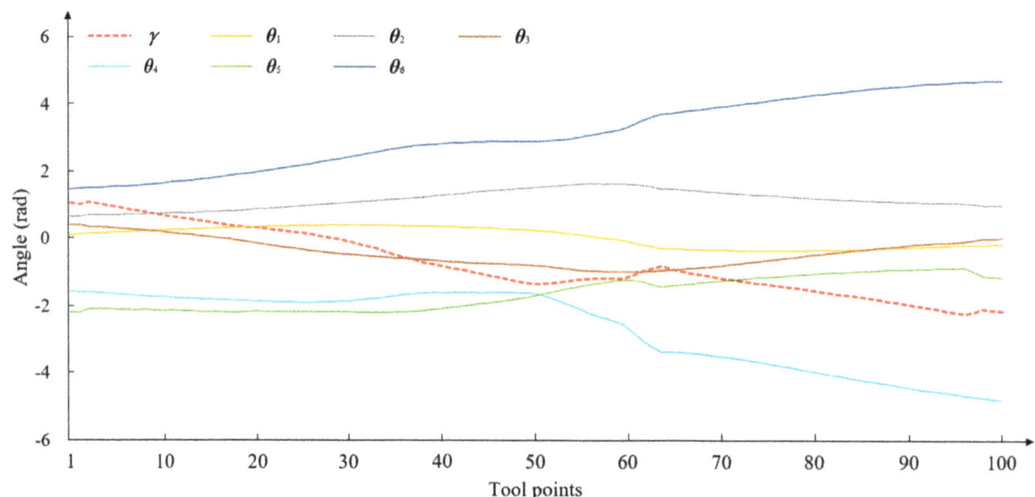

Figure 10. Optimized γ and the corresponding joint angles by PGOS.

5. Conclusions

In this paper, a novel redundant posture optimization method is proposed, and the main conclusions are as follows:

- The joint-limit is regarded as a constraint and the singularity and stiffness performances are the optimization target. Correspondingly, the effective and symmetrical judgment method of joint-limit and the performance indices of singularity and stiffness are designed;

- The PGOS is proposed and all the given tool points in their corresponding segment are considered simultaneously. Meanwhile, the computational load of the optimization solution is limited by the designed segmentation strategy;
- As can be seen from the simulation results, the proposed method has better planning quality and can avoid the gradual deterioration caused by SSPOS, which is suitable for the machining process with a continuous path.

Author Contributions: Conceptualization, H.N. and S.J.; methodology, H.N.; software, H.N.; validation, H.N., Y.Y. and S.J.; formal analysis, S.J.; investigation, Y.Y.; resources, S.J.; data curation, H.N.; writing—original draft preparation, H.N.; writing—review and editing, H.N., S.J. and Y.Y.; visualization, S.J.; supervision, S.J.; project administration, S.J.; funding acquisition, S.J. All authors have read and agreed to the published version of the manuscript.

Funding: This work is supported by the Natural Science Foundation of Shandong Province (Grant No. ZR2021QE128, ZR2019QEE042), Major Scientific and Technological Innovation Project of Shandong Province (Grant No. 2022CXGC010101).

Data Availability Statement: All data generated or analyzed during this study are included in this manuscript.

Conflicts of Interest: The authors declare no conflict of interest.

References

1. Mikolajczyk, T. Manufacturing using robot. In *Advanced Materials Research*; Trans Tech Publications Ltd.: Wollerau, Switzerland, 2012; pp. 1643–1646.
2. Ji, W.; Wang, L. Industrial robotic machining: A review. *Int. J. Adv. Manuf. Technol.* **2019**, *103*, 1239–1255. [CrossRef]
3. Hu, Y.; Chen, Y. Implementation of a robot system for sculptured surface cutting. Part 1. Rough machining. *Int. J. Adv. Manuf. Technol.* **1999**, *15*, 624–629. [CrossRef]
4. Zhu, Z.; Tang, X.; Chen, C.; Peng, F.; Yan, R.; Zhou, L.; Li, Z.; Wu, J. High precision and efficiency robotic milling of complex parts: Challenges, approaches and trends. *Chin. J. Aeronaut.* **2021**, *35*, 22–46. [CrossRef]
5. Huo, L.; Baron, L. The self-adaptation of weights for joint-limits and singularity avoidances of functionally redundant robotic-task. *Robot. Comput. Integr. Manuf.* **2011**, *27*, 367–376.
6. Bigras, P.; Lambert, M.; Perron, C. Robust force controller for industrial robots: Optimal design and real-time implementation on a KUKA robot. *IEEE Trans. Control. Syst. Technol.* **2012**, *20*, 473–479. [CrossRef]
7. Guo, Y.; Dong, H.; Ke, Y. Stiffness-oriented posture optimization in robotic machining applications. *Robot. Comput. Integr. Manuf.* **2015**, *35*, 69–76. [CrossRef]
8. Léger, J.; Angeles, J. Off-line programming of six-axis robots for optimum five-dimensional tasks. *Mech. Mach. Theory* **2016**, *100*, 155–169.
9. Zhang, C.; Ahn, C.K.; Wu, J.; He, W. Online-learning control with weakened saturation response to attitude tracking: A variable learning intensity approach. *Aerosp. Sci. Technol.* **2021**, *117*, 106981.
10. Zhang, C.; Xiao, B.; Wu, J.; Li, B. On low-complexity control design to spacecraft attitude stabilization: An online-learning approach. *Aerosp. Sci. Technol.* **2021**, *110*, 106441.
11. Xie, H.; Li, W.; Yin, Z. Posture optimization based on both joint parameter error and stiffness for robotic milling. In Proceedings of the International Conference on Intelligent Robotics and Applications, Yantai, China, 22–25 October 2021; Springer: Berlin/Heidelberg, Germany, 2018; pp. 277–286.
12. Peng, J.; Ding, Y.; Zhang, G.; Ding, H. Smoothness-oriented path optimization for robotic milling processes. *Sci. China Technol. Sci.* **2020**, *63*, 1751–1763. [CrossRef]
13. Zhu, W.; Qu, W.; Cao, L.; Yang, D.; Ke, Y. An off-line programming system for robotic drilling in aerospace manufacturing. *Int. J. Adv. Manuf. Technol.* **2013**, *68*, 2535–2545. [CrossRef]
14. Jiao, J.; Tian, W.; Liao, W.; Zhang, L.; Bu, Y. Processing configuration off-line optimization for functionally redundant robotic drilling tasks. *Robot. Auton. Syst.* **2018**, *110*, 112–123. [CrossRef]
15. Bu, Y.; Liao, W.; Tian, W.; Zhang, J.; Zhang, L. Stiffness analysis and optimization in robotic drilling application. *Precis. Eng.* **2017**, *49*, 388–400. [CrossRef]
16. Lin, Y.; Zhao, H.; Ding, H. Posture optimization methodology of 6R industrial robots for machining using performance evaluation indexes. *Robot. Comput. Integr. Manuf.* **2017**, *48*, 59–72. [CrossRef]
17. Dufour, K.; Suleiman, W. On maximizing manipulability index while solving a kinematics task. *J. Intell. Robot. Syst.* **2020**, *100*, 3–13. [CrossRef]

18. Li, Y.; Yang, X.; Wu, H.; Chen, B. Optimal design of a six-axis vibration isolator via Stewart platform by using homogeneous Jacobian matrix formulation based on dual quaternions. *J. Mech. Sci. Technol.* **2018**, *32*, 11–19. [CrossRef]
19. Li, G.; Zhu, W.; Dong, H.; Ke, Y. Stiffness-oriented performance indices defined on two-dimensional manifold for 6-DOF industrial robot. *Robot. Comput. Integr. Manuf.* **2021**, *68*, 102076. [CrossRef]
20. Yin, B.; Wenhe, L.; Wei, T.; Zhang, L.; Dawei, L. Modeling and experimental investigation of Cartesian compliance characterization for drilling robot. *Int. J. Adv. Manuf. Technol.* **2017**, *91*, 3253–3264.
21. Mikolajczyk, T. System to surface control in robot machining. In *Advanced Materials Research*; Trans Tech Publications Ltd.: Wollerau, Switzerland, 2012; pp. 708–711.
22. Chen, C.; Peng, F.; Yan, R.; Li, Y.; Wei, D.; Fan, Z.; Tang, X.; Zhu, Z. Stiffness performance index based posture and feed orientation optimization in robotic milling process. *Robot. Comput. Integr. Manuf.* **2019**, *55*, 29–40. [CrossRef]
23. Xiong, G.; Ding, Y.; Zhu, L. A feed-direction stiffness based trajectory optimization method for a milling robot. In Proceedings of the International Conference on Intelligent Robotics and Applications, Wuhan, China, 16–18 August 2017; Springer: Berlin/Heidelberg, Germany, 2017; pp. 184–195.
24. Xiong, G.; Ding, Y.; Zhu, L. Stiffness-based pose optimization of an industrial robot for five-axis milling. *Robot. Comput. Integr. Manuf.* **2019**, *55*, 19–28. [CrossRef]
25. Liao, Z.; Li, J.R.; Xie, H.; Wang, Q.; Zhou, X. Region-based toolpath generation for robotic milling of freeform surfaces with stiffness optimization. *Robot. Comput. Integr. Manuf.* **2020**, *64*, 101953. [CrossRef]

 *symmetry*

Article

Flight Conflict Detection Algorithm Based on Relevance Vector Machine

Senlin Wang [1,2] and Dangmin Nie [1,2,*]

1   Air Traffic Control and Navigation School, Air Force Engineering University, Xi'an 710051, China
2   National Key Laboratory, Air Traffic Collision Prevention, Xi'an 710051, China
*   Correspondence: ndm863@163.com

Abstract: In response to the problems of slow running speed and high error rates of traditional flight conflict detection algorithms, in this paper, we propose a conflict detection algorithm based on the use of a relevance vector machine. A set of symmetrical historical flight data was used as the training set of the model, and we used the SMOTE resampling method to optimize the training set. We obtained relatively symmetrical training data and trained it with the relevance vector machine, improving the kernels through an intelligent algorithm. We tested this method with new symmetrical flight data. The improved algorithm greatly improved the running speed and was able to effectively reduce the missed alarm rate of in-flight conflict detection symmetrically, thus effectively ensuring flight safety.

Keywords: flight conflict detection; relevance vector machine; Bayesian optimization

1. Introduction

With the rapid development of civil aviation and the rapid growth in flight volumes, ensuring flight safety has become an essential guarantee in the rapid development of the aviation industry. Flight conflict detection is a crucial method to ensure flight safety. According to aircraft tracking information, flight intention information and external environmental information provided by navigation equipment are used identify flight conflicts between two aircrafts during the conflict detection time. Two methods are mainly used to solve the problem of flight conflict detection: the geometric method and the probability method [1]. The geometric conflict detection method is based on the use of geometric calculations to judge whether the aircraft has potential flight conflicts within the geometric range of the encounter. However, the problem with this method is that the actual situation cannot be represented by a strict geometric relationship between the aircrafts, and many factors need to be considered, such as the influence of wind, pilot operation, and the external environment.

As a result, the geometric conflict detection method suffers from issues including low certainty and large error values [2]. The study of probability approaches with some fault tolerance is growing in popularity due to the numerous issues with geometric methods that are caused by the mistakes in flight data. Traditional probabilistic conflict detection methods include those based on probability flow theory, complex networks, Markov chains, game theory, and other methods [3–5]. In 2000, Prandini and others used the probability analysis method to analyze the measurement index of complex flight trade-offs and established a flight collision avoidance model. Almost all subsequent studies have improved and expanded upon this model [6]. Shi Lei and others proposed a probabilistic short-term conflict detection algorithm based on the idea of a hybrid system and position space discretization, which uses the tracking and intention information in relation to monitoring information to predict the short-term track of a flight based on the state-related random linear hybrid system [7]. Daalen et al. used probability flow theory to solve the upper probability bound of overall flight conflicts. Jacquemart et al. proposed the simulation of aircraft motion rollout with the use of a Markov chain, and used an important

Citation: Wang, S.; Nie, D. Flight Conflict Detection Algorithm Based on Relevance Vector Machine. *Symmetry* **2022**, *14*, 1992. https://doi.org/10.3390/sym14101992

Academic Editor: Theodore E. Simos

Received: 23 August 2022
Accepted: 19 September 2022
Published: 23 September 2022

Publisher's Note: MDPI stays neutral with regard to jurisdictional claims in published maps and institutional affiliations.

sampling method to calculate conflict probabilities [8]. Compared with the above methods, intelligent methods, such as support vector machines, have the advantages of strong real-time performance, high accuracy, relatively small calculation, and small error values, so they have become a new hot spot in flight conflict detection research [9,10].

In the research on the solution of conflict detection problems through a support vector machine, Jiao Yuliang and others first used a support vector machine to propose the use of the horizontal two-dimensional conflict detection method for high-altitude routes [11]. Han Dong and others used the support vector machine method to detect flight conflicts based on an ellipsoidal protected area, and used simulation data to train the classifier in advance to carry out efficient conflict detection suitable for low-altitude flights [12]. Wu Minggong and others combined the support vector machine classification algorithm, the sigmoid function, and an elliptical protected area to establish a flight conflict detection model that can simultaneously output the conflict or lack thereof and its probability of occurrence [13]. On this basis, Wang Ershen used the improved ID3 decision tree algorithm to reduce the search space to a local method to screen the aircraft with potential flight conflicts, used the random forest method to select the appropriate training set, and used the tanh-function optimization to easily saturate the sigmoid function. A probability map of SVM classification obtains the following results [14]: the above methods have all been used to study flight conflict detection problems with the use of a support vector machine, but support vector machines have many disadvantages. For example, an SVM cannot calculate the posterior probability distribution of a sample output, an SVM is not suitable for multi-classification problems, and SVM hyperparameters need to be obtained through cross-validation, which is very time-consuming. Furthermore, the kernel of an SVM must be positive and definite, whereas RVM avoids these disadvantages. Compared to SVM, a relevance vector machine can better meet the mission needs of flight conflict detection. In this paper, we propose a flight conflict detection method based on the use of a relevance vector machine. Using the SMOTE resampling method to optimize the training set, we establish an ellipsoidal protected area model, analyze historical flight data, establish an RVM classification model, and optimize the kernel through various intelligent algorithms, as well as comprehensively analyzing and selecting the best kernels. In terms of the running speed, this method only takes 0.001 s to run, an extremely fast running speed. In terms of classification accuracy, the optimized relevance vector machine model exhibits a 20% higher accuracy than the SVM model. In terms of innovation, this study represents the first time an RVM has been employed in flight conflict detection. In applications, by loading the trained classifier before take-off, the flight conflict detection speed can be effectively improved, and flight safety can thus be guaranteed.

2. Methods

Flight conflict detection is a two-classification problem, with only two results: conflict or non-conflict. In previous studies, the excellent two-classification machine learning algorithm of the SVM (support vector machine) was used to solve flight conflict detection problems. Although SVM has good promotion ability and avoids the local optimum, there are still problems such as its slow speed on the testing set and its low classification accuracy. Compared with support vector machines, the algorithm used in this paper can give a probabilistic output and it has the following advantages: better generalization ability, better sparsity, more flexible kernel selection, no mandatory positive definite, and simpler parameter settings. Conflict detection requires a high running speed, and RVM has better sparsity than SVM, so its speed on testing sets is faster. Therefore, in this paper, we propose a flight conflict detection method based on the use of a relevance vector machine to solve this problem.

2.1. Data Preprocessing

We first obtain the current position information for the two aircrafts and the speed vectors through the ADS-B system. The position information of the two aircrafts is $X_a = (x_a, y_a, z_a)$, $X_b = (x_b, y_b, z_b)$, and the speed vectors of the two aircrafts are $V_a = (v_{xa}, v_{ya}, v_{za})$ and $V_b = (v_{xb}, v_{yb}, v_{zb})$. We extract the feature quantity, reduce the sample dimension, and obtain the sample information $X = X_a - X_b = (x, y, z)$, $V = V_a - V_b = (v_x, v_y, v_z)$ and use a detection time t, to obtain a simplified sample with a sample feature dimension of 7 [15–17], that is:

$$F = (x, y, z, v_x, v_y, v_z, t) \tag{1}$$

In airspace, the probability of free flight conflicts is a small event in regard to the total sample, so there is an overwhelming advantage for the collected samples without conflicts. In order to avoid this imbalance in the number of samples affecting the conflict detection process, the number of positive and negative samples should be roughly balanced. We improve the robustness of the model to the sample set through the a relatively symmetrical training set. We improve the classification ability through the SMOTE resampling method, that is, by inserting virtual samples between adjacent negative samples to reduce the occurrence of over-adaptation [18]. At the same time, normalization of the data is performed:

$$Y = \frac{X - X_{\min}}{X_{\max} - X_{\min}} \tag{2}$$

wherein $Y \in [0, 1]$ is the normalized value, X is the sample eigenvalue, and $X_{\max}$, and $X_{\min}$ are the maximum and minimum eigenvalues in the total sample, respectively.

2.2. Definition of Relevance Vector Machine

The given training set is $G = \{(x_i, h_i)\}_{i=1}^N$, where $\{x_i\}_{i=1}^N$ is the input sample vector, $\{h_i\}_{i=1}^N$ represents the corresponding target value, and N is the total number of samples. The basic form of the relevance vector machine model output is [19–27]:

$$y(x, w) = \sum_{i=0}^N w_i K(x, x_i) + w = \varphi(x)w \tag{3}$$

Among them, w is the weight value vector, $K(x, x_i)$ is the kernel, and $\varphi(x)$ is the $M \times (M+1)$ order kernel matrix. Assuming that m_i it obeys the normal distribution with a mean value of $y(x_i, w)$ and a variance of σ^2, it is expressed by probability as

$$p(h_i) = N\left(h_i \middle| y(x_i; w), \sigma^2\right) \tag{4}$$

The likelihood function of the sample is

$$p\left(h \middle| w, \sigma^2\right) = \prod_{i=0}^N N\left(h_i \middle| y(x_i; w), \sigma^2\right) = (2\pi\sigma^2)^{-\frac{N}{2}} \exp\{-\frac{\|h - \Phi w\|^2}{2\sigma^2}\} \tag{5}$$

The conditional probability is

$$P(h_*|h) = \int P\left(h_* \middle| w, \sigma^2\right) p\left(w, \sigma^2 \middle| h\right) dw d\sigma^2 = \int P\left(h_* \middle| w, \sigma^2\right) \frac{p(h|w, \sigma^2) p(w, \sigma^2)}{p(h)} dw d\sigma^2 \tag{6}$$

In order to avoid overfitting, the relevance vector machine adds prerequisites to the weight vector w, so that w is a standard normal distribution, so there are:

$$p(w|\mu) = \prod_{i=0}^N N\left(w_i \middle| 0, \mu_i^{-1}\right) = \prod_{i=0}^N \frac{\mu_i}{\sqrt{2\pi}} \exp\left\{-\frac{\mu_i w_i^2}{2}\right\} \tag{7}$$

In this case, the required conditional probability formula is

$$P(h_*|h) = \int P\left(h_*\Big|w,\mu,\sigma^2\right)p\left(w,\mu,\sigma^2\Big|h\right)dwd\mu d\sigma^2 \tag{8}$$

In the formula $\mu = [\mu_0,\mu_1,\mu_2,\cdots,\mu_N]^T$ is a $N+1$-dimensional hyperparameter vector, which obeys the Gamma distribution. Through the Bayesian theory, we can find:

$$p\left(w,\mu,\sigma^2\Big|h\right) = P\left(w\Big|h,\mu,\sigma^2\right)p\left(\mu,\sigma^2\Big|h\right) \tag{9}$$

In the formula:

$$p\left(w\Big|h,\mu,\sigma^2\right) = \frac{p\left(h\big|w,\sigma^2\right)p(w|\mu)}{p(h|\mu,\sigma^2)} = (2\pi) - \frac{N+1}{2}|\Phi|^{-\frac{1}{2}}\exp\left\{-\frac{(w-\alpha)^T\Sigma^{-1}(w-\alpha)}{2}\right\} \tag{10}$$

$$\Phi = \left(\sigma^2\varphi^T\varphi + A\right)^{-1} \tag{11}$$

$$\alpha = \sigma^{-2}\Phi\varphi^T h \tag{12}$$

$$A = \mathrm{diag}(\mu_0,\mu_1,\mu_2,\cdots,\mu_N) \tag{13}$$

From the delta approximation function $P\left(\mu,\sigma^2\big|h\right) \approx \delta\left(\mu_{MP},\sigma_{MP}{}^2\right)$, we can get:

$$P(h_*|h) \approx \int P\left(h_*\Big|w,\mu_{MP},\sigma^2_{MP}\right)P(w|h,\mu_{MP},\sigma^2_{MP}|h)dw \tag{14}$$

Integrating the above formula yields:

$$p\left(h_*\Big|h,\mu_{MP},\sigma_{MP}{}^2\right) = N\left(h_*\Big|y_*,\sigma_*{}^2\right) \tag{15}$$

$$y_* = \varphi(x_*)\alpha \tag{16}$$

$$\sigma_*{}^2 = \sigma_{MP}{}^2 + \varphi\left(\begin{matrix}x\\ *\end{matrix}\right)T\Phi\varphi(x_*) \tag{17}$$

The solution of the model can be further transformed into how to obtain μ_{MP} and $\sigma_{MP}{}^2$, which can be obtained by using the maximum likelihood method:

$$P\left(h\Big|w,\sigma^2\right)P(w,\mu)dw = 2\pi^{-\frac{N}{2}}|\Omega|^{-\frac{1}{2}}\exp\left(-\frac{1}{2}h^T\Omega^{-1}h\right) \tag{18}$$

where $\Omega = \sigma^2 I + \varphi A^{-1}\varphi^T$. In the above formula, the solution of a and b with zero partial derivatives can be obtained:

$$\mu_i^{new} = \frac{\gamma_i}{\mu^2}$$

$$\left(\sigma^2\right)new = \frac{\|h-\Phi\mu\|^2}{N-\sum\limits_{i=0}^{N}\gamma_i} \tag{19}$$

$$\gamma_i = 1 - \alpha_i\Sigma_{i,i}$$

Among them, in the process of parameter change, a part of μ tends to infinity, and its corresponding weight value vector w will tend to zero. This means that a part of the corresponding basis function is "eliminated", and when the final result converges, the w corresponding to the remaining μ is the relevance vector we need.

2.3. Flight Conflict Detection Model

Flight conflict detection involves detecting the aircraft within the scope of the airspace through the use of navigation equipment to determine whether the aircraft can continuously meet the requirements of the minimum safety interval within a certain period of time. If the minimum safety interval is not met at the time t, it is determined that there is a flight conflict. The ellipsoid protected area is established with the aircraft as the center. The protected area can be formulated as follows [28]:

$$\frac{x - x_0}{a^2} + \frac{y - y_0}{b^2} + \frac{z - z_0}{c^2} \leq 1 \tag{20}$$

wherein a b, and c represent x, y, and z, respectively, three half-axis focal lengths.

According to China's "Civil Aviation Air Traffic Management Rules [29]", $a = b = 1000$ m, and $c = 150$ m. These values are used to establish a protected area in Figure 1.

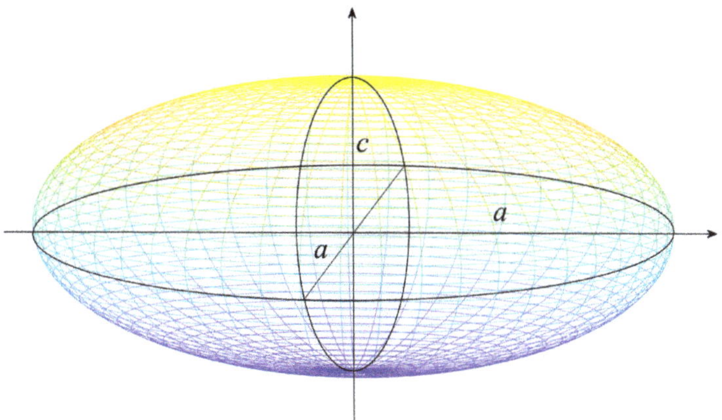

Figure 1. Protected area model.

3. Results

In this section, 10,000 sets of samples from the historical ADS-B dataset were first used as a training set for RVM training to analyze the feasibility of the proposed conflict detection method. The Gauss kernel with a strong nonlinearity mapping capability was selected as the kernel of the RVM, and 100 groups of samples were used to verify the accuracy of the conflict detection method. Secondly, the genetic algorithm, particle swarm algorithm, and Bayesian optimization were used to optimize the hyperparameters of the model. Finally, a comparison of the three optimization methods was carried out to prove the effectiveness of the relevance vector machine in flight conflict detection.

3.1. Accuracy Analysis of Conflict Detection Model

A simulation experiment was carried out on 100 pairs of dual-aircraft flight data in the training set, and the confusion matrix of the classification result and the result prediction diagram of the classification method were obtained. By comparing the predicted value and the actual conflict value, the accuracy of the method was analyzed.

Figure 2 shows a confusion matrix of the testing set without optimization. In the confusion matrix, TP in the upper left corner indicates that there was no conflict risk in real cases and there was no conflict in the prediction, while FP in the lower left corner indicates that there was a conflict risk in real cases, but there was no conflict in the prediction, and FN in the upper right corner indicates that there was no conflict risk in real cases, but there was a conflict in the prediction. Lastly, TN in the lower right corner indicates that there was a conflict risk in real cases and there was a conflict in the prediction. In the result prediction diagram (Figure 3), the blue line represents the actual conflict situation of the training set, and the red line represents the prediction result of the training set.

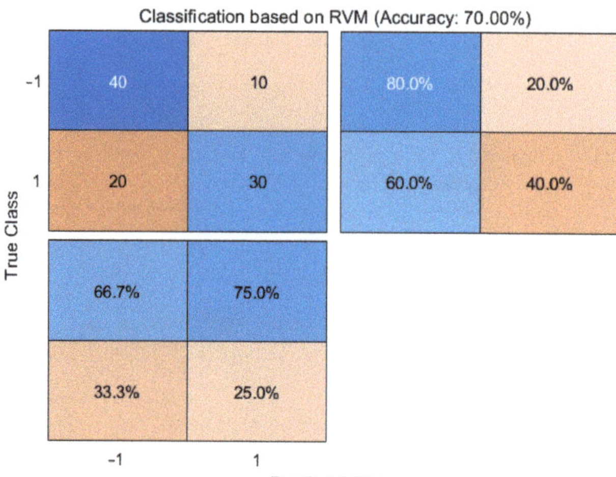

Figure 2. Confusion matrix of the relevance vector machine.

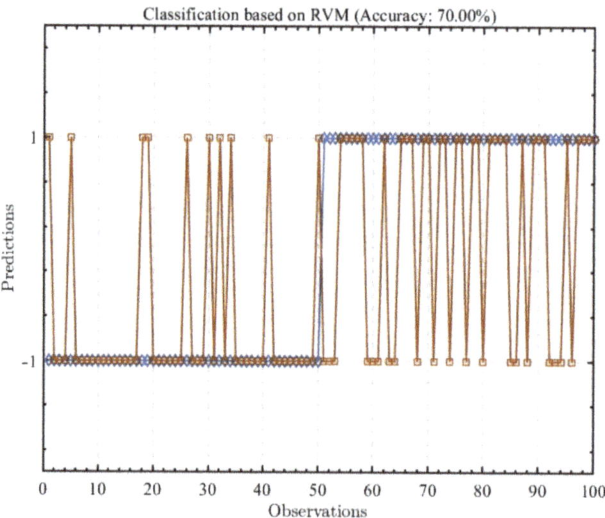

Figure 3. Prediction result graph of the relevance vector machine.

As can be seen in Figure 2, the training model without hyperparameter optimization still demonstrated a 70% accuracy. At the same time, judging from the ratio of TP and TN, after data preprocessing, the leakage missed alarm rate and false alarm rate of conflict risk maintained a relatively consistent level. The flight conflict detection method based on the RVM could be considered to have a certain feasibility, but it can also be seen that the accuracy of this method had great room for improvement. Thus, the detection accuracy of this method was improved through the optimization of the kernel.

3.2. Accuracy Analysis after Optimization of the Genetic Algorithm

In order to improve the accuracy in conflict detection and obtain a better RVM model, the kernel of the relevance vector machine was optimized by using a genetic algorithm [30].

As shown in Figures 4 and 5, in this case, the RVM training model optimized by means of a genetic algorithm displayed an 80% accuracy and was significantly improved before the phase angle optimization. At the same time, the missed alarm rate and false alarm rate of conflict risk maintained relatively consistent levels.

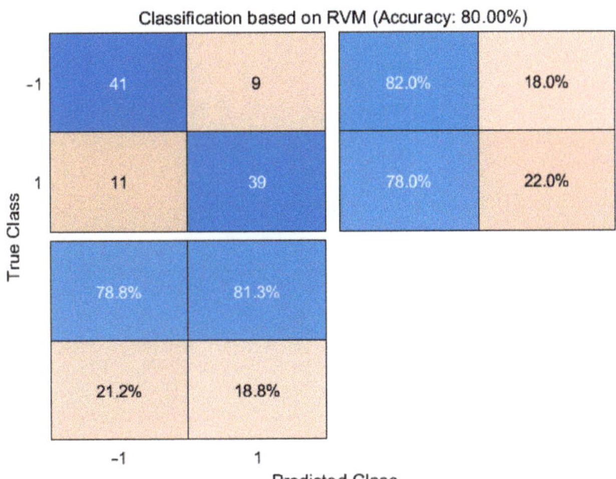

Figure 4. Confusion matrix after genetic algorithm optimization.

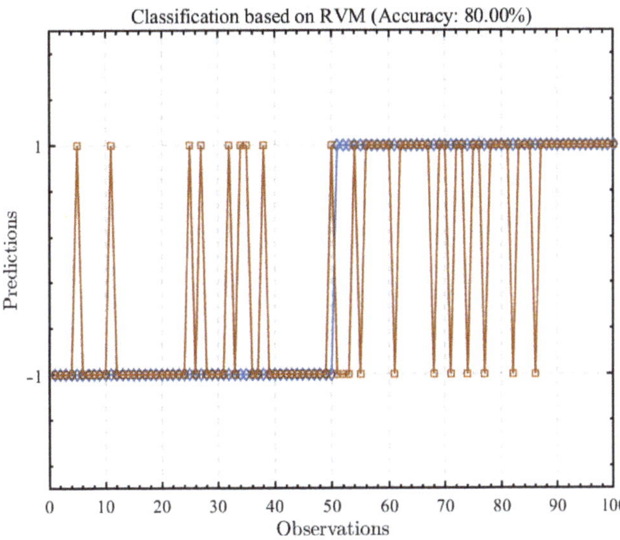

Figure 5. Improved genetic algorithm results prediction graph.

As shown in Figure 6, the optimization of the kernel was close to convergence after eight generations and had a good convergence rate. The disadvantage of this optimization approach is that the optimization speed cannot meet the requirements, and there is still room for improvement in its accuracy.

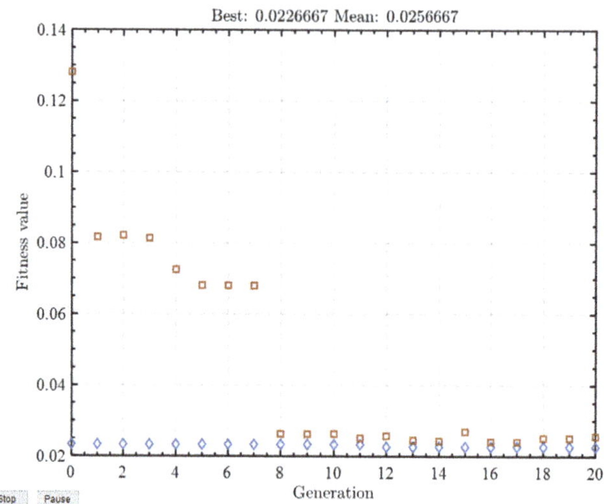

Figure 6. Genetic algorithm optimization process diagram.

3.3. Accuracy Analysis after Optimization of PSO Algorithm

The particle swarm optimization algorithm (PSO) is a swarm intelligence optimization algorithm that simulates the predation behavior of birds and fish. PSO has the advantages of a simple principle, a small amount of calculation, and fewer control parameters, so it is widely used in scheduling problems, optimization problems, path planning, and other practical problems. However, PSO still has some shortcomings, such as the fact that the algorithm easily falls into the local optimum, can easily become "precocious", and has a slow convergence rate and low convergence accuracy [31–33]. As shown in Figures 7 and 8, the RVM kernel was optimized by means of particle swarm optimization, the optimized RVM model demonstrated an 81% accuracy in the classification of the training sets, and the optimized performance of the model was dramatically improved compared to the unoptimized model.

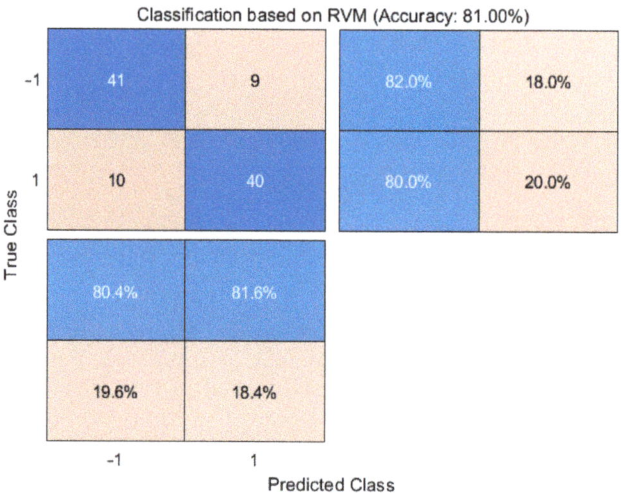

Figure 7. Confusion matrix after particle swarm optimization.

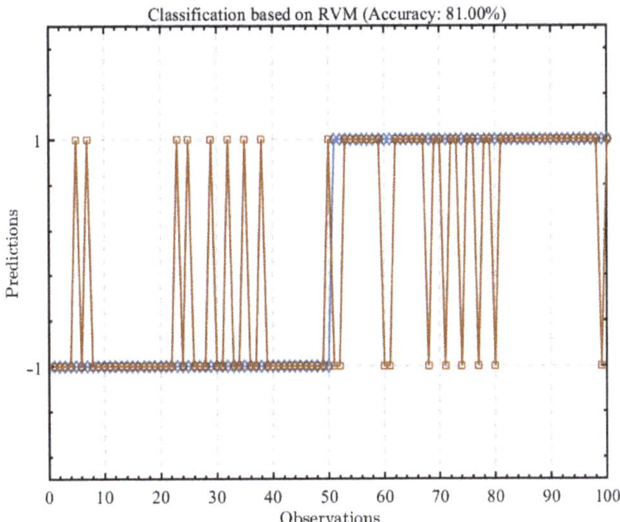

Figure 8. Results prediction chart based on improved particle swarm optimization.

3.4. Accuracy Analysis after Bayesian Optimization

Bayesian optimization (BO) is also called active optimization. This method is essentially model-based sequential optimization, and the next round can only be carried out after the end of the current round of evaluation. The subsequent evaluation position can be selected according to the information obtained by unknown objective functions to obtain the optimal solution at the least cost [34,35].

As shown in Figures 9 and 10, the model optimized by means of the Bayesian approach demonstrated the highest accuracy for the training set, and the balance between missed alarm rate and the false alarm rate was also the best of the four models, with extremely high symmetry for the classification results.

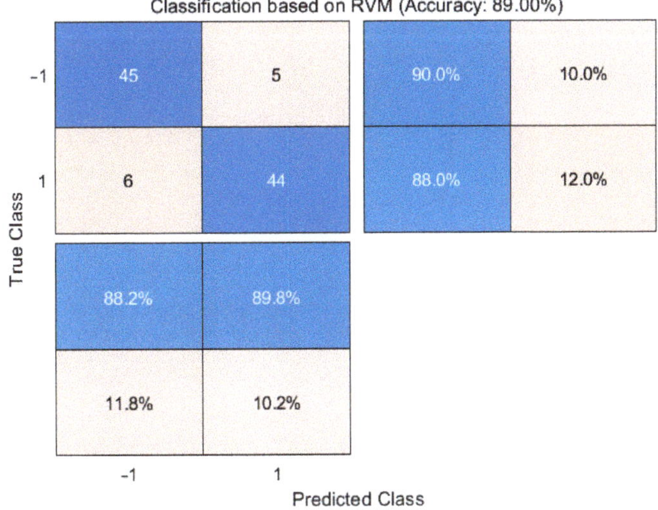

Figure 9. Confusion matrix after Bayesian optimization.

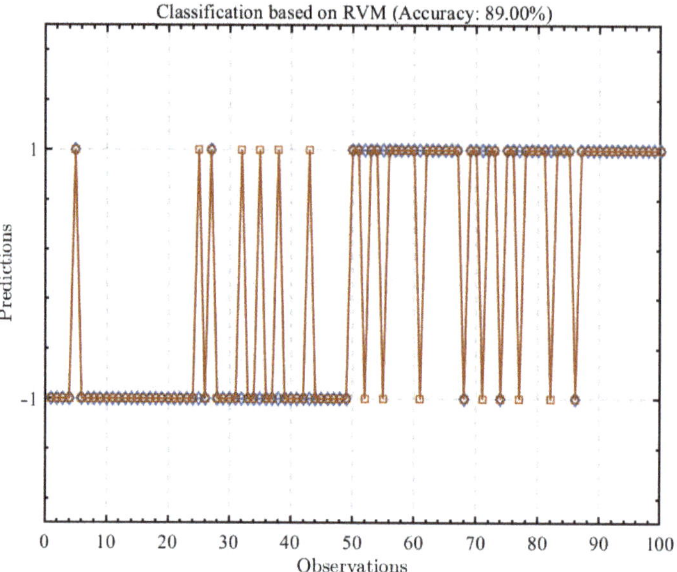

Figure 10. Results prediction chart based on Bayesian optimization.

As can be seen in Figures 11 and 12, in the process of Bayesian optimization, the fourth generation was already close to the optimal kernel, and the convergence rate was also the fastest of all the optimization methods.

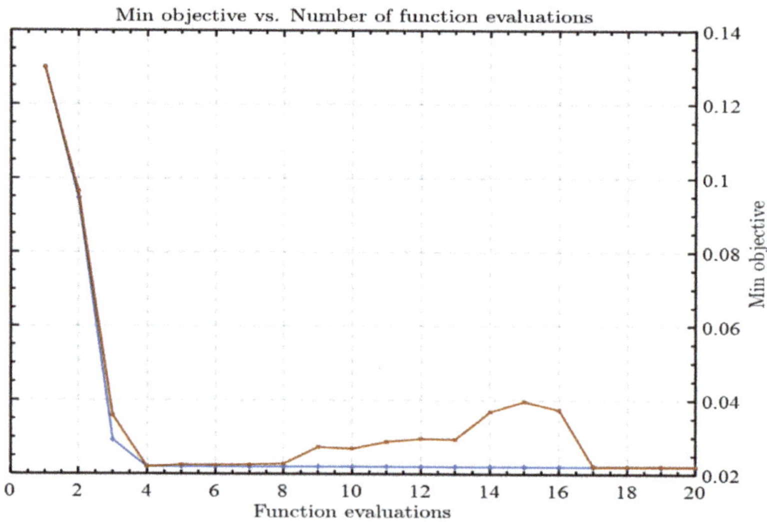

Figure 11. Bayesian optimization process diagram.

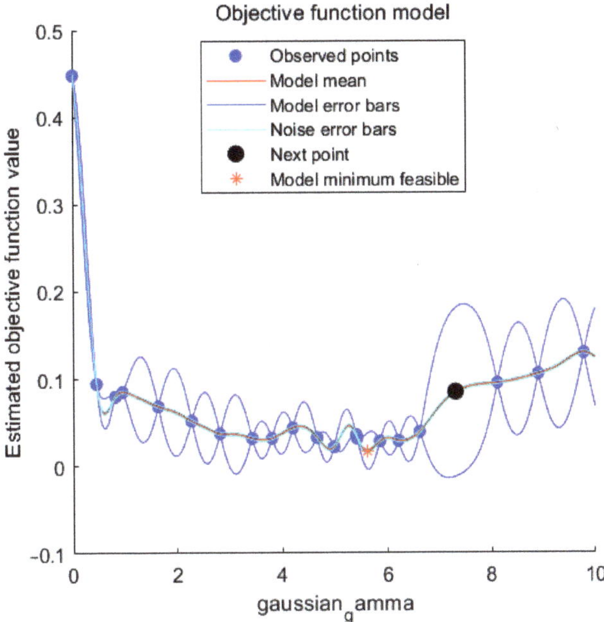

Figure 12. Variation diagram of Bayesian optimization function.

3.5. Optimization Effect Comparison

In order to verify the feasibility of this method for flight conflict detection and the optimized detection accuracy, the accuracy and running speed of the unoptimized flight conflict detection method based on the relevance vector machine were compared with those optimized using different methods in Table 1.

Table 1. Optimization effect comparison.

	SVM	RVM	GA-RVM	PSO-RVM	BO-RVM
Running speed	0.3 s	0.0048 s	0.0012 s	0.0019 s	0.0013 s
Optimization time			3418 s	3184 s	388 s
Accuracy	69%	70%	80%	81%	89%

Through this comparison, we observed that the flight conflict detection methods based on the use of the relevance vector machine were faster, and were thus able to meet the tight time demands of the urgent task of flight conflict detection. From the perspective of optimization speed, the Bayesian optimization method was much faster than the other two optimization methods. It is worth noting that, in fact, the three kinds of optimization refer to the optimization of the kernel. According to the different optimization effects, the obtained kernel was different, and its classification performance was also different. From the perspective of accuracy, Bayesian optimization demonstrated the best optimization effect, with a relatively balanced and low missed alarm and false alarm rates. Thus, it was able to meet the accuracy requirements of flight conflict detection during flight.

4. Conclusions

In this paper, we have proposed a flight conflict detection method based on the use of a relevance vector machine. The conflict risk in the flight process was judged by constructing a flight protection zone. The relevance vector machine was used to construct the model, and the RVM model was trained with 10,000 data elements. The accuracy of the model was improved by optimizing the kernel. The model's performance was optimized by means of a genetic algorithm, a particle swarm algorithm, and a Bayesian optimization method. After optimization, the accuracy of flight conflict detection after Bayesian optimization was observed to be the highest, at 89%. At the same time, the results of this method exhibit a symmetrical missed alarm rate and false alarm rate. In addition, comparing the support vector machine and the relevance vector machine with the optimized kernel, the results show that the optimized conflict detection method based on the relevance vector machine displayed a better accuracy and a faster running speed, which verifies the effectiveness of the method. In practical applications, with its extremely fast running speed and high conflict detection accuracy, the flight conflict detection method based on RVM can reduce flight collision risks through its use as a pre-training model. It cannot be ignored that the flight conflict detection method based on the correlation vector machine still presents problems, such as a poor training ability for large samples and an accuracy which can still be improved. Our future work will focus on further improving the method's accuracy, and the influence of the hybrid kernel on the conflict detection effect will be discussed.

Author Contributions: Conceptualization, S.W.; methodology, S.W.; writing—original draft preparation, S.W; writing—review and editing, S.W. and D.N.; visualization, D.N. All authors have read and agreed to the published version of the manuscript.

Funding: This research received no external funding.

Institutional Review Board Statement: Not applicable.

Informed Consent Statement: Not applicable.

Data Availability Statement: Not applicable.

Conflicts of Interest: The authors declare no conflict of interest.

References

1. Shi, L.; Wu, R.-B. Multi-route mid-term conflict detection algorithm based on discretization of predicted position space. *Signal Process.* **2012**, *28*, 1521–1528.
2. Goss, J.; Rajvanshi, R.; Subbarao, K. Aircraft conflict detection and resolution using mixed geometric and collision cone approaches. In Proceedings of the AIAA Guidance, Navigation, and Control Conference and Exhibit, Providence, RI, USA, 16–19 August 2004; p. 4879. [CrossRef]
3. Van Daalen, C.E.; Jones, T. Fast conflict detection using probability flow. *Automatica* **2009**, *45*, 1903–1909. [CrossRef]
4. Huang, Y.; Tang, J.; Lao, S. UAV flight conflict resolution algorithm based on complex network. *Acta Aeronaut. Astronaut. Sin.* **2018**, *39*, 262–274.
5. Zhang, X.; Guan, X.; Zhu, Y.; Lei, J. Strategic Flight Assignment Approach Based on Multi-Objective Parallel Evolution Algo-rithm with Dynamic Migration Interval. *Chin. J. Aeronaut.* **2015**, *28*, 556–563. [CrossRef]
6. Prandini, M.; Hu, J.; Lygeros, J.; Sastry, S. A probabilistic approach to aircraft conflict detection. *IEEE Trans. Intell. Transp. Syst.* **2000**, *1*, 199–220. [CrossRef]
7. Lei, S.; Ren-Biao, W.; Xiao-Xiao, H. Conflict Detection Algorithm Based on Overall Conflict Probability and Three Dimen-sional Brownian Motion. *J. Electron. Inf. Technol.* **2015**, *37*, 360–366.
8. Jacquemart, D.; Morio, J. Conflict probability estimation between aircraft with dynamic importance splitting. *Saf. Sci.* **2013**, *51*, 94–100. [CrossRef]
9. Liu, H.; Li, B.; Xiao, B.; Ran, D.; Zhang, C. Reinforcement learning-based tracking control for a quadrotor unmanned aerial vehicle under external disturbances. *Int. J. Robust Nonlinear Control* **2022**. [CrossRef]
10. Peral, J.; Gil, D.; Rotbei, S.; Amador, S.; Guerrero, M.; Moradi, H. A Machine Learning and Integration Based Architecture for Cognitive Disorder Detection Used for Early Autism Screening. *Electronics* **2020**, *9*, 516. [CrossRef]
11. Jiao, Y.L.; Zhang, X.J.; Guan, X.M. An Algorithm for Airborne Conflict Detection Based on Support Vector Machine. *Proc. Appl. Mech. Mater. Trans. Technol. Publ.* **2012**, *229*, 1140–1145.
12. Han, D.; Zhang, X.; Nie, Z.; Guan, X. A Conflict Detection Algorithm for Low-Altitude Flights Based on SVM. *J. Beijing Univ. Aeronaut. Astronaut.* **2018**, *44*, 576.
13. Jiang, X.; Wu, M.; Wen, X.; Wang, Z. Application of ensemble learning algorithm in aircraft probabilistic conflict detection of free flight. In Proceedings of the 2018 International Conference on Artificial Intelligence and Big Data (ICAIBD), Chengdu, China, 26–28 May 2018; pp. 10–14.

14. Wang, S.; Song, Y. Improved Model of Conflict Detection in Low Altitude Flight Based on SVM. *J. Beijing Univ. Aeronaut. Astronaut.* **2022**, *48*, 8–14. [CrossRef]
15. Zanin, M. Assessing Airport Landing Efficiency through Large-Scale Flight Data Analysis. *IEEE Access* **2020**, *8*, 170519–170528. [CrossRef]
16. Pan, Z.; Chi, C.; Zhang, J. A model of fuel consumption estimation and abnormality detection based on airplane flight data analysis. In Proceedings of the 2018 IEEE/AIAA 37th Digital Avionics Systems Conference (DASC), London, UK, 23–27 September 2018; pp. 1–6. [CrossRef]
17. Min, S. *A Proactive Safety Enhancement Methodology for General Aviation Using a Synthesis of Aircraft Performance Models and Flight Data Analysis*; Georgia Institute of Technology: Atlanta, GA, USA, 2018.
18. Jeatrakul, P.; Wong, K.W.; Fung, C.C. Classification of imbalanced data by combining the complementary neural network and SMOTE algorithm. In Proceedings of the International Conference on Neural Information Processing, Sydney, Australia, 22–25 November 2010; pp. 152–159. [CrossRef]
19. Yang, B.-S.; Gu, F.; Ball, A. Thermal Image Enhancement Using Bi-Dimensional Empirical Mode Decomposition in Combina-tion with Relevance Vector Machine for Rotating Machinery Fault Diagnosis. *Mech. Syst. Signal Process.* **2013**, *38*, 601–614.
20. Jia, Y.; Kwong, S.; Wu, W.; Gao, W.; Wang, R. Generalized relevance vector machine. In Proceedings of the 2017 Intelligent Systems Conference (IntelliSys), London, UK, 7–8 September 2017; pp. 638–645.
21. Li, H.; Pan, D.; Chen, C.L.P. Intelligent Prognostics for Battery Health Monitoring Using the Mean Entropy and Relevance Vector Machine. *IEEE Trans. Syst. Man Cybern. Syst.* **2014**, *44*, 851–862. [CrossRef]
22. Mianji, F.A.; Zhang, Y. Robust Hyperspectral Classification Using Relevance Vector Machine. *IEEE Trans. Geosci. Remote Sens.* **2011**, *49*, 2100–2112. [CrossRef]
23. Deo, R.C.; Samui, P.; Kim, D. Estimation of monthly evaporative loss using relevance vector machine, extreme learning machine and multivariate adaptive regression spline models. *Stoch. Hydrol. Hydraul.* **2015**, *30*, 1769–1784. [CrossRef]
24. Demir, B.; Erturk, S. Hyperspectral Image Classification Using Relevance Vector Machines. *IEEE Geosci. Remote Sens. Lett.* **2007**, *4*, 586–590. [CrossRef]
25. Elarab, M.; Ticlavilca, A.M.; Torres-Rua, A.F.; Maslova, I.; McKee, M. Estimating chlorophyll with thermal and broadband multispectral high resolution imagery from an unmanned aerial system using relevance vector machines for precision agriculture. *Int. J. Appl. Earth Obs. Geoinf. ITC J.* **2015**, *43*, 32–42. [CrossRef]
26. Caesarendra, W.; Widodo, A.; Yang, B.-S. Application of relevance vector machine and logistic regression for machine degradation assessment. *Mech. Syst. Signal Process.* **2010**, *24*, 1161–1171. [CrossRef]
27. Widodo, A.; Kim, E.Y.; Son, J.-D.; Yang, B.-S.; Tan, A.C.; Gu, D.-S.; Choi, B.-K.; Mathew, J. Fault diagnosis of low speed bearing based on relevance vector machine and support vector machine. *Expert Syst. Appl.* **2009**, *36*, 7252–7261. [CrossRef]
28. Xu, Y.; Yu, G.; Wang, Y.; Wu, X.; Ma, Y. A Hybrid Vehicle Detection Method Based on Viola-Jones and HOG + SVM from UAV Images. *Sensors* **2016**, *16*, 1325. [CrossRef] [PubMed]
29. Civil Aviation Administration of China. *Rules for Air Traffic Management of Civil Aircraft: CCAR-93TM-R5*; Civil Aviation Administration of China: Beijing, China, 2017. (In Chinese)
30. Escamilla-Serna, N.J.; Seck-Tuoh-Mora, J.C.; Medina-Marin, J.; Barragan-Vite, I.; Corona-Armenta, J.R. A Hybrid Search Us-ing Genetic Algorithms and Random-Restart Hill-Climbing for Flexible Job Shop Scheduling Instances with High Flexibility. *Appl. Sci.* **2022**, *12*, 8050. [CrossRef]
31. Li, Y.; Zhou, X.; Gu, J.; Guo, K.; Deng, W. A Novel K-Means Clustering Method for Locating Urban Hotspots Based on Hy-brid Heuristic Initialization. *Appl. Sci.* **2022**, *12*, 8047. [CrossRef]
32. Ahmadian, A.; Elkamel, A.; Mazouz, A. An Improved Hybrid Particle Swarm Optimization and Tabu Search Algorithm for Expansion Planning of Large Dimension Electric Distribution Network. *Energies* **2019**, *12*, 3052. [CrossRef]
33. Garcia-Guarin, J.; Rodriguez, D.; Alvarez, D.; Rivera, S.; Cortes, C.; Guzman, A.; Bretas, A.; Aguero, J.R.; Bretas, N. Smart Microgrids Operation Considering a Variable Neighborhood Search: The Differential Evolutionary Particle Swarm Optimiza-tion Algorithm. *Energies* **2019**, *12*, 3149. [CrossRef]
34. Upreti, D.; Pignatti, S.; Pascucci, S.; Tolomio, M.; Huang, W.; Casa, R. Bayesian Calibration of the Aquacrop-OS Model for Durum Wheat by Assimilation of Canopy Cover Retrieved from VENµS Satellite Data. *Remote Sens.* **2020**, *12*, 2666. [CrossRef]
35. Otero-Casal, C.; Patlakas, P.; Prósper, M.A.; Galanis, G.; Miguez-Macho, G. Development of a High-Resolution Wind Forecast System Based on the WRF Model and a Hybrid Kalman-Bayesian Filter. *Energies* **2019**, *12*, 3050. [CrossRef]

symmetry

Article

Performance Analysis on the Small-Scale Reusable Launch Vehicle

Zheng Gong [1], Zian Wang [2,*], Chengchuan Yang [1], Zhengxue Li [2], Mingzhe Dai [3] and Chengxi Zhang [4]

[1] Department of Aerospace Engineering, Nanjing University of Aeronautics and Astronautics, Nanjing 210016, China
[2] China Academy of Launch Vehicle Technology, Beijing 100076, China
[3] School of Automation, Central South University, Changsha 410083, China
[4] School of Electronic and Information Engineering, Harbin Institute of Technology, Harbin 150006, China
* Correspondence: wangzian@nuaa.edu.cn

Abstract: According to the symmetrical characteristics of a new type of Reusable Launch Vehicle (RLV) in the recovery phase, we studied the basic aerodynamic model data of Starship and the aerodynamic data with rudder deflection, and the causes of its aerodynamic coefficients are expounded. At the same time, we analyzed its stability and maneuverability. According to the flying quality requirements, the lateral-directional model of Starship in the return phase at a high angle of attack is analyzed. Finally, we analyzed the lateral heading stability and control deviation of Starship by using the criterion and nonlinear open-loop simulations. The results show that the Starship has pitching and rolling stability, but it only has heading stability in some ranges of angle of attack, and there is no heading stability at a conventional large angle of attack. At the same time, after modal analysis and comparison of flight quality, it can be seen that the longitudinal long-period model of the starship degenerates into a real root and it is stable and convergent. The lateral heading roll mode is at level 2 flight quality, the helical mode is at level 1 flight quality, and the Dutch roll mode diverges, which needs to be stabilized and controlled later.

Keywords: Reusable Launch Vehicle; stability; manipulativeness; flying qualities; criterion analysis

Citation: Gong, Z.; Wang, Z.; Yang, C.; Li, Z.; Dai, M.; Zhang, C. Performance Analysis on the Small-Scale Reusable Launch Vehicle. *Symmetry* **2022**, *14*, 1862. https://doi.org/10.3390/sym14091862

Academic Editors: Jan Awrejcewicz and Sergei D. Odintsov

Received: 9 July 2022
Accepted: 30 August 2022
Published: 6 September 2022

Publisher's Note: MDPI stays neutral with regard to jurisdictional claims in published maps and institutional affiliations.

1. Introduction

Since the former Soviet Union cosmonaut Gagarin [1] entered space for the first time, after more than 60 years of continuous exploration and development in the field of manned space flight, so far, there are two types of manned spacecraft in the world that can carry out missions between space and earth: the space shuttle of the United States [2] and manned spacecraft represented by Soyuz of Russia and Shenzhou of China [3,4]. However, except for a small part of the space shuttle, other manned spacecraft are not reusable, and the maintenance cost of space shuttles is very expensive, so the cost of space travel has been exceedingly high, which greatly restricts the pace of human exploration of outer space.

Starship and Super Heavy are the next generation of reusable space transportation systems proposed by Musk, founder of SpaceX, based on the vision of Mars colonization. According to the company's assumption, a wide range of missions can be accomplished through a variety of combinations of the two core spacecraft: interplanetary missions such as manned landing on Mars, near-earth missions such as space station transportation, satellite deployment, and globally ultra-fast passenger transportation. A recent lunar version of the starship program also won NASA's bid for the Moon landing mission. As a result, this system can theoretically meet the requirements of large-span transportation activities between different spaces, ranging from near-earth activities to Mars colonization.

In recent years, with the development of civil aerospace enterprises, new opportunities and challenges have been brought to the aerospace field. Since SpaceX publicized the ITS program in 2016, the Starship program has undergone several major design changes and evolutions. In 2019, the first starship prototype was publicly displayed. Since then, SpaceX has accelerated the research and manufacturing of starships by adopting the strategy of rapid testing and iterative verification of prototypes: In 2019, the free suspension test and safe landing test of the star worm preliminary verifier were completed [5]. Since 2020, through intensive flight tests of prototypes SN5–SN15 [6–10], SpaceX gradually mastered the key technologies of suspension at a low altitude of 150 m, flight at a high altitude of 10 km, roll maneuver, engine restarting powered braking, vertically soft landing at a fixed point, and so on [11–15]. According to the current progress, the orbital flight test of the Starship–Super Heavy system with high integration and comprehensive assessment is expected to be realized soon, and the system is planned to be used for carrying out manned missions such as landing on the Moon and Mars in the future [16–18].

The Starship–Super Heavy transportation system uses a two-stage fully reusable vehicle scheme with a designed loading capacity of 100 t. After the superheavy booster completes the first-stage powered flight separation, the starship continues the second-stage powered flight and continues to accelerate to enter orbit. The design of the starship is a combination of a two-stage rocket, orbiter, and reentry vehicle. The crew and payload are placed in the load cabin at the front of the starship, which has a reentry and return capability similar to that of the space shuttle orbiter [18–20]. The starship can carry a 50 t payload on return and uses a power-braking vertical fixed-point recovery scheme during the landing phase, which is similar to the Falcon 9 rocket [21,22]. With a simple shape and body of cone-column combination, the starship adopts a unique tailless canard aerodynamic configuration, and in order to meet the requirements of reentry flight, thermal tiles are laid on the windward side to deal with the thermal environment during reentry flight [23]. The current aerodynamic layout scheme of the starship is different not only from the manned spacecraft and space shuttle schemes of the traditional space transportation system, but also from the radical air-space shuttle scheme, and it is even significantly different from the earlier scheme, thus attracting huge attention once proposed [24,25]. Different from both the conventional manned spacecraft that is recovered by parachutes after semi-ballistic reentry into the atmosphere, and the space shuttle that lands horizontally on the airport, the landing method that the starship adopts is more similar to the recovery landing method of the rocket "Falcon", which realizes the vertical landing by the coordinated control of rudder surfaces and vector thrust. A new rudder surface control that is different from the traditional lift-body aircraft is adopted in the starship for this special takeoff and landing way [26]. Traditional lift-body aircraft realize the control of attitude and path by adopting the ailerons and vertical and horizontal tails, while the starship controls its body through two pairs of wings scattered on the nose and tail, which can deflect along the axial direction [27]. Zuo [27] made a detailed analysis of aerodynamic characteristics of the shape of the early starships (2019) in the landing and low-speed stages. Combined with aerodynamic characteristics such as lift/drag obtained from the simulation of subsonic separation flow field under a large angle of attack and the changing rules of the vertex moment along the deflection angle of leading and rear wings, a conclusion that four wings of the starship layout are subject to the three-channel control was given. While during the hypersonic and supersonic flight of reentry process, how about the wide speed-domain characteristics of this configuration, whether reentry trimming at full speed domain can be realized, how about the characteristics of the center of mass, whether the three channels are stable, what outstanding characteristics and advantages this configuration have, why such a unique design is taken, and many other problems remain to be further analyzed and researched.

At present, domestic and foreign studies on manipulativeness stability characteristics analysis are only limited to aircraft or taxiway takeoff and landing vehicles, and the structure is relatively simple, such as studies on multivariable stability margin of reentry aircraft [28], definitions of static stability margin of aircraft [29], and state feedback control and stability analysis of hypersonic aircraft [30]. There is also research on modal stability analysis of hypersonic aircraft with lift-body configuration [31], aerodynamic characteristics analysis of X-33-like vehicles [32], longitudinal and lateral flight quality research of saucer aircraft [26,27], flight quality research of short take-off and vertical landing aircraft, criterion analysis of Robert Weissman, etc. However, there is little research on the manipulativeness and stability characteristics and flight quality of RLVs that can take off and land vertically.

This paper focuses on the starships, in view of the large angle of attack flight characteristics during the recovery phase. The stability characteristics of aerodynamic derivatives are analyzed, classical theory of flight dynamics of linearized small perturbation method is applied to work out the motion characteristic root of longitudinal and lateral direction, and analysis is carried out. At the same time, the principle of criteria is used to analyze the lateral-directional stability and control the deviation of the starship. Finally, time-history open-loop simulation is used to verify the above analysis.

2. Aerodynamic Configuration of Starships

This paper models the starship according to the size parameters publicized on the official website of SpaceX, as shown in Figure 1. The wings are arranged according to the canard layout, a pair of front wings are arranged at the cone section, and a pair of rear wings are arranged at the end of the column section, both of which adopt trapezoidal wings. The whole ship is 50 m in length, 9 m in diameter, about 18 m in rear wingspan, and 15 m in front wingspan. The projected area of the whole plane is about 545 square meters. The weight of the whole ship is 10^5 kg, and the fuselage is made of stainless steel. In light of the shape and the distribution of the inner fuel tank and engine, the center of gravity of the whole ship is estimated at 40 m from the nose, and the pitching axial inertia moment coming through the center of gravity is 3.7×10^7 kg·m^2.

Figure 1. Configuration of Starship.

Overall, as a lift reentry vehicle, its simple shape gives itself distinctive characteristics, but it also brings several questions.

The simple and coordinated shape of the cone-column -wing, which is facially symmetrical, intuitively facilitates the series combination with superheavy boosters with the same diameter to form a simple two-stage rocket configuration, and this is much more compact than the complicated parallel layout of the orbiter-fuel tank-booster of the space shuttle, and the corresponding aerodynamic characteristics, flight control, and design of booster separation during the active period are also much simpler. Is this simple configuration suitable for lift reentry and return flight in a superwide speed domain?

A canard layout with a front-rear wings combination is adopted. The canard layout is common in the design of tactical missiles and highly maneuverable fighters, but there is no precedent in the design of reentry vehicles. The canard layout with relaxed static stability technology can realize that all wings generate positive lift at the trimming state and improve the aerodynamic efficiency of the aircraft. However, is it necessary for the returning stage? In addition, the canard is located very near the front, which means that the starship may face a severe aerodynamic heating environment during supersonic flight, and will it cause a serious problem for thermal protection?

It is a significant change in comparison to the earlier starship schemes (both the September 2018 and December 2018 versions had vertical tails) that the new version adopts a tailless layout without vertical tails and ventral fin. Tailless design and canard configuration will lead to the directional pressure center moving forward significantly. Intuitively, it can be judged that the starship's directional pressure center will be too forward in most ranges of flight velocity and angle of attack. Will this pose a serious risk to the lateral-directional static stability?

3. Performance Analysis

The new small-scale RLV starship that can vertically take off and land has a wing-body fusion design with four new rudder surfaces (hereinafter referred to as "elevator-like", "aileron-like", and "rudder-like") at the front and rear for attitude motion. In order to evaluate its stability and maneuverability, it is necessary to analyze its manipulativeness and stability characteristics. Figure 2 shows the scaling model obtained by modeling the shape layout published by SpaceX. The basic aerodynamic data and corresponding aerodynamic derivatives were obtained by Computational Fluid Dynamics (CFD) calculation of the layout, and the following analysis was made.

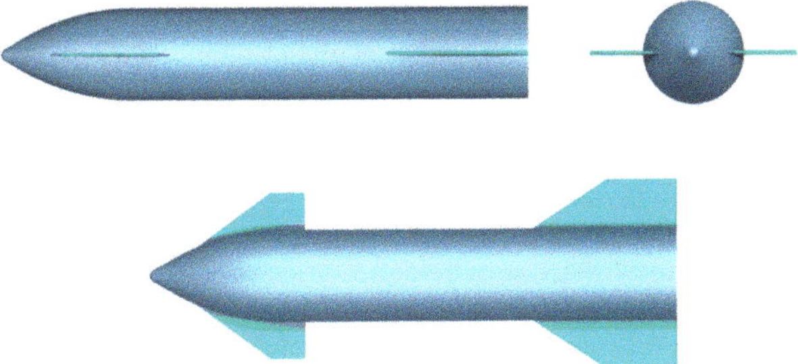

Figure 2. Configuration of small-scale Starship.

3.1. Analysis of Static Force and Moment Coefficients

3.1.1. Analysis of Polar Curve and Lift-Drag Ratio

In the absence of roll, yaw, and pitch, the changes of pole curve and lift–drag ratio with angle of attack are analyzed. The results are shown in Figures 3 and 4. The tangent line of the polar curve is drawn from the origin of the coordinate, the lift–drag ratio achieves its maximum at the tangent point, and the corresponding angle of attack is the favorable angle of attack α_{opt}, which is 20°.

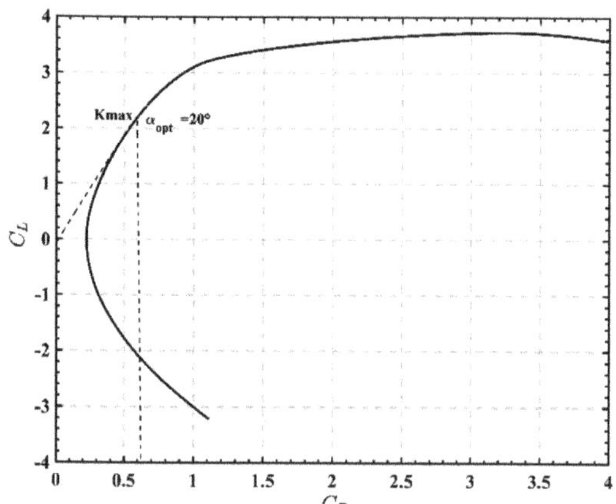

Figure 3. Polar curve.

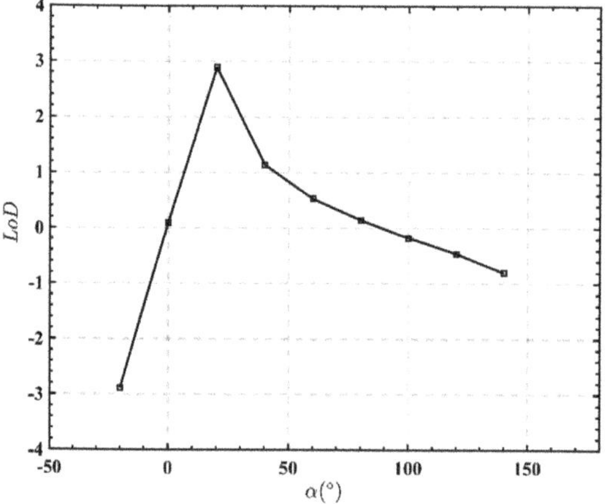

Figure 4. Lift–drag ratio.

3.1.2. Analysis of Longitudinal Forces and Moments

Keep the elevator-like, aileron-like, and rudder-like still, $\delta_e = \delta_a = \delta_r = 0$, and the rate of roll, pitch, and yaw angle is 0, namely, under the condition of $p = q = r = 0$, the starship's lift coefficient, drag coefficient, and roll moment coefficient curve is shown in Figures 5–7. In the figures, C_L, C_D, and C_m are, respectively, lift coefficient, drag coefficient, and roll moment coefficient, and α is angle of attack. It can be seen from the figures:

(1) The lift coefficient increases with the increase in the angle of attack when $-20° < \alpha < 40°$, and decreases with the increase in the angle of attack when $\alpha > 40°$. $\alpha = 40°$ is the critical angle of attack.

(2) When $0° < \alpha < 80°$, the drag coefficient increases with the increase in the angle of attack; at $-20° < \alpha < 0°$ and $\alpha > 80°$, the drag coefficient decreases with the increase in the angle of attack. When the angle of attack is 0, the drag coefficient is the minimum; when the angle of attack is $80°$, the drag coefficient is the maximum. So when the starship returns, choose a horizontal descent to minimize the speed of descent with maximum resistance.

(3) When $-20° < \alpha < 0°$, $C_m > 0$, a positive pitching moment is generated to make the starship raise its nose to reduce the angle of attack. Within the range of angle of attack

$\alpha > 0°$, $C_m < 0$, a negative pitching moment is generated to make the starship bow its nose to reduce the angle of attack.

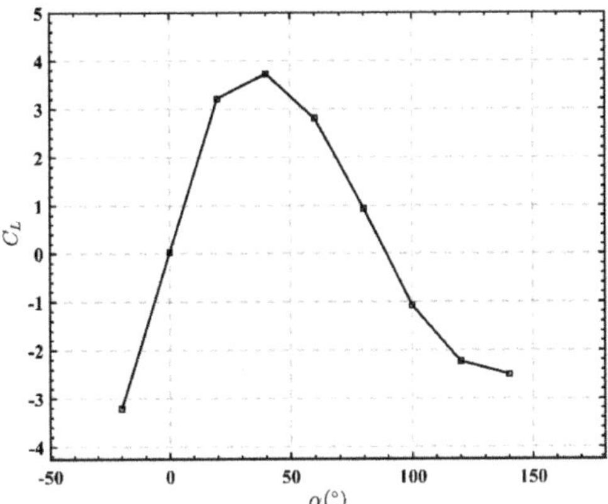

Figure 5. Lift coefficient.

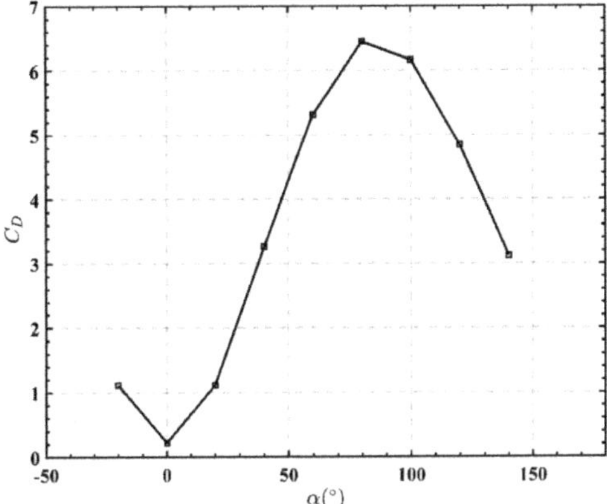

Figure 6. Drag coefficient.

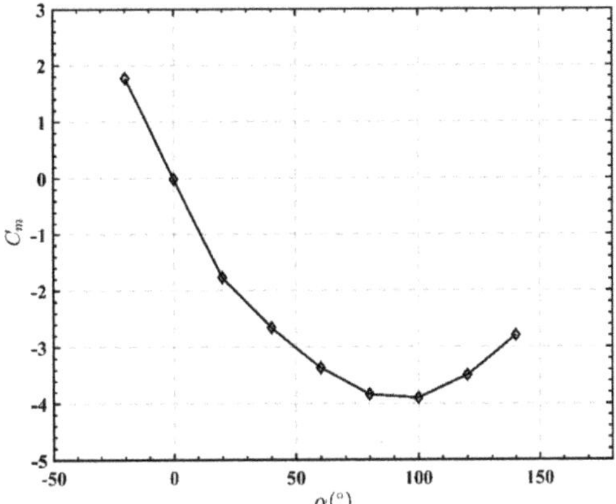

Figure 7. Pitch moment coefficient.

The above analysis shows that the changing rules of longitudinal static force and moment coefficient along the angle of attack are consistent with actual flight characteristics.

3.1.3. Lateral-Directional Force and Moment Analysis

Curves of the starship's lateral force coefficient, roll moment coefficient, and yaw moment coefficient are shown in Figures 8–10. In the figures, C_s, C_l, and C_n are, respectively, the lateral force coefficient, roll moment coefficient, and yaw moment coefficient. It can be seen from the figures:

(1)   The variation of static lateral force coefficient with angle of attack is as follows: negative lateral force is generated if the side slip is positive, while if negative sideslip occurs, positive lateral force occurs.

(2)   When the positive side slip occurs, negative side force is generated, and then negative roll moment is generated, so the roll moment coefficient is negative; when negative sideslipping occurs, positive sideslipping force is generated, and then positive rolling moment is generated, so the rolling moment coefficient is positive.

(3)   When positive sideslip occurs, yaw moment coefficient is positive; when negative sideslip occurs, yaw moment coefficient is negative.

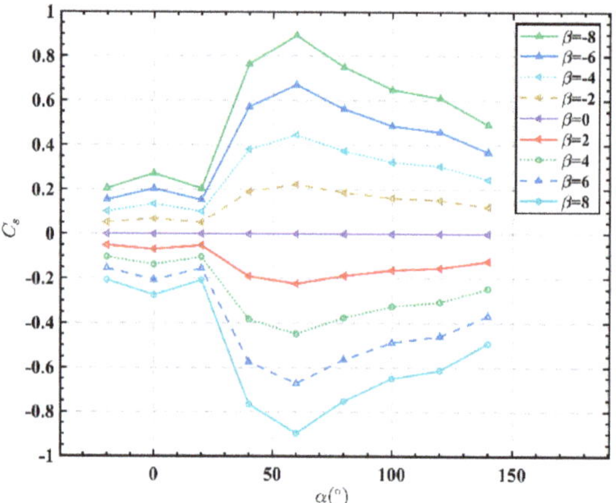

Figure 8. Lateral force coefficient.

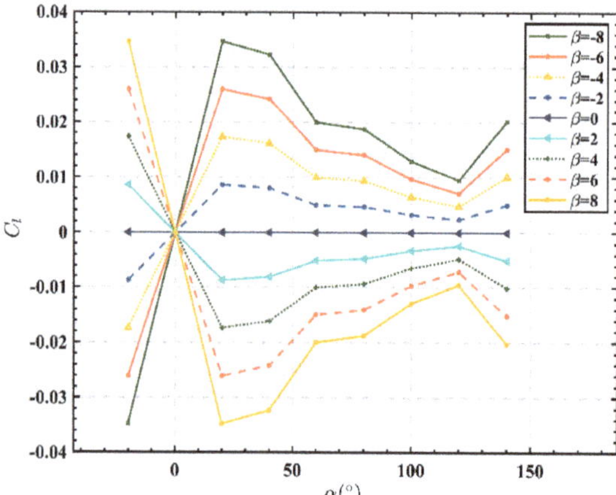

Figure 9. Roll torque coefficient.

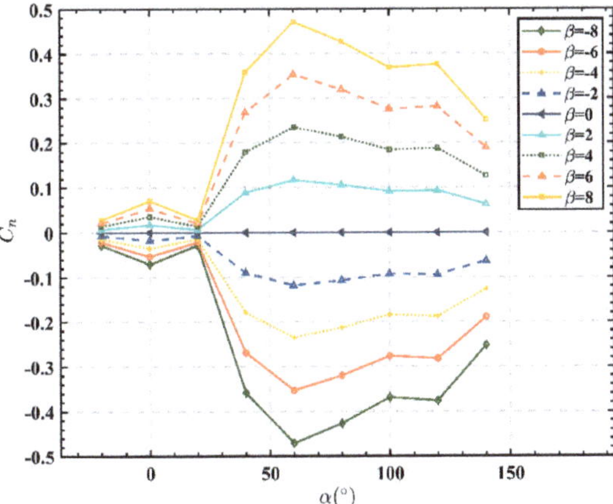

Figure 10. Yaw moment coefficient.

The above analysis shows that the lateral-directional static force and moment coefficient vary with the angle of attack in accordance with the actual flight characteristics.

3.1.4. Static Stability Analysis

The pitch static stability coefficient curve, roll static stability coefficient curve, and yaw static stability coefficient curve are shown in Figures 11–13. In the figures, $C_{m\alpha}$ is the roll static stability coefficient, $C_{l\beta}$ is the roll static stability coefficient, and $C_{n\beta}$ is the yaw static stability coefficient. It can be seen from the figure that the pitching static stability of the starship decreases with the increase in the angle of attack. When $-20° < \alpha < 95°$, $C_{m\alpha} < 0$, the starship is longitudinally stable; when $\alpha = 95°$, $C_{m\alpha} = 0$, the starship is in neutral static stability, and at this point, the center of gravity and the focus coincide; when $\alpha > 95°$, $C_{m\alpha} > 0$, is statically unstable longitudinally. At this time, the longitudinal static stability margin of the starship body needs to be analyzed, which needs to be estimated in combination with the lift derivative $C_{L\alpha}$ (reflected in the following sections). When $-20° < \alpha < 10°$ and $C_{l\beta} > 0$, the static roll is unstable. When $\alpha > 10°$ and $C_{l\beta} < 0$, static roll is stable. In the measured angle of attack range, $C_{n\beta} > 0$, the starship has directional static stability.

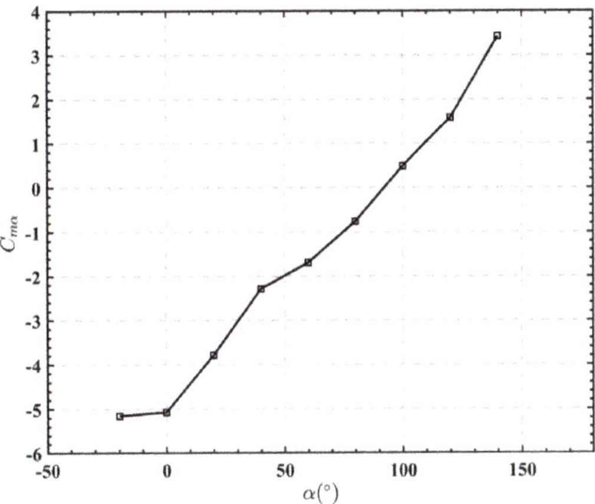

Figure 11. Pitch static stability coefficient.

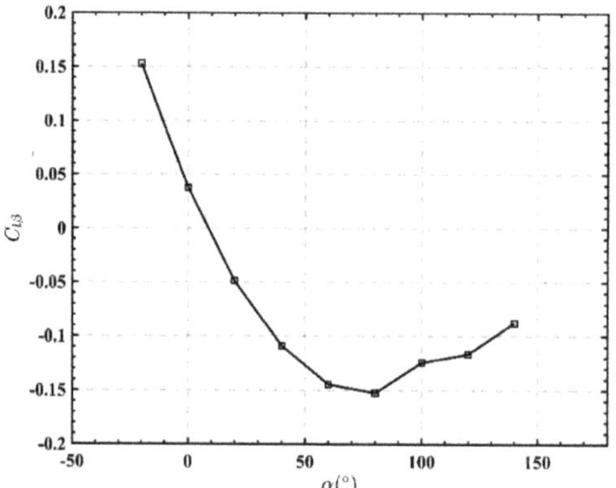

Figure 12. Roll static stability coefficient.

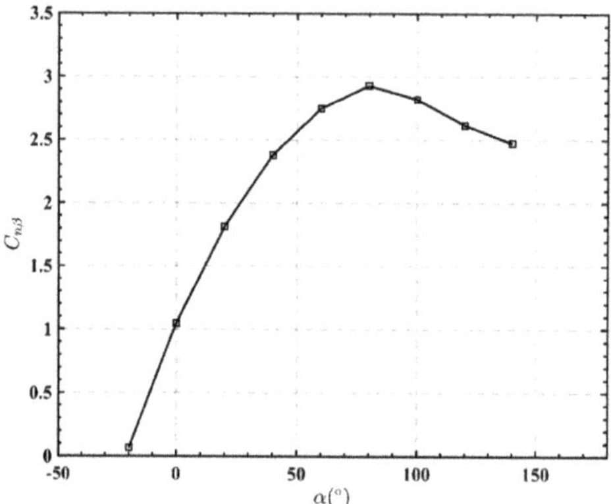

Figure 13. Yaw static stability coefficient.

4. Analysis of Dynamic Characteristics

4.1. Analysis of Dynamic Derivatives

The dynamic derivative is mainly the variation of the derivative of the dimensionless moment with angle of attack measured at different angles of attack when all rudder surfaces are neutral through the oscillating balance experiment, as shown in Figures 14–16.

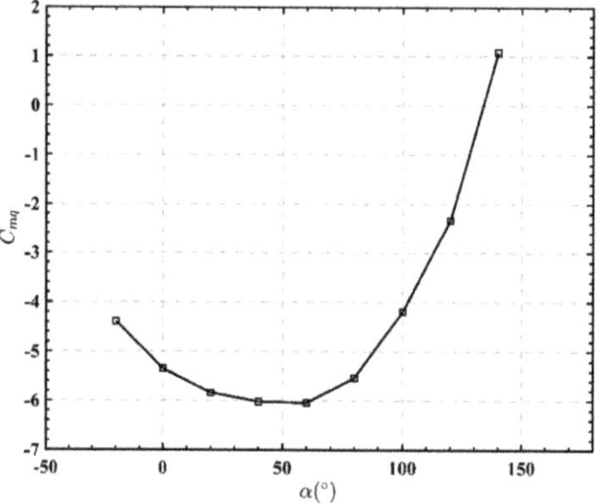

Figure 14. Pitch damping derivative.

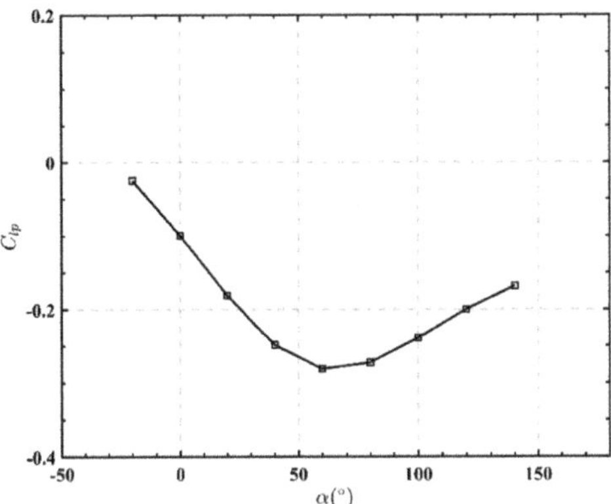

Figure 15. Roll damping derivative.

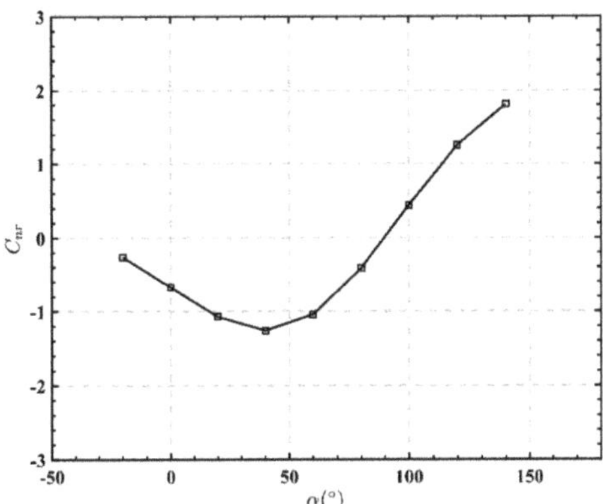

Figure 16. Yaw damping derivative.

It can be seen from the figures that in the range of angle of attack $-20° < \alpha < 90°$ in the starship recovery stage, the dynamic derivative conforms to the actual flight characteristics.

(1) When a starship raises its head and generates a positive pitch angular rate, $q > 0$, a starship will also generate a negative pitch moment to prevent it from rotating upwards. Therefore, in Figure 13, within the angle of attack $-20° < \alpha < 135°$, the pitch damping derivatives are all negative, preventing the starship from rotating. When $\alpha > 135°$, the pitching damping is positive. This is because under a large angle of attack, because the air flow is in chaos, the polarity of the roll moment changes. As a result, during the horizontal descent in the returning stage, the largest angle of attack of the starships is $90°$ according to the analysis of its movement mode and path, so within its range of angle of attack during its movement process, its pitch damping derivative is negative all the time, which conforms to the actual flight characteristics.

(2) If the starship rolls to the right, $p > 0$, the left and right rear wings are asymmetrically deflecting at this time, and a negative roll moment will be generated, which will hinder the roll. Within the whole range of angle of attack, the negative roll moment derivative is exactly caused by the roll angular rate, so the roll damping derivative is consistent with the actual flight characteristics.

(3) The yaw damping derivative is negative in the range of angle of attack $-20° < \alpha < 90°$, but positive in the range of $\alpha > 90°$, which promotes the yaw of the starship. Therefore, this phenomenon is easy to cause the risk of excessive yaw in the recovery stage. According to the analysis, the yaw damping derivative conforms to the actual flight

characteristics in the range of a small angle of attack, but in the process of a large angle of attack, it is underdamped.

4.2. Analysis of Manipulativeness Derivative

4.2.1. Pitching Manipulativeness Derivative

When the deflection of elevator-like is positive, $\delta_e > 0$, this will destroy the balance of the original star trek. Elevator-like is subject to an upward force, which lies behind the center of gravity, which will produce a negative pitch moment, so regular pitch control derivative angle of attack $C_{m\delta_e}$ should be negative, and based on analysis, the pitch control derivative shown in Figure 17 is in line with actual flight characteristics.

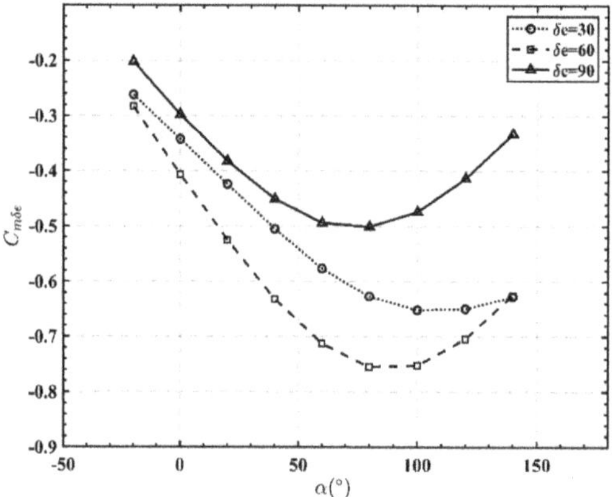

Figure 17. Pitch manipulation derivative.

4.2.2. Rolling Manipulativeness Derivative

It can be seen from Figures 18 and 19 that when the aileron-like wing deflects along the positive direction, $\delta_a > 0$, a negative rolling moment is generated, so the rolling control derivative $C_{l\delta_a}$ is negative. When the rudder-like is deflecting along the positive direction, $\delta_r > 0$, a positive side force is generated. Because the rudder-like is located above the axis of the arrow body, a positive rolling moment is generated. Therefore, the cross-control derivative of rudder-like $C_{l\delta_r}$ is positive, which is consistent with the actual flight situation.

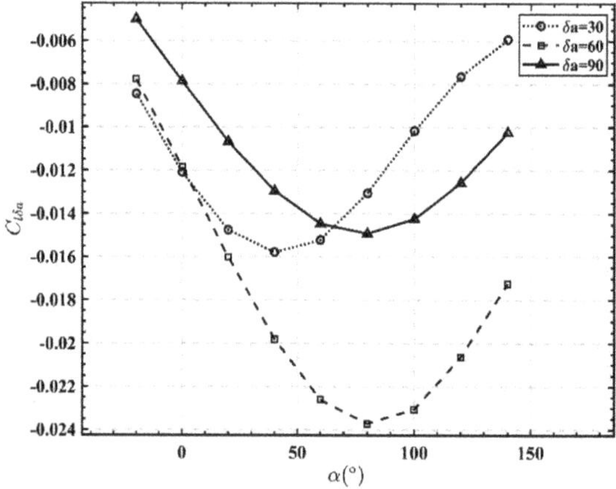

Figure 18. Roll manipulation derivative.

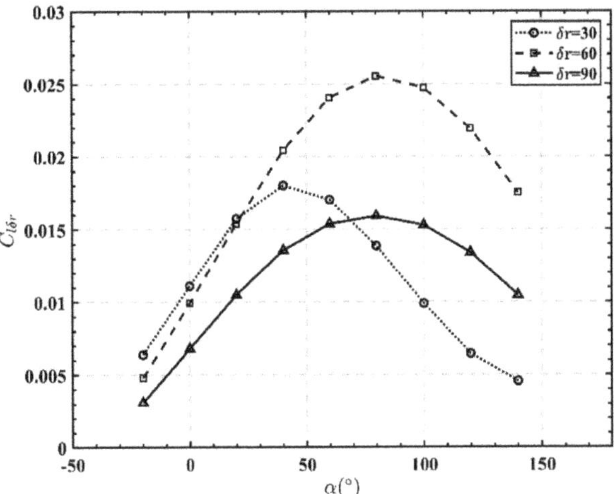

Figure 19. Roll crossover derivative.

4.2.3. Yawing Manipulativeness Derivative

It can be seen from Figures 20 and 21 that when the rudder-like is positively deflected, $\delta_r > 0$, a positive side force is generated, and then a negative yaw moment is generated, so the yaw control derivative $C_{n\delta_r}$ is negative. When the aileron is positively deflected, because the starship is facially symmetrical in aerodynamic layout, it will produce the yaw moment by coupling rudder-like and making it deflect. The positive deflection of aileron-like produces negative roll torque, and the resultant force of lift force and gravity makes it sideslip negatively. In order to offset the negative sideslip, rudder-like will produce a positive yaw moment, so the yaw cross-manipulation derivative is $C_{n\delta_a}$ positive, which is consistent with the actual flight situation.

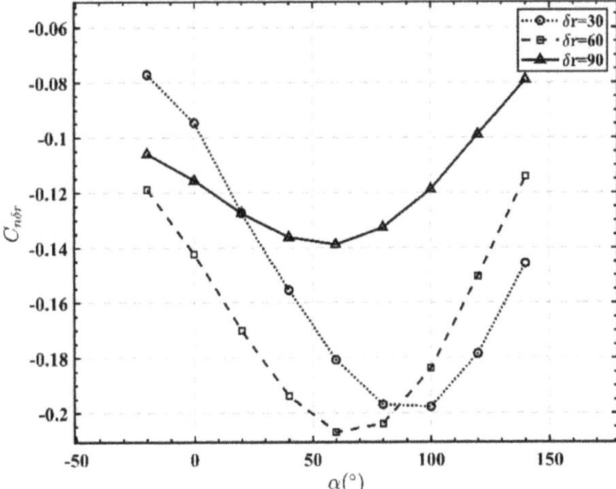

Figure 20. Yaw manipulation derivative.

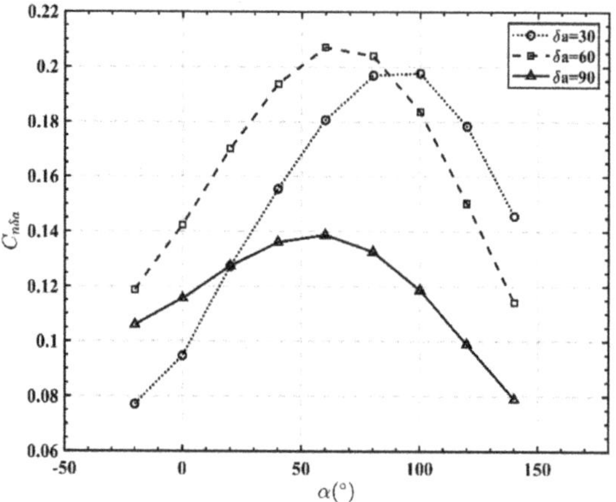

Figure 21. Yaw crossover derivative.

4.3. Analysis of Modal Characteristics and Flight Quality

Similar to the aircraft with the flying wing layout, the starship, with the design of wing-fuselage fusion is a facially symmetrical aircraft from which the vertical tails are canceled, so the dynamic stability analysis method of conventional aircraft is adopted in its dynamic stability analysis [31]. The longitudinal and lateral-directional linearized small disturbance state equations are shown in (1)–(5) and (6), respectively.

$$\dot{x} = Ax + Bu \tag{1}$$

$$x = \begin{bmatrix} \Delta V & \Delta\alpha & \Delta q & \Delta\theta \end{bmatrix}^T \tag{2}$$

$$u = \begin{bmatrix} \Delta\delta_e & \Delta\delta_T \end{bmatrix}^T \tag{3}$$

$$A = \begin{bmatrix} X_V & X_\alpha + g & 0 & -g \\ -Z_V & -Z_\alpha & 1 & 0 \\ M_V - M_{\dot\alpha}Z_V & M_\alpha - M_{\dot\alpha}Z_\alpha & M_q - M_{\dot\alpha} & 0 \\ 0 & 0 & 1 & 0 \end{bmatrix} \tag{4}$$

$$B = \begin{bmatrix} X_{\delta_e} & X_{\delta_T} \\ -Z_{\delta_e} & -Z_{\delta_T} \\ M_{\delta_e} - M_{\dot\alpha}Z_{\delta_e} & M_{\delta_p} - M_{\dot\alpha}Z_{\delta_T} \\ 0 & 0 \end{bmatrix} \tag{5}$$

$$\begin{bmatrix} \dot\beta \\ \dot p \\ \dot r \\ \dot\phi \end{bmatrix} = \begin{bmatrix} Y_\beta & \alpha_* - Y_p & Y_r - 1 & \frac{g\cos\theta_*}{V_*} \\ L_\beta & L_p & L_r & 0 \\ N_\beta & N_p & N_r & 0 \\ 0 & 1 & \tan\theta_* & 0 \end{bmatrix} \begin{bmatrix} \beta \\ p \\ r \\ \phi \end{bmatrix} + \begin{bmatrix} 0 & L_{\delta_a} \\ L_{\delta_a} & L_{\delta_a} \\ N_{\delta_a} & N_{\delta_r} \\ 0 & 0 \end{bmatrix} \begin{bmatrix} \delta_a \\ \delta_r \end{bmatrix} \tag{6}$$

The definition and expression of each parameter in the formula can be found in the literature [31]. The above equation was used to calculate the characteristic roots of the starship at the trimming angle of attack and compared with the motion characteristic roots of B747 and a flying wing aircraft. Then, the analysis in coordination with the flight quality of GJB-185-86 is shown below.

4.3.1. Longitudinal Analysis

The characteristic equation of the longitudinal overall motion of the starship is calculated by Equations (1)–(5):

$$s^4 + 2.119\,s^3 + 5.102\,s^2 + 0.09367\,s + 5.601e^{-14} = 0 \tag{7}$$

The lateral-directional motion characteristic roots of the starship were obtained as shown in Table 1 and compared with the longitudinal characteristic roots of B747 [26] and a flying wing aircraft [26].

Table 1. Longitudinal motion characteristic root comparison.

B747	Flying Aircraft	Starship
$-0.4650 + 1.2456i$	-1.6	$-1.05 + 1.99i$
$-0.4650{-}1.2456i$	0.7	$-1.05{-}1.99i$
$-0.0097 + 0.0445i$	$-0.01 + 0.12i$	-5.98×10^{-13}
$-0.0097{-}0.0445i$	$-0.01{-}0.12i$	-0.0185

It can be seen from Table 1 that the characteristic roots are all located in the negative half-plane, indicating that the longitudinal motion is stable. In addition, the characteristic roots are composed of a pair of conjugate complex roots and two negative real roots. It is easy to find that the long-period mode degenerates into the negative real roots of the third mode, which is between the long and short periods. When the starship is in longitudinal motion, due to the shaking fuel, the center of gravity moves backward, and static stability is decreased, which makes the pitching moment and frequency increase, the damping ratio increase, the imaginary part of the long-period mode slowly change into zero, and the characteristic roots of long-period mode degenerate into two different negative real roots. The starship is overdamped, and the motion response is exponentially and monotonously convergent, indicating that the longitudinal motion is stable.

There is also a method [33] to estimate the characteristic roots of long-period mode according to the balance state quantity without considering the influence of compressibility and the change of thrust velocity. In engineering, there is little error between the result of this method and that of the fourth-order equation. The formula is shown as follows:

$$\begin{cases} \omega_{n.p} = \sqrt{2}\dfrac{g}{u_0} \\ \zeta_p = \dfrac{1}{\sqrt{2}(C_L/C_D)} \\ \lambda_p = -\zeta_p\omega_{n.p} \pm i\omega_{n.p}\sqrt{1 - \zeta_p^2} \end{cases} \tag{8}$$

After substituting the balance state quantity, the long-period modal characteristic roots $-0.06 \pm 0.24475i$ are obtained as Table 2 and the standard characteristics are listed as Table 3.

From the above analysis, it can be seen that the third mode of the starship converges exponentially, the short period mode meets the requirement of level 1 flight quality, and the estimated long-period modes before modal degradation also meet the requirement of flight quality of level 1.

Table 2. Modal characteristics of longitudinal course motion.

Mode	$T_{1/2}/s$	T/s	$N_{1/2}/$ Circle	ζ	ω_n
Third mode	37.46	—-	0	1	0.0185
Short mode	0.66	3.157	0.209	0.4667	2.25
Estimation long-period	11.55	25.671	0.4499	0.2381	0.252

Table 3. GJB-185-86 longitudinal modal characteristic standard.

Mode	Level of Flight Quality	Type of Aircraft	Maximum of ζ	Minimum of ζ
Short mode	1		1.30	0.35
	2	C	2.00	0.25
	3		—	0.15
Long mode	1			$\zeta > 0.04$
	2	C		$\zeta > 0$
	3			$T_2 > 55s$

4.3.2. Lateral-Directional Analysis

Through calculation of Equation (6), the characteristic equation of the starship's lateral-directional overall motion is:

$$s^4 - 0.00956s^3 - 0.04901s^2 + 0.02285s + 0.0001024 = 0 \tag{9}$$

The characteristic roots of lateral-directional motion of the starship are shown in Table 4 and are compared with those of B747 [27] and a flying wing aircraft [27].

Table 4. Lateral heading motion characteristic root comparison.

B747	Flying Aircraft	Starship
$-0.1507 + 0.9431i$	$0.025 + 0.35i$	$0.175 + 0.195i$
$-0.1507 - 0.9431i$	$0.025 - 0.35i$	$0.175 - 0.195i$
-0.3725	-2.0000	-0.336
0.0058	0.0006	-0.00444

It can be seen from the table that the starship and the flying wing aircraft have no vertical tails, so their $C_{n\beta}$, which plays a role in the recovery of Dutch rolling motion, is small, and it is the same thing with C_{nr} and $C_{s\beta}$, which play a role in damping. Moreover, because the $C_{l\beta}$ is also small, $C_{l\beta}$ will be too large, so the directional damping is further reduced, resulting in the instability of the Dutch roll mode. Because the starship's spiral mode, lying in the left of the imaginary axis is close to its imaginary axis, and its roll mode is also located at the left of the imaginary axis, it could be approximately assumed that its roll and spiral modes are stable.

According to the comparison between Tables 5 and 6, the rolling mode of the starship meets level 2 flight quality, and the spiral mode meets level 1 flight quality. A pair of conjugate complex roots corresponding to the Dutch roll mode is in the right plane and in an unstable state. Considering the aerodynamic derivative affecting the Dutch roll mode in the above analysis, $C_{n\beta}$, $C_{n\beta}$, $C_{s\beta}$, $C_{n\delta_r}$, and $C_{n\delta_r}$ should be added to improve the Dutch roll mode, but it should be considered that the impact of $C_{l\beta}$ on the Dutch rolling is relatively small, and the ratio relationship of $C_{l\beta}$ and $C_{n\beta}$ should be considered when improving its value to avoid affecting the stability of the spiral mode.

Table 5. Modal characteristics of lateral course motion.

Parameters	Rolling Time Constant T_R/s	Spiral Amplitude Doubling Time T_2/s	Natural Frequency of Dutch Roll ω_d	Damping Ratio of Dutch Roll ζ_d	$\omega_d \cdot \zeta_d$
Value	2.98	225	0.669	0.261	0.175

Table 6. GJB-185-86 lateral modal characteristic standard.

Parameters	Standards of Flight Quality	Type of Flight Stage	Type of Aircraft	Maximum of T_R/s	Minimum of T_2/s	Minimum of ω_d	Minimum of ζ_d	Minimum of $\omega_d \cdot \zeta_d$
Value	I	C	II	1.4	20	0.4	0.08	0.15
	II		III	3.0	12	0.4	0.02	0.05
	III			10	4	0.4	0.02	—

4.4. Characteristics Analysis of the Ratio between Roll and Swing

In the literature [34], the ratio between roll and swing $|\phi/\beta|_d$ is used to represent the proportion of rolling motion and yaw motion in the Dutch roll mode, and a large ratio between roll and swing means that the proportion of rolling motion in the Dutch roll mode is large. A small ratio between roll and swing indicates that the yaw motion accounts for a large proportion of the Dutch roll mode.

$|\phi/\beta|_d$ is the amplitude of the eigenvector ratio of roll angle and sideslip angle that corresponds to the eigenvalue of Dutch roll, and the approximate expression is shown as follows:

$$\left|\frac{\phi}{\beta}\right| \cong \left|\frac{L'_\beta}{\omega_d}\right| \frac{1}{\sqrt{1 + L'_p{}^2/\omega_d{}^2}} \tag{10}$$

For the flight state of high angle of attack, $\omega_d{}^2 \approx N'_\beta \cos\alpha - L'_\beta \sin\alpha$, the approximate expression is transformed into:

$$\left|\frac{\phi}{\beta}\right| \cong \left|\frac{L'_\beta}{N'_\beta \cos\alpha - L'_\beta \sin\alpha}\right| \frac{1}{\sqrt{1 + L'_p{}^2/(N'_\beta \cos\alpha - L'_\beta \sin\alpha)}} \tag{11}$$

where L'_β is the dominant torque of sideslip angle on the body axis roll, N'_β is the dominant number of yaw moment of body axis caused by the sideslip angle, and L'_p is the dominant torque of the roll angle rate on the body axis.

During the glide stage at a high angle of attack, L'_β/N'_β is smaller, and L'_p is large and has a smaller ratio between roll and swing. Shown in Figure 22 is the roll and swing ratio of each state quantity calculated according to the linear model with small disturbance at a high angle of attack, and it is about 0.0001 under a large angle of attack. Under a small roll and swing ratio, the Dutch roll mode of the starship is mainly reflected by yaw motion.

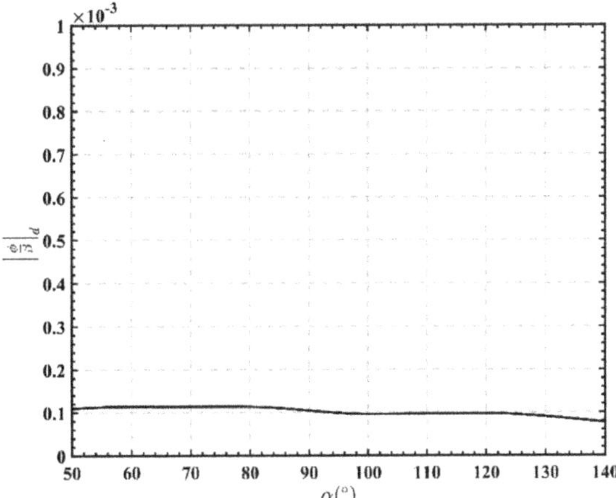

Figure 22. Roll−Sideslip ratio.

5. Analysis Based on the Principle of Criteria

5.1. Criteria of the Margin of Longitudinal Static Stability

Since the layout of starships is similar to that of flying wing aircraft and winged missiles, the longitudinal static stability margin of a traditional aircraft can be used to judge whether it is longitudinally stable or not. Its formula is defined as follows:

$$K_\alpha = -\frac{\partial C_m}{\partial C_L} = -\frac{C_{m\alpha}}{C_{L\alpha}} = x_F - x_G \tag{12}$$

where C_m and $C_{m\alpha}$ are the pitching moment coefficient and the pitching static stability derivative, respectively; C_L and $C_{L\alpha}$ are the lift coefficient and the lift derivative, respectively; x_F and x_G are the dimensionless focus position and the center of gravity position, respectively.

That K_α is positive ($K_\alpha > 0$) indicates that the focus is behind the center of gravity, which is statically stable in longitudinal direction; conversely, $K_\alpha < 0$ indicates that the focus is in front of the center of gravity, which is statically unstable in longitudinal direction. Figure 23 shows that K_α is positive all the time when $-20° < \alpha < 92°$, and it is negative when $\alpha > 92°$, turning statically unstable. In actual flight, because the body of the arrow is elastic and with the consumption and shaking of the fuel [32], with the increase in the angle of attack, in the condition of a small angle of attack, the center of gravity shifts back and forth, and the static stability margin coefficient of the center of gravity also fluctuates, which is shown as follows. At a high angle of attack, the center of gravity moves backward, and the starship changes from statically stable in the longitudinal direction to statically unstable. However, at a high angle of attack, the static instability coefficient is small and meets the requirements of controllability [18]. Therefore, the longitudinal stability augmentation control law and controller should be designed, and the constraint range should be given to assist the longitudinal stability augmentation of the starship when it moves at a high angle of attack.

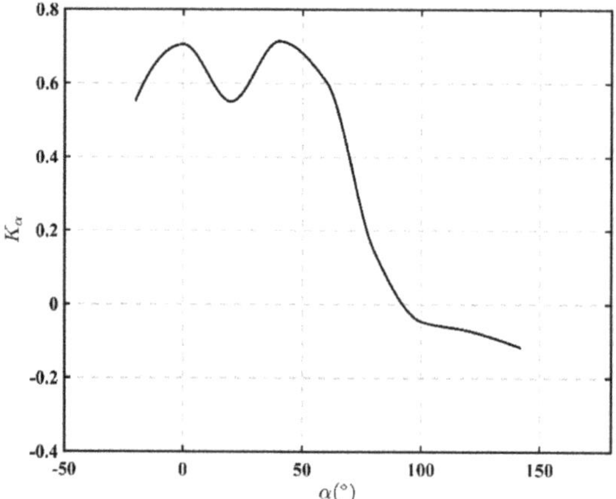

Figure 23. Longitudinal static stability margin.

5.2. Criteria of Comprehensive Analysis of Lateral Direction

This criterion can effectively evaluate the lateral-directional deviation and motion characteristics of a starship by the comprehensive analysis of dynamic directional stability criterion $C_{n\beta dyn}$ and lateral steering deviation stability criterion $LCDP$. Figure 24 shows the graph of analysis results of dynamic deviation and lateral control stability criteria.

$$C_{n\beta dyn} = C_{n\beta} \cos \alpha - \frac{I_z}{I_x} C_{l\beta} \sin \alpha \tag{13}$$

where $C_{n\beta dyn}$ is the dynamic direction stability deviation parameter; I_x, I_z are, respectively, the inertial moment along the x-axis and the inertial moment along the z-axis. $C_{n\beta dyn} > 0$ indicates that the aircraft has dynamic directional stability; otherwise, it does not.

$$LCDP = C_{n\beta} - C_{l\beta} \frac{C_{n\delta a}}{C_{l\delta a}} \tag{14}$$

where $LCDP$ is the lateral manipulation deviation parameter. $LCDP > 0$ indicates that there is no deviation in lateral-directional control; on the contrary, it means that there is deviation in lateral-directional control.

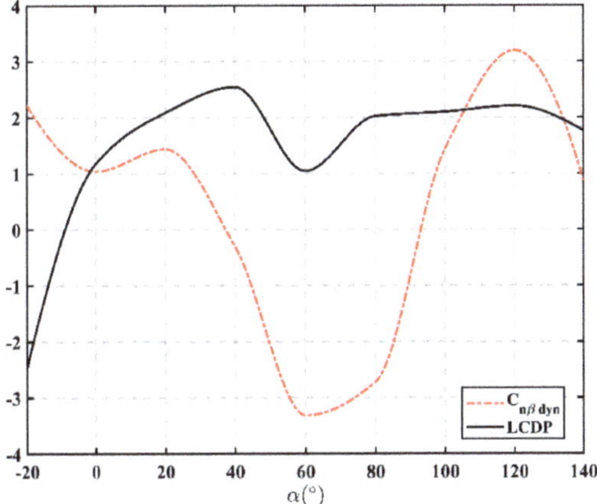

Figure 24. Comprehensive analysis of dynamic deviation and lateral control stability criteria.

As can be seen from the figures, when the angle of attack is at $-20° < \alpha < -10°$, $C_{n\beta dyn} > 0$, $LCDP < 0$, because the absolute value of the deviation parameter of dynamic yaw stability is larger than that of the lateral control deviation parameter, sideslip can also be eliminated by rolling motion of aileron-like without deviation. When the angle of attack is $-10° < \alpha < 38°$ and $\alpha > 93°$, $C_{n\beta dyn} > 0$ m and $LCDP > 0$, if sideslip occurs due to external interference at this time, the starship is stable laterally and the reverse pressure bar can also eliminate sideslip, so there will not be sideslip deviation phenomenon to endanger the starship. When $38° < \alpha < 93°$, $C_{n\beta dyn} < 0$, and $LCDP > 0$, there will be the phenomenon of control backlash. However, in the range of $38° < \alpha < 45°$ and $88° < \alpha < 93°$, because the absolute value of the lateral control deviation parameter is larger than that of the deviation parameter of dynamic yaw stability, yaw coupling brought by the rolling maneuver of aileron-like roll can also eliminate sideslip. When $45° < \alpha < 88°$, the absolute value of the deviation parameter of dynamic yaw stability is larger than that of the lateral control deviation parameter, or there is little difference between them. If sideslip occurs at this time, sideslip cannot be quickly eliminated; if the operation stick is reversed, there will be an anticontrol phenomenon.

5.3. Analysis Criteria of Coordinated Control Deviation of Rudder-like and Aileron-like

In the $LCDP$ criterion, only the operation of aileron-like is considered, and the rudder-like is in neutral state, while in the actual flight condition, aileron-like and rudder-like are in coordination to realize horizontal directional movement. As a result, the parameters quantity of rudder-like control are also introduced into the analysis of control deviation, forming a new deviation analysis criterion of coordinated control of rudder-like and aileron-like, which is shown as follows:

$$LCDP_{ARI} = C_{n\beta} - C_{l\beta} \frac{C_{n\delta_a} + kC_{n\delta_r}}{C_{l\delta_a} + kC_{l\delta_r}} \tag{15}$$

$$k = \frac{\delta_r}{\delta_a} \tag{16}$$

where $LCDP_{ARI}$ is the coordinated control deviation parameter of aileron-like and rudder-like. Similar to the lateral steering deviation stability criterion, when $LCDP_{ARI} > 0$, the rudder-like and aileron-like coordinated control is stable in operation; otherwise, it is not stable.

According to Figure 25, when the value of k is larger, the range of angle of attack with the coordinated control deviation parameter, which is more than zero increases, indicating that the stability of rudder-like and aileron-like coordinated control is much better than that of rudder-like or aileron-like control alone. When $k = 1/3$, within the angle of attack range of $-20° < \alpha < 0°$ and $25° < \alpha < 55°$, the coordinated control deviation parameter is negative, while in the angle of attack range of $0° < \alpha < 25°$ and $\alpha > 55°$, the coordinated control deviation parameter is positive. When $k = 0.5$, the coordinated control deviation parameter is negative within the angle of attack range of $-20° < \alpha < -5°$ and $32° < \alpha < 50°$, while the coordinated control deviation parameter is positive within the range of $-5° < \alpha < 32°$ and $\alpha > 50°$. When $k = 2/3$, the coordinated control deviation parameter is negative within the range of $-20° < \alpha < -15°$, and the parameter is positive within the range of $\alpha > -15°$. When $k = 1$, the deviation parameters of coordinated control are all positive in all angle of attack ranges, indicating that when the deflection angles of rudder-like and aileron-like are the same, the best control effect of coordinated operation is achieved.

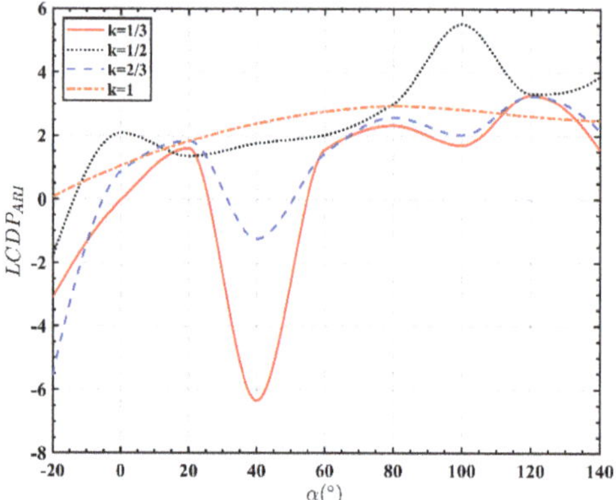

Figure 25. Evolution of parameters $LCDP_{ARI}$ versus angle of attack (AoA) when k takes different positive values.

As can be seen from Figure 26, when different negative values of k are taken, the deviation parameters change little, and the effect is of little difference to that of the $LCDP$ criterion. Therefore, when different negative values of k are taken, the rudder-like cannot compensate the aileron-like.

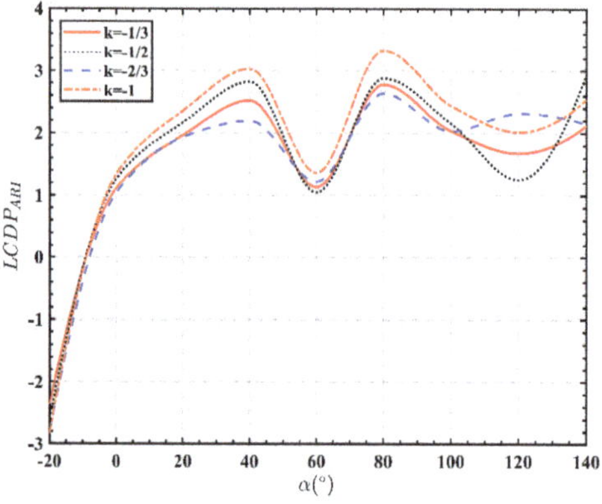

Figure 26. Evolution of parameters $LCDP_{ARI}$ versus AoA when k takes different positive values.

5.4. Weissman Graph Criteria

According to reusable carriers Weissman criterion that is introduced in the literature [35,36], the distribution of typical lateral-directional aerodynamic characteristics state points of the coordinated control of rudder-like and aileron-like of the starship in horizontal slip returning on the Weissman graph was analyzed. What is shown in Figure 27a is the Weissman diagram after the boundary was updated in 1980, and Figure 27b is the distribution diagram of the lateral-directional aerodynamic characteristics points on Weissman.

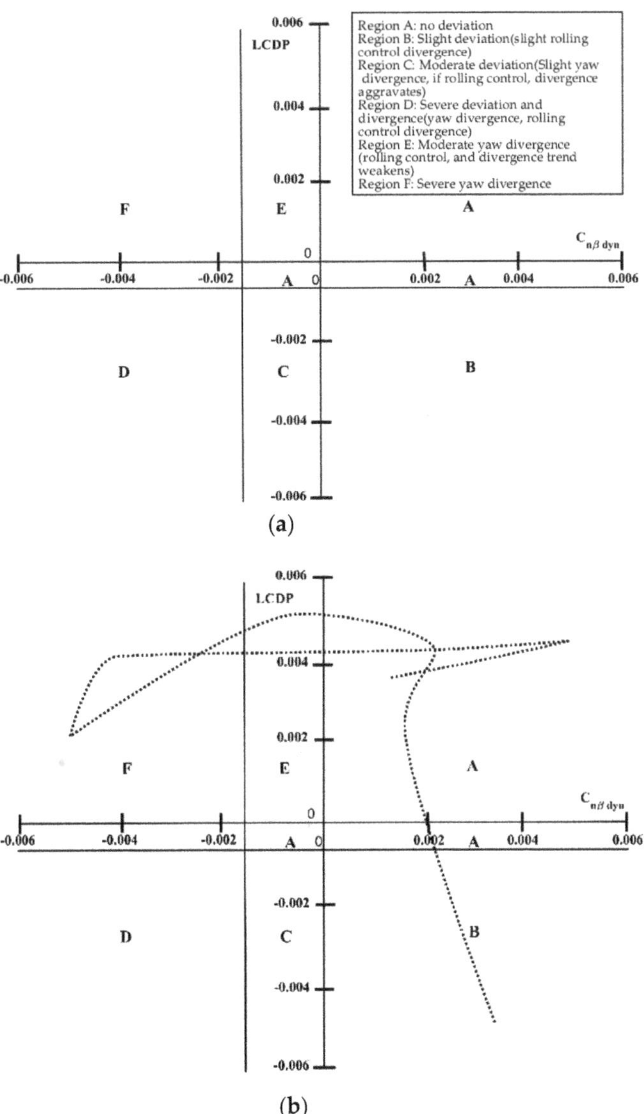

Figure 27. Weissman. (**a**) The Weissman diagram with updated boundary. (**b**) the distribution diagram of the lateral-directional aerodynamic characteristics points on Weissman.

In the figures, Region A is the no-deviation zone, Region B is a slight-deviation zone, Region C is a moderate-deviation zone, Region D is a heavy-deviation zone, Region E is a moderate -yaw -divergence zone (roll control is carried out, and divergence trend is weakened), and Region F is the strong -yaw -divergence zone.

Figure 27b shows that most states of starship lie in Area A, without lateral-directional deviation, and a portion of the state points falls in Area F, because during the returning stage with a large angle of attack, horizontal directional movement may be unstable, so the directional stabilization augmentation controller should be designed to move Area F overall to the right to Area E or the controllable part of Area A. There are only a few cases of slight yaw or divergence of rolling in the figure, and the divergence can be weakened by mutual compensation control.

6. Simulation Results of Nonlinear Open Loop

6.1. Longitudinal Simulation Results

Now the accuracy of the above analysis results has been verified by the time response simulation of the starship scaling model under $V = 60\ m/s, H = 3000\ m$. A unit step response is given to the rudder-like when the starship is in the trimming state. The response curves of angle of attack, pitch rate, and pitch angle over time are shown in Figures 28–30.

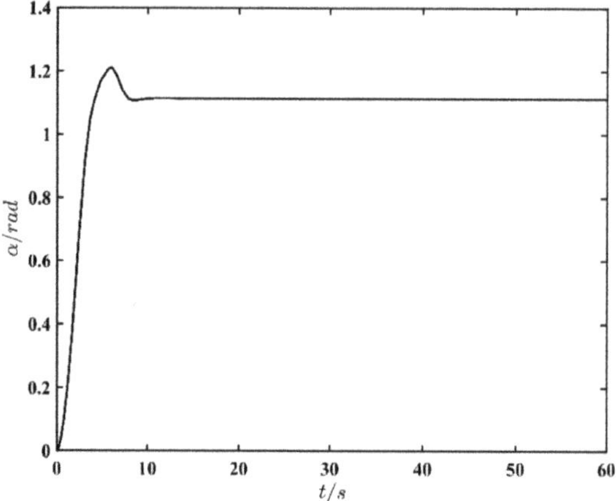

Figure 28. Angle of attack response over time.

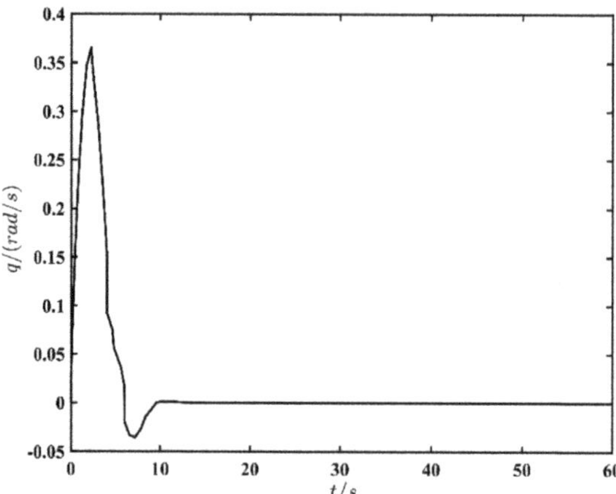

Figure 29. Pitch rate response over time.

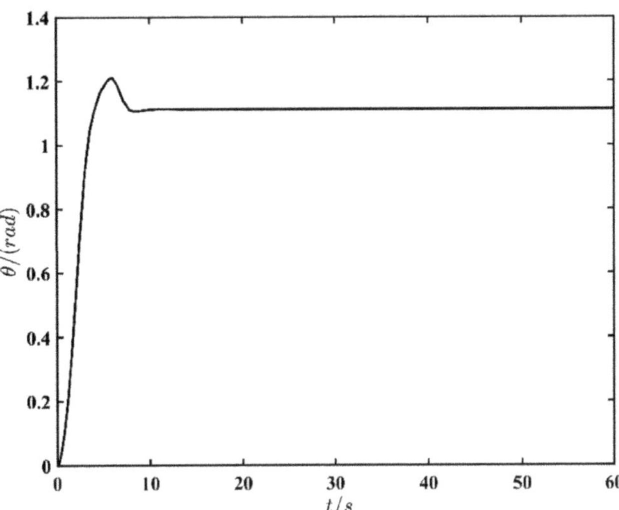

Figure 30. Pitch angle response with time.

It can be seen from the figures that short period characteristics of the starship are obvious during the unit step deflection of elevator-like, whose reflection on the angle of attack and pitch angular rate is that the starship could turn to the stable state quickly. At the same time, from the response curve of the pitch angle over time, the pitch angle also quickly returned to a stable state without accompanying oscillation, indicating that the long period degenerated into the third mode, which is an exponentially monotonous and convergent motion. Therefore, in the perspective of simulation results, it is consistent with the above analysis of control stability characteristics, and it verifies the accuracy of the above analysis.

6.2. Lateral-Directional Simulation Results

Now a unit step response is given to the aileron-like under $V = 60\ m/s$, H = 3000 m, $\beta = 0$, and the original $p = q = r = 0$ is ensured. The response curves along time of sideslip angle, roll angle, roll angular rate, and yaw angle are shown as follows.

Figures 31–34 show that after a step response is given to aileron-like, because the roll mode belongs to level 2 flying quality, this makes it slow for the roll angle to recover to the stable state. According to Figures 31 and 32, when the aileron-like steps, it takes the rolling angle 30 s to recover to a stable state. Because the lateral force caused by aileron-like deflection is small, coupling yaw motions are also small, and the yaw angle returns faster to the stable state than the roll angle. The starship is short of vertical tails, so its $C_{n\beta}$, which plays a role in the recovery of Dutch rolling motion, is small. Moreover, the $C_{l\beta}$ is also small, which leads the directional damping further reduced, resulting in the instability of the Dutch roll mode. Because the starship's spiral mode is close to its imaginary axis, and its roll mode is also located at the left of the imaginary axis, it could be approximately assumed that its roll and spiral modes are stable. Therefore, the half-life of the roll response is about 0.3 s.

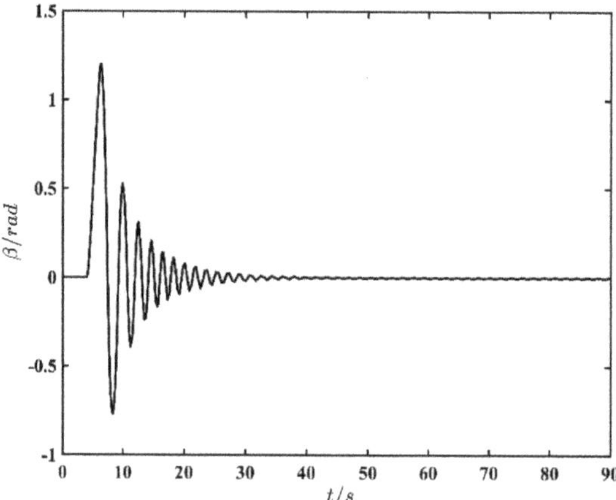

Figure 31. Sideslip angle response over time.

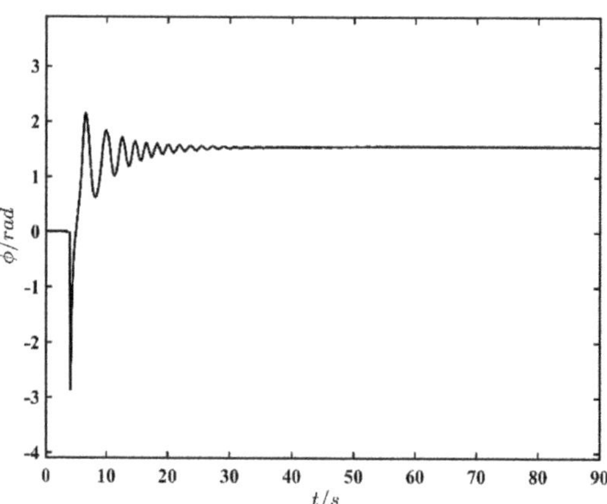

Figure 32. Roll angle response over time.

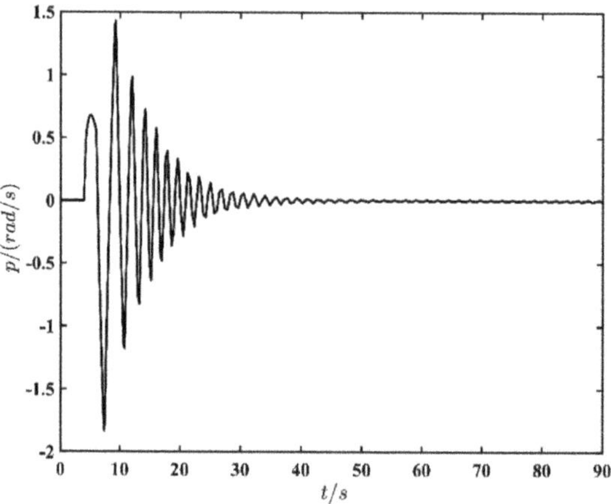

Figure 33. Roll rate response over time.

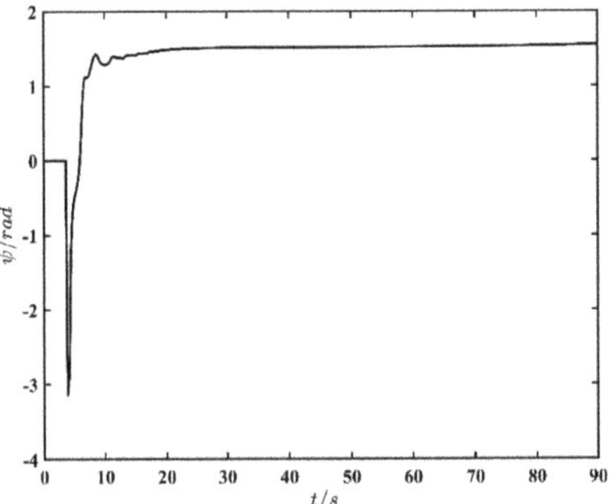

Figure 34. Yaw angle response with time.

Meanwhile, it can be seen from the analysis in the last section that the dynamic instability of the yaw motion of the starship is divergent, and by controlling the aileron, the sideslip could be eliminated, and the divergence trend could be weakened. As can be seen from Figures 30 and 33, the simulation results are consistent with the analysis mentioned in the last section.

Now a unit step response is given to the rudder-like under $V = 60\,\text{m/s}$, $H = 3000\,\text{m}$, $\beta = 0$, and at the same time, $p = q = r = 0$, the response curves of sideslip angle, yaw angle, roll angle, and yaw angle rate along the time are shown below.

According to the analysis in the previous section, sideslip can be eliminated by the combined control of aileron-like and rudder-like, so the sideslip angle in Figure 34 can recover to a stable state. However, when the rudder-like steps, the lateral-directional motion wholly diverges, which is caused by the instability of the Dutch roll mode of the starship itself. From the analysis of the roll and swing ratio characteristics in the last segment, it can be seen that the yaw motion accounts for a large proportion of the Dutch roll, so when the rudder-like steps, the yaw motion diverges. Meanwhile, according to the Weissman criterion, some of the state points of the starship fall in the strong yaw divergence zone during lateral–directional motions. It can be concluded that when the rudder-like steps, the simulation results are also consistent with the characteristics analysis. And the response curves, namely, sideslip angle, yaw, roll angle, yaw rate, versus time are showed as Figures 35–38 below.

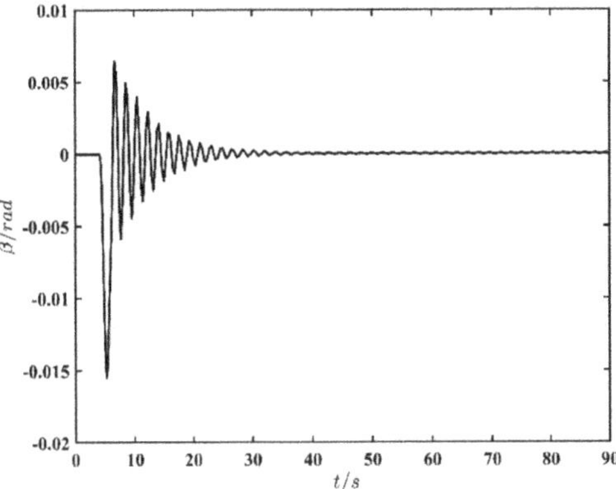

Figure 35. Sideslip angle response over time.

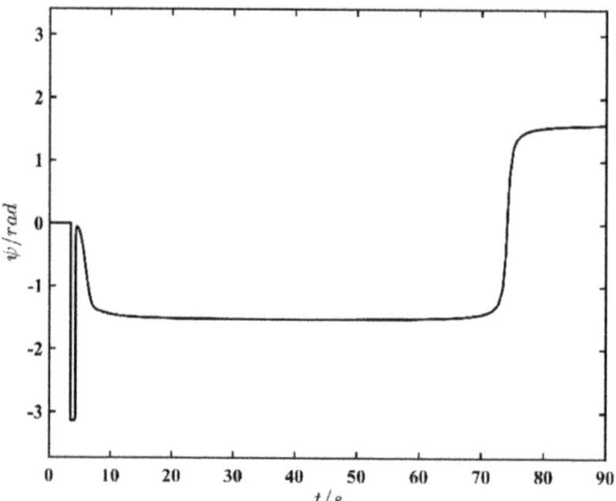

Figure 36. Yaw angle response with time.

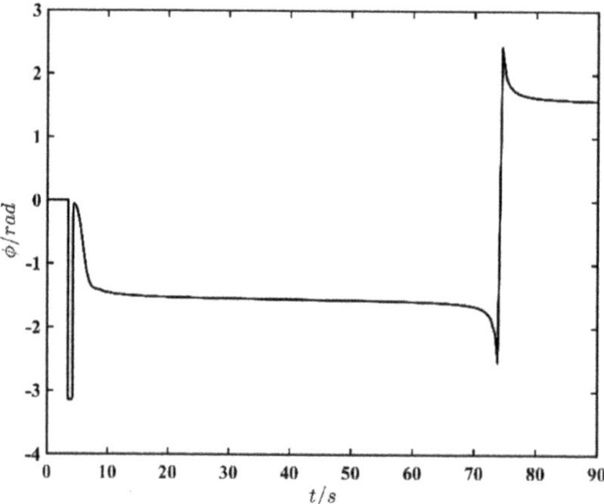

Figure 37. Roll angle response over time.

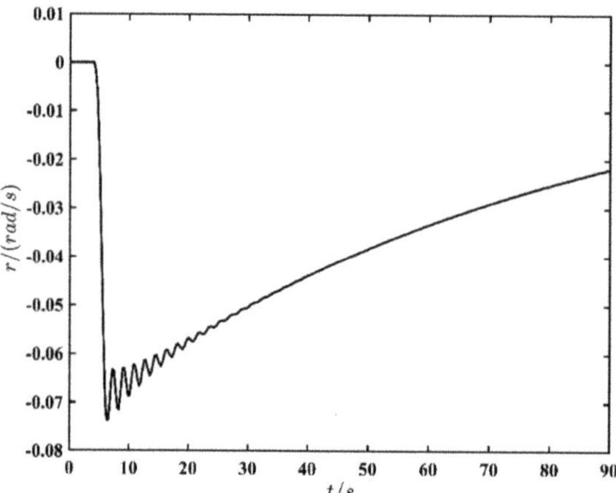

Figure 38. Yaw rate response over time.

As can be seen from the lateral-directional simulation results, the characteristics of the lateral-directional motions are consistent with the lateral-directional characteristics analysis in the previous section.

7. Conclusions

After the aerodynamic characteristics analysis, modal analysis, and deviation criterion analysis of the starship, the following conclusions are drawn:

The starship is longitudinally stable when $-20° < \alpha < 92°$, and becomes longitudinally statically unstable when $\alpha > 92°$, at which time the center of gravity moves to the focus. It does not have rolling static stability when $-20° < \alpha < 10°$. When $\alpha > 10°$, it has rolling static stability; The starship is always yaw statically stable within the range of angle of attack.

Both long and short longitudinal period mode of the starship meet the level 1 flight quality, and the motion response of the third mode after degradation is exponentially monotonous convergent. The lateral-directional roll mode meets level 2 flight quality, and the spiral mode meets level 1 flight quality. A pair of conjugate complex roots corresponding to the Dutch roll mode is in the right plane, which is in an unstable state. The Dutch roll mode is mainly reflected in yaw motion, which is divergent.

Through the comprehensive analysis of dynamic deviation and lateral control stability criteria, it can be seen that when $45° < \alpha < 88°$, if sideslip occurs, the anticontrol phenomenon is easy to occur, thus the design of directional control systems deserves attention.

The effect of combined control of rudder-like and aileron-like is better than that of single control of either rudder-like or aileron-like, and the best control effect is achieved when the compensation gain of rudder-like to aileron-like is 1. When the compensation gain is negative, the rudder-like cannot compensate the aileron-like.

The simulation result of open-loop ontology shows that when a unit step response is given to elevator-like and rudder-like and aileron-like, after the elevator-like and aileron-like step, the starship ontology can recover to a stable state, but after the rudder-like step, the lateral-directional motion of the whole starship diverges, which indicates that the direction of the starship ontology is unstable, and a directional stability augmentation controller is necessary to be designed.

Author Contributions: Methodology, Z.L.; formal analysis, C.Y. and M.D.; data curation, C.Z.; writing—original draft preparation, Z.W.; writing—review and editing, Z.G. All authors have read and agreed to the published version of the manuscript.

Funding: This research received no external funding.

Institutional Review Board Statement: Not applicable.

Informed Consent Statement: Not applicable.

Data Availability Statement: Not applicable.

Conflicts of Interest: The authors declare no conflict of interest.

Abbreviations

RLV	Reusable Launch Vehicle
CFD	Computational Fluid Dynamics
LCDP	Lateral Control Departure Parameter
AoA	Angle of Attack

References

1. Britt, R.T.; Arthurs, T.D.; Jacobson, S.B. Aeroservoelastic analysis of the B-2 bomber. *J. Aircr.* **2000**, *37*, 745–752. [CrossRef]
2. Jacobson, S.; Britt, R.; Freim, D.; Kelly, P. Residual pitch oscillation (rpo) flight test and analysis on the b-2 bomber. *Ices J. Mar. Sci.* **2003**, *67*, 1260–1271.
3. Su, W.; Gao, Z.; Xia, L. Multiobjective optimization design of aerodynamic configuration constrained by stealth performance. *Acta Aerodyn. Sin.* **2006**, *24*, 137–140.

4. Johnson, J.; Colbo, H. Space Shuttle main engine progress through the first flight. In Proceedings of the 17th Joint Propulsion Conference, Colorado Springs, CO, USA, 27–29 July 1981.

5. Klotz, I. SpaceX Aims for Starship Crew Flights By 2023. *Aerosp. Dly. Def. Rep.* **2020**, *272*, 3.

6. SpaceX Starship SN8 Takes Flight, 2020. Available online: https://www.spacex.com/updates/starship-sn8-takes-flight/index.html (accessed on 10 June 2022).

7. SpaceX. Starship SN9 High-Altitude Flight Test, 2021. Available online: https://www.spacex.com/updates/starship-sn9-flight-test/index.html (accessed on 10 June 2022).

8. SpaceX. Starship SN10 High-Altitude Flight Test, 2021. Available online: https://www.spacex.com/updates/starship-sn10-flight-test/index.html (accessed on 10 June 2022).

9. SpaceX. Starship SN11 High-Altitude Flight Test, 2021. Available online: https://www.spacex.com/updates/starship-sn11-flight-test/index.html (accessed on 10 June 2022).

10. SpaceX. StarShip SN15, 2021. Available online: https://www.spacex.com/updates/starship-sn15-flight-test/index.html (accessed on 10 June 2022).

11. Klotz, I. Starship Hop on Hold Pending Demo-2. *Aerosp. Dly. Def. Rep.* **2020**, *272*, 2.

12. Norris, G. SpaceX Plans Pacific Splashdown for First Starship Orbital Test. *Aerosp. Dly. Def. Rep.* **2021**, *276*, 6.

13. Goman, M.G.; Khramtsovsky, A.V.; Kolesnikov, E.N. Evaluation of Aircraft Performance and Maneuverability by Computation of Attainable Equilibrium Sets. *J. Guid. Control Dyn.* **2008**, *31*, 329–339. [CrossRef]

14. Klotz, I. Starship Lands, Then Bursts into Fireball. *Aviat. Week Space Technol.* **2021**, *183*, 30.

15. Klotz, I. SpaceX Starship Prototype Passes Static Test Fire. *Aerosp. Dly. Def. Rep.* **2020**, *272*, 4.

16. Palmer, C. SpaceX Starship Lands on Earth, but Manned Missions to Mars will Require More. *Engineering* **2021**, *7*, 1345–1347. [CrossRef]

17. Owen, D.B. The starship. *Commun. Stat.-Simul. Comput.* **1988**, *17*, 315–323. [CrossRef]

18. Klotz, I. Third Starship Prototype Lost During Cryogenic Tanking Test. *Aerosp. Dly. Def. Rep.* **2020**, *272*, 2.

19. Norris, G. SpaceX Sticks Starship Landing After High-Altitude Flight Test. *Aerosp. Dly. Def. Rep.* **2021**, *276*, 1–2.

20. Weiss, D. Creating Materials for the Starship. *J. Br. Interplanet. Soc. (JBIS)* **2015**, *68*, 211–213.

21. Galea, P. Machine Learning and the Starship—A Match Made in Heaven. *J. Br. Interplanet. Soc. (JBIS)* **2012**, *65*, 278–282.

22. Norris, G. SpaceX's Star Hopper Verifies Raptor Performance for Starship. *Aviat. Week Space Technol.* **2019**, *181*, 28.

23. Klotz, I. SpaceX Starship Crashes after High-Altitude Flight Test. *Aerosp. Dly. Def. Rep.* **2020**, *274*, 3.

24. Seedhouse, E. Starship. In *SpaceX*; Springer: Cham, Switzerland, 2022; pp. 171–188.

25. Klotz, I. SpaceX Prototype Starship Survives Low-Altitude Hop. *Aerosp. Dly. Def. Rep.* **2020**, *273*, 3.

26. Hayes, C.; Starship Troopers. ACM SIGGRAPH 98 Conference Abstracts and Applications. 1998. p. 311. Available online: https://dl.acm.org/doi/pdf/10.1145/280953.282447 (accessed on 10 June 2022).

27. Wang, Z.; Mao, S.; Gong, Z. Energy Efficiency Enhanced Landing Strategy for Manned eVTOLs Using L1 Adaptive Control. *Symmetry* **2021**, *13*, 2125. [CrossRef]

28. Zhang, Y.J.; Zuo, G. Numerical simulation on aerodynamic characteristics of new type control surface of Starship. *Acta Aeronaut. Astronaut. Sin.* **2021**, *42*, 624058. (In Chinese)

29. Parsons, D.G.; Levin, D.E.; Panteny, D.J.; Wilson, P.N.; Rask, M.R.; Morris, B.L. F-35 STOVL Performance Requirements Verification. In *The F-35 Lightning II: From Concept to Cockpit*; American Institute of Aeronautics and Astronautics: Reston, VA, USA, 2022; pp. 641–680.

30. Walker, G.; Allen, D. X-35B STOVL Flight Control Law Design and Flying Qualities. In Proceedings of the 2002 Biennial International Powered Lift 641 Conference and Exhibit, Williamsburg, VA, USA, 5–7 November 2002; pp. 1–13.

31. Osder, S.; Caldwell, D. Design and robustness issues for highly augmented helicopter controls. *J. Guid. Control Dyn.* **1992**, *15*, 1375–1380. [CrossRef]

32. Carter, J.; Stoliker, P. Flying quality analysis of a JAS 39 Gripen ministick controller in an F/A-18 aircraft. In Proceedings of the AIAA Guidance, Navigation, and Control Conference and Exhibit, Dever, CO, USA, 14–17 August 2000; pp. 1–20.

33. Yang, X.; Fan, Y.; Zhu, J. Transition Flight Control of Two Vertical/Short Takeoff and Landing Aircraft. *J. Guid. Control Dyn.* **2008**, *31*, 371–385.

34. Gerontakos, P.; Lee, T. Trailing-edge flap control of dynamic pitching moment. *AIAA J.* **2007**, *45*, 1688–1694. [CrossRef]

35. Andrews, J.; Andrews, D. Designing reusable launch vehicles for future space markets. In Proceedings of the AIAA Space 2001 Conference and Exposition, Albuquerque, NM, USA, 28–30 August 2001; p. 4544.

36. Zhong, Y.; Liu, D.; Wang, C. Research progress of key technologies for typical reusable launcher vehicles. In *IOP Conference Series: Materials Science and Engineering*; IOP Publishing: Bristol, UK, 2018; Volume 449, p. 012008.

Article

Design of Thrust Vectoring Vertical/Short Takeoff and Landing Aircraft Stability Augmentation Controller Based on L1 Adaptive Control Law

Zan Zhou [1], Zian Wang [2,*], Zheng Gong [1], Xiong Zheng [2], Yang Yang [2] and Pengcheng Cai [1]

[1] Department of Aerospace Engineering, Nanjing University of Aeronautics and Astronautics, Nanjing 210016, China
[2] China Academy of Launch Vehicle Technology, Beijing 100076, China
* Correspondence: wangzian@nuaa.edu.cn

Abstract: Aiming at the conversion process of thrust vectoring vertical/short takeoff and landing (V/STOL) aircraft with a symmetrical structure in the transition stage of takeoff and landing, there is a problem with the coupling and redundancy of the control quantities. To solve this problem, a corresponding inner loop stabilization controller and control distribution strategy are designed. In this paper, a dynamic system model and a dynamic model are established. Based on the outer loop adopting the conventional nonlinear dynamic inverse control, an L1 adaptive controller is designed based on the model as the inner loop stabilization control to compensate the mismatch and uncertainty in the system. The key feature of the L1 adaptive control architecture is ensuring robustness in the presence of fast adaptation, so as to achieve a unified performance boundary in transient and steady-state operations, thus eliminating the need for adaptive rate gain scheduling. The control performance and robustness of the controller are verified by inner loop simulation and the shooting Monte Carlo approach. The simulation results show that the controller can still track the reference input well and has good robustness when there is a large parameter perturbation.

Keywords: L1 adaptive control; V/STOL aircraft; Monte Carlo simulations

Citation: Zhou, Z.; Wang, Z.; Gong, Z.; Zheng, X.; Yang, Y.; Cai, P. Design of Thrust Vectoring Vertical/Short Takeoff and Landing Aircraft Stability Augmentation Controller Based on L1 Adaptive Control Law. *Symmetry* **2022**, *14*, 1837. https://doi.org/10.3390/sym14091837

Academic Editor: Jan Awrejcewicz

Received: 10 July 2022
Accepted: 26 August 2022
Published: 4 September 2022

Publisher's Note: MDPI stays neutral with regard to jurisdictional claims in published maps and institutional affiliations.

1. Introduction

Thrust vectoring technology can directly change the thrust magnitude and thrust direction of aircraft, which is an important technical scheme to achieve the high maneuverability of modern aircraft. Unlike conventional fighter jets, vertical/short takeoff and landing (V/STOL) aircraft is a new type of aircraft [1–3]. It can not only realize the vertical takeoff and landing, but also carry out a high-speed cruise in a conventional aircraft configuration and will be widely used in the military field in the future. Therefore, the thrust vectoring vehicle has the characteristics of a large flight space, complex flight action, and diverse flight tasks, resulting in the strong nonlinearity of its control system and severe changes in the external environment. How to design a control scheme that can deal with this large uncertainty is a key problem in the control design of a thrust vectoring vehicle.

Regarding the aspect of mathematical model establishment and simulation, Songlin Ma and Weijun Wang [4] established a longitudinal model in a transition flight phase for a new concept of vertical/short takeoff and landing aircraft. Xiaomeng Zhang and Weijun Wang [5] analyzed the dynamics of the vertical/short takeoff and landing of unmanned aerial UAVs and obtained a nonlinear dynamics model. References [6,7] investigated the thrust vector control of thrust vectoring V/STOL aircraft using a six-degree-of-freedom flight dynamics simulation. For controller design, Yang, Xili et al. [8] proposed two nonlinear approaches for the autonomous transition control of two vertical/short takeoff and landing aircraft. Walker, G. and Allen, D. [9] presented an overview of the X-35B control law requirements, design, analysis, and summary of the STOVL flight test results. In Reference [10], a control scheme comprising the dynamic characteristics of the thrust vectoring system was developed for V/STOL aircraft. Simulation and experimental results were presented. Zhiqiang Cheng et al. [11] proposed an optimal trajectory transition controller. Reference [12] showed an application of L1 adaptive control theory for the attitude control of UAVs. Chiang R Y, et al. [13] presented an H-∞ flight control system design case study for a supermaneuverable fighter flying quality of the Herbst maneuver, which may provide some reference for hovering state flying quality. Zian, Wang et al. [14] used an L1 adaptive inner loop controller and designed a roll-horizon landing deceleration and landing strategy for hybrid-wing vehicles. Seshagiri S, et al. [15] considered the application of a conditional integrator-based sliding mode control design for robust regulation of minimum-phase nonlinear systems for the control of the longitudinal flight dynamics of an F-16 aircraft. Liu, N, et al. [16] used the nonlinear active disturbance rejection controller (ADRC) to control the tilt wings. In terms of control allocation, Min, B.M, et al. [17] focused on applying various control allocation schemes to the SAT-II UAV system. Tan J, et al. [18] studied attitude tracking UAVs with the terminal sliding mode based on the extended state observer and with the multi-objective nonlinear control allocation. For the problem of simulation verification, References [19,20] provided a practical Monte Carlo method to verify the robustness of the controller.

In the transition stage of the thrust vectoring V/STOL aircraft, the tilt angle of the vectoring nozzle and the opening of the lift fan will change greatly. In this power conversion process, both the aerodynamic rudder surface and the power system of the aircraft can control the six degrees of freedom of the aircraft. At this stage, the aircraft will face the problems of strong nonlinearity, control quantity coupling, and redundancy. The design of the control law is a great challenge. Under the physical constraints of the thrust vectoring vehicle, the aerodynamic control inputs and thrust vectoring control inputs are allocated according to the virtual control variables. To solve this problem, based on the F35B scale model, the dynamic equation modeling is given in this paper. On this basis, the inner loop controller and control distribution method are designed. Finally, the performance and robustness of the controller are verified by inner loop simulation and the shooting Monte Carlo approach.

The remainder of this paper is organized as follows. Section 2 introduces the composition of the thrust vector V/STOL aircraft power system and establishes the dynamic equation modeling by combining the whole vehicle dynamics equation, kinematic equation, and moment equation. Section 3 introduces the design of the inner loop and outer loop controllers in detail, and the control allocation method is given. The robustness of the inner loop controller is simulated and verified by Monte Carlo simulation in Section 4, and conclusions and recommendations for future work are stated in Section 5

2. Dynamic Equation Modeling

The thrust vectoring vertical/short takeoff and landing aircraft used in this paper is a self-made F35B scale model with a total weight of about 13 kg, a cruising speed of 30 m/s, and a cruising altitude of less than 100 m.

The F35B consists of a fuselage, wings, tail, and power system as shown in Figure 1. The F35B mainly has the fixed-wing mode and vertical takeoff and landing (VTOL) mode. The platform's aerodynamic control surfaces include a full-motion horizontal tail, a full-motion vertical tail, and flaperons. The power system consists of two subsystems: the lift fan and auxiliary motor subsystem installed in the front section of the fuselage; the main ducted fan and the main motor subsystem installed in the rear section of the fuselage; a three-bearing swivel duct (3BSD) nozzle is connected behind the main ducted fan.

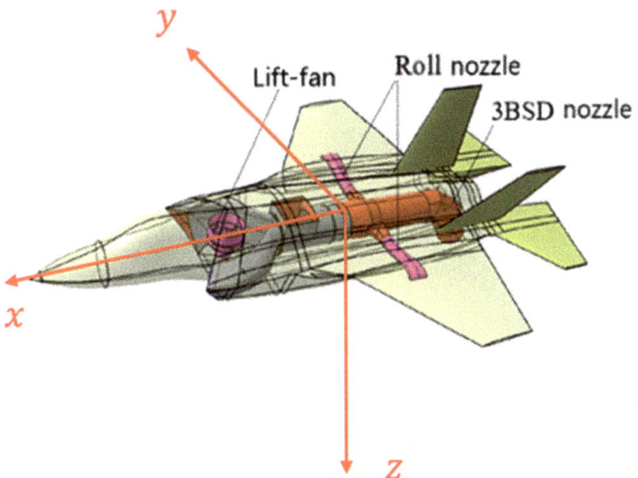

Figure 1. F35 model.

2.1. Whole Vehicle Dynamics Model

We use $[T_x\ T_y\ T_z]$ to denote the thrust components of the engine under the three shafts. The aerodynamic force A is defined in the airflow coordinate system, and considering the lift loss in the transition section, the total aerodynamic force can be expressed as Equation (1), where D is the drag force, C is the side force, and L is the lift force.

$$
\begin{bmatrix} A_x \\ A_y \\ A_z \end{bmatrix} = \begin{bmatrix} -D\cos\alpha\cos\beta - C\cos\alpha\sin\beta + L\sin\alpha \\ -D\sin\beta + C\cos\beta \\ -D\sin\alpha\cos\beta - C\sin\alpha\sin\beta - L\cos\alpha \end{bmatrix} \tag{1}
$$

The direction of the aircraft's gravity is given in the z-axis of the ground coordinate system, which is also converted to the airframe coordinate system.

$$
m\begin{bmatrix} g_x \\ g_y \\ g_z \end{bmatrix} = L_{bg}m\begin{bmatrix} 0 \\ 0 \\ g \end{bmatrix} = m\begin{bmatrix} -g\sin\theta \\ g\sin\phi\cos\theta \\ g\cos\phi\cos\theta \end{bmatrix} \tag{2}
$$

Combined with Equations (1) and (2), the linear motion equation of the aircraft in the body coordinate system can be determined from Equation (3).

$$
\begin{cases} m\frac{du}{dt} = T_x - D\cos\alpha\cos\beta - C\cos\alpha\sin\beta \\ \quad + L\sin\alpha - mg\sin\theta + mvr - mwq \\ m\frac{dv}{dt} = T_y - D\sin\beta + C\cos\beta \\ \quad + mg\sin\phi\cos\theta - mur + mwp \\ m\frac{dw}{dt} = T_z - D\sin\alpha\cos\beta - C\sin\alpha\sin\beta \\ \quad - L\cos\alpha + mg\cos\phi\cos\theta + muq - mvp \end{cases} \tag{3}
$$

where $[u, v, w]$ are the velocity components of the three coordinate axes in the body coordinate system and $[p, q, r]$ are roll angle velocity, pitch angle velocity, and yaw angle velocity in the body coordinate system, respectively.

2.2. Kinematic Equations and Moment Equation

The kinematic equations and moment equation of the aircraft are [9]:

$$\begin{cases} \dot{\phi} = p + (r\cos\phi + q\sin\phi)\tan\theta \\ \dot{\theta} = q\cos\phi - r\sin\phi \\ \dot{\psi} = \frac{1}{\cos\theta}(r\cos\phi + q\sin\phi) \end{cases} \tag{4}$$

$$\begin{cases} \dot{p} = (c_1 r + c_2 p)q + c_3 L + c_4 N \\ \dot{q} = c_5 pr - c_6(p^2 - r^2) + c_7 M \\ \dot{r} = (c_8 p - c_2 r)q + c_4 L + c_9 N \end{cases} \tag{5}$$

where $[\phi, \theta, \psi]$ are the roll angle, pitch angle, and yaw angle of the aircraft, respectively; $c_1 \sim c_9$ are the moment equation coefficients; $[L, M, N]$ are roll moment, pitch moment, and yaw moment in the body coordinate system, respectively.

Based on Equations (3)–(5), the nonlinear dynamics model of the aircraft can be derived:

$$\begin{cases} m\dot{u} = T_x - D\cos\alpha\cos\beta - C\cos\alpha\sin\beta \\ \quad +L\sin\alpha - mg\sin\theta + mvr - mwq \\ m\dot{v} = T_y - D\sin\beta + C\cos\beta \\ \quad +mg\sin\phi\cos\theta - mur + mwp \\ m\dot{w} = T_z - D\sin\alpha\cos\beta - C\sin\alpha\sin\beta \\ \quad -L\cos\alpha + mg\cos\phi\cos\theta + muq - mvp \\ \dot{p} = (c_1 r + c_2 p)q + c_3 L + c_4 N \\ \dot{q} = c_5 pr - c_6(p^2 - r^2) + c_7 M \\ \dot{r} = (c_8 p - c_2 r)q + c_4 L + c_9 N \\ \dot{\phi} = p + (r\cos\phi + q\sin\phi)\tan\theta \\ \dot{\theta} = q\cos\phi - r\sin\phi \\ \dot{\psi} = \frac{1}{\cos\theta}(r\cos\phi + q\sin\phi) \end{cases} \tag{6}$$

3. L1 Stabilization Controller Design

The stability augmentation controller in this paper constitutes a control and stability augmentation system (CSAS) by controlling the roll, pitch, and yaw of the aircraft. The CSAS can generate angular velocity and acceleration control commands and provides torque commands to the control distributor. Then, a single-input single-output (SISO) L1 adaptive controller is designed for each of the three angular velocity channels to compensate for mismatch uncertainties in the dynamics [21–23]. For outer-loop designs such as pitch and roll, nonlinear dynamic inversion (NDI) provides the tracking of the required dynamics. Figure 2 shows the adaptive control flow diagram using NDI and the L1 adaptive control law.

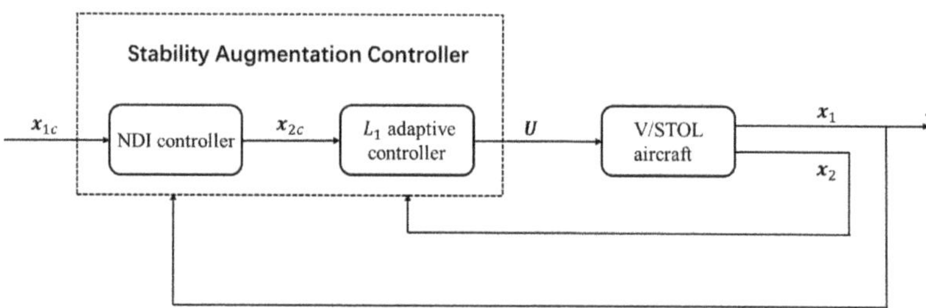

Figure 2. Flow chart of L1 adaptive controller.

In Figure 2, the outer loop state variable $x_1 = [\phi\,\theta\,\psi]^T$ and the inner loop state variable $x_2 = [p\,q\,r]^T$; the control inputs of the V/STOL aircraft are $U_1 = [\delta_a\,\delta_e\,\delta_r\,\delta_T\,\delta_L\,\delta_{az}\,\delta_{ac}\,\delta_{Tl}\,\delta_{Tr}]^T$;

$\delta_a, \delta_e, \delta_r$ are the aileron deflection, elevator deflection, and rudder deflection, respectively; δ_L is the ratio of the lift fan thrust to the maximal lift fan thrust; δ_T is the ratio between the maximum three-bearing swivel duct nozzle thrust and the three-bearing swivel duct nozzle thrust; δ_{ac} is the three-bearing swivel duct nozzle's pitch angle, when the aircraft is in fixed-wing mode, $\delta_{ac} = 0°$, and in vertical takeoff and landing mode, $\delta_{ac} = 90°$; δ_{az} is the yaw angle of the three-bearing swivel duct nozzle, as shown in Figure 3; δ_{Tl}, δ_{Tr} are the thrust of the left and right roll nozzles, respectively. The actuator characteristics are described in Table 1.

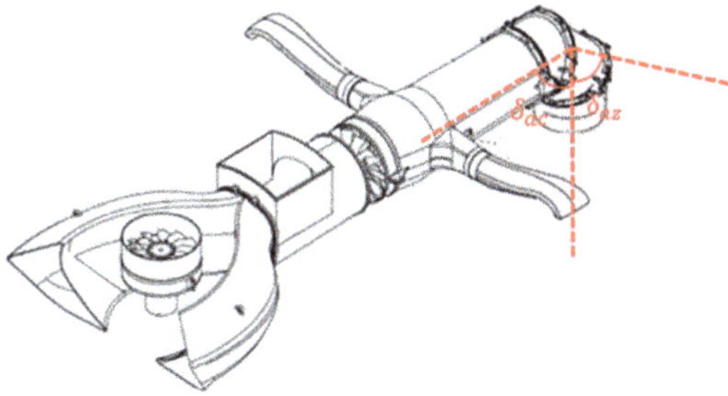

Figure 3. Power system diagram.

Table 1. Effector characteristics of the F35B.

Effector	Position Limit	Rate Limit
$\delta_a, \delta_e, \delta_r$	$[-30, 30]$ deg	± 30 deg/s
δ_T, δ_L	$[0, 1]$	± 0.4
δ_{ac}	$[0, 90]$	± 40 deg/s
δ_{az}	$[-12, 12]$ deg	± 40 deg/s
δ_{Tl}, δ_{Tr}	$[0, 1000]$ N	± 4000 N/s

The pitch and roll channels realize the decoupling of the angle and angular velocity according to the timescale separation principle. The attitude angle is controlled by the NDI method, while the L1 adaptive controller controls the angular velocity. The L1 adaptive law can be divided into four parts: control object, state predictor, adaptive law, and control law.

3.1. Inner Loop Controller Design

3.1.1. NDI Controller Design of Roll Loop

First, the NDI controller needs to be designed for the roll loop. The kinematic equation of the outer loop is:

$$\dot{x}_1 = f_1(x_1, t) + g_1(x_1, t)x_2 \tag{7}$$

where $f_1(x_1, t)$, $g_1(x_1, t)$ are affine functions satisfying $f_1(x_1, t) = 0_{3\times 1}$ and $g_1(x_1, t) = \begin{bmatrix} 1 & \sin\phi\tan\theta & \cos\phi\tan\theta \\ 0 & \cos\phi & \sin\phi \\ 0 & \sin\phi/\cos\theta & \cos\phi/\cos\theta \end{bmatrix}$.

The error is defined as $\Delta x_1 = x_{1c} - x_1$. Based on the NDI method, the angular velocity command p_c can be obtained:

$$x_{2c} = g_1^{-1}(x_1, t)(K_1 \Delta x_1 - f_1(x_1, t)) \tag{8}$$

where $K_1 \in R^{3\times 3}$ is the bandwidth coefficient and is a diagonal matrix, which is determined by the flight quality of the aircraft and the outer loop bandwidth.

3.1.2. L1 Adaptive Controller Design of Roll Angular Velocity Loop

According to the small perturbation theory, the kinematic equation of the outer loop Equation (7) can be simplified as

$$
\dot{x}_2 = \begin{bmatrix} L_\beta & L_p \\ M_\alpha & M_q \\ N_\beta & N_r \end{bmatrix} \begin{bmatrix} \beta & \alpha & \beta \\ p & q & r \end{bmatrix} + \begin{bmatrix} L_{\dot{p}_{uc}} & 0 & 0 \\ 0 & M_{\dot{q}_{uc}} & 0 \\ 0 & 0 & N_{\dot{r}_{uc}} \end{bmatrix} \dot{x}_{2uc} \tag{9}
$$

where $\dot{x}_{2uc} = \begin{bmatrix} \dot{p}_{uc} & \dot{q}_{uc} & \dot{r}_{uc} \end{bmatrix}^T$ is the virtual angular acceleration, $\begin{bmatrix} L_\beta & L_p \\ M_\alpha & M_q \\ N_\beta & N_r \end{bmatrix}$ are the

aerodynamic coefficients, and $L_{\dot{p}_{uc}}, M_{\dot{q}_{uc}}, N_{\dot{r}_{uc}}$ are the virtual control torque generated by the virtual output signal $\dot{x}_{2uc}$. $\dot{x}_{2uc}$ can be solved by Equation (9) and expressed as:

$$
\dot{x}_{2uc} = \dot{x}_{2Tc} + \dot{x}_{2ac} \tag{10}
$$

where $\dot{x}_{2ac}$ and $\dot{x}_{2Tc}$ are the roll angular accelerations generated by the air surface and the vector nozzle, respectively. At low airspeeds, the control torque provided by the vectoring nozzle dominates. The vector nozzle provides the required pitch acceleration, i.e., $\dot{x}_{2Tc} \triangleq \dot{x}_{2uc}, \dot{x}_{2ac} = 0$. At high airspeeds, the aerodynamic surfaces subject to the required pitch angular acceleration, i.e., $\dot{x}_{2ac} \triangleq \dot{x}_{2uc}, \dot{x}_{2Tc} = 0$. During the transitional flight, both the aerodynamic surfaces and the vectoring nozzle achieve the commanded roll angular acceleration.

Considering the uncertainty of moment, Equation (9) can be expressed as:

$$
\begin{aligned}
\dot{x}_2 =\ & \begin{bmatrix} L_\beta + \hat{L}_\beta & L_p + \hat{L}_p \\ M_\alpha + \hat{M}_\alpha & M_q + \hat{M}_q \\ N_\beta + \hat{N}_\beta & N_r + \hat{N}_r \end{bmatrix} \begin{bmatrix} \beta & \alpha & \beta \\ p & q & r \end{bmatrix} \\
& + \begin{bmatrix} L_{\dot{p}_{uc}} + \hat{L}_{\dot{p}_{uc}} & 0 & 0 \\ 0 & M_{\dot{q}_{uc}} + \hat{M}_{\dot{q}_{uc}} & 0 \\ 0 & 0 & N_{\dot{r}_{uc}} + \hat{N}_{\dot{r}_{uc}} \end{bmatrix} \begin{bmatrix} \dot{p}_{uc} \\ \dot{q}_{uc} \\ \dot{r}_{uc} \end{bmatrix} + \sigma
\end{aligned} \tag{11}
$$

where $\hat{L}_\beta, \hat{L}_p, \hat{L}_{\dot{p}_{uc}}, \hat{M}_\alpha, \hat{M}_q, \hat{M}_{\dot{q}_{uc}}, \hat{N}_\beta, \hat{N}_r, \hat{N}_{\dot{r}_{uc}}$ are the uncertain aerodynamic factors and $\sigma = \begin{bmatrix} \sigma_p & \sigma_q & \sigma_r \end{bmatrix}^T$ is the disturbance factor. The structure of the first-order reference model is as follows:

$$
\dot{x}_2 = K_2 \Delta x_2 = K_2 (x_{2c} - x_2) \tag{12}
$$

Combining Equations (9)–(12), we can write the control model of the rolling loop as:

$$
\begin{cases}
\dot{x}_2 = -K_2 x_2 + K_2 \eta \\
\eta = \omega_{x_2} \dot{x}_2 + f_2(t, x_2) \\
f_2(t, x_2) = \lambda x_2 + \sigma_{x_2}
\end{cases} \tag{13}
$$

where $K_2(k_p, k_q, k_r) \in R^{3 \times 3}$ is the feedback gain matrix, $\omega_{x_2} = \begin{bmatrix} 1 + \dfrac{\hat{L}_{\dot{p}_{uc}}}{L_{\dot{p}_{uc}}} & 1 + \dfrac{\hat{M}_{\dot{q}_{uc}}}{M_{\dot{q}_{uc}}} & 1 + \dfrac{\hat{N}_{\dot{r}_{uc}}}{N_{\dot{r}_{uc}}} \end{bmatrix}$

is the virtual control factor, $\lambda = diag(\hat{L}_p - \dfrac{\hat{L}_{\dot{p}_{uc}}}{L_{\dot{p}_{uc}}} L_p, \hat{M}_q - \dfrac{\hat{M}_{\dot{q}_{uc}}}{M_{\dot{q}_{uc}}} M_q, \hat{N}_r - \dfrac{\hat{N}_{\dot{r}_{uc}}}{N_{\dot{r}_{uc}}} N_r)$ is the aero-

dynamic factor, and $\sigma_{x_2} = diag(\hat{L}_\beta - \dfrac{\hat{L}_{\dot{p}_{uc}}}{L_{\dot{p}_{uc}}} L_\beta, \hat{M}_\alpha - \dfrac{\hat{M}_{\dot{q}_{uc}}}{M_{\dot{q}_{uc}}} M_\alpha, \hat{N}_\beta - \dfrac{\hat{N}_{\dot{r}_{uc}}}{N_{\dot{r}_{uc}}} N_\beta) x_2 + \sigma$ is the aerodynamic disturbance.

Three assumptions and a lemma are given.

Assumption 1. *The unknown constant ω_{x_2} is uniformly bounded, i.e., $\omega_{x_2} \in \Omega \subset R^{3 \times 3}$, Ω is a compact convex set.*

Assumption 2. $f_2(t, x_2)$ is uniformly bounded in Equation (19), such as $\|f_2(t,0)\|_\infty \leq b, b > 0,$ where $\|\bullet\|_\infty$ is the ∞-norm.

Assumption 3. The partial derivatives of f_2 are semi-globally uniformly bounded: for $\delta > 0$, there exist $d_{fp}(\delta) > 0$ and $d_{ft}(\delta) > 0$ independent of time t to ensure that the partial derivatives of $f_2(t, x_2)$ are piecewise continuous and bounded, as follows:

$$\begin{cases} \|\frac{\partial f_2(t,x_2)}{\partial x_2}\|_\infty \leq d_{fp}(\delta) \\ \|\frac{\partial f_2(t,x_2)}{\partial t}\|_\infty \leq d_{ft}(\delta) \end{cases} \tag{14}$$

The assumptions can be met since the uncertainty for the inner loop control system is of a given magnitude.

Lemma 1. For $\tau > 0$, if $\|x_\tau\|_{L_\infty} \leq \rho$ and $\|\dot{x}_\tau\|_{L_\infty} \leq d_x$, where ρ and d_x are positive constants, λ and σ_{x_2} are continuous. Furthermore, their derivatives with respect to $t \in [0, \tau]$ are:

$$\begin{cases} f_2(t, x_2) = \lambda \|x\|_{L_\infty} + \sigma_{x_2} \\ \|\lambda\| < d_{fq}(\rho), \|\dot{\lambda}\| \leq d_\lambda \\ \|\sigma_{x_2}\| < b, \|\dot{\sigma}_{x_2}\| \leq d_\sigma \end{cases} \tag{15}$$

where d_λ and d_σ are computable finite values; $\|\bullet\|_{L_\infty}$ is the L_∞-norm.

Based on the above assumptions and lemma, an adaptive law controller is designed, which consists of a state predictor, an adaptive law, and a control law:

- State predictor:

According to Equation (13), the state predictor can be expressed as:

$$\begin{cases} \dot{\hat{x}}_2 = -K_2\hat{x}_2 + K_2\hat{\eta} \\ \hat{\eta} = \dot{x}_{2c}\hat{\omega}_{x_2} + \hat{\lambda}x_2 + \hat{\sigma}_{x_2} \\ \hat{y} = \hat{x}_2 \end{cases} \tag{16}$$

where $\hat{\omega}_{x_2}$ is the predicted level of control factor uncertainty, $\hat{\lambda}$ is the expected degree of the aerodynamic factor's uncertainty, and $\hat{\sigma}_{x_2}$ is the predicted degree of the aerodynamic disturbance's uncertainty; $\hat{\omega}_{x_2}$ denotes the estimated level of control factor uncertainty, B the estimated level of aerodynamic factor uncertainty, and C the estimated level of aerodynamic disturbance uncertainty.

- Law of adaptation:

$$\begin{cases} \dot{\hat{\lambda}} = \Gamma K_{proj}(\hat{\lambda}, -\tilde{x}_2 PK_2 \|x_2\|_\infty) \\ \dot{\hat{\sigma}}_{x_2} = \Gamma K_{proj}(\hat{\lambda}, -\tilde{x}_2 PK_2) \\ \dot{\hat{\omega}}_{x_2} = \Gamma K_{proj}(\hat{\omega}_{x_2}, -\tilde{x}_2 PK_2\dot{x}_2) \end{cases} \tag{17}$$

where Γ is the adaptive gain, $\tilde{x}_2 = \hat{x}_2 - x_2$ is the prediction error, and P is the solution of the Lyapunov equation $K_2^T P + PK_2 = -Q$ $(Q = Q^T > 0)$. K_{proj} is the projection operator.

- Control law:

The control law is as follows:

$$\dot{x}_{2c} = K_d D(K_g x_2 - \hat{\eta}) \tag{18}$$

where K_g is the adaptive feedback gain, D is the low-pass filter, and K_d is the adaptive feedforward gain. The L1 adaptive control scheme is shown in Figure 3. Control laws should be designed to ensure that the following transfer functions are strictly regular:

$$C(s) = \omega_{x_2} K_d D(s)(I + \omega_{x_2} K_d D(s))^{-1} \tag{19}$$

where $C(s)$ satisfies $C(0) = I$ and I is a 3×3 identity matrix.

When obtaining K_d and D, the following conditions must be satisfied to ensure the stability of adaptive control:

$$\|G(s)\|_{L_1} < \frac{\rho_r - \|H(s)C(s)K_g\|_{L_1}\|x_{2c}\|_{L_\infty} - \rho_{in}}{L_{p_r}\rho_r + b} \tag{20}$$

where ρ_{in}, $H(s)$, $G(s)$, and L_{p_r} are defined as:

$$\begin{cases} \rho_{in} = \|s(sI + K_p)^{-1}\|_{L_1}\rho_0 \\ H(s) = (sI + K_p)^{-1}K_p \\ G(s) = H(s)[I - C(s)] \\ L_{p_r} = \frac{\rho_r + \overline{\gamma}_1}{\rho_r} d_{fp}[\rho_r + \overline{\gamma}_1] \end{cases} \tag{21}$$

where $\overline{\gamma}_1$ is an arbitrary positive number.

In the case of satisfying Equation (20), The adaptive controller of the inner loop consists of Equations (16)–(18). The block diagram of the adaptive controller system is shown in Figure 4.

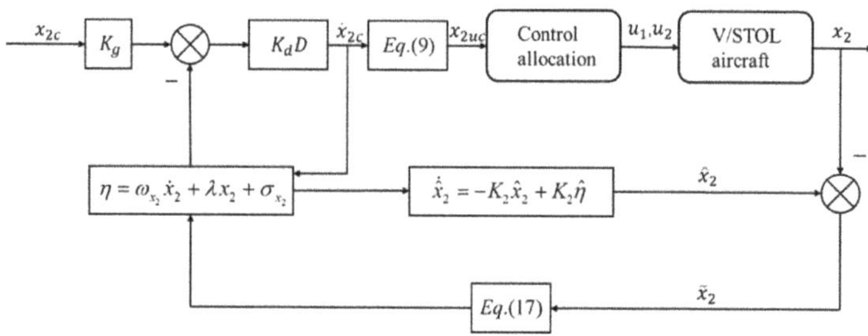

Figure 4. Adaptive controller system block diagram.

3.2. Control Allocation

Equation (6) can be expressed as:

$$\dot{X} = f(X, U_2, t) \tag{22}$$

where $\dot{X}$ are the system state variables, U_2 are the system control variables, and $\dot{X} = [u\,v\,w\,p\,q\,r\,\phi\,\theta\,\psi]$, $U_2 = [\delta_a\,\delta_e\,\delta_r\,\delta_T\,\delta_L\,\delta_{az}\,\delta_{Tl}\,\delta_{Tr}]^T$.

Expanding Equation (22) by Taylor series at the equilibrium point, keeping the linear part, and ignoring the high-order part, we can obtain:

$$\Delta\dot{X} = A\Delta X + B\Delta U_2 \tag{23}$$

where $A = \left.\frac{\partial f}{\partial X}\right|_{X=X_{trim}}$ is the system state matrix, which consists of the partial derivatives of forces and moments with respect to the state variables. $B = \left.\frac{\partial f}{\partial U_2}\right|_{U_2=U_{trim}}$ is the control derivative matrix, which is the aerodynamic change generated by the unit control amount, which can reflect the control efficiency of the aircraft by the control input. The subscript trim indicates the trim value. System control variables ΔU_2 can be solved by Equation (23).

4. Inner Loop Controller Simulation Experiment

According to the inner loop controller and control distribution method designed in the previous section, the simulation verification results are given in this section. The difficulty in controlling the V/STOL aircraft is that there is a flight mode conversion process during the takeoff and landing process. At this stage, the two sets of control mechanisms of the aircraft are involved in the work of jointly controlling the position and attitude of the aircraft. Therefore, in this state, the working point is selected for inner loop simulation and the shooting Monte Carlo approach to verify the performance of the inner loop stabilization controller and control distribution method designed in this paper.

4.1. Monte Carlo Targeting at Nominal State

An L1 controller with $\omega_p = 8$ and $D = 1/s$ receives a $5°/s$ frequent square roll rate command as the input. Figures 5 and 6 show the outcomes for various combinations of Γ and k_p.

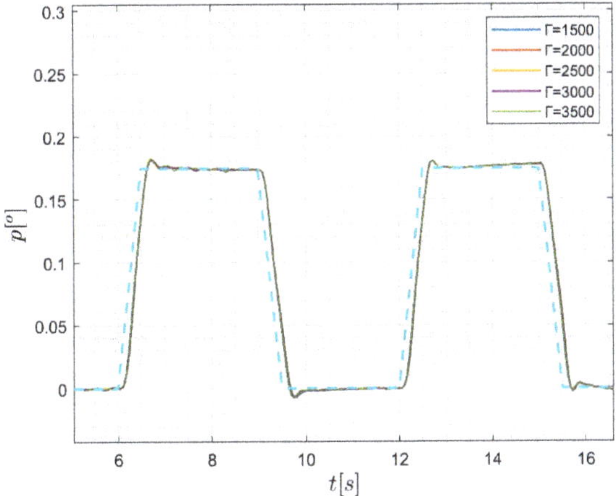

Figure 5. Roll rate command tracking results at $K_p = 8$.

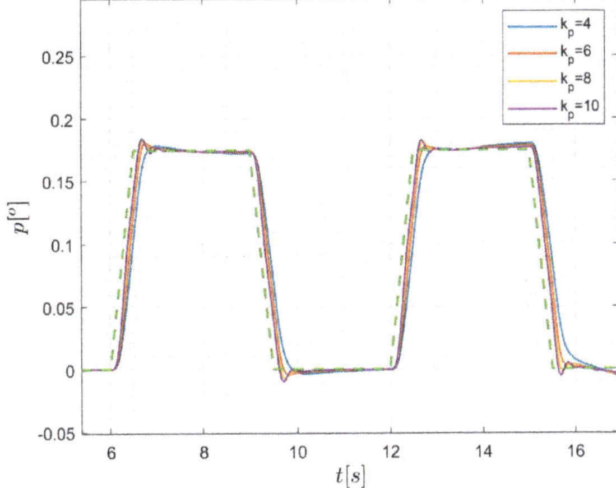

Figure 6. Roll rate command tracking results at $\Gamma = 2500$.

We settled on $k_p = 8$ and $\Gamma = 2500$ as the virtual control coefficient and virtual state coefficient, respectively, because they strike a balance between speed and stability, have a brief risetime, and exhibit zero overshoots.

According to the inner loop controller designed in the previous section, in fixed–wing mode, set the height to 25 m and the airspeed to 24 m/s. In this state, given step signals for the roll and pitch angles, the simulation results are shown in Figure 7.

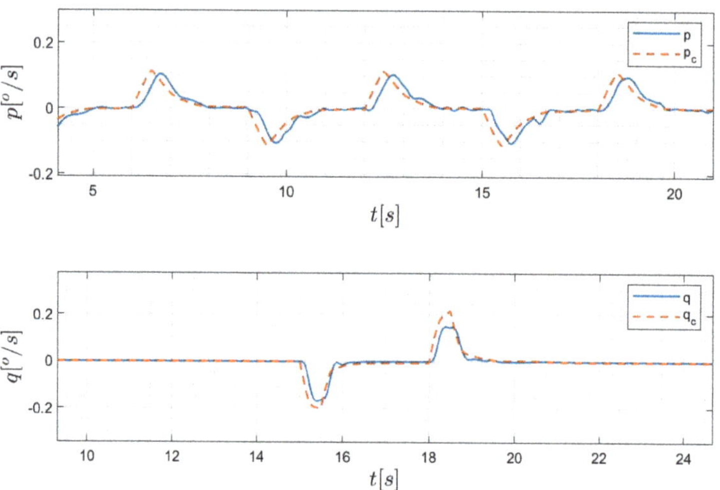

Figure 7. Roll rate and pitch rate command tracking results in fixed-wing mode.

It can be seen from the figure that this controller can track the pitch angle command well; the adjustment time of the inner loop pitch angle and roll angle command is about 1 s, and there is no obvious overshoot.

In VTOL mode, set the height to 25 m and the airspeed to 16 m/s. In this state, given step signals for the roll and pitch angles, the simulation results are shown in Figure 8:

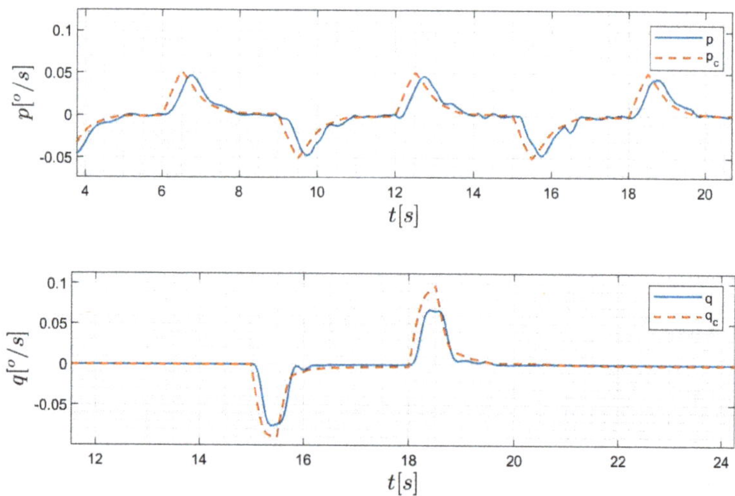

Figure 8. Roll rate and pitch rate command tracking results in VTOL mode.

The findings reveal that the curve's shape is the same in both modes, and the L1 controller's tracking performance is superb with a steady-state error of zero and a minor temporal lag.

4.2. Target Shooting Monte Carlo Approach in Level Flight

The robustness of the inner loop controller was simulated and verified by Monte Carlo simulation. Its parameter perturbation variables mainly include aerodynamic characteristics, mass characteristics, and the dynamic system. The change of perturbation parameters is shown in Table 2.

Table 2. Main parameters of the V/STOL vehicle.

Parameters	Perturbations
$C_{L_{\delta e}}, C_{m_{\delta e}}, C_{S_{\delta r}}, C_{n_{\delta r}}, C_{l_{\delta a}}$	$\pm 10\%$
$C_{m_\alpha}, C_{S_\beta}, C_{l_\beta}, C_{n_\beta}$	$\pm 10\%$
C_{l_p}, C_{n_r}	$\pm 10\%$
C_{l_r}, C_{n_p}	$\pm 10\%$
J_{xx}, J_{yy}, J_{zz}	$\pm 10\%$

$C_{L_{\delta e}}, C_{m_{\delta e}}, C_{S_{\delta r}}, C_{n_{\delta r}}, C_{l_{\delta a}}$ are the control derivatives; $C_{m_\alpha}, C_{S_\beta}, C_{l_\beta}, C_{n_\beta}$ are the stability derivatives; C_{l_p}, C_{n_r} are the damping derivatives; C_{l_r}, C_{n_p} are the cross-damping derivatives; J_{xx}, J_{yy}, J_{zz} are the inertia moments.

The above parameters are randomly changed, and the shooting Monte Carlo approach is performed on the aircraft in constant level flight and acceleration conditions, respectively. At an altitude of 100 m, the aircraft is in constant level flight at a speed of 20 m/s. At 15 s, a 10° pitch angle command is given to conduct a target shooting Monte Carlo approach simulation test. Then, comparing it with the PID controller [24–26], the simulation results are shown in Figures 9–13.

As can be seen from the above figure, the control channels of each direction of the aircraft are decoupled. The controller designed in this paper is obviously better than the PID controller for instruction tracking. This shows that the controller designed in this paper can still track the command well and has good robustness under the premise that the parameters of the aircraft are taken.

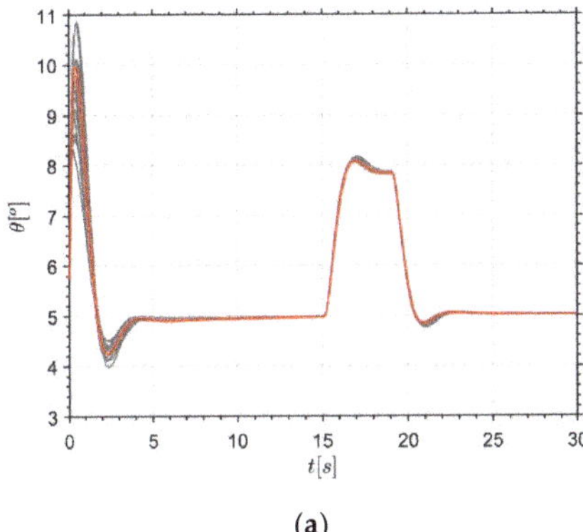

(a)

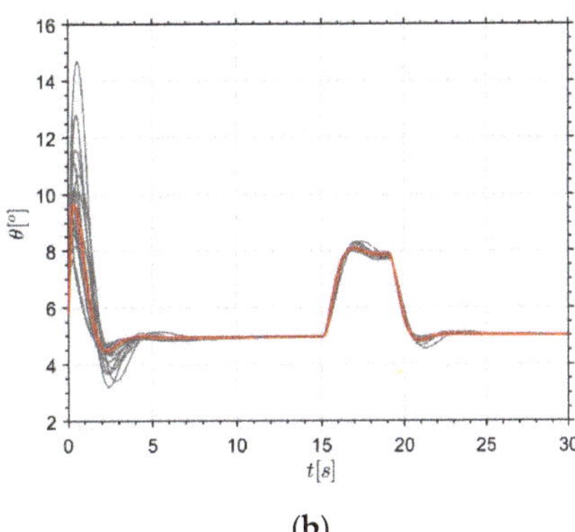

(b)

Figure 9. The shooting Monte Carlo approach at the pitch angle. (a) L1 adaptive controller; (b) PID controller.

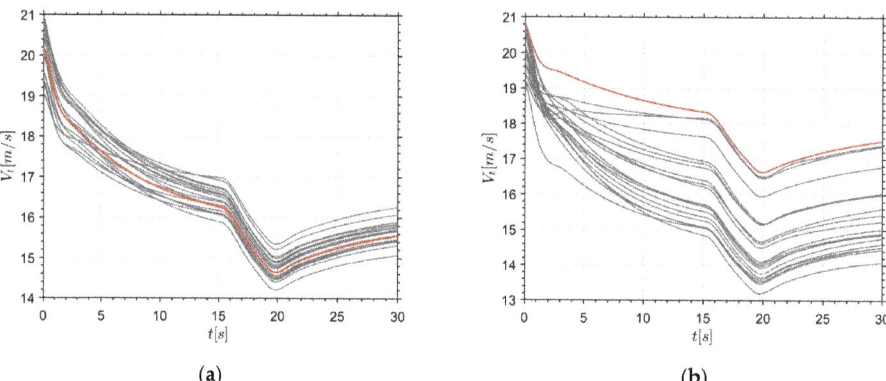

(a)   (b)

Figure 10. The shooting Monte Carlo approach at the airspeed; (**a**) L1 adaptive controller; (**b**) PID controller.

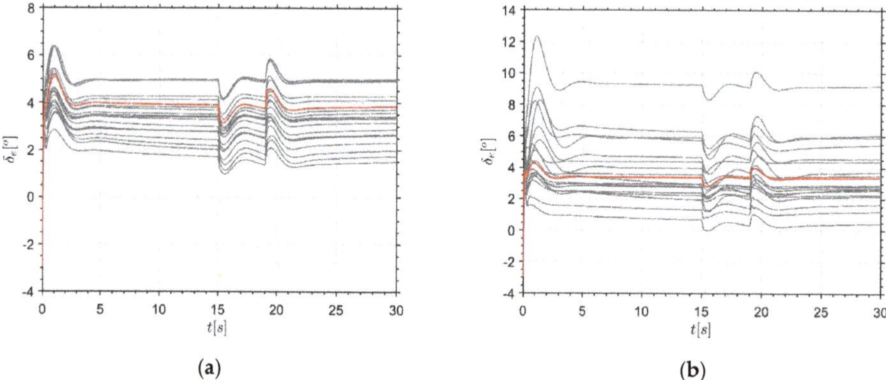

(a)   (b)

Figure 11. The target shooting Monte Carlo approach with the elevator; (**a**) L1 adaptive controller; (**b**) PID controller.

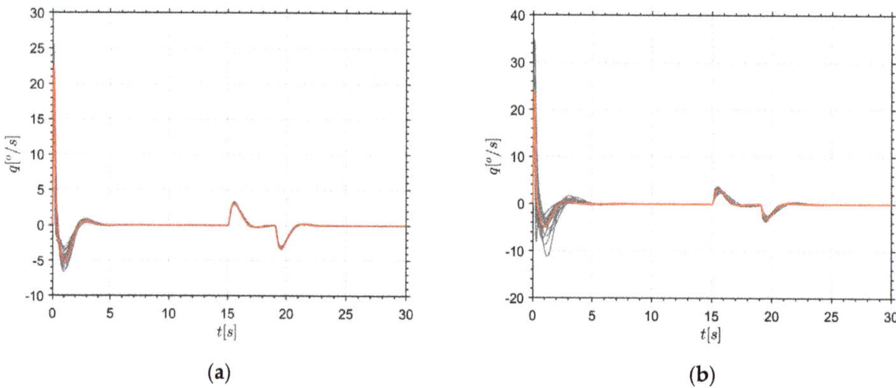

(a)   (b)

Figure 12. The shooting Monte Carlo approach at the pitch angle velocity; (**a**) L1 adaptive controller; (**b**) PID controller.

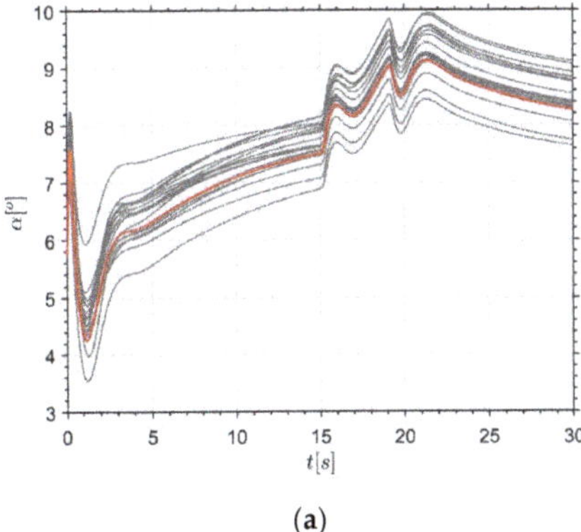

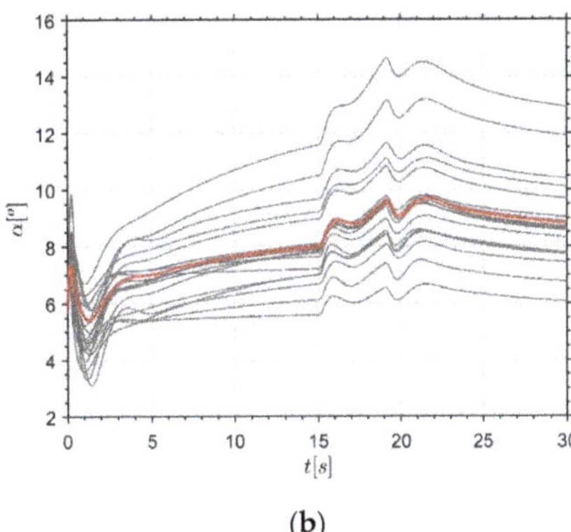

Figure 13. The shooting Monte Carlo approach at the angle of attack; (**a**) L1 adaptive controller; (**b**) PID controller.

5. Conclusions

Based on the conventional nonlinear dynamic inverse control of the outer loop, an L1 adaptive controller was designed as the inner loop stability augmentation control to compensate the uncertainty in the system. The expected verification results were obtained from the Monte Carlo simulations. The main contributions of this paper are as follows:

(1) This paper introduced the composition of the power system of the thrust vectoring V/STOL aircraft and established the dynamic equation of the F35B scale model prototype.

(2) For the thrust vector V/STOL aircraft with strong coupling and nonlinearity, based on the conventional dynamic inverse control in the outer loop, an L1 adaptive stabilization controller was designed on the inner loop to compensate for the mismatch uncertainty. The designed control structure integrates the fixed-wing mode and the VTOL mode.

(3) The uncertainty of modeling and possible input disturbances were fully considered and compared with the PID controller. It was verified by simulation that the controller quickly responds to the command and has good robustness when there is a large parameter perturbation.

Author Contributions: Conceptualization, Z.G. and Z.Z.; methodology, Z.G.; software, Z.G. and Z.Z.; validation, Z.Z. and Z.G.; formal analysis, Z.Z. and Z.W.; investigation, Z.W.; resources, X.Z.; data curation, Z.W.; writing—original draft preparation, Z.G. and Z.Z.; writing—review and editing, Z.W. and Z.Z.; visualization, Y.Y.; supervision, Z.W.; project administration, P.C. All authors have read and agreed to the published version of the manuscript.

Funding: This research received no external funding.

Data Availability Statement: Data available on request due to restrictions eg privacy or ethical. The data presented in this study are available on request from the corresponding author. The data are not publicly available due to commercial use.

Conflicts of Interest: The authors declare no conflict of interest.

References

1. Su, W.; Gao, Z.H.; Xia, L. Multiobjective optimization design of aerodynamic configuration constrained by stealth performance. *Acta Aerodyn. Sin.* **2006**, *24*, 137–140.
2. Jacobson, S.; Britt, R.; Freim, D.; Kelly, P. Residual pitch oscillation (rpo) flight test and analysis on the b-2 bomber. *ICES J. Mar. Sci.* **2003**, *67*, 1260–1271.
3. Britt, R.T.; Arthurs, T.D.; Jacobson, S.B. Aeroservoelastic analysis of the B-2 bomber. *J. Aircr.* **2000**, *37*, 745–752. [CrossRef]

4. Ma, S.; Wang, W. The Longitudinal Modeling of a New Concept V/STOL UAV in Transition Flight by Using the Method of System Identification. In Proceedings of the 2018 10th International Conference on Intelligent Human-Machine Systems and Cybernetics (IHMSC), Hangzhou, China, 25–26 August 2018.

5. Zhang, X.; Wang, W. Dynamic modelling of the hovering phase of a new V/STOL UAV and verification of the PID control strategy. *IOP Conf. Ser. Mater. Sci. Eng.* **2018**, *408*, 12–17. [CrossRef]

6. Wang, X.; Zhu, J.; Zhang, Y. Dynamics modeling and analysis of thrust-vectored V/STOL aircraft. In Proceedings of the 32nd Chinese Control Conference, IEEE, Xi'an, China, 26–28 July 2013.

7. Nagabhushan, B.L.; Faiss, G.D. Thrust vector control of a v/stol airship. *J. Aircr.* **1984**, *21*, 408–413. [CrossRef]

8. Yang, X.; Fan, Y.; Zhu, J. Transition flight control of two vertical/short takeoff and landing aircraft. *J. Guid. Control. Dyn.* **2008**, *31*, 371–385.

9. Walker, G.; Allen, D. X-35B STOVL Flight Control Law Design and Flying Qualities. In Proceedings of the 2002 Biennial International Powered Lift Conference and Exhibit, Williamsburg, VA, USA, 5–7 November 2002.

10. Wang, X.; Zhu, B.; Zhu, J.; Cheng, Z. Thrust vectoring control of vertical/short takeoff and landing aircraft. *Sci. China* **2020**, *63*, 218–220. [CrossRef]

11. Cheng, Z.; Zhu, J.; Yuan, X.; Wang, X. Design of optimal trajectory transition controller for thrust-vectored v/stol aircraft. *Sci. China Inf. Sci.* **2021**, *64*, 139201. [CrossRef]

12. Mallikarjunan, S.; Nesbitt, B.; Kharisov, E. L1 adaptive controller for attitude control of multirotors. In Proceedings of the Guidance, Navigation, and Control Conference, Minneapolis, MN, USA, 13–16 August 2012.

13. Chiang, R.Y.; Safonov, M.G.; Haiges, K.; Madden, K.; Tekawy, J. A fixed H∞ controller for a supermaneuverable fighter performing the herbst maneuver. *Automatic* **1993**, *29*, 111. [CrossRef]

14. Wang, Z.; Mao, S.; Gong, Z.; Zhang, C.; He, J. Energy Efficiency Enhanced Landing Strategy for Manned eVTOLs Using L1 Adaptive Control. *Symmetry* **2021**, *13*, 21–25. [CrossRef]

15. Seshagiri, S.; Promtun, E. Sliding mode control of F-16 longitudinal dynamics. In Proceedings of the 2008 American Control Conference, Seattle, WA, USA, 11–13 June 2008.

16. Liu, N.; Cai, Z.; Wang, Y. Fast level-flight to hover mode transition and altitude control in tiltrotor's landing operation. *Chin. J. Aeronaut.* **2020**, *34*, 181–193. [CrossRef]

17. Min, B.M.; Kim, E.T.; Tahk, M.J. Application of Control Allocation Methods to SAT-II UAV. In Proceedings of the Aiaa Guidance, Navigation, & Control Conference & Exhibit, San Francisco, CA, USA, 15–18 August 2005.

18. Tan, J.; Zhou, Z.; Zhu, X.; Xu, M. Attitude control of flying wing uav based on terminal sliding mode and control allocation. *J. Northwestern Polytech. Univ.* **2014**, *32*, 505–510.

19. Shakarian, A. Application of Monte-Carlo Techniques to the 757/767 Autoland Dispersion Analysis by Simulation. *Guid. Control Conf.* **1983**, *83*, 181–194.

20. Chao, Z.; Chen, L.; Chen, Z. Monte Carlo Simulation for Vision-based Autonomous Landing of Unmanned Combat Aerial Vehicles. *J. Syst. Simul.* **2010**, *22*, 2235–2240.

21. Xargay, E.; Hovakimyan, N.; Dobrokhodov, V.; Statnikov, R.; Kaminer, I.; Cao, C.; Gregory, I. L1 Adaptive Flight Control System: Systematic Design and Verification and Validation of Control Metrics. In Proceedings of the AIAA Guidance, Navigation, and Control Conference, Toronto, ON, Canada, 2–5 August 2010.

22. Leman, T.; Xargay, E.; Dullerud, G.; Hovakimyan, N.; Wendel, T. L1 Adaptive Control Augmentation System for the X-48b AircrafAIA Guidance. In Proceedings of the Navigation, and Control Conference, Chicago, IL, USA, 10–13 August 2009; AIAA, Urbana-Champaign: Champaign, IL, USA, 2009.

23. Jiang, G.; Liu, G.; Yu, H. A Model Free Adaptive Scheme for Integrated Control of Civil Aircraft Trajectory and Attitude. *Symmetry* **2021**, *13*, 347. [CrossRef]

24. Faruk, T.; Ali, A.; Naim, A.; Firas, A.; Basil Al, K.; Özyavaş, A. Robust Nonlinear Non-Referenced Inertial Frame Multi-Stage PID Controller for Symmetrical Structured UAV. *Symmetry* **2022**, *14*, 689. [CrossRef]

25. Ma, J.; Xie, H.; Song, K.; Liu, H. Self-Optimizing Path Tracking Controller for Intelligent Vehicles Based on Reinforcement Learning. *Symmetry* **2022**, *14*, 31. [CrossRef]

26. Zhong, C.Q.; Wang, L.; Xu, C.F. Path Tracking of Permanent Magnet Synchronous Motor Using Fractional Order Fuzzy PID Controller. *Symmetry* **2021**, *13*, 1118. [CrossRef]

 *symmetry*

Article

A Multi-Scale and Lightweight Bearing Fault Diagnosis Model with Small Samples

Shouwan Gao [1,2,*], Jianan He [1], Honghua Pan [3] and Tao Gong [4]

[1] School of Computer Science and Technology, China University of Mining and Technology, Xuzhou 221116, China; wjhjn@cumt.edu.cn
[2] Mine Digitization Engineering Research Center of the Ministry of Education, China University of Mining and Technology, Xuzhou 221116, China
[3] Shandong Provincial Weifang Big Data Center, Weifang 261000, China; panhonghua@wf.shandong.cn
[4] School of Mechatronic Engineering, China University of Mining and Technology, Xuzhou 221116, China; gongtao@cumt.edu.cn
* Correspondence: gaoshouwan@cumt.edu.cn

Abstract: Currently, deep-learning-based methods have been widely used in fault diagnosis to improve the diagnosis efficiency and intelligence. However, most schemes require a great deal of labeled data and many iterations for training parameters. They suffer from low accuracy and over fitting under the few-shot scenario. In addition, a large number of parameters in the model consumes high computing resources, which is far from practical. In this paper, a multi-scale and lightweight Siamese network architecture is proposed for the fault diagnosis with few samples. The architecture proposed contains two main modules. The first part implements the feature vector extraction of sample pairs. It is composed of two lightweight convolutional networks with shared weights symmetrically. Multi-scale convolutional kernels and dimensionality reduction are used in these two symmetric networks to improve feature extraction and reduce the total number of model parameters. The second part takes charge of calculating the similarity of two feature vectors to achieve fault classification. The proposed network is validated by multiple datasets with different loads and speeds. The results show that the model has better accuracy, fewer model parameters and a scale compared to the baseline approach through our experiments. Furthermore, the model is also proven to have good generalization capability.

Keywords: convolutional neural network; fault diagnosis; few shot; Siamese network; lightweight

Citation: Gao, S.; He, J.; Pan, H.; Gong, T. A Multi-Scale and Lightweight Bearing Fault Diagnosis Model with Small Samples. *Symmetry* **2022**, *14*, 909. https://doi.org/10.3390/sym14050909

Academic Editor: Christos Volos

Received: 16 March 2022
Accepted: 26 April 2022
Published: 29 April 2022

Publisher's Note: MDPI stays neutral with regard to jurisdictional claims in published maps and institutional affiliations.

1. Introduction

At present, bearings are an essential component of machine manufacturing equipment. The good or bad running conditions of bearings directly affects the operation of the equipment. However, complex real environments, including abnormal humidity, temperatures and current magnitudes, cause different degrees of damage to the bearings, resulting in the occurrence of faults. This produces high maintenance costs as well as delays of production progress to the factory and even threatens the personal safety of personnel. Therefore, the safety of bearings has become a crucial concern. The research on the bearing fault diagnosis algorithm is of great significance to the safety of equipment [1,2].

Thus far, the traditional bearing fault diagnosis technology is to manually analyze the vibration signal obtained by the accelerometer [3]. The corresponding methods are used to extract the characteristic information from the vibration signal, which mainly include fast Fourier transform (FFT) [4], wavelet transformation (WT) [5], empirical mode decomposition (EMD) [6], short-time Fourier transform (STFT) [7] and Wigner–Ville distribution (WVD) [8]. Furthermore, the advent of Hilbert transformation (HT) [9] made it possible to diagnose transient bearing faults. These methods have been shown to be effective in practice. In recent years, machine learning has been utilized in the study of bearing fault diagnosis.

The main methods are artificial neural networks (ANN) [10], principal component analysis (PCA) [11], K-Nearest Neighbors (K-NN) [12] and support vector machines (SVM) [13]. Machine learning as a branch of artificial intelligence is widely used in various fields. The use of machine learning has taught computers how to process data efficiently compared to traditional methods. The computer can find more subtle features to analyze, which improves the accuracy and intelligence of bearing fault diagnosis. However, with the rapid changes of current technology, the amount and types of data have also ushered in rapid growth. Feature selection, which we need to rely on experts to perform, becomes time-consuming and laborious. Deep learning not only has better accuracy and processing speed but also can solve problems end-to-end. Therefore, deep learning is gradually being widely adopted.

Deep learning has made great breakthroughs in the fields of computer vision, natural language and data mining. Typical methods, such as convolutional neural networks (CNN) [14,15], Recurrent Neural Networks (RNN), Long Short-Term Memory (LSTM) [16] and Generative Adversarial Networks (GAN) [17], have obvious effects in dealing with problems in these fields. These methods simplify the step of feature extraction at the same time. Furthermore, deep learning has good application prospects in the field of bearing faults. Compared with the traditional diagnosis methods, the deep-learning method realizes the automatic extraction of features and has a good effect on the accuracy of diagnosis.

A fault diagnosis method based on CNN multi-sensor fusion was proposed in the literature [18]. An automatic recognition architecture for rolling bearing fault diagnosis based on reinforcement learning was also proposed in the literature [19]. With the use of real-life scenarios, the problem of insufficient training samples has been noticed and studied. In recent years, excellent progress has been made in the study of neural networks based on small samples [20,21]. Fang, Q. et al., proposed a denoised fault diagnosis algorithm with small samples that can solve the problem of bearing fault diagnosis under small samples [22].

However, a model with complex structures often requires a large number of parameters. This leads to a higher level of operational equipment. Too many parameters make it far from practical in real world scenarios. Furthermore, this may also affect the computational speed. Hence, controlling the number of model parameters is extremely important in practical applications. Fang, H. et al. proposed a lightweight fault diagnosis model that can solve the problem of too many model parameters [23]. However, it cannot perform fault diagnosis when there are insufficient samples. This shows that the recently proposed models are unable to achieve a better trade-off between accuracy and lightweight [24,25].

To overcome the problems of few samples and a huge amount of parameters, an end-to-end multi-scale and lightweight Siamese network with symmetrical architecture (MLS-net) is proposed in this paper. MLS-net not only maintains good accuracy in bearing fault diagnosis under small samples but also has fewer parameters to reduce resource consumption and a good generalization ability. The main contributions are summarized below.

- We construct a novel fault diagnosis network architecture by combining an improved Siamese network and few-shot learning for the case of small samples.
- A multi-scale feature extraction module is designed to improve the feature extraction capability of the model. Furthermore, we use the dimensionality reduction method to compress the parameters of the model to conserve device resources.
- Extensive experiments are conducted on multiple datasets to demonstrate the efficiency and generalization of the proposed architecture.

The rest of the paper is organized as follows: Section 2 introduces the mentioned basic theory. Section 3 describes the proposed network structure. Section 4 presents the details and results of the experiments. Section 5 concludes the paper.

2. Preliminaries

2.1. Inception

The inception module has an important role in model compression and feature extraction [26]. The key to the module improvement is the introduction of 1×1 convolution and the construction of a multi-scale convolution structure. The accuracy of the model on image classification was proven to be significantly improved through experiments. In addition, the model can utilize the computational resources more efficiently. More features are acquired with the same amount of parameters. The accuracy is also significantly improved. At the same time, the problems of overfit, gradient explosion and gradient disappearance due to the increased depth of the model are also improved.

There are four branches in the inception module. The first three branches use 1×1 convolution kernels for dimensionality reduction, which serves to optimize the problem of too many parameters caused by convolution operations. The reduction in dimensionality also brings a reduction in calculations. It is beneficial to the full utilization of computational resources. In addition, convolution kernels with different sizes are used to obtain different perceptual fields in the inception module. The branch contains 1×1, 3×3 and 5×5 sizes.

The different sizes of the convolution kernels allow for richer information to be extracted from the features. At the same time, multi-scale convolution uses the principle of decomposing sparse matrices into dense matrices to speed up the convergence. A comparison of the traditional convolution and inception modules is shown in Figure 1.

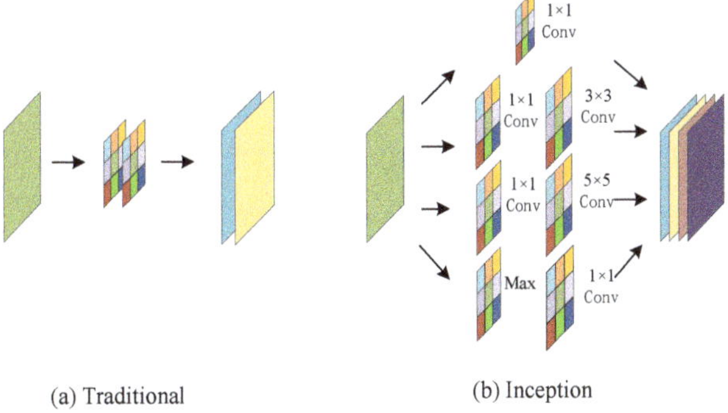

(a) Traditional   (b) Inception

Figure 1. Traditional convolution vs. inception module convolution.

2.2. Siamese Network

A Siamese network [27] is a symmetrical architecture built from two neural networks. They are mainly applied in small sample cases. The inputs of the model are two samples from the same or different datasets. The main body consists of feature extraction and a similarity calculation module. The function of the feature extraction module outputs the feature vectors of the input samples. The similarity calculation module calculates the similarity of the two feature vectors. The similarity is compared between the predicted data and the prior knowledge using the model obtained from training. Thus, fault classification with small samples is achieved. Its emergence solves the problem of deep neural networks to obtain high accuracy and overfitting in the absence of a large number of data samples.

2.3. Few-Shot Learning

Few-shot learning [28] is the use of few samples for classification or regression. It is different from traditional supervised learning methods. Few-shot learning does not generalize the training set to the test set. It aims to make the model understand the similarities and differences of things and learn to distinguish between different things.

Few-shot learning generally consists of a support set (S), a query set (Q), a training set (T) and a judgment rule (R), where S contains a small amount of supervised information in Q.

The combination of S and Q is used for predictive classification. R is the procedure equipped to determine the similarities and differences of things. We use T as prior knowledge to train R. R is then used to determine the similarities and differences between samples in S and Q to achieve small sample classification.

3. Methodology

3.1. Structure of MLS-Net

The overall architecture of MLS-net is shown in Figure 2. We fuse the improved sub-network with a Siamese network. A multi-scale and lightweight bearing fault diagnosis architecture applied to the small sample situation is constructed. The whole structure body consists of two symmetrical branches. As we can see from the figure, the overall architecture contains four parts: data pre-processing, the sub-network, similarity classification and the few-shot learning test.

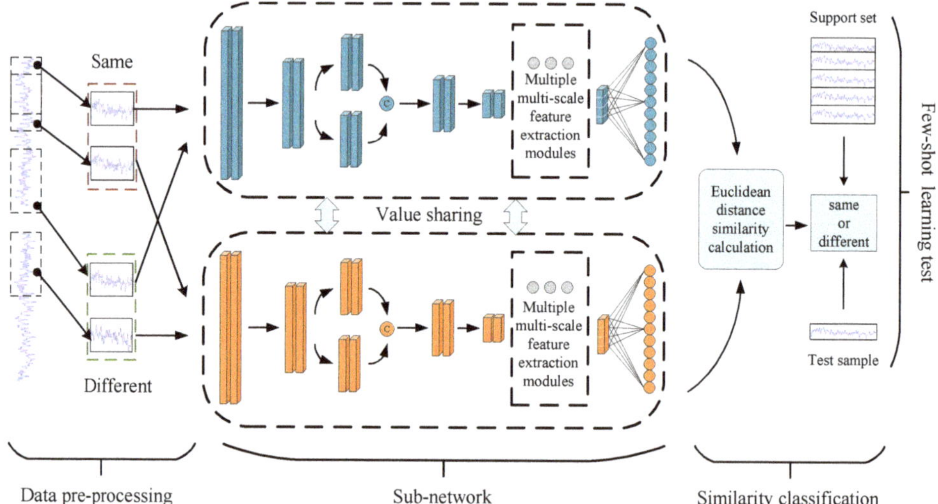

Figure 2. Architecture of the MLS-net consisting of four modules.

3.1.1. Data Pre-Processing

The data pre-processing part focuses on the construction of the dataset that needs to be input into the model. The intercepted fault data from the same class and different classes are randomly combined to form the same and different class sample pairs. In the input sample pairs, we input the bearing vibration data from the sample pairs into the two symmetrical feature extraction branches separately.

The whole network needs to input pairs of samples in the format $(x_1, x_2, \mathcal{Y})$. Each sample pair contains a label $\mathcal{Y}$. When $\mathcal{Y}$ is 1, it means that the input sample pairs are fault data of the same category. However, when $\mathcal{Y}$ is 0, it means that the input sample pairs are the fault data of different categories. The corresponding x_1 and x_2 represent the two vibration data to be input. Further details of the data pre-processing can be found in Section 4.1.

3.1.2. Sub-Network

The sub-network part has two feature extraction modules with shared weights. The shared weights ensure that the results obtained from the two branches are comparable during similarity classification. The main purpose of this part is to extract the feature vector of the bearing fault data after convolutional processing using an optimized sub-network. The structure of the sub-network consists of multiple multi-scale and reduced dimensional feature extraction modules. The fused multi-scale feature information is used to enhance the model's ability to obtain information from the samples. The function of the dimensionality reduction module is as a reducing parameter.

3.1.3. Similarity Classification

The similarity classification part mainly uses the distance formula to calculate the similarity between the feature vectors of the two branches. A similarity percentage is given after normalization. This similarity percentage is used to determine whether the two input bearing fault vibration data are of the same class. We use the trained similarity calculation model in combination with the few-shot learning test method to realize bearing fault diagnosis.

To implement the similarity calculation module, we first use the distance formula to obtain the distance between two feature vectors. The closer the two feature vectors are, the more likely that we can assume that they are the same class. When far apart, they are different classes. We chose the Euclidean distance as the metric formula for the two feature vectors. The formula is as follows.

$$s_t(x_1, x_2) = ||t(x_1) - t(x_2)|| \tag{1}$$

where x_1 and x_2 are the input samples. $t(x_1)$ and $t(x_2)$ are the feature vectors obtained after sub-network processing. $s_T(x_1, x_2)$ is the Euclidean distance.

The output of the entire network indicates the similarity of the sample pairs. In fact, this problem has been transformed into a binary classification problem in the similarity classification module. This is to give a probability to determine whether two input samples are of the same class or different classes. We use the sigmoid function to map the distance between two feature vectors to the range of (0, 1). The probability is used to intuitively predict the magnitude of the distance between the two vectors. The formula to calculate the output is as follows.

$$p(x_1, x_2) = sigm(FC(s_T(x_1, x_2))) \tag{2}$$

where FC is full connection processing for Euclidean distance output. *sigm* is sigmoid function. $p(x_1, x_2)$ is the probability of the similarity of sample pairs.

As the whole similarity calculation module is transformed into a binary classification problem. When the entire network is trained, binary cross entropy is used as the loss function of the network. The corresponding formula is as follows:

$$Loss = -\mathcal{Y}(x_1, x_2)log(p(x_1, x_2)) + (1 - \mathcal{Y}(x_1, x_2))log(1 - p(x_1, x_2)) \tag{3}$$

where $\mathcal{Y}(x_1, x_2)$ represents the corresponding label. The same is "1", and different is "0".

Once the loss function is determined, a gradient descent function can be used to train the Siamese network. The model weights are fine-tuned over multiple cycles by using forward and backward propagation. A network model that can determine the similarity of the fault samples is trained. We can use the trained similarity classification model and few-shot learning test method for bearing fault diagnosis.

3.1.4. Few-Shot Learning Testing

The C-shot K-way testing is generally used to test the model under the situation of small samples. K classes are extracted from the existing dataset, and C samples from each class are taken to build the support set as the test criteria. The test set is called the query set. We use the similarity classification module to calculate the similarity probability between the support set and the query set. With the help of the similarity probability, we can easily determine the category of the test sample. The general testing methods include one-shot K-way testing and C-shot K-way testing.

In one-shot K-way testing, each class in the support set S contains only one sample. This test aims to calculate the probability of similarity for each sample pair consisting of a support set and a query set. The sample pairs with the highest probability are of the same kind. The definition of S and the formula for the highest score T are as follows.

$$S = \{(x_i, y_i) | i = 1, 2...K\}$$ (4)

$$T(\hat{x}, S) = argmax(P(\hat{x}, x_k)), x_k \in S$$ (5)

In the C-shot K-way testing, each class in the support set S contains only C samples. Unlike one-shot K-way testing, this is performed by comparing the maximum of the sum of the probability. The specific formula is as follows.

$$S_C = \{(S_i | i = 1, 2...C)\}$$ (6)

$$T(\hat{x}, S_C) = argmax\left\{ \sum_{i=1}^{n} P(\hat{x}, x_C k) \right\}, x_C k \in S_C$$ (7)

3.2. Sub-Net Optimization

Siamese networks have been proven to be effective in dealing with small sample size problems. However, when the structure of feature extraction network is simple. The model cannot effectively extract enough features from the samples for classification. This article improves on the branches in the Siamese network to overcome the above problem. The sub-network is improved to a multi-scale and lightweight structure. The model extracts rich features through convolution kernels of different sizes. We use 1×1 convolution kernels to compress the model and reduce the calculations. This method can improve the model feature extraction capability and model accuracy. The optimization of sub-network module is shown in Figure 3.

The optimization for sub-network module is mainly inspired by the Inception module. The improvement of the sub-network is based on a convolutional network with a first layer of wide convolution [29]. As shown in Figure 3, the new model retains the wide convolutional layer of the first layer in the original network. This is to extract more feature information from the one-dimensional bearing vibration data, while other layers introduce multi-scale modules and 1×1 for sub-network optimization. The 3×3 and 5×5 convolution kernels generate a huge amount of computation when there are too many parameters in the input. We introduce a 1×1 convolution kernel to achieve dimensionality reduction of the data, thus, reducing the amount of calculation and model parameters.

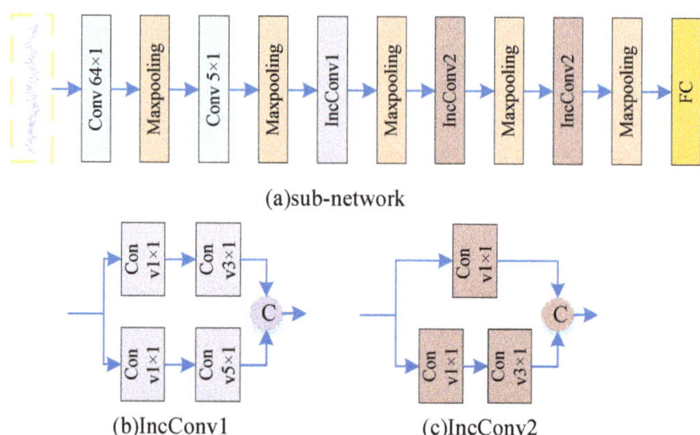

Figure 3. Structure of the sub-network with two types of multi-scale convolutional modules.

After the model is processed by the first layer of wide convolution, it continues to be processed by the multi-scale module to obtain richer feature information. The multi-scale module can have different perceptual fields compared to the conventional convolution module. There are two types of multi-scale convolution modules in the model. One is a combination of 3×3 convolution and 5×5 convolution named IncConv1. The other is the combination of 1×1 convolution and 3×3 convolution named IncConv2. Due to the larger scale of the data in the first stage. The IncConv1 is chosen for processing in the model. As the scale of the processed data decreases and the depth of the model increases, the IncConv2 is gradually chosen for processing. This is to reduce the parameters and computational effort.

The parameters of the convolutional layers in the sub-network are as follows: the input part is 2048×1 size data. The size of the convolutional layer in the first layer is 64×1 and contains a total of 16 convolutional kernels. The step size of the convolutional layers after this one is 1. The size of the second layer is 5×1 and contains 32 convolutional kernels. The third layer is the IncConv1 module mentioned above. It is a combination of convolutional kernels of sizes 3×3 and 5×5. The fourth and fifth convolutional layers are the IncConv1 modules. These are a combination of 1×1 and 3×3 size convolutional kernels.

Before the fully connected layer is a dropout layer with a parameter set to 0.5. It is used to prevent overfitting during model training and to accelerate the convergence of the model. The final output is a fully-connected layer with an output of 100. We add a maximum pooling layer after each convolutional layer. The size of each maximum pooling layer is 2×1, and the stride is 1. The maximum pooling is to reduce the model parameters and increase the computational speed while extracting features robustly.

3.3. Processing of the Network

The specific operations can be seen in Figure 4 in the following modules.

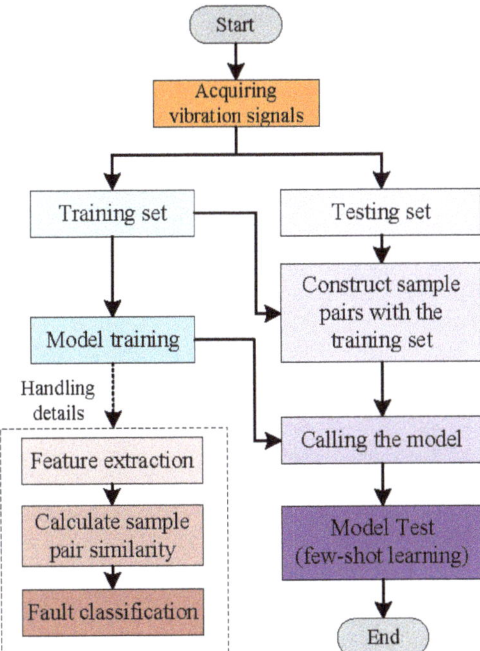

Figure 4. Model workflow of MLS-net.

- Preprocessing: The bearing fault vibration data is segmented using sliding windows to obtain the bearing fault samples. We construct a training set and a test set according to the requirement that the model input is a sample pair.
- Model training: The training set is divided into sample pairs with equal proportions of the same and different classes of faults. We feed the training samples into the network. Then, the network is trained by using the Adam gradient descent algorithm and a binary cross entropy loss function. Finally, we save the model with the best training results.
- Model testing: We first combine the samples from the test set and the training set in order to form a support set. The trained optimal model is used in the similarity probability calculation. The sample pair with the highest similarity probability among the obtained multiple similarities is selected as the fault class.

4. Experimentation and Analysis

4.1. Data Set Preparation

We must understand the performance of the proposed network structure in the case of insufficient samples. Three datasets are used for validation in this experiment. They are the Case Western Reserve University (CWRU) bearing fault dataset [30], Mechanical fault Prevention Technology Institute (MFPT) bearing fault dataset [31] and Laboratory simulated bearing fault dataset.

(1) CWRU bearing fault dataset

For this experiment, the 12 kHz bearing fault on the drive side from the Case Western Reserve University bearing dataset is used as the experimental data. The fault types are divided into four categories: normal, ball fault, inner ring fault and outer ring fault. Each fault, in turn, contains three fault categories of 0.007, 0.014 and 0.021 inch dimensions; therefore, the total number of fault categories is 10. The specific classification is in Table 1.

(2) MFPT bearing fault dataset

The MFPT dataset is provided by the Mechanical Prevention Technology Association. The dataset contains data from the experimental bench and three real-world fault data. Fault types are divided into three categories: baseline conditions, outer race fault conditions and inner race fault conditions. The sampling frequency of the data set is 25 Hz. We selected seven types of data from MFPT to construct the experimental dataset. The fault types are classified into three categories: normal, outer ring fault and inner ring fault. Each fault class data is selected with load conditions of 50, 200 and 300 lbs. The total number of classes of the fault categories in the experimental dataset is seven. The specific classification is in Table 2.

(3) Laboratory bearing fault dataset

The main structure of the test bench is shown in the diagram below. The components are the following: accelerometer, bearings, motors, acquisition cards, frequency converter and external computers and other key devices. The positions of the individual devices are marked in Figure 5. The entire experimental equipment is rotated by motors driving the bearing parts. The accelerometers collect the vibration signal in real time. The vibration signal is then transferred to the computer for storage and analysis by means of an acquisition card.

The experiments conducted in this case are set up for three fault situations. The faults are outer race fault, inner race fault and ball fault. All three faults are set as scratch faults. Three faults are set to penetrate in the axial. The width of the fault is 1.2 mm, and the depth is 0.5 mm. All three faults are tested twice at 1800 and 3000 r/min, respectively. Therefore, all fault categories are divided into six categories. Details of the corresponding health conditions are shown in the following Table 3.

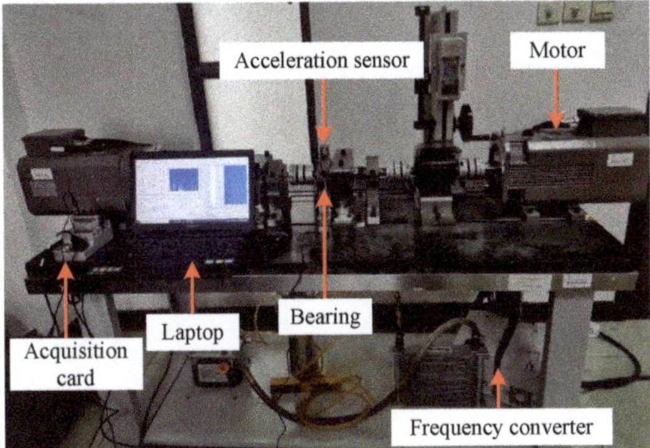

Figure 5. Schematic of the bearing fault diagnosis simulation test bench.

Table 1. CWRU bearing dataset classification description.

Fault Location (inch)	Status Labels	Number of Training Samples	Number of Test Samples
Normal	0	1980	75
Rolling ball (0.007)	1	1980	75
Rolling ball (0.014)	2	1980	75
Rolling ball (0.021)	3	1980	75
Inner race (0.007)	4	1980	75
Inner race (0.014)	5	1980	75
Inner race (0.021)	6	1980	75
Out race (0.021)	7	1980	75
Out race (0.021)	8	1980	75
Out race (0.021)	9	1980	75

Table 2. MFPT bearing dataset classification description.

Fault Location (lbs)	Status Labels	Number of Training Samples	Number of Test Samples
normal	0	660	25
Out race (50)	1	660	25
Out race (200)	2	660	25
Out race (300)	3	660	25
Inner race (50)	4	660	25
Inner race (200)	5	660	25
Inner race (300)	6	660	25

Table 3. Laboratory bearing dataset classification description.

Fault Location (r/min)	Status Labels	Number of Training Samples	Number of Test Samples
Out race (1800)	0	660	25
Out race (3000)	1	660	25
Inner race (1800)	2	660	25
Inner race (3000)	3	660	25
Rolling ball (1800)	4	660	25
Rolling ball (3000)	5	660	25

Each type of fault data is a vibration signal collected by an accelerometer. To ensure consistent conditions with the comparison schemes, the dataset is constructed based on the method in [29]. The detailed schematic diagram for building the training and test sets is shown in Figure 6. We build the training set from the first half of the vibration signal and the test set from the second. Each training sample is 2048 points in length. We use a sliding window with a step size of 80 to intercept the training samples sequentially backwards. The data intercepted by the sliding window is the training set. The second half of the vibration signal is divided into multiple non-overlapping test samples, and each test sample also contains 2048 points.

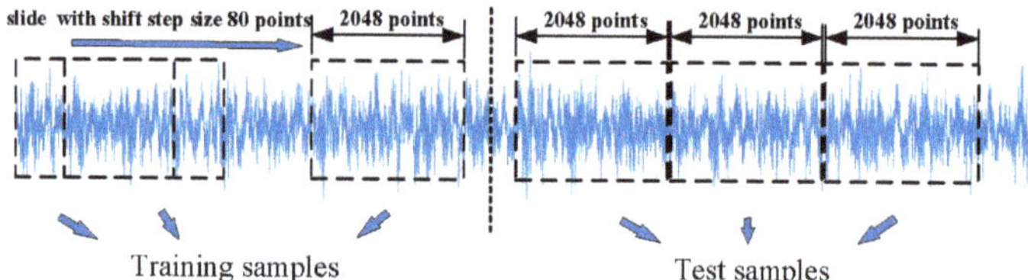

Figure 6. Schematic of the bearing fault vibration signal cut and constructed data set.

4.2. Experimental Setup

The training samples are divided into the training set and validation set. By comparing the loss rate under different ratios in Figure 7, the ratio of the training set and validation set is configured to be 4:1 for better convergence performance. In addition, the model is implemented using the Keras library and Python 3.6. The total epoch of model training is 15,000, and the small batch size is 32. The optimal model is saved after 20 training sessions have been conducted in the experiment.

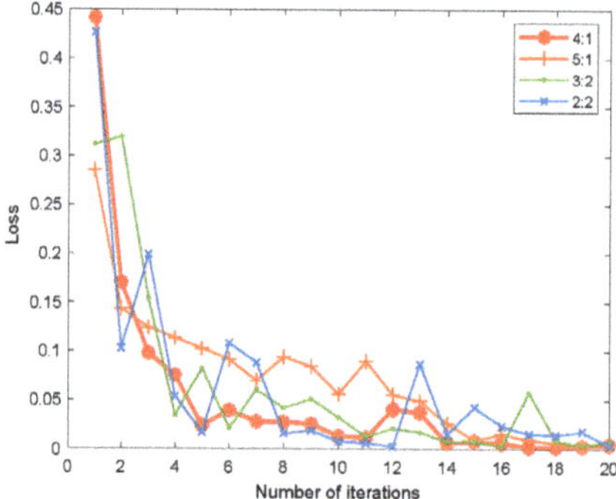

Figure 7. Comparison of the model training loss rate under different ratios.

To validate the performance of the models obtained by training under different samples, the quantities 60, 90, 120, 200, 300, 600 and 900 are randomly selected on the CWRU and Laboratory datasets. The number of fault types selected in MFPT is seven. For the sake of balance of the training data, we randomly select the quantities 70, 105, 140, 210, 280, 490 and 700.

The sample pairs we input each time are randomly selected from the above training set. When they belong to the same class, they are labeled as positive samples; otherwise, they are negative samples. We also need to ensure that the number of positive and negative sample pairs is equal to ensure a balanced sample.

In Experiment 1, we vary the number of multi-scale modules in the model to determine the optimal model structure. In Experiment 2, we test the model on three datasets to verify the performance of MLS-net. In Experiment 3, we visualize the model performance using visualization tools. In Experiment 4, we calculate the model size and parameters.

The following three methods will be tested on the three datasets to compare with the new proposed model.

1. Support Vector Machine (SVM): SVM is a classical machine learning method for dichotomous classification problems. We use an SVM between any two classes to implement a multi-classification task.

2. One-Dimensional Convolutional Neural Network (1D-CNN) [29]: 1D-CNN, which is a five-layer DCNN. It uses 64 × 1 convolutional kernels for the first layer and 3 × 1 convolutional kernels for the next four layers. The corresponding number of convolutional kernels is 16, 32, 64, 64 and 64. We add a maximum pooling layer of size 2 × 1 after each convolutional layer. The final layer is a fully connected layer with an output of 100.

3. The baseline Siamese network (BS-net) [32]: BS-net has the same structure as the proposed Siamese network. However, the structure of sub-network in the Siamese network is the 1D-CNN mentioned above.

4.3. Determination of the Number of Multi-Size Modules

We want to determine the optimal number of multi-size modules in the model. The multi-size modules are divided into two categories by the introduction of the sub-network. The larger size is a fusion of 5 × 5 and 3 × 3, which we call IncConv1. The smaller size is a fusion of 3 × 3 and 1 × 1, which we call IncConv2. Under the premise that the sample size is set to 60, we will vary the number of these two modules to determine the optimal number of modules.

The comparison between Tables 4 and 5 shows that the trend of the accuracy of the model decreases as the number of the IncConv1 increases. This shows that the number of modules for IncConv1 should be 1. Figure 8a,b shows the variation of accuracy on each data set and the mean of the three types of data. The mean lines in both plots show that, as the number of modules increases, the accuracy rate decreases. However, our previous analysis shows that the number of IncConv1 should be 1.

Therefore, we only need to observe Table 4 to determine the number of IncConv2. We find that the accuracy rate decreases as the number of modules increases.The accuracy of the models is similar at number 1 and 2; however, the total number of parameters is different. To balance the accuracy and the total number of parameters, we finally decided to set the number of IncConv2 to 2.

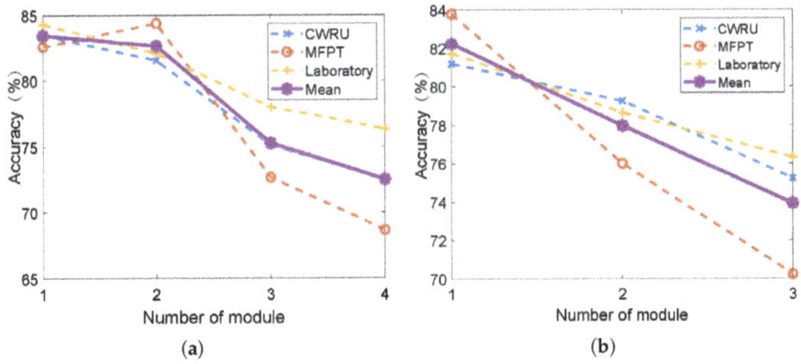

Figure 8. Accuracy comparison with different number of modules: (a) number of IncConv1 is 1 and (b) number of IncConv1 is 2.

Table 4. Comparison of IncConv2 numbers (number of IncConv1 is 1).

IncConv2 Numbers	Datasets	Accuracy (%)	Mean (%)	Total Number of Parameters
1	CWRU	83.50	83.47	63,897
	MFPT	82.60		
	Laboratory	84.32		
2	CWRU	81.55	82.67	41,449
	MFPT	84.39		
	Laboratory	82.08		

Table 4. *Cont.*

IncConv2 Numbers	Datasets	Accuracy (%)	Mean (%)	Total Number of Parameters
3	CWRU	75.12		
	MFPT	72.67	75.26	31,801
	Laboratory	78.00		
4	CWRU	72.56		
	MFPT	68.64	72.51	28,452
	Laboratory	76.34		

Table 5. Comparison of IncConv2 numbers (number of IncConv1 is 2).

IncConv2 Numbers	Datasets	Accuracy (%)	Mean (%)	Total Number of Parameters
1	CWRU	81.21		
	MFPT	83.83	82.25	43,497
	Laboratory	81.73		
2	CWRU	79.26		
	MFPT	76.00	77.96	33,849
	Laboratory	78.64		
3	CWRU	75.23		
	MFPT	70.23	73.92	30,601
	Laboratory	76.32		

4.4. Model Effects with Different Sample Sizes

In this section, we want to verify that the proposed method performs well in the case of insufficient samples. We chose the methods described above: (SVM), 1D-CNN, BS-net and MLS-net for performance comparison. Several models are tested on three bearing fault datasets. Table 6 and Figure 9 show the results of the experiments.

Table 6. Comparison of sample sizes.

Datasets	Models	Number of Samples						
CWRU		60	90	120	200	300	600	900
	SVM	18.93	26.56	31.20	38.67	43.89	50.05	52.35
	1D-CNN	73.97	77.39	84.19	93.97	96.59	97.03	98.69
	BS-net	79.33	88.41	90.15	94.00	95.45	96.00	98.07
	ANS-net [22]	88.64	90.60	-	-	98.24	98.59	99.05
	MLS-net	81.55	90.02	91.44	93.04	95.59	97.97	98.74
MFPT		70	105	140	210	280	490	700
	SVM	45.14	53.14	54.26	56.57	73.14	74.85	78.85
	1D-CNN	60.76	70.28	78.47	91.42	95.09	96.23	97.13
	BS-net	74.97	81.6	89.6	94.97	95.12	96.25	96.57
	MLS-net	84.39	89.32	91.44	95.12	95.79	96.53	97.91
Laboratory		60	90	120	180	300	600	900
	SVM	41.33	51.53	66.66	73.33	74	74.66	76
	1D-CNN	54.13	60.93	71.66	82.46	83.73	84.26	88.26
	BS-net	72.80	81.20	82.00	82.93	85.00	87.06	88.00
	MLS-net	82.08	83.99	86.23	87.56	89.05	90.41	92.02

It is clear that the MLS-net shows the most excellent results. We analyze the results of each dataset and see that the SVM method has a significant difference in accuracy compared to the other methods. The accuracy of the SVM differs from other methods by nearly 20% or more when the sample is insufficient. There is also a 10% difference in accuracy with a large number of samples. It can be seen that the deep-learning approach is far superior to SVM. Compared with 1D-CNN, the Siamese network model is more complex in structure. The model cleverly uses metrics for similarity calculation and incorporates few-shot learning methods.

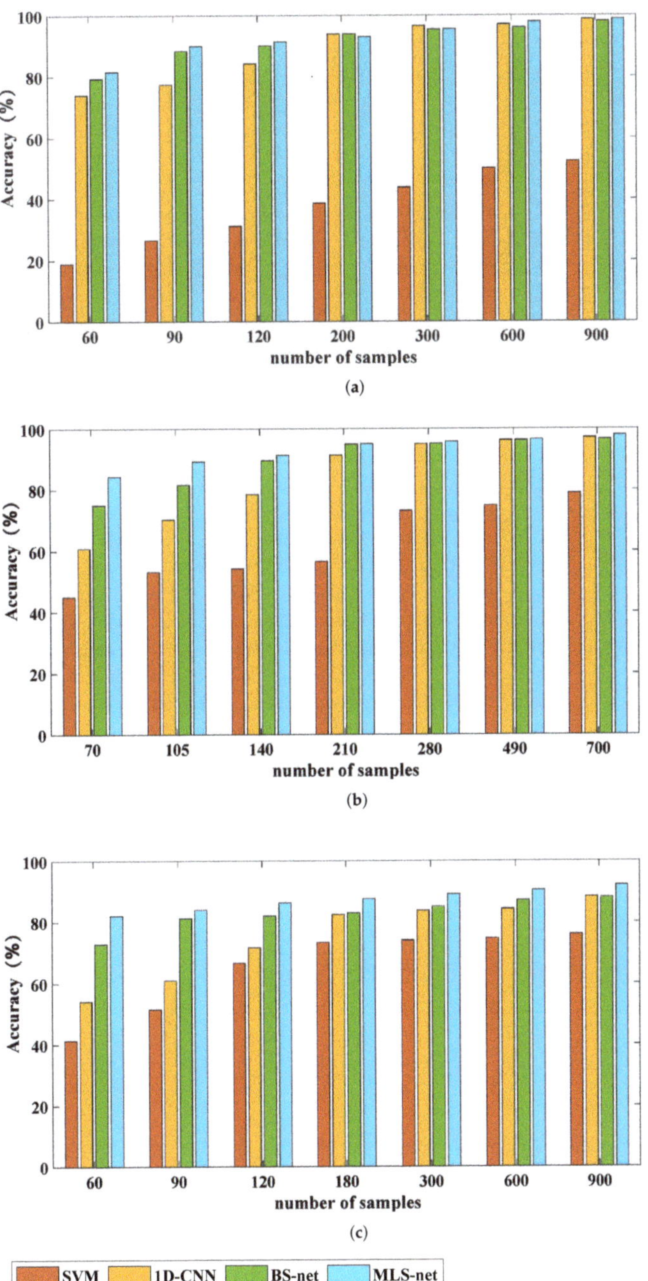

Figure 9. Comparison of sample sizes: (**a**) CWRU dataset, (**b**) MFPT dataset and (**c**) Laboratory dataset.

This makes the ability of fault classification significantly better than 1D-CNN in the case of small samples. The MLS-net is compared with BS-net by experimental data. When the samples are insufficient, the accuracy of the MLS-net is improved in all cases. It can be seen that the introduced multi-size convolutional module can obtain richer information. With the increase of samples, the accuracy of both tends to be the same. Sometimes the accuracy of the old model is higher than that of the MLS-net. The difference between the two models is within 1%. It can be seen that there is minimal loss of accuracy when the sample is sufficient.

4.5. Visualization Analysis

We attempted to obtain a better understanding of how well the model performs in the presence of insufficient samples. We would like to make further proof by using the feature visualization method of t-SNE and the confusion matrix of the test results. In Figure 10, we show the visualization of the last layer of the fully connected layer on the CWRU dataset and Laboratory dataset. The number of samples for model training is set to 90. In Figure 11, the confusion matrix plot of the test results on these two datasets is also shown. The comparison methods used in both plots are the BS-net and the MLS-net proposed in this paper.

In Figure 10, the Figure 10a,b are of the CWRU dataset. Figure 10c,d are the Laboratory dataset. As can be seen in the figure on the CWRU dataset, the MLS-net can be clearly seen on categories 1, 2 and 3 with a good distinction. Whereas, on the BS-net it shows that the three categories are mixed together and cannot be clearly distinguished. It can be seen that the BS-net is not as good at classifying as the MLS-net. This problem is more apparent in the Laboratory data set. Multiple classes are mixed together and all data distribution is discrete on the BS-net. This problem is well resolved in the plots of the MLS-net. It can be seen that the MLS-net has a better ability to classify samples with small samples.

In Figure 11, Figure 11a,b are the CWRU bearing dataset and Figure 11c,d are the Laboratory bearing dataset. As we can see in both Figure 11a,b, the number of accurate judgements for each category in Figure 11a is greater. Whereas, in Figure 11b, it is clear that the accuracy of per category is much lower. In Figure 11c,d, the comparison of the two models is also the same. This shows that the new model also has a superior performance in prediction compared to the BS-net. At the same time, the performance of the MLS-net is consistent across the different dataset. It means that the MLS-net can be applied to practical bearing fault diagnosis.

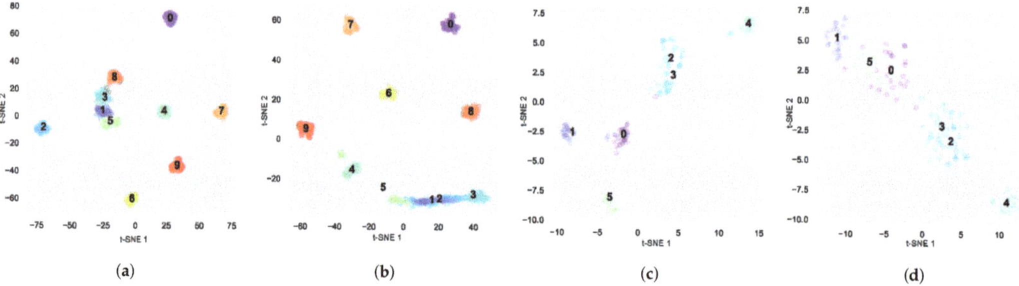

Figure 10. Visualization via t-SNE: (**a**,**b**) CWRU dataset, (**c**,**d**) Laboratory dataset.

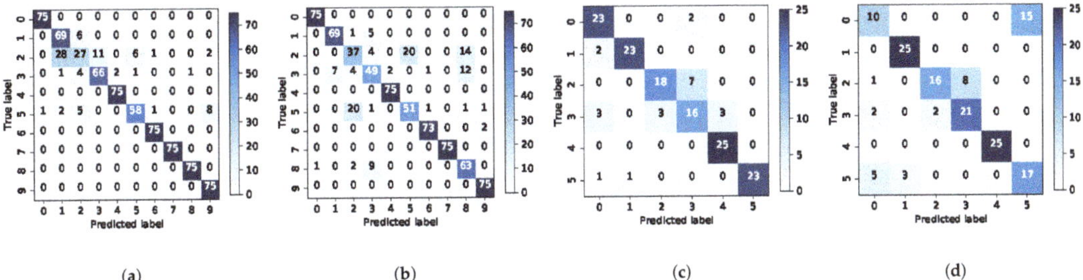

Figure 11. confusion matrix: (**a**,**b**) CWRU dataset, (**c**,**d**) Laboratory dataset.

4.6. Model Lightweight Comparison

In this subsection, the main purpose is to analyze the comparison of model size under different models and datasets. The results of the experiment mainly contain the total number of model parameters and model sizes for SVM, 1D-CNN, BS-net and MLS-net under the three dataset. The recently proposed bearing fault diagnosis models ANS-net [22] and LEFE-net [23] are also compared. We jointly determine the merit of a model based on the parameters and the accuracy rate. A model that has fewer parameters while having the higher accuracy will have superior performance. The system cost and the speed of computation will be greatly increased under fewer parameters. The details are shown in the following Figure 12 and Table 7.

In Figure 12, we mainly depict the relationship between model accuracy and total number of parameters. The horizontal coordinate indicates the model parameters. The vertical coordinate indicates the accuracy rate. The accuracy of each model is obtained from the experiment when the sample size is set as 900 for the CWRU dataset. As we can see from Figure 12, MLS-net shows the better performance in terms of the model parameters and accuracy compared with 1D-CNN and BS-net.

Although ANS-net has similar accuracy with MLS-net, the number of MLS-net parameters is only 41449, which is greatly reduced. The ANS-net, on the other hand, has far more than 100,000 parameters. LEFE-net has fewer parameters than ANS-net. However, its accuracy is lower than ANS-net and MLS-net when the training sample size is 900. When the sample drops to 60, the accuracy of LEFE-net will be further greatly reduced. Overall, MLS-net is able to run efficiently with less computation cost as well as guaranteed accuracy.

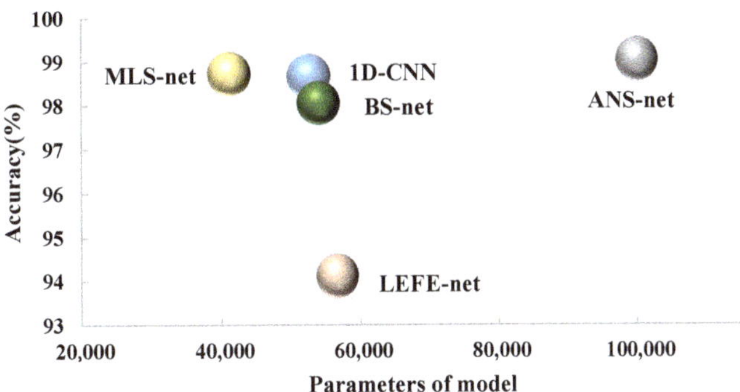

Figure 12. Relationship between model accuracy and parameters.

From Tables 6 and 7 and Figure 12, it is clear that MLS-net has a significant improvement in model size and accuracy. Specifically, SVM is 100-times larger than MLS-net in terms of model size, while its accuracy is only 50% of MLS-net under small samples. MLS-net is also more advantageous in terms of the parameters of the model. It compresses about 20% parameters in comparison with BS-net and 1D-CNN. However, MLS-net under small samples improves the accuracy compared to 1D-CNN with an improvement of 10–15% and improves the accuracy by about 2–5% compared with BS-net.

ANS-net was recently proposed as a bearing fault diagnosis model for the small sample case. Although it has a high accuracy rate under small samples, a large number of parameters (more than 100,000) are needed to ensure the accuracy. In addition, MLS-net performs better in accuracy and lightweight than the lightweight bearing fault diagnosis model LEFE-net. Through the above experiments, MLS-net is proven to have a lighter model structure and better accuracy under small samples, which can greatly improve the efficiency of bearing fault diagnosis.

Table 7. Comparison of model parameters and sizes.

Models	Datasets	Total Number of Parameters	Model Size
SVM	CWRU	-	110,837 KB
	MFPT	-	108,460 KB
	Laboratory	-	108,460 KB
1D-CNN	CWRU	52,806	663 KB
	MFPT	52,503	660 KB
	Laboratory	52,402	659 KB
BS-net	CWRU		
	MFPT	53,945	680 KB
	Laboratory		
ANS-net	CWRU		
	MFPT	>100,000	-
	Laboratory		
MLS-net	CWRU		
	MFPT	41,449	566 KB
	Laboratory		
LEFE-net	CWRU	56,640	-

5. Conclusions

In this paper, we proposed the MLS-net for the end-to-end bearing fault diagnosis problem. The model has a great ability to classify in the case of small samples. It also has a multi-scale feature fusion module to enable further feature information to be acquired. With dimensionality reduction, the model is also able to obtain comparable accuracy with fewer parameters. The model was mainly designed based on the idea of metrics. Two symmetrical sample feature extraction modules with shared parameters are contained. These are mainly used to extract the feature vectors of the two sample pairs of the input. The similarity calculation module is used to calculate the similarity of the two extracted feature vectors. Thus, the trained model has the ability to compare the similarity probability between the standard samples and predicted samples. This enables the classification of the bearing fault.

To better validate the proposed model MLS-net, we tested it on three datasets to demonstrate its performance. The results show that the model had higher accuracy with fewer parameters when the sample was insufficient compared to recently proposed methods. This proves that MLS-net as proposed in our paper makes a good tradeoff between the accuracy and computing cost. In addition, the results were consistent across the three datasets tested. This indicates that the whole model has good generalization ability for different fault datasets.

The model showed good performance by retraining the method in this paper on multiple datasets. However, the need of retraining the model each time makes the operation cumbersome. In future work, we can focus our research more on the transfer scenarios of the model and fault diagnosis in noisy environments.

Author Contributions: Conceptualization, S.G. and J.H.; methodology, S.G. and J.H.; software, J.H.; validation, J.H.; formal analysis, J.H.; investigation, H.P.; resources, S.G.; data curation, T.G.; writing—original draft preparation, J.H; writing—review and editing, S.G.; visualization, J.H.; supervision, S.G. and H.P.; project administration, S.G.; funding acquisition, S.G. All authors have read and agreed to the published version of the manuscript.

Funding: This research was funded by the Major Project of Experimental Technology Research and Development of China University of Mining and Technology: S2021Z004; the Fundamental Research Funds for the Central Universities: 2021ZDPY0204.

Institutional Review Board Statement: Not applicable.

Informed Consent Statement: Not applicable.

Data Availability Statement: Not applicable.

Conflicts of Interest: The authors declare no conflict of interest.

References

1. Wei, Y.; Li, Y.; Xu, M.; Huang, W. A review of early fault diagnosis approaches and their applications in rotating machinery. *Entropy* **2019**, *21*, 409. [CrossRef] [PubMed]
2. Zhang, S.; Zhang, S.; Wang, B.; Habetler, T.G. Deep learning algorithms for bearing fault diagnostics—A comprehensive review. *IEEE Access* **2020**, *8*, 29857–29881. [CrossRef]
3. Mohammed, S.A.; Ghazaly, N.M.; Abdo, J. Fault Diagnosis of Crack on Gearbox Using Vibration-Based Approaches. *Symmetry* **2022**, *14*, 417. [CrossRef]
4. Sikder, N.; Bhakta, K.; Al Nahid, A.; Islam, M.M. Fault diagnosis of motor bearing using ensemble learning algorithm with FFT-based preprocessing. In Proceedings of the 2019 International Conference on Robotics, Electrical and Signal Processing Techniques (ICREST), Dhaka, Bangladesh, 10–12 January 2019; pp. 564–569.
5. Zhang, K.; Ma, C.; Xu, Y.; Chen, P.; Du, J. Feature extraction method based on adaptive and concise empirical wavelet transform and its applications in bearing fault diagnosis. *Measurement* **2021**, *172*, 108976. [CrossRef]
6. Cheng, Y.; Wang, Z.; Chen, B.; Zhang, W.; Huang, G. An improved complementary ensemble empirical mode decomposition with adaptive noise and its application to rolling element bearing fault diagnosis. *ISA Trans.* **2019**, *91*, 218–234. [CrossRef] [PubMed]
7. Yu, G. A concentrated time–frequency analysis tool for bearing fault diagnosis. *IEEE Trans. Instrum. Meas.* **2019**, *69*, 371–381. [CrossRef]
8. Quinde, I.R.; Sumba, J.C.; Ochoa, L.E.; Morales-Menendez, R. Bearing fault diagnosis based on optimal time-frequency representation method. *IFAC-Pap.* **2019**, *52*, 194–199. [CrossRef]
9. Elbouchikhi, E.; Choqueuse, V.; Amirat, Y.; Benbouzid, M.E.H.; Turri, S. An efficient Hilbert–Huang transform-based bearing faults detection in induction machines. *IEEE Trans. Energy Convers.* **2017**, *32*, 401–413. [CrossRef]
10. Qian, S.; Yang, X.; Huang, J.; Zhang, H. Application of new training method combined with feedforward artificial neural network for rolling bearing fault diagnosis. In Proceedings of the 2016 23rd International Conference on Mechatronics and Machine Vision in Practice (M2VIP), Nanjing, China, 28–30 November 2016; pp. 1–6.
11. Xue, X.; Zhou, J. A hybrid fault diagnosis approach based on mixed-domain state features for rotating machinery. *ISA Trans.* **2017**, *66*, 284–295. [CrossRef]
12. Wang, T.; Liu, Z.; Lu, G.; Liu, J. Temporal-spatio graph based spectrum analysis for bearing fault detection and diagnosis. *IEEE Trans. Ind. Electron.* **2020**, *68*, 2598–2607. [CrossRef]
13. Goyal, D.; Choudhary, A.; Pabla, B.; Dhami, S. Support vector machines based non-contact fault diagnosis system for bearings. *J. Intell. Manuf.* **2020**, *31*, 1275–1289. [CrossRef]
14. Wang, X.; Mao, D.; Li, X. Bearing fault diagnosis based on vibro-acoustic data fusion and 1D-CNN network. *Measurement* **2021**, *173*, 108518. [CrossRef]
15. Hsueh, Y.M.; Ittangihal, V.R.; Wu, W.B.; Chang, H.C.; Kuo, C.C. Fault diagnosis system for induction motors by CNN using empirical wavelet transform. *Symmetry* **2019**, *11*, 1212. [CrossRef]
16. Chen, X.; Zhang, B.; Gao, D. Bearing fault diagnosis base on multi-scale CNN and LSTM model. *J. Intell. Manuf.* **2021**, *32*, 971–987. [CrossRef]
17. Goodfellow, I.; Pouget-Abadie, J.; Mirza, M.; Xu, B.; Warde-Farley, D.; Ozair, S.; Courville, A.; Bengio, Y. Generative adversarial nets. *arXiv* **2014**, arXiv:1406.2661.
18. Xia, M.; Li, T.; Xu, L.; Liu, L.; De Silva, C.W. Fault diagnosis for rotating machinery using multiple sensors and convolutional neural networks. *IEEE/ASME Trans. Mechatron.* **2017**, *23*, 101–110. [CrossRef]

19. Wang, R.; Jiang, H.; Li, X.; Liu, S. A reinforcement neural architecture search method for rolling bearing fault diagnosis. *Measurement* **2020**, *154*, 107417. [CrossRef]
20. Sung, F.; Yang, Y.; Zhang, L.; Xiang, T.; Torr, P.H.; Hospedales, T.M. Learning to compare: Relation network for few-shot learning. In Proceedings of the IEEE conference on computer vision and pattern recognition, Salt Lake City, UT, USA, 18–23 June 2018; pp. 1199–1208.
21. Snell, J.; Swersky, K.; Zemel, R.S. Prototypical networks for few-shot learning. *arXiv* **2017**, arXiv:1703.05175.
22. Fang, Q.; Wu, D. ANS-net: Anti-noise Siamese network for bearing fault diagnosis with a few data. *Nonlinear Dyn.* **2021**, *104*, 2497–2514. [CrossRef]
23. Fang, H.; Deng, J.; Zhao, B.; Shi, Y.; Zhou, J.; Shao, S. LEFE-Net: A lightweight efficient feature extraction network with strong robustness for bearing fault diagnosis. *IEEE Trans. Instrum. Meas.* **2021**, *70*, 1–11. [CrossRef]
24. Xiong, S.; Zhou, H.; He, S.; Zhang, L.; Shi, T. Fault diagnosis of a rolling bearing based on the wavelet packet transform and a deep residual network with lightweight multi-branch structure. *Meas. Sci. Technol.* **2021**, *32*, 085106. [CrossRef]
25. Zhang, S.; Ye, F.; Wang, B.; Habetler, T.G. Few-Shot Bearing Fault Diagnosis Based on Model-Agnostic Meta-Learning. *IEEE Trans. Ind. Appl.* **2021**, *57*, 4754–4764. [CrossRef]
26. Szegedy, C.; Liu, W.; Jia, Y.; Sermanet, P.; Reed, S.; Anguelov, D.; Erhan, D.; Vanhoucke, V.; Rabinovich, A. Going deeper with convolutions. In Proceedings of the IEEE Conference on Computer Vision and Pattern Recognition, Boston, MA, USA, 7–12 June 2015; pp. 1–9.
27. Chicco, D. Siamese neural networks: An overview. *Artif. Neural Netw.* **2021**, *2190*, 73–94. [CrossRef]
28. Wang, Y.; Yao, Q.; Kwok, J.T.; Ni, L.M. Generalizing from a few examples: A survey on few-shot learning. *ACM Comput. Surv.* **2020**, *53*, 1–34. [CrossRef]
29. Zhang, W.; Peng, G.; Li, C.; Chen, Y.; Zhang, Z. A new deep learning model for fault diagnosis with good anti-noise and domain adaptation ability on raw vibration signals. *Sensors* **2017**, *17*, 425. [CrossRef]
30. Loparo, K. Case western reserve university bearing data center. *Bearings Vibration Data Sets*; Case Western Reserve University: Cleveland, OH, USA, 2012; pp. 22–28. Available online: https://engineering.case.edu/bearingdatacenter/download-data-file (accessed on 1 October 2021).
31. Dataset, M. Society for Machinery Failure Prevention Technology. 2012. Available online: https://www.mfpt.org/exhibitor-info/proceedings/ (accessed on 1 October 2021).
32. Zhang, A.; Li, S.; Cui, Y.; Yang, W.; Dong, R.; Hu, J. Limited data rolling bearing fault diagnosis with few-shot learning. *IEEE Access* **2019**, *7*, 110895–110904. [CrossRef]

Article

Velocity-Free State Feedback Fault-Tolerant Control for Satellite with Actuator and Sensor Faults

Mingjun Liu [1], Aihua Zhang [1] and Bing Xiao [2],*

1 School of Control Science and Engineering, Bohai University, Jinzhou 121013, China; ys_mingjun_liu@163.com (M.L.); zhangaihua@qymail.bhu.edu.cn (A.Z.)
2 School of Automation, Northwestern Polytechnical University, Xi'an 710129, China
* Correspondence: xiaobing@nwpu.edu.cn

Abstract: A velocity-free state feedback fault-tolerant control approach is proposed for the rigid satellite attitude stabilization problem subject to velocity-free measurements and actuator and sensor faults. First, multiplicative faults and additive faults are considered in the actuator and the sensor. The faults and system states are extended into a new augmented vector. Then, an improved sliding mode observer based on the augmented vector is presented to estimate unknown system states and actuator and sensor faults simultaneously. Next, a velocity-free state feedback attitude controller is designed based on the information from the observer. The controller compensates for the effects of actuator and sensor faults and asymptotically stabilizes the attitude. Finally, simulation results demonstrate the effectiveness of the proposed scheme.

Keywords: attitude stabilization; fault reconstruction; fault-tolerant control; sliding mode observer; state feedback

Citation: Liu, M.; Zhang, A.; Xiao, B. Velocity-Free State Feedback Fault-Tolerant Control for Satellite with Actuator and Sensor Faults. *Symmetry* **2022**, *14*, 157. https://doi.org/10.3390/sym14010157

Academic Editor: Jan Awrejcewicz

Received: 23 December 2021
Accepted: 11 January 2022
Published: 13 January 2022

Publisher's Note: MDPI stays neutral with regard to jurisdictional claims in published maps and institutional affiliations.

1. Introduction

As an important component of the satellite, the attitude control system plays a key role in practical aerospace missions, such as space on-orbit services and spacecraft pointing and turning. Numerous studies on attitude control methods have emerged, such as adaptive variable structure control (VSC) [1,2], robust control [3–7], output feedback control [8,9], time-delayed control [10], and finite-time control [11–13]. The premise of these control methods is the assumption that there exists no actuator or sensor fault during satellite maneuvers. However, due to the harsh space environment in satellite operation, actuator and sensor faults are inevitable. If the designed attitude control system does not have the ability to deal with the faults, it may lead to the failure of the target space missions or even the destruction of the satellite [14]. Inspired by this problem, this work mainly studies the fault tolerant control (FTC) of attitude stabilization guaranteed in the case of actuator and sensor faults.

For actuator faults or sensor faults, some scholars have used observer methods to estimate fault values. In [15], a fault-tolerant control method based on the iterative learning observer was proposed. Fault-tolerant control and closed-loop control assignment were achieved. In [16], a fixed-time observer was presented to estimate the lumped disturbances, including actuator faults and external disturbances. At the same time, a fixed-time attitude controller was presented according to the homogeneity, estimated disturbances, and integral sliding mode. For gyroscope constant deviation, a coupled quaternion filter and a bias observer were employed to achieve attitude control in [17]. For the linear parameter varying (LPV) system, a reduced-order LPV observer was considered to estimate unmeasured states and sensor faults in [18], reducing the computation of full-order estimation. Refs. [19,20] proposed an adaptive fault-tolerant attitude controller based on VSC. Compared with the observer method, this method does not need accurate fault information and compensates for fault effects by adaptive law. Different from model-based observers, some scholars have used the neural network algorithm to solve fault problems. In [21], the uncertain terms and the fault boundary of the system were estimated by using the neural network and the online update law, respectively. Based on estimations, a modified fault-tolerant control law was designed to achieve global asymptotic stability of attitude. In [22], the recursive neural network was considered to detect and isolate the actuator and sensor faults of the satellite attitude subsystem. The impacts of component faults on the system were well solved in the above literature, but either actuator faults or sensor faults were considered, without discussing the simultaneous faults of both actuators and sensors. For the systems with simultaneous actuator and sensor faults, some observation schemes have been proposed in [23–26] and applied to the circuit model and the vehicle model.

The above methods require the satellite attitude and angular velocity to be measurable. However, in practical applications, angular velocity measurements may not be available due to sensor faults or reduced satellite costs [27]. Therefore, a velocity-free attitude control system becomes the trend in satellite development. Considering the unavailability of angular velocity measurements, a velocity-free attitude stabilization control scheme relying solely on attitude information was presented in [28]. In [29–31], finite-time observers were designed to estimate angular velocity, which was applied to attitude stability control [29,30] and attitude synchronization control [31]. In [32], the finite-time observer based on the neural network was used to obtain the unknown angular velocity. Compared with the method in [29–31], it did not require accurate knowledge of the system model. To obtain the angular velocity faster and more stably, based on the fixed-time theory in [33], a fixed-time angular velocity observer was designed to complete satellite formation control in [34]. If the actuator faults are also considered, the design of the control system encounters greater challenges. In the face of actuator faults and velocity-free measurements, [35] designed two finite-time observers and presented a fault-tolerant controller with attitude information only. In [36], an adaptive fault-tolerant controller was proposed based on neural networks using the information from the finite-time observer.

Although there are credible results for single fault and velocity-free measurements in the above literature, the simultaneous occurrence of actuator faults, sensor faults, and velocity-free measurements are not considered. When the above problems occur simultaneously, the controller design will face major challenges: (1) Velocity-free measurements lead to the reduction of measurable information. (2) The simultaneous faults of actuator and sensor lead to the complexity of fault information, which increases the difficulty of fault detection and compensation. (3) Velocity-free measurements and sensor faults lead to the lack of accurate attitude information. To solve these problems, a velocity-free state feedback fault-tolerant control scheme is proposed in this paper. The main contributions of this work are summarized as follows:

(a) An improved sliding mode observer is proposed to estimate system states and faults simultaneously. Compared with the observer in [26], the steady-state performance is improved.

(b) The multiplicative faults and additive faults of actuator and sensor are considered. The designed scheme is able to tolerate the lumped faults. The controller presented has a strong fault-tolerance ability such that the closed-loop attitude system is asymptotically stable.

(c) The proposed fault-tolerant control scheme does not require angular velocity measurements, which reduces satellite mass and the cost of airborne sensors.

The remainder of this paper is organized as follows. In Section 2, satellite attitude dynamics and actuator and sensor faults models are described. The required mathematical preliminaries are also given in this part. In Section 3, the proposed improved sliding mode observer and the state feedback fault-tolerant attitude controller are presented, respectively. Numerical simulation is provided to demonstrate the effectiveness of the proposed scheme in Section 4. Conclusions are given in Section 5.

2. Preliminaries

2.1. Notations and Lemmas

$I_n \in \Re^{n \times n}$ represents the identity matrix with a dimension n. $0_{n \times m} \in \Re^{n \times m}$ is an n by m zero matrix. $\| \cdot \|$ stands for the induced norm of a matrix or the Euclidean norm of a vector. $\lambda_{\min}(\cdot)$ and $\lambda_{\max}(\cdot)$ denote the minimum and maximum eigenvalues of a matrix, respectively. For a given scalar $\alpha > 0$ and a vector $x = [\, x_1 \quad x_2 \quad \cdots \quad x_n \,]^T \in \Re^n$, the notation can be defined as follows: $\mathrm{sig}^\alpha(x) = [\, \mathrm{sig}^\alpha(x_1) \quad \mathrm{sig}^\alpha(x_2) \quad \cdots \quad \mathrm{sig}^\alpha(x_n) \,]^T$, where $\mathrm{sig}^\alpha(x_i) = |x_i|_\alpha \mathrm{sgn}(x_i)$ and $\mathrm{sgn}(\cdot)$ denotes the sign function.

The function $\Gamma(x, \alpha, \beta)$ is defined as

$$\Gamma(x, \alpha, \beta) = \begin{cases} \mathrm{sgn}^\alpha(x) \; \|x\| \leq 1 \\ \mathrm{sgn}^\beta(x) \; \|x\| > 1 \end{cases} \tag{1}$$

where $x \in \Re^n, 0 < \alpha < 1$, and $\beta > 1$.

For a given vector $a = [\, a_1 \quad a_2 \quad a_3 \,]^T \in \Re^3$, the notation $a^\times$ indicates the skew-symmetric matrix:

$$a^\times = \begin{bmatrix} 0 & -a_3 & a_2 \\ a_3 & 0 & -a_1 \\ -a_2 & a_1 & 0 \end{bmatrix} \tag{2}$$

Lemma 1 [33]. *Consider the following nonlinear system of differential equations:*

$$\dot{x} = f(t, x), x(0) = x_0 \tag{3}$$

where $x \in \Re^n$ is the system state and $f : \Re^+ \times \Re^n \to \Re^n$ is a continuous function defined in an open neighborhood U of the origin. If there is a positive definite function $V(x) : \Re^n \to \Re$ satisfying $\dot{V}(x) \leq -(\alpha V^p(x) + \beta V^q(x)) + \varsigma$, where $\alpha, \beta \in \Re^+, 0 < p < 1, q > 1$, and $0 < \varsigma < \infty$, the trajectory of Equation (3) is practical fixed-time stable. The settling time T for the system to reach a steady state satisfies $T \leq 1/\alpha(1 - p) + 1/\beta(q - 1)$.

2.2. Satellite Attitude Dynamics Model

To describe the satellite attitude, three coordinate systems are commonly used: the inertial fixed reference coordinate frame $\mathcal{F}_I$, the orbital coordinate frame $\mathcal{F}_o$, and the body-fixed coordinate frame $\mathcal{F}_b$. The angular velocity ω_r of the body-fixed coordinate frame relative to the orbital coordinate frame is obtained by a yaw–pitch–roll sequence of rotations. It can be described by [35]

$$\omega_r = R(\Theta)\dot{\Theta} = \begin{bmatrix} 1 & 0 & -\sin\theta \\ 0 & \cos\phi & \cos\theta\sin\phi \\ 0 & -\sin\phi & \cos\theta\cos\phi \end{bmatrix} \begin{bmatrix} \dot\phi \\ \dot\theta \\ \dot\psi \end{bmatrix} \tag{4}$$

where $\Theta = [\, \phi \quad \theta \quad \psi \,]^T \in \Re^3$ is the attitude Euler angle vector. The angular velocity $\omega = [\, \omega_x \quad \omega_y \quad \omega_z \,]^T \in \Re^3$ of the body frame with respect to the inertial frame in the body frame is defined as

$$\omega = \omega_r + \omega_O \tag{5}$$

where

$$\omega_O = - \begin{bmatrix} \cos\theta\sin\psi \\ \cos\phi\cos\psi + \sin\phi\sin\theta\sin\psi \\ -\sin\phi\cos\psi + \cos\phi\sin\theta\sin\psi \end{bmatrix} \omega_0 \tag{6}$$

Here, ω_0 is the orbital angular velocity. Substituting Equations (4) and (6) into Equation (5) and taking into account small Euler angle rotations, the attitude kinematics can be given by [21]

$$\omega = \begin{bmatrix} \omega_x & \omega_y & \omega_z \end{bmatrix}^T = \begin{bmatrix} \dot{\phi} - \omega_0\psi & \dot{\theta} - \omega_0 & \dot{\psi} + \omega_0\phi \end{bmatrix}^T \tag{7}$$

The angular momentum is expressed as $H = I_b\omega$. In the case of considering the gravity gradient torque and the external disturbance torque, the attitude dynamics model of the rigid satellite can be described by [35]

$$I_b\dot{\omega} + \omega^\times I_b\omega = \tau + \tau_d + \tau_g \tag{8}$$

where $\tau = \begin{bmatrix} \tau_1 & \tau_2 & \tau_3 \end{bmatrix}^T \in \Re^3$ is the total torque generated by the actuator, $\tau_d = \begin{bmatrix} \tau_{d1} & \tau_{d2} & \tau_{d3} \end{bmatrix}^T \in \Re^3$ is the external disturbance torque, and $I_b = \mathrm{diag}(\begin{bmatrix} I_{b1} & I_{b2} & I_{b3} \end{bmatrix})$ is the inertia matrix. In the case of a small attitude angle maneuver, the gravity gradient torque $\tau_g = \begin{bmatrix} \tau_{g1} & \tau_{g2} & \tau_{g3} \end{bmatrix}^T \in \Re^3$ can be approximated as

$$\tau_g = \begin{bmatrix} -3\omega_0^2(I_{b2} - I_{b3})\phi & -3\omega_0^2(I_{b1} - I_{b3})\theta & 0 \end{bmatrix}^T \tag{9}$$

Substituting Equations (7) and (9) into Equation (8) yields

$$I_b\ddot{\Theta} + a_1\dot{\Theta} + a_0\Theta = \tau + \tau_d \tag{10}$$

Define a new state $x = \begin{bmatrix} x_1^T & x_2^T \end{bmatrix}^T = \begin{bmatrix} \Theta^T & \dot{\Theta}^T \end{bmatrix}^T$. Then, Equation (10) can be rewritten as:

$$\begin{cases} \dot{x} = Ax + B\tau + D\tau_d \\ y = Cx \end{cases} \tag{11}$$

where $x \in \Re^6$, $\tau \in \Re^3$, $\tau_d \in \Re^3$, and $y \in \Re^6$ denote system state, total torque, external disturbance torque, and measurement output, respectively. The system parameter matrixes $A \in \Re^{6\times6}$, $B \in \Re^{6\times3}$, $D \in \Re^{6\times3}$, and $C \in \Re^{6\times6}$ are defined as

$$a_0 = \omega_0^2 \begin{bmatrix} 4(I_{b2} - I_{b3}) & 0 & 0 \\ 0 & 3(I_{b1} - I_{b3}) & 0 \\ 0 & 0 & I_{b2} - I_{b1} \end{bmatrix}, a_1 = \omega_0 \begin{bmatrix} 0 & 0 & -I_{b1} + I_{b2} - I_{b3} \\ 0 & 0 & 0 \\ I_{b1} - I_{b2} + I_{b3} & 0 & 0 \end{bmatrix},$$

$$A = \begin{bmatrix} 0_3 & I_3 \\ -I_b^{-1}a_0 & -I_b^{-1}a_1 \end{bmatrix}, B = D = \begin{bmatrix} 0_3 \\ I_b^{-1} \end{bmatrix}, c_1 = \begin{bmatrix} 0 & 0 & -\omega_0 \\ 0 & 0 & 0 \\ \omega_0 & 0 & 0 \end{bmatrix}, C = \begin{bmatrix} I_3 & 0_3 \\ c_1 & I_3 \end{bmatrix}.$$

2.3. Faults Model

Referring to the definition of actuator faults in [36], the actuator faults can be divided into multiplicative faults and additive faults with the form

$$\tau = \rho u + u_f \tag{12}$$

where $u = \begin{bmatrix} u_1 & u_2 & u_3 \end{bmatrix}^T \in \Re^3$ represents the actuator commanded control torque; $\rho = \mathrm{diag}(\begin{bmatrix} \rho_1 & \rho_2 & \rho_3 \end{bmatrix}^T) \in \Re^{3\times3}$ denotes the degree of actuator failure, $0 < \rho_i \leq 1$, where $i = 1, 2, 3$; and $u_f \in \Re^3$ stands for actuator additive faults. Equation (12) can be rewritten as

$$\tau = u + f_a \tag{13}$$

where the lumped actuator faults $f_a \in \Re^3$ can be defined as

$$f_a = (\rho - I_3)u + u_f \tag{14}$$

In this paper, the fault-tolerant control of the satellite is studied under the condition of velocity-free measurements so that the actual measurement output only contains the attitude angle information. Considering partial failures and additive faults of the sensor, the expression of measurement output can be defined as

$$y_a = \delta y + y_f \tag{15}$$

where $y_a \in \Re^3$ is the actual output of the sensor; $y_f \in \Re^3$ is sensor additive faults; and $\delta = \mathrm{diag}([\ \delta_1 \quad \delta_2 \quad \delta_3\ ]^T) \in \Re^{3\times 3}$ denotes the degree of sensor failure, $0 < \delta_i \le 1$, where $i = 1, 2, 3$. Equation (14) can be rewritten as

$$y_a = Cx + f_s \tag{16}$$

where the lumped sensor faults $f_s \in \Re^3$ can be defined as

$$f_s = (\delta - I_3)y + y_f \tag{17}$$

Combined with Equations (13) and (16), the system dynamics with actuator and sensor faults can be written as

$$\begin{cases} \dot{x} = Ax + B(u + f_a) + D\tau_d \\ y_a = Cx + f_s \end{cases} \tag{18}$$

where $f_a \in \Re^3$ and $f_s \in \Re^3$ denote the lumped faults of the actuator and the lumped faults of the sensor, respectively. The parameter matrix $C = [\ I_3 \quad 0_3\ ] \in \Re^{3\times 6}$. According to Equation (18), the following assumptions are made:

Assumption 1. *There exist positive constants k_Θ and k_ω such that $||\Theta|| \le k_\Theta$ and $||\omega|| \le k_\omega$ for all $t \ge 0$.*

Assumption 2. *The external disturbances and actuator and sensor faults satisfy $||\tau_d|| \le r_d$, $||f_s|| \le r_s$, $||f_a|| \le r_a$, and $||\dot{f}_a|| \le r_{a1}$, where $r_d > 0$, $r_s > 0$, $r_a > 0$, and $r_{a1} > 0$ are known constants. There is a known constant $\eta > 0$ such that $r_d + r_a + r_s + r_{a1} \le \eta$.*

Remark 1. *For the feedback control problem, the assumption of bounded system states and disturbances is necessary. Due to the physical limitation of the equipment, the attitude of the satellite and the output torque of the actuator are limited in practical engineering. If the external disturbance is infinite, the attitude system will not be controllable. Similar assumptions also can be found in related literature, such as Assumption 1 in [29] and Assumption 2 in [26].*

2.4. Problem Statement

The objective of this work is stated as follows: for the satellite with simultaneous actuator faults, sensor faults, and velocity-free measurements, the designed observer is required to estimate the unknown system states and faults in real time. The velocity-free fault-tolerant controller is provided to asymptotically stabilize the attitude, i.e., $\Theta \to 0$ and $\omega \to 0$, even in the presence of external disturbances and actuator and sensor faults.

3. Observer-Based State Feedback Attitude Controller Design

In this section, a velocity-free state feedback fault-tolerant attitude controller (VSFTC) will be proposed to stabilize the attitude of the satellite. The structure of the closed-loop system is shown in Figure 1. This control structure includes two modules: an improved sliding mode observer is designed to estimate Θ, ω, f_a, and f_s and a velocity-free controller is proposed by using the estimate information of the observer.

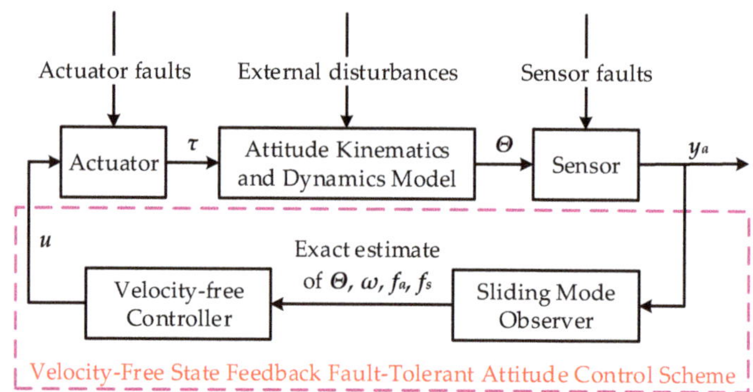

Figure 1. The structure of the proposed fault-tolerant attitude control system without velocity measurements.

3.1. Improved Sliding Mode Observer Design

According to Equation (18), the augmented system is constructed as follows:

$$\begin{cases} \overline{E}\dot{\overline{x}} = \overline{A}\overline{x} + \overline{B}u + \overline{D}\overline{d} \\ y_a = \overline{C}\overline{x} \end{cases} \tag{19}$$

where

$$\overline{x} = \begin{bmatrix} x \\ f_a \\ f_s \end{bmatrix}, \overline{d} = \begin{bmatrix} \tau_d \\ f_a + d_a \\ f_s \end{bmatrix}, \overline{E} = \begin{bmatrix} I_6 & 0_{6\times3} & 0_{6\times3} \\ 0_{3\times6} & I_3 & 0_3 \\ 0_{3\times6} & 0_3 & 0_3 \end{bmatrix}, \tag{20}$$

$$\overline{B} = \begin{bmatrix} B \\ 0_{6\times3} \end{bmatrix}, \overline{C} = \begin{bmatrix} C & 0_3 & I_3 \end{bmatrix}, \tag{21}$$

$$\overline{A} = \begin{bmatrix} A & B & 0_{6\times3} \\ 0_{3\times6} & -I_3 & 0_3 \\ 0_{3\times6} & 0_3 & -I_3 \end{bmatrix}, \overline{D} = \begin{bmatrix} D & 0_{6\times3} & 0_{6\times3} \\ 0_3 & I_3 & 0_3 \\ 0_3 & 0_3 & I_3 \end{bmatrix} \tag{22}$$

Let $S = \overline{E} + \overline{F}\overline{C} \in \Re^{12\times12}$; $\overline{F} \in \Re^{12\times3}$ is selected as

$$\overline{F} = \begin{bmatrix} 0_3^T & 0_3^T & 0_3^T & F^T \end{bmatrix}^T \tag{23}$$

where $F = \text{diag}(\begin{bmatrix} \sigma_1 & \sigma_2 & \sigma_3 \end{bmatrix}^T) \in \Re^{3\times3}$ and $\sigma_i > 0, i = 1, 2, 3$, such that S is non-singular.

Define $\overline{\overline{A}} = S^{-1}\overline{A}, \overline{\overline{B}} = S^{-1}\overline{B}, \overline{\overline{D}} = S^{-1}\overline{D}$, and $\overline{\overline{F}} = S^{-1}\overline{F}$. Then Equation (19) can be rewritten as

$$\begin{cases} \dot{\overline{x}} = \overline{\overline{A}}\overline{x} + \overline{\overline{B}}u + \overline{\overline{D}}d + \overline{\overline{F}}\dot{y}_a \\ y_a = \overline{C}\overline{x} \end{cases} \tag{24}$$

Let $\hat{\overline{x}}$ and $\hat{y}_a$ be estimates of $\overline{x}$ and y_a, respectively. The improved sliding mode observer is proposed as follows:

$$\dot{\hat{\overline{x}}} = \overline{\overline{A}}\hat{\overline{x}} + \overline{\overline{B}}u + \overline{\overline{D}}\Delta + \overline{\overline{F}}\dot{y}_a - L\Gamma(\overline{C}e, \alpha, \beta) - kL(\dot{\hat{y}}_a - \dot{y}_a) \tag{25}$$

where $e = \hat{\overline{x}} - \overline{x} = \begin{bmatrix} e_1^T & e_2^T & e_3^T & e_4^T \end{bmatrix}^T$ is the system state error vector; $0 < \alpha < 1, \beta > 1, k > 0$, and $L = \begin{bmatrix} L_1^T & L_2^T & L_3^T & L_4^T \end{bmatrix}^T \in \Re^{12\times3}$ are the gain matrixes; $L_i = l_i I_3$; $l_i > 0$, where $i = 1, 2, 3, 4$; and Δ is the compensation input. According to Assumption 2, Δ is designed as

$$\Delta = -(\eta + \varepsilon)\text{sgn}(\overline{\overline{D}}^T N^{-T} Pe) \tag{26}$$

where $\varepsilon > 0$ is a small scalar quantity and $N, P \in \Re^{12 \times 12}$ is the parameter matrix to be designed.

By defining $z = \hat{\bar{x}} - \bar{\bar{F}} y_a + kL(\hat{y}_a - y_a)$, Equation (25) can be rewritten as

$$
\begin{cases}
\dot{z} = \bar{\bar{A}} z + \bar{\bar{B}} u + \bar{\bar{D}} \Delta + \bar{\bar{A}}(\bar{F} + kL) y_a - k\bar{\bar{A}} L \hat{y}_a - L\Gamma(\bar{C}e, \alpha, \beta) \\
\qquad \hat{\bar{x}} = z + (\bar{F} + kL) y_a - kL\hat{y}_a
\end{cases}
\tag{27}
$$

where $z \in \Re^{12}$ is the system state and $\hat{\bar{x}}$ is the system output.

According to Equations (24) and (25), the estimation errors can be given by

$$
\dot{e} = \bar{\bar{A}} e + \bar{\bar{D}} \Delta - \bar{\bar{D}} d - L\Gamma(\bar{C}e, \alpha, \beta) - kL\bar{C}\dot{e}
\tag{28}
$$

Define a matrix $N = I_{12} + kL\bar{C} \in \Re^{12 \times 12}$. According to the definition of L and Equation (21), N is a non-singular matrix. Then, Equation (28) can be rewritten as

$$
\dot{e} = N^{-1}(\bar{\bar{A}} e + \bar{\bar{D}} \Delta - \bar{\bar{D}} d - L\Gamma(\bar{C}e, \alpha, \beta))
\tag{29}
$$

Theorem 1. *Consider the plant (18) subject to the actuator faults (14), sensor faults (17), and the external disturbances under Assumption 2. If the observer is designed according to Equation (27), there exist appropriate gain matrixes $L \in \Re^{12 \times 3}$, $P \in \Re^{12 \times 12} > 0$, and $Q \in \Re^{12 \times 12} > 0$, satisfying:*

$$
(\bar{\bar{A}} - L\bar{C})^T N^{-T} P + PN^{-1}(\bar{\bar{A}} - L\bar{C}) + Q \leq 0
\tag{30}
$$

such that the estimated states $\hat{\bar{x}}$ converges asymptotically to its actual states $\bar{x}$ and the error dynamic system of Equation (29) is ultimately uniformly bounded.

Proof of Theorem 1. Define a Lyapunov function $V_1 = e^T P e$. The time derivative of V_1 along the trajectories of the error dynamics in Equation (29) leads to

$$
\dot{V}_1 = 2e^T PN^{-1}\bar{\bar{A}} e + 2e^T PN^{-1}\bar{\bar{D}} \Delta - 2e^T PN^{-1}\bar{\bar{D}} d - 2e^T PN^{-1}L\Gamma(\bar{C}e, \alpha, \beta)
\tag{31}
$$

According to Equation (1), $\Gamma(x, \alpha, \beta)$ is a piecewise function. We discuss $\dot{V}_1$ in two parts: Part 1 ($||\bar{C}e|| \leq 1$) and Part 2 ($||\bar{C}e|| > 1$).

Part 1: $||\bar{C}e|| \leq 1$. Based on Equation (1), Equation (31) can be rewritten as

$$
\dot{V}_1 = 2e^T PN^{-1}\bar{\bar{A}} e + 2e^T PN^{-1}\bar{\bar{D}} \Delta - 2e^T PN^{-1}\bar{\bar{D}} d - 2e^T PN^{-1}L\mathrm{sig}^\alpha(\bar{C}e)
\tag{32}
$$

Since $||\bar{C}e|| \leq 1$ and $0 < \alpha < 1$, we have $||\bar{C}e||_\alpha \geq ||\bar{C}e||$. Substituting it into Equation (32) yields

$$
\dot{V}_1 \leq 2e^T PN^{-1}\bar{\bar{A}} e - 2e^T PN^{-1}L\bar{C}e + 2e^T PN^{-1}\bar{\bar{D}} \Delta - 2e^T PN^{-1}\bar{\bar{D}} d
\tag{33}
$$

According to Assumption 2, $||\bar{d}|| \leq \eta$. If the inequality in Equation (30) holds, substituting Equation (26) into Equation (33), we have

$$
\begin{aligned}
\dot{V}_1 &\leq 2e^T PN^{-1}\bar{\bar{A}} e - 2e^T PN^{-1}L\bar{C}e + 2e^T PN^{-1}\bar{\bar{D}} \Delta + 2\eta ||e^T PN^{-T}\bar{\bar{D}}|| \\
&\leq e^T((\bar{\bar{A}} - L\bar{C})^T N^{-T} P + PN^{-1}(\bar{\bar{A}} - L\bar{C}))e + 2||e^T PN^{-T}\bar{\bar{D}}||(-(\varepsilon + \eta) + \eta) \\
&\leq -e^T Q e - 2\varepsilon ||e^T PN^{-T}\bar{\bar{D}}||
\end{aligned}
\tag{34}
$$

From the standard inequality for quadratic forms, we obtain $\lambda_{\min}(Q)||e||_2 \leq V_1 \leq \lambda_{\max}(Q)||e||^2$. For $\varepsilon > 0$, the inequality in Equation (34) can also be rewritten as

$$
\dot{V}_1 \leq -\lambda_{\min}(Q)||e||_2 \leq 0
\tag{35}
$$

Part 2: $||\overline{C}e|| > 1$. Based on Equation (1), Equation (31) can be rewritten as

$$\dot{V}_1 = 2e^T P N^{-1} \overline{\overline{A}} e + 2e^T P N^{-1} \overline{\overline{D}} \Delta - 2e^T P N^{-1} \overline{\overline{D}} d - 2e^T P N^{-1} L \text{sig}^\beta(\overline{C}e) \tag{36}$$

Since $||\overline{C}e|| > 1$ and $\beta > 1$, we have $||\overline{C}e||_\beta \geq ||\overline{C}e||$. Substituting it into Equation (36) yields

$$\dot{V}_1 \leq 2e^T P N^{-1} \overline{\overline{A}} e - 2e^T P N^{-1} L \overline{C}e + 2e^T P N^{-1} \overline{\overline{D}} \Delta - 2e^T P N^{-1} \overline{\overline{D}} d \tag{37}$$

Following the same procedure in Part 1, $\dot{V}_1$ satisfies

$$\dot{V}_1 \leq -\lambda_{\min}(Q)||e||_2 \leq 0 \tag{38}$$

With the combination of Part 1 and Part 2, we have $\dot{V}_1 \leq 0$ when Equation (30) holds. According to the Lyapunov stability theorem, the errors system in Equation (29) is asymptotically stable. This completes the proof of Theorem 1. $\square$

3.2. Velocity-Free State Feedback Fault-Tolerant Attitude Controller Design

The parameter matrix of the observer in Equation (25) is expressed as follows:

$$\overline{\overline{A}} = \begin{bmatrix} \overline{\overline{A}}_{11} & \overline{\overline{A}}_{12} & \overline{\overline{A}}_{13} & \overline{\overline{A}}_{14} \\ \overline{\overline{A}}_{21} & \overline{\overline{A}}_{22} & \overline{\overline{A}}_{23} & \overline{\overline{A}}_{24} \\ \overline{\overline{A}}_{31} & \overline{\overline{A}}_{32} & \overline{\overline{A}}_{33} & \overline{\overline{A}}_{34} \\ \overline{\overline{A}}_{41} & \overline{\overline{A}}_{42} & \overline{\overline{A}}_{43} & \overline{\overline{A}}_{44} \end{bmatrix}, \overline{\overline{B}} = \begin{bmatrix} \overline{\overline{B}}_1 \\ \overline{\overline{B}}_2 \\ \overline{\overline{B}}_3 \\ \overline{\overline{B}}_4 \end{bmatrix}, \overline{\overline{D}} = \begin{bmatrix} \overline{\overline{D}}_1 \\ \overline{\overline{D}}_2 \\ \overline{\overline{D}}_3 \\ \overline{\overline{D}}_4 \end{bmatrix}$$

where $\overline{\overline{A}}_{i,j} \in R^{3\times3}$, $\overline{\overline{B}}_i \in R^{3\times3}$, and $\overline{\overline{D}}_i \in R^{3\times3}$, where $i,j = 1,2,3,4$. According to the parameter setting in Equations (11) and (19), $\overline{\overline{A}}_{1,2} = I_3$ and $\overline{\overline{A}}_{2,3} = \overline{\overline{B}}_2 = I_b^{-1}$. By decomposing Equation (25), the differential forms of attitude angle estimation $\hat{x}_1$ and attitude angle velocity estimation $\hat{x}_2$ can respectively be described as

$$\dot{\hat{x}}_1 = \hat{x}_2 - L_1 \Gamma(\overline{C}e, \alpha, \beta) - kL_1 \overline{C}e \tag{39}$$

$$\dot{\hat{x}}_2 = \overline{\overline{A}}_{2,1} \hat{x}_1 + \overline{\overline{A}}_{2,2} \hat{x}_2 + I_b^{-1} \hat{f}_a + I_b^{-1} u + \overline{\overline{D}}_2 \Delta - kL_2 \overline{C}e - L_2 \Gamma(\overline{C}e, \alpha, \beta) \tag{40}$$

Based on Theorem 1, the following assumption is proposed:

Assumption 3. *The feedback term* $\Gamma(\overline{C}e, \alpha, \beta) + k(\dot{\hat{y}}_a - \dot{y}_a)$ *of Equation (25) can rewritten as* $\Gamma(\overline{C}e, \alpha, \beta) + k\overline{C}e$. *There exists a positive scalar* k_e, *satisfying* $||\Gamma(\overline{C}e, \alpha, \beta) + k\overline{C}\dot{e}|| \leq k_e$, *for all* $t \geq 0$.

Remark 2. *It has been proved in Theorem 1 that the observer errors system is asymptotically stable, which means that both e and $\dot{e}$ can converge to neighbors of zero. Therefore, the observer feedback term $\Gamma(\overline{C}e, \alpha, \beta) + k\overline{C}\dot{e}$ is bounded. Assumption 3 is reasonable.*

Theorem 2. *Consider the satellite attitude model (18) with the actuator, sensor faults, and velocity-free measurements. Based on the observer in Equation (27), the controller is designed as*

$$\begin{aligned} u &= I_b(-\overline{\overline{A}}_{2,1}\hat{x}_1 - \overline{\overline{A}}_{2,2}\hat{x}_2 - \overline{\overline{D}}_2\Delta + L_2\Gamma(\overline{C}e, \alpha, \beta) + kL_2\overline{C}\dot{e} \\ &\quad + g_1(pk_2\xi_1 + qk_3\xi_2)(-\hat{x}_2 + L_1\Gamma(\overline{C}e, \alpha, \beta) + kL_1\overline{C}\dot{e}) \\ &\quad - g_1(\gamma_1\text{sig}^p(s) + \gamma_2\text{sig}^q(s) + \gamma_3 s + g_1\hat{x}_1)) - \hat{f}_a \end{aligned} \tag{41}$$

where $\xi_1 = \text{diag}(|\hat{x}_1|_{p-1})$, $\xi_2 = \text{diag}(|\hat{x}_1|_{q-1})$, $k_i, \gamma_i > 0$ $(i = 1,2,3)$, $p \in (0,1)$, $q > 1$, *and* $g_1 = 1/k_1$; s *is defined as follows:*

$$s = k_1\hat{x}_2 + k_2\text{sig}^p(\hat{x}_1) + k_3\text{sig}^q(\hat{x}_1) \tag{42}$$

Then, the estimated attitude angle and the estimated attitude angular velocity converge to a steady state within a fixed time. According to Theorem 1, the observer dynamic error is asymptotically stable. Therefore, the satellite attitude system asymptotically stabilizes.

Proof of Theorem 2. Using Equations (39) and (40), the time derivative of s becomes

$$
\begin{aligned}
\dot{s} &= k_1\dot{\hat{x}}_2 + pk_2\varsigma_1\dot{\hat{x}}_1 + qk_3\varsigma_2\dot{\hat{x}}_1 \\
&= k_1(\overline{\overline{A}}_{2,1}\hat{x}_1 + \overline{\overline{A}}_{2,2}\hat{x}_2 + I_b^{-1}\hat{f}_a + I_b^{-1}u + \overline{\overline{D}}_2\Delta - L_2\Gamma(\overline{C}e,\alpha,\beta) - kL_2\overline{C}\dot{e}) \\
&\quad + (pk_2\varsigma_1 + qk_3\varsigma_2)(\hat{x}_2 - L_1\Gamma(\overline{C}e,\alpha,\beta) - kL_1\overline{C}\dot{e})
\end{aligned}
\tag{43}
$$

Substituting Equation (41) into Equation (43) yields

$$
\dot{s} = -\gamma_1\mathrm{sig}^p(s) - \gamma_2\mathrm{sig}^q(s) - \gamma_3 s - g_1\hat{x}_1
\tag{44}
$$

Then, define a new state vector $\zeta = [\; s^T \quad \hat{x}^T_1 \;]^T$. Choose a candidate Lyapunov function as $V_2 = \zeta^T\zeta$. Computing its time derivative by using Equations (39) and (44) gives

$$
\begin{aligned}
\dot{V}_2 &= 2\dot{\zeta}^T\zeta = 2s^T s. + 2\hat{x}^T_1\dot{\hat{x}} \\
&= 2s^T(-\gamma_1\mathrm{sig}^p(s) - \gamma_2\mathrm{sig}^q(s) - \gamma_3 s - g_1\hat{x}_1) + 2\hat{x}^T_1(\hat{x}_2 - L_1\Gamma(\overline{C}e,\alpha,\beta) - kL_1\overline{C}\dot{e})
\end{aligned}
\tag{45}
$$

The expression $\hat{x}_2 = g_1(s - k_2\mathrm{sig}^p(\hat{x}_1) - k_3\mathrm{sig}^q(\hat{x}_1))$ can be obtained by Equation (42). Substituting it into Equation (45) yields

$$
\begin{aligned}
\dot{V}_2 &= -2(\gamma_1 s^T\mathrm{sig}^p(s) + \gamma_2 s^T\mathrm{sig}^q(s) + \gamma_3 s^T s + g_1 s^T\hat{x}_1 - g_1\hat{x}^T_1 s + g_1 k_2\hat{x}^T_1\mathrm{sig}^p(\hat{x}_1) \\
&\quad + g_1 k_3\hat{x}^T_1\mathrm{sig}^q(\hat{x}_1) + \hat{x}^T_1 L_1(\Gamma(\overline{C}e,\alpha,\beta) + k\overline{C}\dot{e})) \\
&\leq -2\chi_1\zeta^T\mathrm{sig}^p(\zeta) - 2\chi_2\zeta^T\mathrm{sig}^q(\zeta) - 2\hat{x}^T_1 L_1(\Gamma(\overline{C}e,\alpha,\beta) + k\overline{C}\dot{e}) \\
&\leq -2\chi_1 V_2^{\alpha_1} - 2\chi_2 V_2^{\beta_1} - 2\hat{x}^T_1 L_1(\Gamma(\overline{C}e,\alpha,\beta) + k\overline{C}\dot{e})
\end{aligned}
\tag{46}
$$

where $\alpha_1 = (1+p)/2$, $\beta_1 = (1+q)/2$, $\chi_1 = \min\{\gamma_1, g_1 k_2\}$, and $\chi_2 = \min\{\gamma_2, g_1 k_3\}$. According to Assumption 1 and Theorem 1, there exists $k_4 > 0$ such that $\|\hat{x}_1\| \leq k_4$. By Assumption 3, the upper bound of the term $\hat{x}^T_1 L_1(\Gamma(\overline{C}e,\alpha,\beta) + k\overline{C}\dot{e})$ is obtained:

$$
\hat{x}^T_1 L_1(\Gamma(\overline{C}e,\alpha,\beta) + k\overline{C}\dot{e}) \leq l_1\|\hat{x}^T_1\|\,\|\Gamma(\overline{C}e,\alpha,\beta) + k\overline{C}\,\dot{e}\| \leq l_1 k_4 k_e < \infty
\tag{47}
$$

Substituting Equation (47) into Equation (46) yields

$$
\dot{V}_2 \leq -2\chi_1 V_2^{\alpha_1} - 2\chi_2 V_2^{\beta_1} + \upsilon
\tag{48}
$$

where $\upsilon = 2l_1 k_4 k_e$. From the definitions of p and q, α_1 and β_1 satisfy $0 < \alpha_1 < 1$ and $\beta_1 > 1$, respectively. According to Lemma 1, if the controller is chosen as Equation (41), s and $\hat{x}_1$ will converge in the neighborhood of zero within a fixed time. The systems in Equations (39) and (40) are practical fixed-time stable. Moreover, the setting time T is given by

$$
T \leq T_{\max} = \frac{1}{2\chi_1(1 - \alpha_1)} + \frac{1}{2\chi_2(\beta_1 - 1)}
\tag{49}
$$

Theorem 1 proves that the observer errors are asymptotically stable, which means that the satellite attitude is asymptotically stable. Thus, the argument stated in Theorem 2 holds and the proof is completed. $\square$

4. Simulation Results

In this section, simulation results are presented to verify the effectiveness of the proposed observer (Equation (27)) and VSFTC (Equation (41)). Consider a small angular maneuvering satellite with actuator and sensor faults without velocity information. The inertia matrix is

$$I_b = \mathrm{diag}([\ 18.73 \quad 20.77 \quad 23.63\ ]^T)\mathrm{kg \cdot m^2} \tag{50}$$

The initial attitude angle is chosen as

$$\Theta(0) = x_1(0) = [\ 0.0859 \quad -0.1628 \quad 0.1109\ ]^T \mathrm{rad} \tag{51}$$

The initial velocity is chosen as

$$w(0) = x_2(0) = [\ -0.0415 \quad 0.0496 \quad -0.0557\ ]^T \mathrm{rad/s} \tag{52}$$

The satellite orbital angular velocity is $w_0 = 0.0012$ rad/s. The external disturbance torque is set as follows:

$$\tau_d = 1.5 \times 10^{-5} \times \begin{bmatrix} 3\cos(w_0 t) + 1 \\ 1.5\sin(w_0 t) + 3\cos(w_0 t) \\ 3\sin(w_0 t) + 1 \end{bmatrix} \mathrm{Nm} \tag{53}$$

The initial values of observer system states are $\hat{\bar{x}}(0) = [\ 0^T \quad y_a^T\ ]^T$.

Non-gyroscopic attitude sensors are equipped to measure the attitude angle Θ. A zero-mean Gaussian random noise with the variance of 1×10^{-5} is added to the attitude sensors model. The parameters of Equation (27) are chosen as $\eta = 0.5$, $\varepsilon = 0.0001$, $\sigma_1 = \sigma_2 = \sigma_3 = 40$, $l_1 = 7.68$, $l_2 = 0.66$, $l_3 = 0.09$, $l_4 = 6.44$, $k = 0.001$, $\alpha = 0.6$, and $\beta = 1.4$. The gains for the observer scheme in [26] are set as $\eta = 0.5$, $\varepsilon = 0.0001$, $\sigma_1 = \sigma_2 = \sigma_3 = 40$, and $L = [\ 7.68 I_3 \quad 0.66 I_3 \quad 0.09 I_3 \quad 6.44 I_3\ ]^T$. The gains for the controller in Equation (41) are set as $\gamma_1 = 1.4$, $\gamma_2 = 0.84$, $\gamma_3 = 0.7$, $k_1 = 1$, $k_2 = k_3 = 1.1$, $p = 0.9$, and $q = 1.4$.

4.1. Observer-Based PD Controller Simulation

In this part of the simulation, the performance of the proposed observer (Equation (27)) is compared with the observer method in [26]. The observer parameters in [26] are consistent with the above sets in this paper. To better display the effectiveness of the proposed observer, a PD controller is set as

$$u = -3I_3 \hat{x}_1 - 5I_3 \hat{x}_2 \tag{54}$$

For a satellite without angular velocity measurements, the lumped faults, including multiplicative faults and additive faults, are designed as

$$f_a = \begin{cases} f_{a_1} = \begin{cases} -0.15 & 50 \leq t < 75 \\ 0.2\sin(0.35t) - 0.1 & 75 \leq t < 150 \end{cases} \\ f_{a_2} = \begin{cases} 0.2 & 50 \leq t < 75 \\ 0.2\sin(0.35t) - 0.1\cos(0.15t) & 75 \leq t < 150 \end{cases} \\ f_{a_3} = \begin{cases} 0.1\sin(0.25t) & 0 \leq t < 100 \\ 0.15 & 100 \leq t < 150 \end{cases} \end{cases} \quad (\rho - I_3)u = \begin{cases} -0.15u_1 & 0 \leq t < 75 \\ -0.2u_2 & 0 \leq t < 75 \\ -0.3u_3 & 0 \leq t < 75 \end{cases} \tag{55}$$

$$f_s = \begin{cases} 0.02\sin(0.025\pi t) + 0.1 & 0 \leq t \leq 200 \\ 0.03\sin(0.02\pi t) + 0.05 & 0 \leq t \leq 200 \\ 0.025\cos(0.015\pi t) + 0.06 & 0 \leq t \leq 200 \end{cases} \quad (\delta - I_3)y_a = \begin{cases} -0.25y_{a1} & 0 \leq t \leq 200 \\ -0.4y_{a2} & 0 \leq t \leq 200 \\ -0.3y_{a3} & 0 \leq t \leq 200 \end{cases} \tag{56}$$

where u_i and y_{ai} represent the triaxial component of u and y_a, respectively, and t is in seconds.

The curves in Figures 2 and 3 illustrate the time response of attitude estimation errors e_1 and angular velocity estimation errors e_2 by two observers. It can be clearly seen from Figures 2 and 3 that the observation errors of Θ and w under the observer in Equation (27) can respectively converge to $|e_{1i}| \leq 2 \times 10^{-5}$ and $|e_{2i}| \leq 2 \times 10^{-6}$ within 2.8 s;

the observation errors of Θ and ω under the observer in [26] can respectively converge to $|e_{1i}| \leq 3 \times 10^{-4}$ and $|e_{2i}| \leq 2 \times 10^{-5}$ within 16 s.

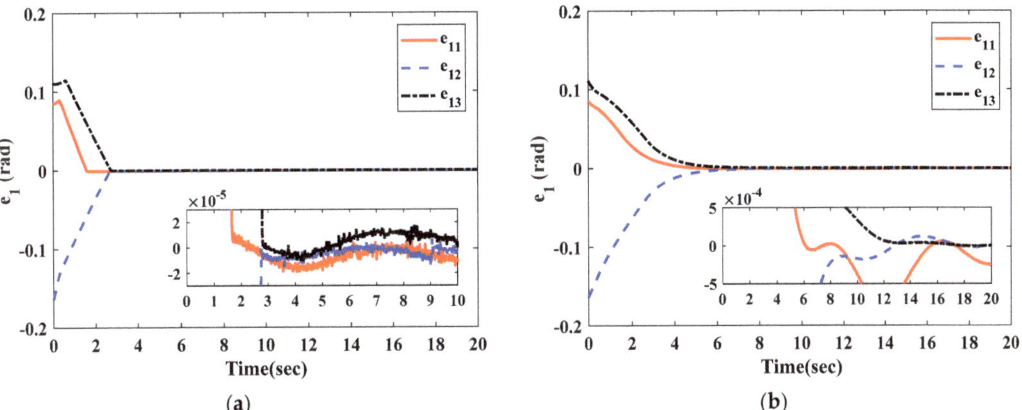

Figure 2. The initial response of attitude angle estimation errors. (**a**) Observer in Equation (27); (**b**) observer in [26].

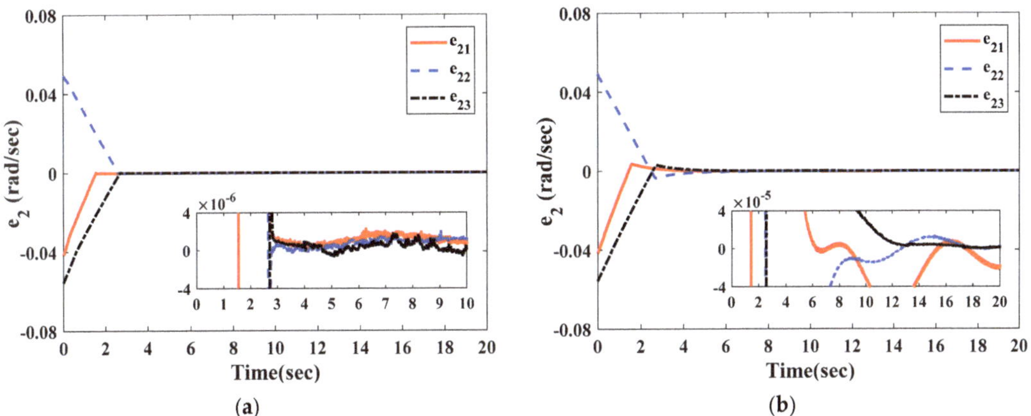

Figure 3. The initial response of angular velocity estimation errors. (**a**) Observer in Equation (27); (**b**) observer in [26].

Figures 4 and 5, respectively, show the reconstruction errors of the observer in Equation (27) and the observer in [26] for actuator faults f_a and sensor faults f_s. It can be seen in Figures 4a and 5a that the reconstruction using the proposed observer is achieved accurately within 2.8 s with steady-state accuracies of $|e_{3i}|, |e_{4i}| \leq 2 \times 10^{-5}$. In Figure 4b, the observer in [26] also shows good reconstruction performance, and the actuator faults converge to $|e_{3i}| \leq 1 \times 10^{-4}$ at 2.8 s. Figure 5b shows that the reconstruction of sensor faults under the observer in [26] can converge to $|e_{4i}| \leq 3 \times 10^{-4}$ within 16 s. The summary can also be found in Table 1.

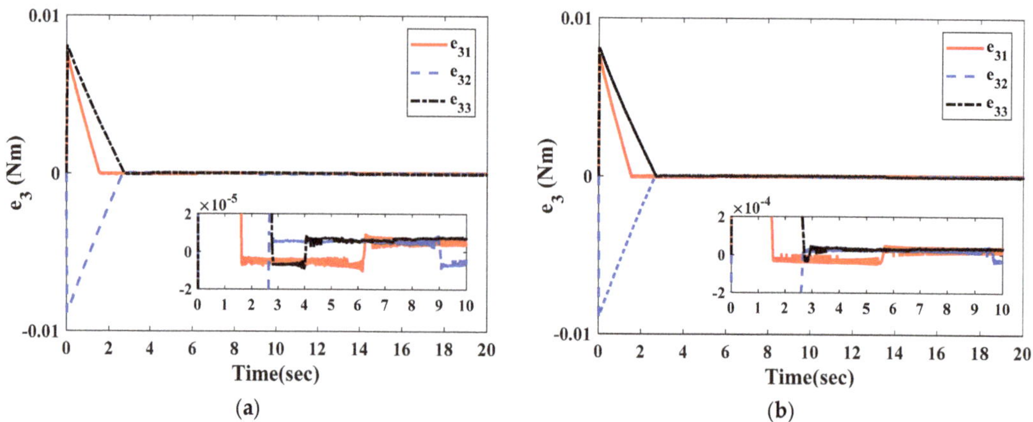

Figure 4. The initial response of actuator fault reconstruction errors. (**a**) Observer in Equation (27); (**b**) observer in [26].

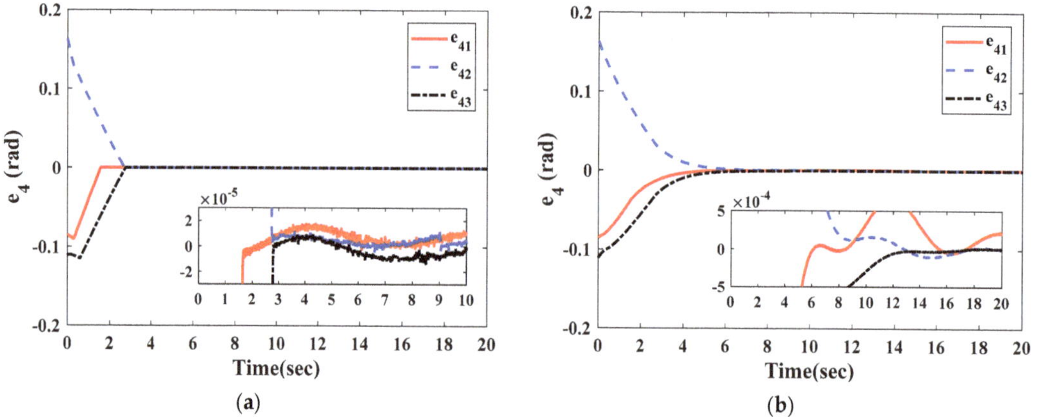

Figure 5. The initial response of sensor fault reconstruction errors. (**a**) Observer in Equation (27); (**b**) observer in [26].

Table 1. The comparison results of the observer in Equation (27) and the observer in [26].

	Observer (Equation (27))		Observer in [26]	
	Ultimate Bound	Settling Time	Ultimate Bound	Settling Time
e_1 (rad)	2×10^{-5}	2.8	3×10^{-4}	16
e_2 (rad/s)	2×10^{-6}	2.8	2×10^{-5}	16
e_3 (Nm)	2×10^{-5}	2.8	1×10^{-4}	2.8
e_4 (rad)	2×10^{-5}	2.8	3×10^{-4}	16

To give a more instructive comparison of the two observers, the function $X(t) = ||e(t)||$ is defined motivated by [29] and its time response is shown in Figure 6. From Figure 6, it is obvious that the proposed observer (Equation (27)) provides a faster convergence rate than the observer in [26]. In Figures 7 and 8, the steady-state error performances of the two observers are shown during the same fault period. The observer proposed has a higher steady-state accuracy than the observation scheme in [26].

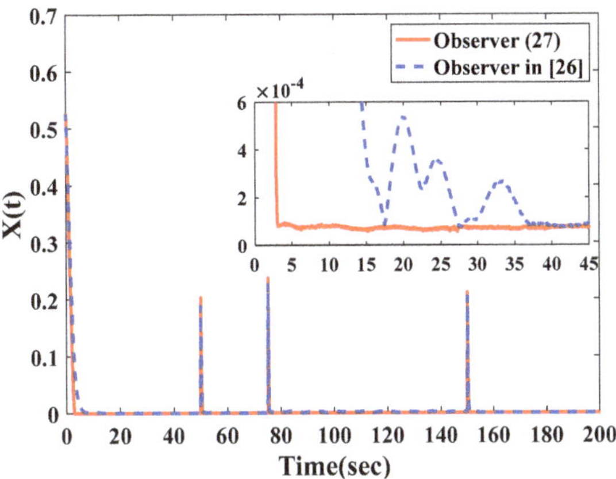

Figure 6. The function $X(t)$ by the observer in Equation (27) and the observer in [26].

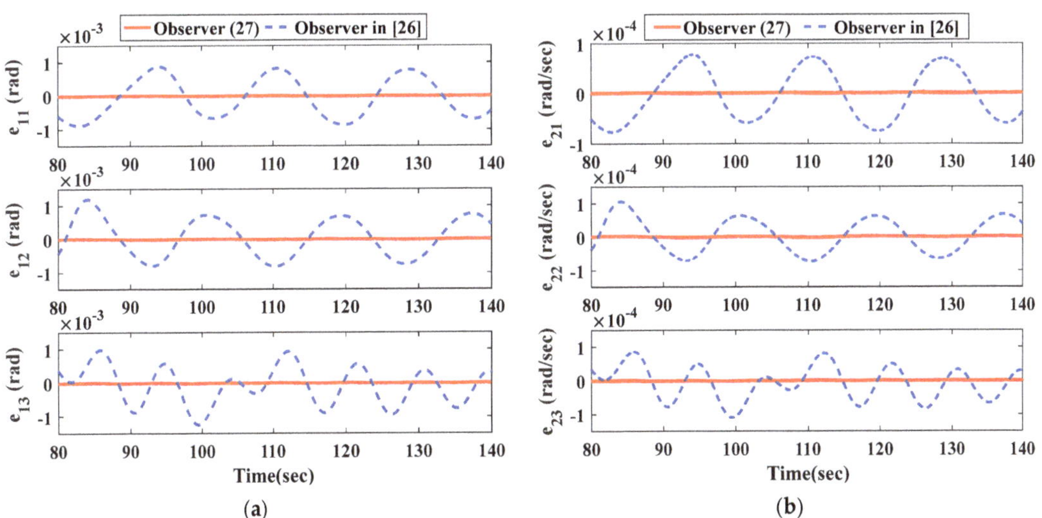

Figure 7. The steady-state behaviors of system state estimation errors by the observer in Equation (27) and the observer in [26]. (**a**) Attitude estimation errors; (**b**) angular velocity estimation errors.

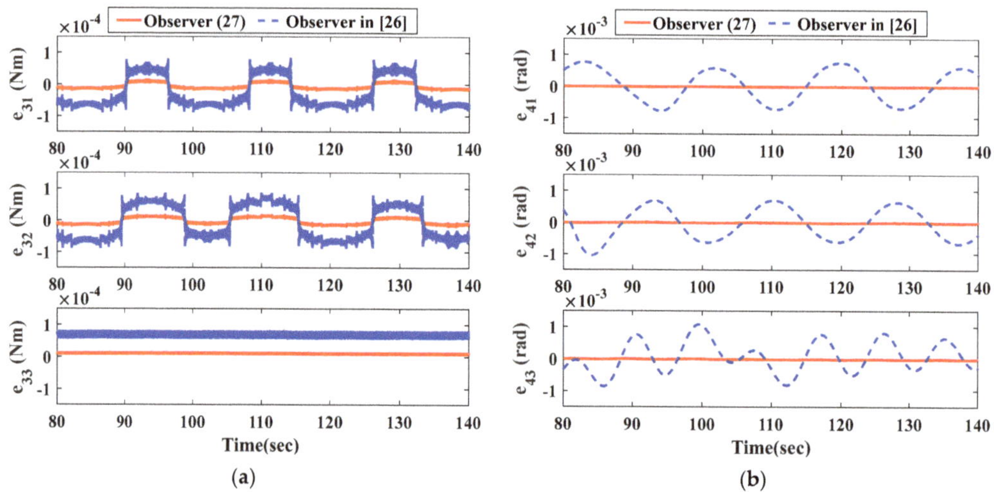

Figure 8. The steady-state behaviors of fault reconstruction errors by the observer in Equation (27) and the observer in [26]. (**a**) Actuator faults reconstruction errors; (**b**) sensor faults reconstruction errors.

4.2. VSFTC Simulation

To verify the applicability of the observation method, the actuator faults are redefined in this section. The sensor faults follow Equation (57). Due to the accurate reconstruction of the actuator and sensor faults supplied by Equation (27), the controller in Equation (41) can completely compensate for the effects of actuator and sensor faults. The accurate attitude information can be obtained. Thus, the controller can guarantee the asymptotic stability of satellite attitude and angular velocity. The actuator faults are selected as follows:

$$
f_a = \begin{cases} f_{a_1} = \begin{cases} -0.15 & 75 \le t < 150 \\ 0.1\sin(0.2t + \pi/3) - 0.1 & 150 \le t < 250 \end{cases} \\ f_{a_2} = \begin{cases} 0.2 & 50 \le t < 150 \\ 0.2\sin(0.1t) - 0.15\cos(0.2t) & 150 \le t < 250 \end{cases} \quad (\rho - I_3)u = \begin{cases} -0.1u_1 & 50 \le t < 200 \\ -0.15u_2 & 50 \le t < 200 \\ -0.2u_3 & 50 \le t < 200 \end{cases} \\ f_{a_3} = \begin{cases} 0.15\sin(0.1t + \pi/3) & 0 \le t < 100 \\ 0.15 & 100 \le t < 250 \end{cases} \end{cases} \tag{57}
$$

For the given faults, the controller in Equation (41) gives the actuator commanded control torque, as shown in Figure 9. It is clearly seen from Figure 9 that the controller effectively compensates for the actuator faults in a short time. As the time response behavior is shown in Figure 10, the total torque τ acting on the satellite attitude control system reaches the steady state within 8 s. Figures 11 and 12 show the responses of the estimated attitude Θ and the estimated angular velocity ω driven by the VSFTC (Equation (41)). The estimations of Θ and ω converge to $|\Theta_i| \le 1 \times 10^{-5}$ and $|\omega_i| \le 2 \times 10^{-5}$ within 9 s. Figures 13 and 14 show that the actual attitude Θ and the angular velocity ω take 9 s to reach steady-state behavior. As per the steady-state behavior shown in Figures 13 and 14, the attitude pointing accuracy achieves a level of $|\Theta_i| \le 1 \times 10^{-5}$ and the actual angular velocity has a pointing accuracy of 2×10^{-5}. By comparing Figures 11–14, it is concluded that the proposed observer can reconstruct the faults and system states quickly and accurately. Therefore, in the face of actuator faults, sensor faults, and rate-free measurements, the satellite attitude control system completes the steady-state control mission and achieves the required accuracy in a short time through the VSFTC method.

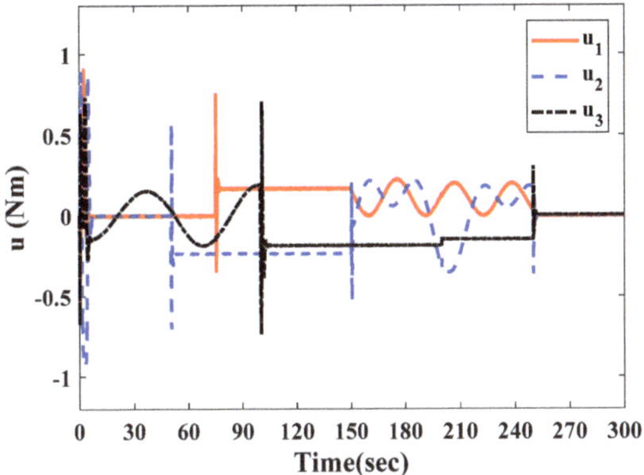

Figure 9. The actuator commanded control torque u.

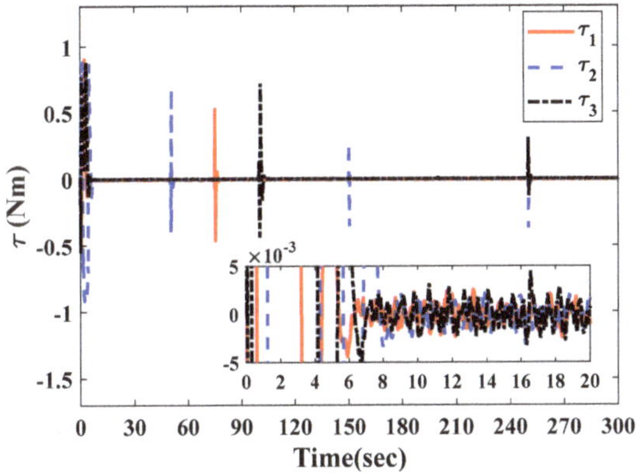

Figure 10. The total torque τ.

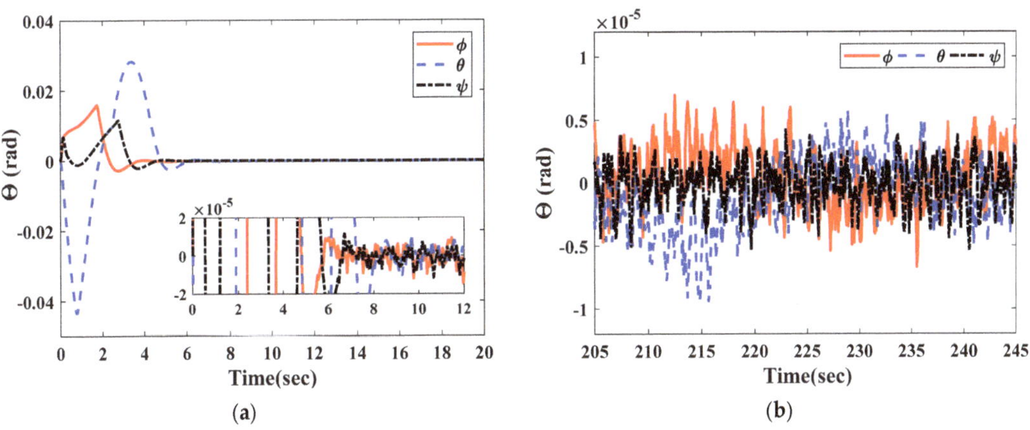

Figure 11. The attitude angle estimation of the satellite. (**a**) The initial response; (**b**) the steady-state behavior.

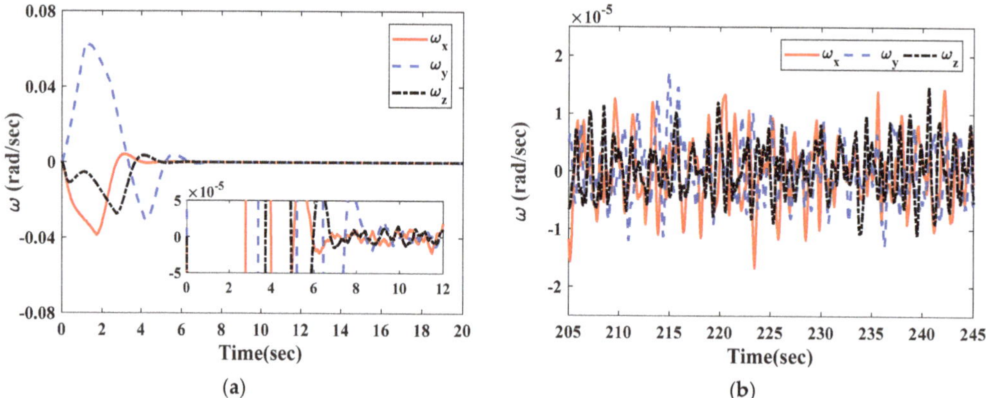

Figure 12. The angular velocity estimation of the satellite. (**a**) The initial response; (**b**) the steady-state behavior.

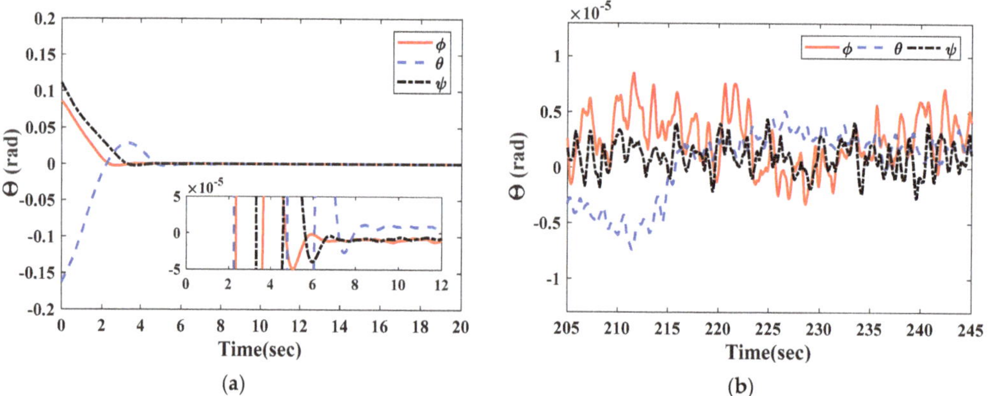

Figure 13. The actual attitude angle of the satellite. (**a**) The initial response; (**b**) the steady-state behavior.

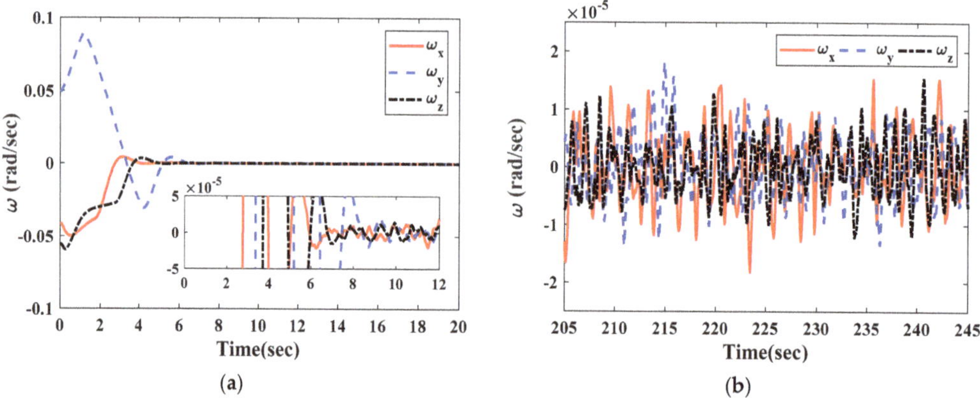

Figure 14. The actual angular velocity of the satellite. (**a**) The initial response; (**b**) the steady-state behavior.

5. Conclusions

For the satellite attitude control system without velocity measurements, the problems of fault reconstruction, state estimation, and stability control were studied when the actuator and sensor fail simultaneously. An improved sliding mode observer was proposed to quickly reconstruct the system states and faults. Based on the improved sliding mode observer, a VSFTC was presented to guarantee that the closed-loop attitude system asymptotically stabilizes in the presence of multiple faults and unknown angular velocity. Numerical simulations illustrated that the proposed observer leads to higher steady-state accuracy and faster settling time than the traditional sliding mode observer in [26] with the same parameters. By using VSFTC, the attitude of the satellite arrived at a stable state in a short time under complex faults, external disturbances, and measurement noises.

However, in this paper, only attitude stabilization was considered under multiple faults. The observer-based fault-tolerant control for attitude tracking is more challenging. Therefore, in future research, it is desirable to design an attitude tracking fault-tolerant controller to achieve fast attitude tracking under multiple faults and disturbances.

Author Contributions: Conceptualization and methodology, M.L., B.X. and A.Z.; mathematical calculations and simulations, M.L.; validation and writing—original draft preparation, M.L. and B.X.; review and editing, B.X. and A.Z. All authors have read and agreed to the published version of the manuscript.

Funding: This research received no external funding.

Conflicts of Interest: The authors declare no conflict of interest.

References

1. Lee, D.; Vukovich, G.; Gui, H.C. Adaptive variable-structure finite-time mode control for spacecraft proximity operations with actuator saturation. *Adv. Space Res.* **2017**, *59*, 2473–2487. [CrossRef]
2. Hu, Q.; Ma, G. Adaptive variable structure controller for spacecraft vibration reduction. *IEEE Trans. Aerosp. Electron. Syst.* **2008**, *44*, 861–876.
3. Roy, S.; Kar, I.N.; Lee, J.; Tsagarakis, N.G.; Caldwell, D.G. Adaptive-robust control of a class of EL systems with parametric variations using artificially delayed input and position feedback. *IEEE Trans. Control Syst. Technol.* **2019**, *27*, 603–615. [CrossRef]
4. Li, J.F.; Wang, Y.B.; Liu, Z.Y.; Jing, X.; Hu, C.W. A new recursive composite adaptive controller for robot manipulators. *Space Sci. Technol.* **2021**, *2021*, 9801421. [CrossRef]
5. Liu, F.S.; Jin, D.P. A high-efficient finite difference method for flexible manipulator with boundary feedback control. *Space Sci. Technol.* **2021**, *2021*, 9874563. [CrossRef]
6. Xiao, B.; Cao, L.; Xu, S.; Liu, L. Robust tracking control of robot manipulators with joint velocity measurement uncertainty and actuator faults. *IEEE/ASME Trans. Mechatron.* **2020**, *25*, 1354–1465. [CrossRef]
7. Wu, B.; Cao, X. Robust attitude tracking control for spacecraft with quantized torques. *IEEE Trans. Aerosp. Electron. Syst.* **2018**, *54*, 1020–1028. [CrossRef]
8. Zou, A.; Dev Kumar, K.; Hou, Z. Quaternion-based adaptive output feedback attitude control of spacecraft using chebyshev neural networks. *IEEE Trans. Neural Netw.* **2010**, *21*, 1457–1471.
9. Ran, D.; De Ruiter AH, J.; Yao, W.; Chen, X. Distributed and reliable output feedback control of spacecraft formation with velocity constraints and time delays. *IEEE/ASME Trans. Mechatron.* **2019**, *24*, 2541–2549. [CrossRef]
10. Roy, S.; Kar, I.N.; Lee, J. Toward position-only time-delayed control for uncertain Euler–Lagrange systems: Experiments on wheeled mobile robots. *IEEE Robot. Autom. Lett.* **2017**, *2*, 1925–1932. [CrossRef]
11. Shi, X.; Zhou, Z.; Zhou, D. Finite-time attitude trajectory tracking control of rigid spacecraft. *IEEE Trans. Aerosp. Electron. Syst.* **2017**, *53*, 2913–2923. [CrossRef]
12. Du, H.; Li, S.; Qian, C. Finite-time attitude tracking control of spacecraft with application to attitude synchronization. *IEEE Trans. Autom. Control* **2011**, *56*, 2711–2717. [CrossRef]
13. Cao, L.; Xiao, B.; Golestani, M. Robust fixed-time attitude stabilization control of flexible spacecraft with actuator uncertainty. *Nonlinear Dyn.* **2020**, *100*, 2505–2519. [CrossRef]
14. Robertson, B.; Stoneking, E. Satellite GN&C anomaly trends. In Proceedings of the Annual AAS Rocky Mountain Guidance and Control Conference, San Diego, CA, USA, 5–9 February 2003.
15. Li, B.; Hu, Q.; Ma, G.; Yang, Y. Fault-tolerant attitude stabilization incorporating closed-loop control allocation under actuator failure. *IEEE Trans. Aerosp. Electron. Syst.* **2019**, *55*, 1989–2000. [CrossRef]
16. Gao, J.; Fu, Z.; Zhang, S. Adaptive fixed-time attitude tracking control for rigid spacecraft with actuator faults. *IEEE Trans. Ind. Electron.* **2019**, *66*, 7141–7149. [CrossRef]

17. Mayhew, C.G.; Sanfelice, R.G.; Teel, A.R. Quaternion-based hybrid control for robust global attitude tracking. *IEEE Trans. Autom. Control* **2011**, *56*, 2555–2566. [CrossRef]
18. Abdullah, A.; Zribi, M. Sensor-fault-tolerant control for a class of linear parameter varying systems with practical examples. *IEEE Trans. Ind. Electron.* **2013**, *60*, 5239–5251. [CrossRef]
19. Xiao, B.; Hu, Q.; Wang, D.; Poh, E.K. Attitude tracking control of rigid spacecraft with actuator misalignment and fault. *IEEE Trans. Control Syst. Technol.* **2013**, *21*, 2360–2366. [CrossRef]
20. Bustan, D.; Sani, S.H.; Pariz, N. Adaptive fault-tolerant spacecraft attitude control design with transient response control. *IEEE/ASME Trans. Mechatron.* **2014**, *19*, 1404–1411.
21. Xiao, B.; Hu, Q.; Zhang, Y. Adaptive sliding mode fault tolerant attitude tracking control for flexible spacecraft under actuator saturation. *IEEE Trans. Control Syst. Technol.* **2012**, *20*, 1605–1612. [CrossRef]
22. Talebi, H.A.; Khorasani, K.; Tafazoli, S. A recurrent neural-network-based sensor and actuator fault detection and isolation for nonlinear systems with application to the satellite's attitude control subsystem. *IEEE Trans. Neural Netw.* **2009**, *20*, 45–60. [CrossRef] [PubMed]
23. Gao, Z. Fault estimation and fault-tolerant control for discrete-time dynamic systems. *IEEE Trans. Ind. Electron.* **2015**, *62*, 3874–3884. [CrossRef]
24. Zhu, F.; Shen, Y.; Zhang, J.; Wang, F. Observer-based fault reconstructions and fault tolerant control designs for uncertain switched systems with both actuator and sensor faults. *IET Control Theory Appl.* **2020**, *14*, 2017–2029. [CrossRef]
25. Lee, T.H.; Lim, C.P.; Nahavandi, S.; Roberts, R.G. Observer-based H-infinity fault-tolerant control for linear systems with sensor and actuator faults. *IEEE Syst. J.* **2019**, *13*, 1981–1990. [CrossRef]
26. Yang, H.; Yin, S. Reduced-order sliding-mode-observer-based fault estimation for markov jump systems. *IEEE Trans. Autom. Control* **2019**, *64*, 4733–4740. [CrossRef]
27. Kruk, J.W.; Class, B.F.; Rovner, D.; Westphal, J.; Ake, T.B.; Moos, H.W.; Roberts, B.; Fisher, L. FUSE in-orbit attitude control with two reaction wheels and no gyroscopes. In Proceedings of the SPIE—The International Society for Optical Engineering, Bellingham, WA, USA, 24 February 2003.
28. Tayebi, A.; Roberts, A.; Benallegue, A. Inertial vector measurements based velocity-free attitude stabilization. *IEEE Trans. Autom. Control* **2013**, *58*, 2893–2898. [CrossRef]
29. Hu, Q.; Jiang, B. Continuous finite-time attitude control for rigid spacecraft based on angular velocity observer. *IEEE Trans. Aerosp. Electron. Syst.* **2018**, *54*, 1082–1092. [CrossRef]
30. Du, H.; Li, S. Semi-global finite-time attitude stabilization by output feedback for a rigid spacecraft. *Proc. Inst. Mech. Eng. Part G J. Aerosp. Eng.* **2012**, *227*, 1881–1891. [CrossRef]
31. Peng, X.; Geng, Z.; Sun, J. The specified finite-time distributed observers-based velocity-free attitude synchronization for rigid bodies on SO(3). *IEEE Trans. Syst. Man Cybern. Syst.* **2020**, *50*, 1610–1621. [CrossRef]
32. Cui, B.; Xia, Y.; Liu, K.; Wang, Y.; Zhai, D. Velocity-observer-based distributed finite-time attitude tracking control for multiple uncertain rigid spacecraft. *IEEE Trans. Ind. Inform.* **2020**, *16*, 2509–2519. [CrossRef]
33. Polyakov, A. Nonlinear feedback design for fixed-time stabilization of linear control systems. *IEEE Trans. Autom. Control* **2012**, *57*, 2106–2110. [CrossRef]
34. Zou, A.; Fan, Z. Fixed-time attitude tracking control for rigid spacecraft without angular velocity measurements. *IEEE Trans. Ind. Electron.* **2020**, *67*, 6795–6805. [CrossRef]
35. Xiao, B.; Huo, M.; Yang, X.; Zhang, Y. Fault-tolerant attitude stabilization for satellites without rate sensor. *IEEE Trans. Ind. Electron.* **2015**, *62*, 7191–7202. [CrossRef]
36. Wang, X.; Tan, C.P.; Wu, F.; Wang, J. Fault-tolerant attitude control for rigid spacecraft without angular velocity measurements. *IEEE Trans. Cybern.* **2021**, *51*, 1216–1229. [CrossRef]

 *symmetry*

Article

Improved Path Planning for Indoor Patrol Robot Based on Deep Reinforcement Learning

Jianfeng Zheng, Shuren Mao, Zhenyu Wu, Pengcheng Kong and Hao Qiang *

School of Mechanical Engineering and Rail Transit, Changzhou University, Changzhou 213164, China;
zjf@cczu.edu.cn (J.Z.); 19085206180@smail.cczu.edu.cn (S.M.); 19085206557@smail.cczu.edu.cn (Z.W.);
19085206920@smail.cczu.edu.cn (P.K.)
* Correspondence: qhao@cczu.edu.cn

Abstract: To solve the problems of poor exploration ability and convergence speed of traditional deep reinforcement learning in the navigation task of the patrol robot under indoor specified routes, an improved deep reinforcement learning algorithm based on Pan/Tilt/Zoom(PTZ) image information was proposed in this paper. The obtained symmetric image information and target position information are taken as the input of the network, the speed of the robot is taken as the output of the next action, and the circular route with boundary is taken as the test. The improved reward and punishment function is designed to improve the convergence speed of the algorithm and optimize the path so that the robot can plan a safer path while avoiding obstacles first. Compared with Deep Q Network(DQN) algorithm, the convergence speed after improvement is shortened by about 40%, and the loss function is more stable.

Keywords: patrol robot; path planning; autonomous navigation; DQN; rewards and punishments function

Citation: Zheng, J.; Mao, S.; Wu, Z.; Kong, P.; Qiang, H. Improved Path Planning for Indoor Patrol Robot Based on Deep Reinforcement Learning. *Symmetry* **2022**, *14*, 132. https://doi.org/10.3390/sym14010132

Academic Editors: Chengxi Zhang, Jin Wu and Chong Li

Received: 10 December 2021
Accepted: 7 January 2022
Published: 11 January 2022

Publisher's Note: MDPI stays neutral with regard to jurisdictional claims in published maps and institutional affiliations.

1. Introduction

Path planning is an essential direction of robot research [1]. It is the key to realizing autonomous navigation tasks, which means a robot can independently explore a smooth and collision-free path trajectory from the starting position to the target position [2]. Traditional path planning algorithms include the A-star algorithm [3], Artificial Potential Field Method [4], and Rapidly Exploring Random Tree [5], and so on, which are used to solve the path planning in a known environment and are easy to implement. Still, robots have poor exploration ability in path planning. Given, because of the problems in traditional algorithms, a deep reinforcement learning algorithm has been introduced to enable robots to make more accurate movement directions in environmental states, which is a combination of deep learning and reinforcement learning [6–8]. Deep learning obtains the observation information of the target state by perceiving the environment. In contrast, reinforcement learning, which uses the reward and punishment functions to guide whether the action is good or not, is a process in which the patrol robot and the environment interact trial and error constantly.

The first deep reinforcement learning model, namely DQN, was proposed by Mnih et al. [9], which combined neural network with Q-learning and used a neural network to replace the Q value table to solve the problem of dimension disaster in Q-learning. Still, the convergence speed was slow in network training. DQN was applied to path planning for model-free obstacle avoidance by Tai et al. [10], which was the over-estimation of state-action value, inevitably resulting in sparse rewards and not the optimal path for robots. The artificial potential field, which accelerated the convergence speed of the network, increased the action step length and adjusted the direction of the robot to improve the precision of the robot's route planning, was introduced in the initialization process of Q value [11], leading to having good effect in the local path planning of the robot but poor implementation in the global path planning.

Because of the fixed and symmetrical patrol route, it is difficult for the patrol robot to avoid obstacles. The A-star algorithm is characterized by large performance consumption when encountering target points with obstacles. Artificial Potential Field Method cannot find a path between similar obstacles. The path planned by the Artificial Potential Field Method may oscillate and swing in narrow channels when new environmental obstacles are detected. Rapidly Exploring Random Tree is challenging to find the path in an environment with limited channels. Therefore, to effectively solve the patrol robot exploration algorithm problems of slow convergence speed and poor ability, an improved path planning for indoor patrol robot based on deep reinforcement learning is put forward in this paper, which use PTZ to perceive the surrounding environment information, combining the patrol robot's position information with the target of a state-space as network input [12–15]. The position and speed of the patrol robot are taken as the output, and a reasonable reward and punishment function is designed to significantly improve the convergence speed of the algorithm and optimize the reward sparsity of the environmental state space in the paper [16–19].

2. Background

Reinforcement learning is a process in which different rewards are obtained by way of "trial and error" when the patrol robot and the environment interact, which means that the patrol robot will not be affected by the initial stage of the environment, which does not guide the patrol robot to move but only rate its interaction [20–22]. High score behavior and low score behavior need to be remembered by the patrol robot, which just needs to use the same behavior to get a high score and avoid a low score when it interacts next time [23–26]. The interaction process of reinforcement learning is shown in Figure 1. DQN, which uses a neural network for fusion, is based on Q-learning for overcoming the defect of "dimension disaster" caused by large memory consumption of Q-learning to store data [27–29].

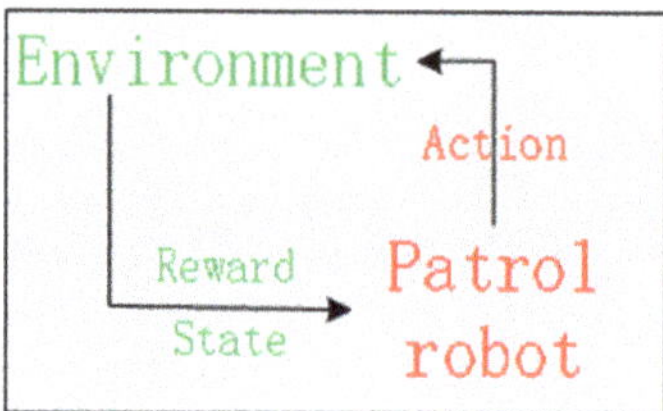

Figure 1. Reinforcement learning flow chart.

Deep reinforcement learning mainly realizes the learning interaction between the patrol robot and the environment, which is composed of deep learning and reinforcement learning. Deep learning uses the patrol robot's built-in sensors to receive and perceive the information of the surrounding environment and obtain the information of the current state of the patrol robot. While reinforcement learning is responsible for the patrol robot to explore and analyze the acquired environmental information, which helps the patrol robot to make correct decisions and makes the patrol robot can achieve navigation tasks come true [30–32].

The DQN algorithm combines a neural network, which needs to model Q table and uses RGB image as input to generate all Q values and Q-learning which uses Markov decision for modeling and uses the current state, action, reward, strategy, and next action in Markov decision for representation. The experience playback mechanism is introduced to improve the sample correlation of the robot and solve the efficiency utilization problem of the robot in DQN, which uses the uniqueness of the target Q value to enhance the smoothness of the action update. DQN includes three steps: establishing target function, target network, and introducing experience replay.

Target function. The target function of DQN is constructed by Q-learning, and the formula is as follows:

$$Q'(s,a) \leftarrow Q(s,a) + \alpha[r + \gamma \max_{a'} Q(s',a') - Q(s,a)] \tag{1}$$

where (s,a) indicates the current status and action. (s',a') indicates the next state and action. Q(s,a) represents the current state-action value. Q'(s,a) represents the updated state-action value. α represents the learning rate which means how many errors in the current state will be updated and learned. The value ranges from 0 to 1. γ represents the attenuation value of future rewards, ranging from 0 to 1. Because the Q value in the DQN algorithm is mostly randomly generated. For the convenience of calculation, we need to use the maximum value max.$\max_{a'} Q(s',a')$ means the maximum value of Q at the next state-action value. Breaking down the expression for Q values yields the following expression:

$$(s1) = r2 + \gamma Q(s2) = r2 + \gamma[r3 + \gamma Q(s3)] = r2 + \gamma\{r3 + \gamma[r4 + \gamma Q(s4)]\} = \cdots \tag{2}$$

$$\text{Namely } Q(s1) = r2 + \gamma \cdot r3 + \gamma^2 \cdot r4 + \gamma^3 \cdot r5 + \gamma^4 \cdot r6 + \cdots \tag{3}$$

It is not difficult to conclude that the value of Q is correlated with the rewards at each subsequent step, but these associated rewards decay over time, and the further away from state s1, the more state decay.

The target state-action value function can be expressed by the Bellman equation as follows:

$$y' = r + \gamma \max Q(s',a',\theta) \tag{4}$$

where y' is the target Q value. θ is the weight parameter trained in the neural network structure model. $\max Q(s',a',\theta)$ means the maximum value of Q ant the next state-action value and θ.

The loss function is the mean square error loss function, and the formula is as follows:

$$L(\theta) = E\left[(y' - Q(s,a,\theta))^2\right] \tag{5}$$

Target network. The current state-action value function is evaluated by DQN through the target network and prediction network. The target network is based on a neural network to get the target Q value, using the target Q value to estimate the Q value of the next action. The prediction network uses the stochastic gradient descent method to update the weight of the network $\Delta\theta$, and the formula of the gradient descent algorithm is shown as follows:

$$\Delta\theta = E\left[y' - Q(s,a,\theta_1)\nabla_\theta Q(s,a,\theta_1)\right] \tag{6}$$

where ∇ is Hamiltonian.

Experience replay. The experiential replay mechanism improves the sample relevance of the patrol robot and solves the problem of efficient utilization of the patrol robot. The patrol robot can obtain the sample database during the information interaction between the patrol robot and the environment. It stores the sample database into the established experience pool and randomly selects a small part of data for training samples, and then sends the training samples into the neural network for training. The experience replay mechanism utilizes the repeatability of the sample itself to improve learning efficiency.

The structure of the DQN algorithm is shown in Figure 2. s (state) represents the current cycle step, r (reward) represents the reward value generated by the current cycle step, a(action) represents the behavior caused according to the current cycle step state, and s_(next State) represents the next cycle step. Target_net and Eval_net refer to Q_target and Q_eval, respectively. Parameters of Eval_net are deferred to parameters in Target_net. Q_target and Q_eval represent the two neural networks of the DQN algorithm, which follows Q-learning, the predecessor of the DQN algorithm. They are not much different, just different parameters. Q_eval contains behavioral parameters, and theoretical behavior is bound to deviate from actual conduct (loss). Thus, in the traditional DQN algorithm, parameters s, loss, and symmetric Q_target and Q_eval will affect the Train value.

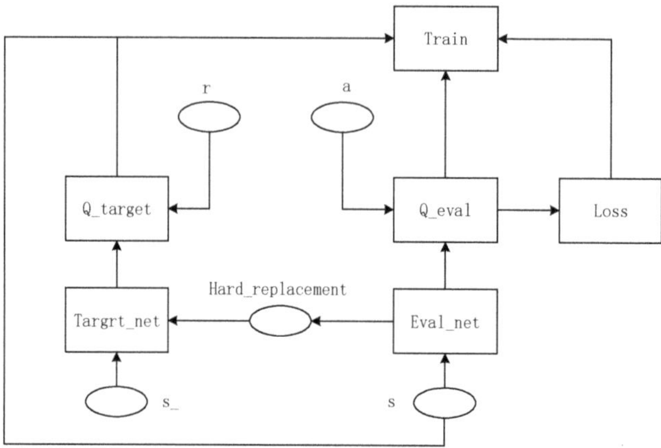

Figure 2. Structure diagram of DQN algorithm.

3. Improved Algorithm

The main goal of the patrol robot is to be able to reach the target point and return to the starting point autonomously in the indoor route environment. So a kind of improved deep reinforcement learning algorithm is put forward in this paper, which takes the acquired image information and target position information as the input of the network, the position and speed of the patrol robot as the output of the next action, and the specified circular route with a boundary as the test. Thus, the patrol robot can realize the process of its running towards the target point in the specified route with limited conditions and finally returning to the starting point to complete the automatic walking task. Since the robot needs to walk on the indoor prescribed route to realize the patrol function, the conventional DQN algorithm is easy to find the optimal path for the open route. Still, for the circular prescribed route, it is easy to fall into the local optimal and cannot complete the path. Therefore, the DQN algorithm needs to be improved.

3.1. Overview of Improved DQN Algorithm

The structure of the improved DQN algorithm is shown in Figure 3, which is transmitted to the evaluation function through the current iteration step s. The evaluation function generates four different data types and sends the data to the target function through four symmetric routes. In addition to the four transmission lines of the objective function, the next iteration step s_ also affects them. The evaluation function and Q function often have a loss value during operation, which, together with the evaluation function, Q function, and the current iterative step s, guides the patrol robot training.

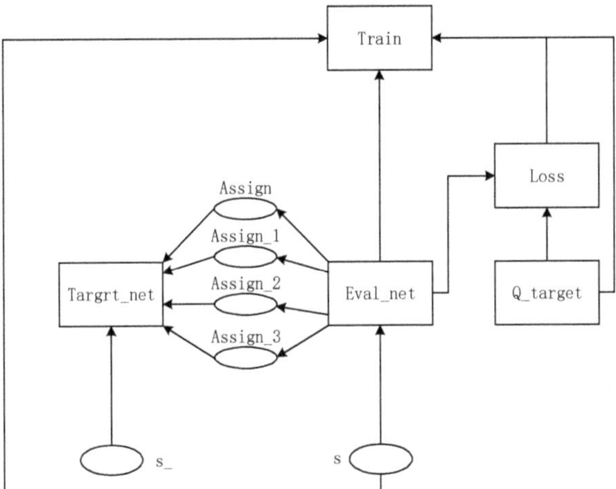

Figure 3. Structure diagram of improved DQN algorithm.

In the simulation environment, the patrol robot takes the directly collected image information as the training sample, then combines its environmental state characteristics and the target point to be reached as the input of the network, and takes the Q value of the current position as the output of the network model, and uses the ε-greedy strategy to select the action to reach the next state. When the next state is reached, a complete data tuple (s,a,r,s′) can be obtained by calculating the corresponding reward value r. Then, data of this series are stored in the experience replay pool D, and small-batch samples are extracted from the experience replay pool D and put into the neural network for training.

In the course of network training, if the patrol robot uses PTZ to identify the obstacle, the improved algorithm can make the robot avoid the block effectively. Otherwise, the patrol robot will continue to navigate until the target point is reached. The improved algorithm design is shown in Algorithm 1.

Algorithm 1. Improved Deep Q-learning Network Based on Patrol Robot.

Initialize replay memory D to capacity N
Initialize action-value function Q with random weights
for episode = 1, M **do**
Initialize sequence s_t from replay memory D and preprocess sequenced θ, $\theta = \theta_1$
    for t = 1, T **do**
    With probability ε select a random action a_t
    Otherwise select optimal action $a_t = \max_a Q'((s_t), a; \theta)$
    Execute action a_t in emulator, observe new reward function r_t and next image s_{t+1}
    Set transition (s_t, a_t, r_t, s_{t+1}) and store it in D
    Preprocess $\theta_{t+1} = \theta_{(s_{t+1})}$
    Sample random mini-batch of transitions $\left(s_j, a_j, r_j, s_{j+1}\right)$ from D

$$
\text{Set } y_j(a|s) = \begin{cases} r_j & for\ terminal_{j+1} \\ r_j + \gamma max_{a'} Q\left(s_{j+1}, a'; \theta\right) & for\ non-terminal_{j+1} \end{cases}
$$

    Perform a gradient descent step on $L(\theta)$

$$
L(\theta) = E\left(y_j - Q(s, a, \theta)\right)^2
$$

    Update policy gradient
    end for
end for

3.2. Improved Target Point Function

The target point is the coordinate of the position that the patrol robot needs to achieve in the motion state and represents the final position of the patrol robot. It needs to drive on a fixed route for the patrol robot when performing tasks and return to the initial position after completing the patrol. If the target point is placed in the starting position, the DQN algorithm will skip to the end without iteration, thus skipping the patrol step. Even if the design function makes the algorithm stop iteration when it reaches the initial position for the second time, due to the characteristics of the DQN algorithm, it will also stop iteration quickly, and the final walking path may only be a small lattice.

In the circular path, the patrol robot has a long distance to walk as the start point of the path is just the endpoint. When encountering obstacles, the conventional DQN algorithm will begin from the starting point again, resulting in a long calculation cycle of the whole circle.

Therefore, the paper improves the target point function, which the circular route used in this paper is abstracted. In a grid of 30*30, the track boundary is set in black to represent obstacle points. The red point represents the starting point which is the initial position of the patrol robot. The yellow point represents the target point. The paper segments the whole ring line, the yellow spot under an initial state for the first target of the patrol robot. While the patrol robot achieves the first target point, the second target point will be set in the forward line of the patrol robot, at the same time, the first target point will be the new starting point, and turn the white grid point closest to the first target point on the previous route into an obstacle point to make it become the barriers for which the patrol robot can accelerate the efficiency of iteration in the subsequent iterations of the patrol robot, and it also prevents the patrol robot from going "backward" in each iteration. The target point of the last stage is set as the starting point of the initial step to ensure that the patrol robot can complete the whole loop route.

3.3. Improved Reward and Punishment Function

The calculation is performed only according to the improved target point function modified in Section 3.2. As the robot has a relatively large space for walking and the degree of freedom is greatly improved, the algorithm calculation time is too long, which violates the original intention of using the DQN algorithm. Therefore, this paper starts from the reward and punishment function r, and the reward and punishment function r of the design change is shown as follows.

$$
r_{(i,j)} = \begin{cases} -1, & (i,j) \text{ is an obstacle} \\ 1, & (i,j) \text{ is a target point} \\ \mu, & r_{(m,n)} \neq -1, m \in [i-1, i+1], n \in [j-1, j+1] \\ 0, & \text{others} \end{cases} \tag{7}
$$

In the original algorithm, the reward value of all points except the obstacle point and target point is 0. However, the reward and punishment function used in this paper adds a new reward and punishment value μ, which ranges from 0 to 1. The point to which this value is assigned must have the following characteristics: the nine points centered on this point, including the eight surrounding points, are not obstacle points. This point is better than other blank points, and the patrol robot should choose it first.

As shown in Figure 4, the blue point is the new punishment and reward point, but the green point is not because the point to the lower left of the green point is the obstacle point.

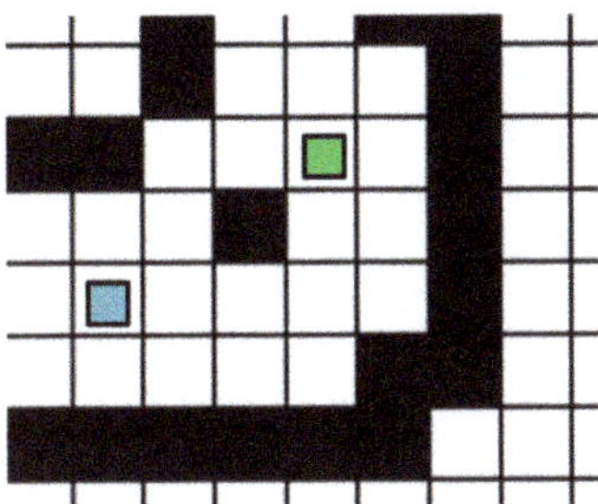

Figure 4. An illustration of the new reward and punishment value.

4. Experimental Analysis and Results

4.1. Experimental Environment and Parameter Configuration

To realize the running experiment of patrol robots and verify the validity of the algorithm in this paper under the specified route, the following experiments comparing with DQN are carried out in this paper. The experimental environment is composed of NVIDIA GeForce GTX 1660 SUPER GPU server, ROS operating system of the robot and Pycharm. The patrol robot training process is trained in the simulation environment built by Pycharm and then transplanted into the patrol robot named Raspblock with PTZ.

The patrol robot takes the state Q value as the input and the action Q value as the output, thus forming a state-action pair. If the patrol robot bumps into obstacles while running, it will get a negative reward. If the patrol robot reaches the target point, it will get a positive reward; the patrol robot also receives fewer positive rewards if it walks on a road point with no obstacles around it. The patrol robot can avoid obstacles in the learning process and keep approaching the target to complete the path planning process through the method of reward and punishment mechanism. Parameter Settings of the improved deep reinforcement learning algorithm are shown in Table 1.

Table 1. Parameter Settings.

Parameter	Value
Batch	32
Episode	10,000
Learning rate α	0.01
Reward decay γ	0.9
ε − greedy	0.9

4.2. Experimental Modeling Procedure

The paper is modeling as described in Section 3.2 and the characteristics of map symmetry are shown in Figure 5a. The red point represents the starting point of the robot, which can change as needed. The yellow point represents the target point, and the next target point will generate after the robot reaches it until it runs a full lap. The black points represent obstacles or track boundaries. The state of the final phase is shown in Figure 5b. The last target point is in the same position as the start of the first stage.

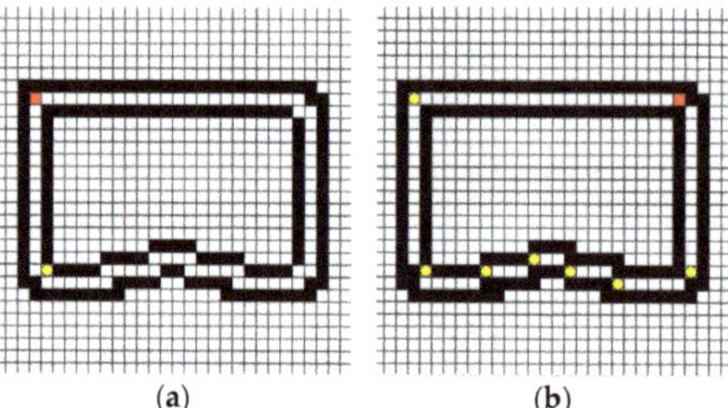

(a)   (b)

Figure 5. Schematic diagram of the initial model. (**a**) The first stage; (**b**) The final stage.

The annular region described above conforms to the general condition. However, when obstacles suddenly appear in the specified path, the patrol robot must prioritize obstacle avoidance and then patrol according to the loop. In this paper, the corridor shape is retained, and the white area in the middle of the two black circles is enlarged to deal with obstacles in the line. The new model that preserves the map symmetry is shown in Figure 6. As the black block points in the corridor are randomly set in Figure 6a, the randomness of barriers and the rationality of the algorithm are ensured. The black spots representing obstacles in the model include the original track boundary and the newly added obstacle spots. The new reward and punishment value μ at the part of Section 3.3 is introduced. The state of the last stage under the new model is shown in Figure 6b.

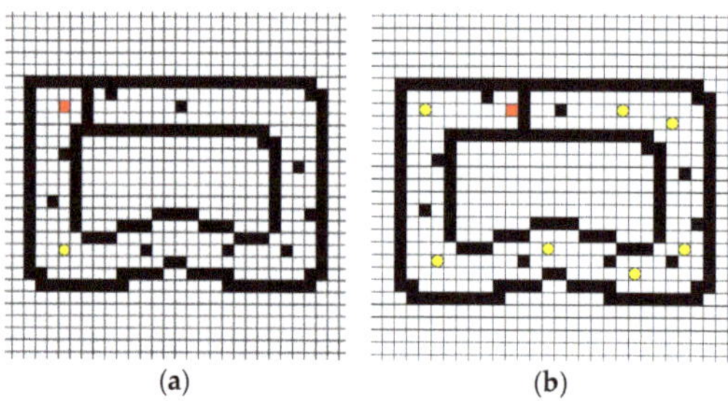

(a)   (b)

Figure 6. Schematic diagram of the new model. (**a**) The first stage; (**b**) The final stage.

4.3. Analysis of Experimental Results

In this paper, the DQN algorithm and the improved algorithm using training times and loss function values are analyzed, and the experimental results are compared. With the same parameters, the convergence speed of the improved DQN algorithm is about 30% faster than that of the DQN algorithm, and the average loss function value of the improved DQN algorithm is about 25% smaller than that of The DQN algorithm under the same training times. As shown in Figure 7, Figure 7a represents the change curve of training times-loss function obtained by the operation of the DQN algorithm, and Figure 7b represents the same by the operation of improved DQN algorithm with a value of 0.5 for the particular reward and punishment value μ. The number of training times is less than 500, which belongs to the initial training stage. The patrol robot is in the stage of exploration and learning and fails to make correct judgments on obstacles, resulting in a significant loss. When the number of training reaches 500 times, the patrol robot is still exploring the learning obstacle avoidance stage, indicating that it has begun to identify obstacles and correctly avoid some obstacles. However, due to the lack of training, it is still learning and interacting with the environment to adjust further actions to avoid more obstacles to reduce losses. The improved DQN algorithm in the current state is more stable. When the training times are between 500 and 1500, the patrol robot is unstable under the DQN algorithm and the improved DQN algorithm. When the training times reach about 1800, the loss function of the improved DQN algorithm tends to be balanced. When the training times reach about 2600, the loss function of the DQN algorithm is regionally balanced. In the testing stage, the training result model is used to test the network in the same environment to verify its effectiveness further. The objective function and reward value function of test and training are consistent. Therefore, the improved algorithm can shorten the network training time and make the patrol robot plan a shorter path.

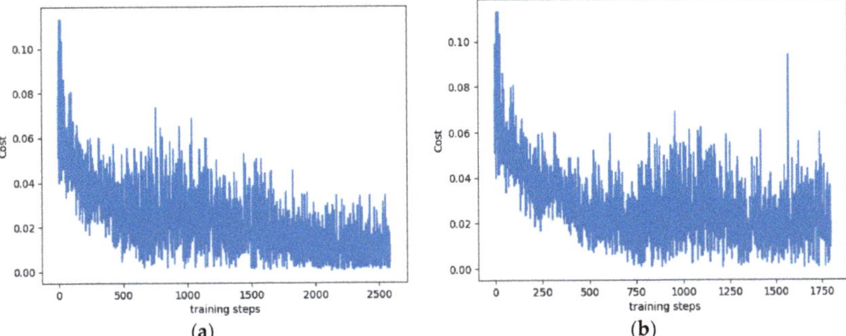

Figure 7. A comparison of the two DQN algorithms. (**a**) Conventional DQN algorithm; (**b**) Improved DQN algorithm ($\mu = 0.5$).

Although the improved objective function and reward value function reduces the convergence speed of the algorithm, the corresponding loss function slightly increases. The analysis results show that because the particular reward and punishment value of 0.5 set in the above experiment is too large, the accumulation speed of the reward and punishment value of the patrol robot is too fast, which makes the patrol robot have serious interference in the judgment of the later stage. Therefore, the following modifications are made in this paper, which is the special reward and punishment value 0.5 is changed to 10^{-6}, and it only affects the initial learning, adjusted to 0 after training. The training-loss function generated by the modified algorithm is shown in Figure 8. Figure 8a represents the change curve of the training-loss function obtained when the value of μ is 10^{-6} forever. Figure 8b illustrates the corresponding change curve obtained when the value of μ is 10^{-6} at the initial stage of the experiment and 0 at the later stage of the investigation. Compared with the improved algorithm before, the training times of the two algorithms are increased, which is caused by the sharp decrease of the particular reward and punishment value μ. However, by comparison, the convergence speed of the optimized algorithm with changing μ value is about 40% less than that of the algorithm with unchanged μ value. The number of training times is less than 500, which belongs to the initial training stage. The patrol robot is in the stage of exploration and learning and fails to make a correct judgment on obstacles. The loss value is still large, but both algorithms have a downward trend, which the loss function of the algorithm with μ value changing declines earlier. When the number of training reaches 500 times, both algorithms tend to balance, and the algorithm with changed μ value has fewer training times, which means that the improved DQN algorithm with changed μ value can complete the training faster, and the patrol robot can avoid obstacles and reach the target point more quickly. The feasibility of the improved algorithm is further verified.

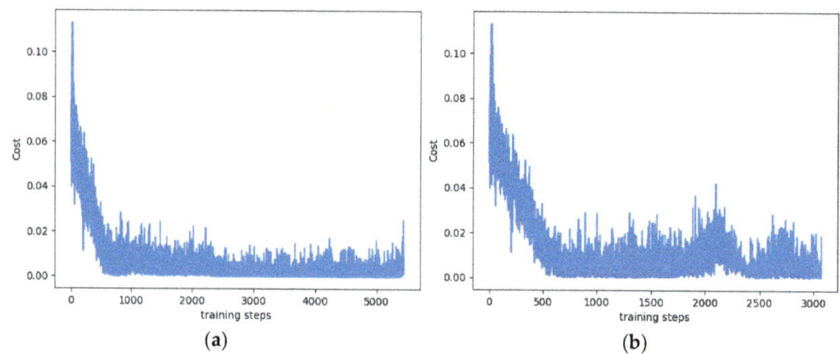

Figure 8. Influence of μ value on improved DQN algorithm. (**a**) $\mu = 10^{-6}$; (**b**) Changeable μ.

After several rounds of learning according to the improved algorithm, the patrol robot patrols according to the route shown in Figure 9. If the improved algorithm is not used, the patrol robot is prone to an infinite loop at the yellow target point in the lower-left corner of Figure 9, thus unable to complete the walking task. Because this algorithm is cyclic, no matter which small segment of the target point to take a screenshot will contain part of the obstacle point. Figure 9 is the final part of the target point of the algorithm in the whole circle. The corresponding obstacle point of the last part will be displayed, and other screenshots will also contain the related obstacle point. To distinguish, the transformation of the obstacle points is especially changed to blue.

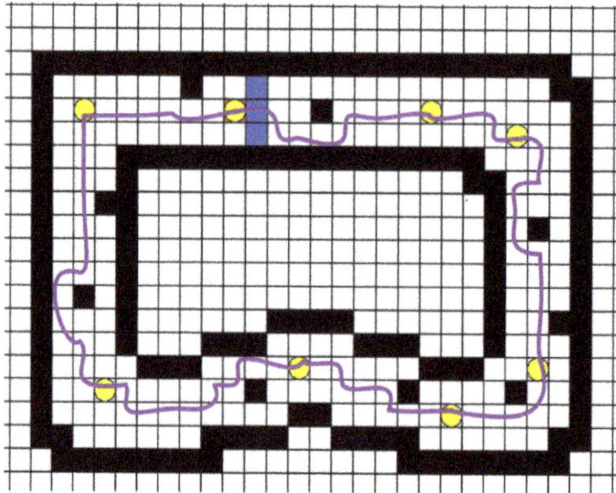

Figure 9. The approximate route of the improved algorithm.

The calculation time of the four algorithms are analyzed, including the convergence steps of their curves to be smooth, the total number of training steps, and the loss function tabulation of convergence. The data is shown in Table 2. Particular reward values are used to distinguish the four algorithms, $\mu = 0$ for Figure 7a, $\mu = 0.5$ for Figure 7b, $\mu = 10^{-6}$ for Figure 8a, and changeable μ for Figure 8b.

Table 2. Some data for four algorithms.

	$\mu = 0$	$\mu = 0.5$	$\mu = 10^{-6}$	Changeable μ
Operation time	300 s	80 s	63 s	35 s
Convergence steps	2000	1500	750	600
Total training steps	2600	1800	5500	3100
Loss function	0.030	0.040	0.010	0.015

5. Conclusions

To solve the problem that the patrol robot can complete the loop route, an improved DQN algorithm based on deep image information is proposed. The depth image information of the obstacle is obtained by using PTZ, and then the information is directly input into the network, which improves the convergence speed of network training. By enhancing the reward and punishment functions and adding new reward value points, the reward value of the robot is improved, the problem of sparse reward in the environment state space is solved by optimizing the state-action space, and the robot's action selection is more accurate to complete the patrol task. Simulation and experimental results show that the training times and loss function values of the DQN algorithm and the improved DQN algorithm are analyzed through comparative experiments, and the effective implementation of the improved algorithm is further verified in the testing stage. The improved algorithm not only enhances the robot's exploration ability and obstacle avoidance ability but also makes the planned path length safer, which verifies the improved DQN algorithm's feasibility in path planning.

In addition to the black ring boundary, the target point, the starting point, and the intermediate obstacle point in this paper are randomly set, so the improved algorithm has strong universality. According to the previous experiments, the DQN algorithm cannot achieve the best stability and the fastest calculation speed. The optimal result of this paper is to select the optimal operating speed based on ensuring stability. In the future, more restrictive conditions can be added to verify the correctness and reliability of the improved DQN algorithm.

Author Contributions: Conceptualization, J.Z. and H.Q.; methodology, S.M.; software, S.M.; validation, S.M., Z.W. and P.K.; formal analysis, S.M.; investigation, Z.W.; resources, P.K. and Z.W.; data curation, J.Z.; writing—original draft preparation, S.M.; writing—review and editing, S.M.; supervision, H.Q.; project administration, J.Z.; funding acquisition, S.M. All authors have read and agreed to the published version of the manuscript.

Funding: This research was funded by the Postgraduate Research & Practice Innovation Program of Jiangsu Province under the grant number [SJCX20_0932].

Institutional Review Board Statement: Not applicable.

Informed Consent Statement: Not applicable.

Data Availability Statement: No new data were created or analyzed in this study. Data sharing does not apply to this article.

Conflicts of Interest: The authors declare no conflict of interest.

References

1. Sun, Y.; Wang, J.; Duan, X. Research on Path Planning Algorithm of Indoor Mobile Robot. In Proceedings of the 2013 International Conference on Mechatronic Sciences, Electric Engineering and Computer (MEC), Shenyang, China, 20–22 December 2013.
2. Wang, C.; Zhu, D.; Li, T.; Meng, M.Q.H.; Silva, C.D. SRM: An Efficient Framework for Autonomous Robotic Exploration in Indoor Environments. *arXiv* **2018**, arXiv:1812.09852.
3. Candra, A.; Budiman, M.A.; Pohan, R.I. Application of A-Star Algorithm on Pathfinding Game. *J. Phys. Conf. Ser.* **2021**, *1898*, 012047. [CrossRef]
4. Rostami, S.M.H.; Sangaiah, A.K.; Wang, J.; Liu, X. Obstacle avoidance of mobile robots using modified artificial potential field algorithm. *EURASIP J. Wirel. Commun. Netw.* **2019**, *2019*, 70. [CrossRef]
5. Zhang, Z.; Qiao, B.; Zhao, W.; Chen, X. A Predictive Path Planning Algorithm for Mobile Robot in Dynamic Environments Based on Rapidly Exploring Random Tree. *Arab. J. Sci. Eng.* **2021**, *46*, 8223–8232. [CrossRef]
6. Lynnerup, N.A.; Nolling, L.; Hasle, R.; Hallam, J. A Survey on Reproducibility by Evaluating Deep Reinforcement Learning Algorithms on Real-World Robots. In Proceedings of the Conference on Robot Learning: CoRL 2019, Osaka, Japan, 30 October–1 November 2019; Volume 100, pp. 466–489.
7. Zhang, C.; Ma, L.; Schmitz, A. A sample efficient model-based deep reinforcement learning algorithm with experience replay for robot manipulation. *Int. J. Intell. Robot. Appl.* **2020**, *4*, 217–228. [CrossRef]
8. Chen, Y.; Leixin, X. Deep Reinforcement Learning Algorithms for Multiple Arc-Welding Robots. *Front. Control Eng.* **2021**, *2*, 1.

9. Mnih, V.; Kavukcuoglu, K.; Silver, D.; Rusu, A.A.; Veness, J.; Bellemare, M.G.; Graves, A.; Riedmiller, M.; Fidjeland, A.K.; Ostrovski, G.; et al. Human-level control through deep reinforcement learning. *Nature* **2015**, *518*, 529–533. [CrossRef]

10. Tai, L.; Li, S.; Liu, M. A Deep-Network Solution towards Model-Less Obstacle Avoidance. In Proceedings of the 2016 IEEE/RSJ International Conference on Intelligent Robots and Systems (IROS), Daejeon, Korea, 9–14 October 2016.

11. Yu, X.; Wang, P.; Zhang, Z. Learning-Based End-to-End Path Planning for Lunar Rovers with Safety Constraints. *Sensors* **2021**, *21*, 796. [CrossRef]

12. Miao, K.; Ma, J.; Li, Z.; Zhao, Y.; Zhu, W. Research on multi feature fusion perception technology of mine fire based on inspection robot. *J. Phys. Conf. Ser.* **2021**, *1955*, 012064. [CrossRef]

13. Shi, X.; Lu, J.; Liu, F.; Zhou, J. Patrol Robot Navigation Control Based on Memory Algorithm. In Proceedings of the 2014 4th IEEE International Conference on Information Science and Technology, Shenzhen, China, 26–28 April 2014; pp. 189–192.

14. Xu, H.; Chen, T.; Zhang, Q.; Lu, J.; Yang, Z. A Deep Learning and Depth Image based Obstacle Detection and Distance Measurement Method for Substation Patrol Robot. *IOP Conf. Ser. Earth Environ. Sci.* **2020**, *582*, 012002.

15. Dong, L.; Lv, J. Research on Indoor Patrol Robot Location based on BP Neural Network. *IOP Conf. Ser. Earth Environ. Sci.* **2020**, *546*, 052035. [CrossRef]

16. Van Nguyen, T.T.; Phung, M.D.; Pham, D.T.; Tran, Q.V. Development of a Fuzzy-based Patrol Robot Using in Building Automation System. *arXiv* **2020**, arXiv:2006.02216.

17. Ji, J.; Xing, F.; Li, Y. Research on Navigation System of Patrol Robot Based on Multi-Sensor Fusion. In Proceedings of the 2019 8th International Conference on Advanced Materials and Computer Science(ICAMCS 2019), Chongqing, China, 6–7 December 2019; pp. 224–227. [CrossRef]

18. Xia, L.; Meng, Q.; Chi, D.; Meng, B.; Yang, H. An Optimized Tightly-Coupled VIO Design on the Basis of the Fused Point and Line Features for Patrol Robot Navigation. *Sensors* **2019**, *19*, 2004. [CrossRef]

19. Zhao, F.; Yang, Z.; Li, X.; Guo, D.; Li, H. Extract Executable Action Sequences from Natural Language Instructions Based on DQN for Medical Service Robots. *Int. J. Comput. Commun. Control* **2021**, *16*, 1–12. [CrossRef]

20. Seok, P.K.; Man, P.J.; Kyu, Y.W.; Jo, Y.S. DQN Reinforcement Learning: The Robot's Optimum Path Navigation in Dynamic Environments for Smart Factory. *J. Korean Inst. Commun. Inf. Sci.* **2019**, *44*, 2269–2279.

21. Sasaki, H.; Horiuchi, T.; Kato, S. Experimental Study on Behavior Acquisition of Mobile Robot by Deep Q-Network. *J. Adv. Comput. Intell. Intell. Inform.* **2017**, *21*, 840–848. [CrossRef]

22. Han, B.; Zhao, Y.; Luo, Q. Walking Stability Control Method for Biped Robot on Uneven Ground Based on Deep Q-Network. *J. Beijing Inst. Technol.* **2019**, *28*, 220–227.

23. Rahman, M.M.; Rashid, S.; Hossain, M.M. Implementation of Q learning and deep Q network for controlling a self balancing robot model. *Robot. Biomim.* **2018**, *5*, 8. [CrossRef]

24. da Silva, I.J.; Perico, D.H.; Homem TP, D.; da Costa Bianchi, R.A. Deep Reinforcement Learning for a Humanoid Robot Soccer Player. *J. Intell. Robot. Syst.* **2021**, *102*, 69. [CrossRef]

25. Peng, X.; Chen, R.; Zhang, J.; Chen, B.; Tseng, H.W.; Wu, T.L.; Meen, T.H. Enhanced Autonomous Navigation of Robots by Deep Reinforcement Learning Algorithm with Multistep Method. *Sens. Mater.* **2021**, *33*, 825. [CrossRef]

26. Tallamraju, R.; Saini, N.; Bonetto, E.; Pabst, M.; Liu, Y.T.; Black, M.J.; Ahmad, A. AirCapRL: Autonomous Aerial Human Motion Capture using Deep Reinforcement Learning. *IEEE Robot. Autom. Lett.* **2020**, *5*, 6678–6685. [CrossRef]

27. Abanay, A.; Masmoudi, L.; Elharif, A.; Gharbi, M.; Bououlid, B. Design and Development of a Mobile Platform for an Agricultural Robot Prototype. In Proceedings of the 2nd International Conference on Computing and Wireless Communication Systems, Larache, Morocco, 14–16 November 2017; pp. 1–5.

28. Budiharto, W.; Santoso, A.; Purwanto, D.; Jazidie, A. A method for path planning strategy and navigation of service robot. *Paladyn* **2011**, *2*, 100–108. [CrossRef]

29. Arvin, F.; Samsudin, K.; Nasseri, M.A. Design of a Differential-Drive Wheeled Robot Controller with Pulse-Width Modulation. In Proceedings of the 2009 Innovative Technologies in Intelligent Systems and Industrial Applications, Kuala Lumpur, Malaysia, 25–26 July 2009; pp. 143–147.

30. Bethencourt, J.V.M.; Ling, Q.; Fernández, A.V. Controller Design and Implementation for a Differential Drive Wheeled Mobile Robot. In Proceedings of the 2011 Chinese Control and Decision Conference (CCDC), Mianyang, China, 23–25 May 2011; pp. 4038–4043.

31. Zeng, D.; Xu, G.; Zhong, J.; Li, L. Development of a Mobile Platform for Security Robot. In Proceedings of the 2007 IEEE International Conference on Automation and Logistics, Jinan, China, 18–21 August 2007; pp. 1262–1267.

32. Sharma, M.; Sharma, R.; Ahuja, K.; Jha, S. Design of an Intelligent Security Robot for Collision Free Navigation Applications. In Proceedings of the 2014 International Conference on Reliability Optimization and Information Technology (ICROIT), Faridabad, India, 6–8 February 2014; pp. 255–257.

Article

Finite-Time Controller for Flexible Satellite Attitude Fast and Large-Angle Maneuver

You Li [1], Haizhao Liang [2,*] and Lei Xing [3]

[1] School of Aerospace Science and Technology, Xidian University, Xi'an 710126, China; liyouhahaha@163.com
[2] School of Aeronautics and Astronautics, Sun Yat-sen University, Shenzhen 518107, China
[3] School of Astronautics, Harbin Institute of Technology, Harbin 150006, China; xinglei@hit.edu.cn
* Correspondence: lianghch5@mail.sysu.edu.cn

Abstract: In order to deal with the fast, large-angle attitude maneuver with flexible appendages, a finite-time attitude controller is proposed in this paper. The finite-time sliding mode is constructed by implementing the dynamic sliding mode method; the sliding mode parameter is constructed to be time-varying; hence, the system could have a better convergence rate. The updated law of the sliding mode parameter is designed, and the performance of the standard sliding mode is largely improved; meanwhile, the inherent robustness could be maintained. In order to ensure the system's state could converge along the proposed sliding mode, a finite-time controller is designed, and an auxiliary term is designed to deal with the torque caused by flexible vibration; hence, the vibration caused by flexible appendages could be suppressed. System stability is analyzed by the Lyapunov method, and the superiority of the proposed controller is demonstrated by numerical simulation.

Keywords: attitude control; fast large-angle maneuver; finite-time control; flexible appendages

Citation: Li, Y.; Liang, H.; Xing, L. Finite-Time Controller for Flexible Satellite Attitude Fast and Large-Angle Maneuver. *Symmetry* **2022**, *14*, 45. https://doi.org/10.3390/sym14010045

Academic Editor: Deming Lei

Received: 19 November 2021
Accepted: 16 December 2021
Published: 30 December 2021

Publisher's Note: MDPI stays neutral with regard to jurisdictional claims in published maps and institutional affiliations.

1. Introduction

Current space missions, such as push-broom imaging and stare imaging, need satellites that have the ability to perform fast large-angle maneuvers. However, standard controllers, such as the PID controller and the sliding mode controller, have the issue of a low convergence rate. It is necessary to develop an attitude controller with a faster convergence rate. Furthermore, the deformation and vibration of flexible appendages would bring unexpected torque on the satellite system; hence, the overall goal of this paper is to develop a satellite attitude controller subject to fast large-angle attitude maneuvers with a better convergence rate compared to standard controllers. Meanwhile, the flexible vibrations could be suppressed.

In the field of satellite attitude control, PID control and sliding mode control are the most mature and widely used methods. They both have the advantage of simple structures and strong robustness; hence, a lot of work has been performed by researchers. However, a low convergence rate is the main drawback of these two methods. Li [1–3] developed PID controllers for satellite's fast maneuvering with flexible appendages, and standard PID controllers are modified in his work. Hence, the system could have a better convergence rate. There is also some work [4,5] focusing on satellite orbit control, and the main goal of these two papers is to optimize energy consumption. The sliding mode controller is also a mature method in the field of satellite attitude control. Chakrabarti [6] and Ye [7] designed sliding mode controllers for satellite attitude maneuvers; the standard sliding mode for satellite attitude control is modified in these works, and the system robustness is enhanced. However, the system convergence rate is not taken into consideration in these works. In order to improve the system convergence rate, Li [8,9] has done some work, and the standard sliding mode is modified by Bang-Bang logic and dynamic sliding mode. The system convergence rate is largely improved by implementing the updated law of the sliding mode parameter. Moreover, Xiao [10–12] has performed some work focusing on the modification of sliding mode controllers. The controllers with fault-tolerant and strong robustness are proposed in these works. Ye [13] also designed a sliding mode control algorithm for an attitude tracking controller, and his main focus is the system convergence rate. In his work, it is pointed out that by optimizing the trajectory of angular velocity, the system could have both a fast convergence rate and low energy consumption. However, the common drawback of these works is that the system convergence rate is exponential, which means the system state would reach its equilibrium point with infinite time. The terminal convergence rate is relatively low in these works, and it is necessary to develop a controller with a better convergence rate, especially for fast attitude maneuverable conditions.

In order to deal with the exponential convergence rate issue, finite-time control theory, which could largely improve the system convergence rate to near its equilibrium point, is developed by researchers. Li [14] developed a finite-time controller with three stage structures, and a braking curve for angular velocity is constructed. The system convergence rate is improved by maintaining angular velocity revers to attitude quaternion; meanwhile, its norm is reaching its upper bound for as long as possible. The singularity issue is solved by using the property when the angular velocity is reversed to the vector of the Euler Axis. Liang [15,16], Wang [17,18] and Wu [19,20] also designed finite-time controllers for satellite attitude control, and their main focus is the structure of the finite-time sliding mode. It is pointed out that the key to achieving finite-time stability is to construct the fraction order of the system state properly. However, these works do not consider the flexible deformation of large flexible appendages, such as solar sails and large antennas, which are very common in satellites. Noting that when a satellite is on a fast attitude maneuver, the flexible vibrations can not be ignored; hence, it is necessary to design controllers that are robust to flexible deformations.

In order to deal with the satellite attitude control issue considering flexible vibration, some fundamental work [21–24] has been done. Wie constructed the basic structure of a PID controller for satellite attitude control and some typical methods for stability analysis is proposed. Some typical sliding mode surfaces are also proposed for satellite attitude control. Generally, the basic idea to deal with flexible vibrations could be concluded as following two aspects: 1. treat it as another kind of disturbance with a normal upper bound; 2. design a state observer to estimate it based on its dynamic model. The former is easier for engineering practice, and the latter has better performance in theoretical research.

In this paper, a finite-time controller for a satellite capable of fast, large-angle maneuvers with flexible appendages will be proposed. The next section will give the dynamic and kinetic models used in this paper, Section 3 is the core of this paper, and the finite-time controller will be given in this section. Section 4 will demonstrate the performance by numerical simulation, and Section 5 will conclude this paper.

2. Explanation of the Symbols Used in This Paper

In order to make it easier to understand this paper, the symbols used in this paper are explained in the following Table 1.

Table 1. Explanation of the symbols.

J	Inertia matrix of satellite (3×3 matrix)
$\hat{J}$	Inertia matrix best estimate (3×3 matrix)
$\tilde{J}$	Error inertia matrix (3×3 matrix)
ω	Angular velocity (3×1 vector)
δ	Coupling matrix between flexible appendages and rigid body
η	Modal coordinate vector
u	Control torque
d	Unknown disturbance torque
$\bar{r}$	Norm upper bound of vector r

Table 1. *Cont.*

ω_{ni}	Natural frequencies of flexible appendages
ξ_i	Associated damping of flexible appendages
N	The elastic modes need to be considered
$r^\times$	Product matrix of the three-dimensional vector r
$\|r\|$	Euclidean two-norm of vector or matrix r
$\lambda_M(A), \lambda_m(A)$	Maximum and minimum eigenvalue of matrix A
q	Attitude quaternion (4×1 vector)
q_v	Vector part of attitude quaternion (3×1 vector)
q_0	Scalar part of attitude quaternion
$\text{sgn}(x)$	Sign function of vector or scalar x

3. Dynamic and Kinetic Model

The dynamic model of the rigid satellite could be written as follows.

$$J\dot{\omega} + \delta^T\ddot{\eta} = -\omega^\times\left(J\omega + \delta^T\dot{\eta}\right) + u + d$$
$$\ddot{\eta} + C\dot{\eta} + K\eta = -\delta\dot{\omega} \tag{1}$$

where ω is the angular velocity, J is the inertia matrix of satellite, which is a symmetric matrix, d is the unknown disturbance torque with a normal upper bound $\|d\| < \bar{d}$. δ is the coupling matrix between the flexible appendages and the rigid body and δ describes how the flexible appendages influence the rigid body, η is the modal coordinate vector, its definition could be found in Table 1, C and K are defined as follows.

$$C = diag(2\xi_i\omega_{ni}), i = 1, 2 \cdots N$$
$$K = diag\left(\omega_{ni}^2\right), i = 1, 2 \cdots N \tag{2}$$

where ω_{ni} is the natural frequency and ξ_i is the associated damping, and N is the number of the elastic modes need to be considered.

Product matrix $r^\times$ of vector r is defined as:

$$r^\times = \begin{bmatrix} 0 & -r_3 & r_2 \\ r_3 & 0 & -r_1 \\ -r_2 & r_1 & 0 \end{bmatrix} \tag{3}$$

Generally, the inertia matrix J could not be accurately known, and it is assumed that:

$$J = \hat{J} + \tilde{J} \tag{4}$$

where $\hat{J}$ is the inertia matrix best estimate and $\tilde{J}$ is the error matrix. Product matrix has an important property, which will be used in the latter part that the eigenvalues of $r^\times$ satisfies:

$$\lambda(r^\times) = 0, \|r\|_2$$
$$\lambda_{\max}(r^\times) = \|r\|_2 \tag{5}$$

Define ψ and ϑ as follows.

$$\psi = \dot{\eta} + \delta\omega, \vartheta = \begin{bmatrix} \eta^T & \psi^T \end{bmatrix}^T \tag{6}$$

The dynamic model (1) could be transformed to:

$$\bar{J}\dot{\omega} = -\omega^\times\left(\bar{J}\omega + H\vartheta\right) + L\vartheta - M\omega + u + d$$
$$\dot{\vartheta} = A\vartheta + B\omega \tag{7}$$

where $\bar{J}, H, L, M, A, B$ are defined as follows.

$$\bar{J} = J - \delta^T\delta, H = \begin{bmatrix} 0 & \delta^T \end{bmatrix}, L = \delta^T \begin{bmatrix} K & C \end{bmatrix}, M = \delta^T C \delta$$
$$A = \begin{bmatrix} 0 & I \\ -K & -C \end{bmatrix}, B = \begin{bmatrix} -I \\ C \end{bmatrix}\delta \tag{8}$$

The kinetic model based on attitude quaternion could be written as follows.

$$\dot{q} = \begin{bmatrix} \dot{q}_0 \\ \dot{q}_v \end{bmatrix} = \begin{bmatrix} -\frac{1}{2}q_v^T\omega \\ \frac{1}{2}(q_0I_3 + q_v^\times)\omega \end{bmatrix} = \frac{1}{2}\begin{bmatrix} -q_v^T \\ F \end{bmatrix}\omega \tag{9}$$

where $F = q_0I_3 + q_v^\times$ and the eigenvalue of F satisfies:

$$\lambda(F) = |q_0|, 1$$
$$\lambda_M(F) = 1 \tag{10}$$

In order to simplify the text, the maximum and minimum eigenvalue of matrix A is described as $\lambda_M(A)$ and $\lambda_m(A)$.

Considering that q and $-q$ describes the same attitude, it is assumed that $q_0 \geq 0$ in this paper.

Furthermore, it is worth noticing that in engineering practice, system angular velocity ω and control torque u has its norm upper bound, and it is assumed to be $\bar{l}$ and $\bar{u}$ in this paper.

4. Finite-Time Controller

4.1. Problem Description

The sliding mode control method has been proposed for a decade, and a lot of work based on this method has been performed on the satellite attitude control issue. The most widely used and standard sliding mode for satellite attitude control could be written as follows.

$$s = \omega + kq_v, (k > 0) \tag{11}$$

After reaching this sliding mode surface, the system state has such properties:

$$\omega = -kq_v$$
$$\dot{q}_v = \frac{1}{2}(q_0I_3 + q_v^\times)\omega = -\frac{1}{2}kq_0q_v \tag{12}$$

When the system maneuvers along (11), the angular velocity vector is reversed to the attitude quaternion vector, and a lot of work has been done based on this sliding mode. The model uncertainty and unknown disturbance issue could be effectively solved using sliding mode (11), and it could be concluded that the reverse property could improve system robustness. However, based on Equation (11), it could be easily found that the convergence rate of q_v is exponential, which means the system would reach the equilibrium point with infinite time and the convergence rate needs to be improved.

In order to improve the system convergence rate, a finite-time controller is an effective method. Generally, in order to achieve the finite-time stability, a fraction order feedback is used as follows to construct the sliding mode.

$$\dot{x} = -ksign(x)|x|^r, 0 < r < 1 \tag{13}$$

Sliding mode (13) would bring another issue, i.e., the singularity issue. Since the control torque is always related to $\ddot{x}$, i.e., the second derivative of x, the singularity term x^{r-1} would be brought into the controller. In order to deal with the singularity issue, some typical finite-time controllers are designed [14]. However, the system robustness issue is not taken into consideration, and the reverse property does not hold in these works. System robustness needs to be improved to suppress the perturbations, such as inertia matrix uncertainty and unknown disturbance. In summary, the robustness issue and singularity issue should be both taken into consideration to design the robust finite-time controller.

Based on the discussion above, the goal of this paper could be: design a finite-time controller for satellite stabilization issues and the following properties should be satisfied:

1. Compared with the standard sliding mode, the system convergence rate near the equilibrium point should be largely improved;
2. Finite-time stability should be satisfied, i.e., there exist positive scalar $\varepsilon, \varepsilon'$ and T to satisfy following inequality;

$$\|q_v\| \leq \varepsilon, \|w\| \leq \varepsilon' \quad fot \quad \forall t \geq T \tag{14}$$

3. The singularity issue should be solved, i.e., $q_v, \dot{q}_v, w, \dot{w}$ are all bound during the whole control process;
4. The controller should be robust to inertia matrix uncertainty and unknown disturbance torque.

4.2. Finite-Time Sliding Mode

In paper [9], the author pointed out that the fixed sliding mode caused the low convergence rate, and a dynamic sliding mode is constructed in this paper. The maneuver stage with constant angular velocity and converge stage with constant angular acceleration are designed based on the updated law of sliding mode parameter k, and the system convergence rate is largely improved compared to the standard sliding mode. Inspired by the method in [3], the finite-time sliding mode proposed in this paper could be written as follows.

$$s = w + kq_v$$

$$\dot{k} = \begin{cases} 0 & \|s\| > \varepsilon_1 \\ k(1-\alpha)\beta q_0 \|q_v\|^{\alpha-1} & \|s\| \leq \varepsilon_1 \end{cases} \tag{15}$$

$$1/2 < \alpha < 1, \beta = k(t_0)/\|q_v(t_0)\|^{\alpha-1} \tag{16}$$

where the initial value of k satisfies $k(t_0) > 0$, ε_1 is a small positive scalar, and α, β are all positive scalars.

A sliding mode (15) has the same structure as a standard sliding mode; hence. the reversed property could be maintained. Moreover, the same structure could make it possible to design a robust finite-time controller based on standard sliding mode methods. Based on (15), it could be found that the maneuvering process is constructed in two stages: in the first stage, i.e., $\|s\| > \varepsilon_1$, the system performance is totally the same as that of a standard sliding mode, and the sliding mode parameter k is fixed; in the second stage, i.e., $\|s\| \leq \varepsilon_1$, it could be treated as a system that has reached the sliding mode, and angular velocity vector has been reversed to the attitude quaternion vector. In this stage, the sliding mode parameter k begins to update. Moreover, based on the updated law of k, it could be found that k is monotonically increasing to affect the exponential convergence rate. The key work of this paper is the updated law of the sliding mode parameter k and when the system convergences along (15), i.e., $s = 0$, system (5) would converge to its equilibrium point within finite-time, and during this process, w and $\dot{w}$ are all norm upper bound.

The next step is to discuss the finite-time stability of the sliding mode (15). When the system reaches the sliding mode (15), it is defined as follows, and its derivative could be calculated as:

$$V_q = q_v^T q_v = \|q_v\|^2 \tag{17}$$

$$\dot{V}_q = 2q_v^T \dot{q}_v = -kq_0 q_v^T q_v = -kq_0\|q_v\|^2 \tag{18}$$

In order to achieve the goal of finite-time stability, the derivative of Lyapunov function should satisfy the following inequality:

$$\dot{V}_q \leq -\gamma q_0 \|q_v\|^{\alpha+1}, \quad with \quad \alpha \in (0,1), \gamma > 0 \tag{19}$$

Compared with (14) and (15), it could be that if there exists a positive scalar γ to satisfy the following inequality, the finite-time stability could be ensured.

$$k = \gamma \|q_v\|^{\alpha-1} \tag{20}$$

In order to satisfy the finite-time condition (20), the fixed parameter k is not feasible since the right part of (20) tends towards infinite, and a very large k would cause the control torque of an angular velocity to exceed the system upper bound drastically. Hence, it is necessary to design a time-variable parameter k and its update law to satisfy (20), and that is how the dynamic sliding mode (15) is found. In fact, selecting parameters as follows, it could be found that:

$$\gamma = k(t_0)/\|q_v(t_0)\|^{\alpha-1}, \beta = \gamma \tag{21}$$

Noting that the structure of the sliding mode parameter update law in (15), it could be found that:

$$k(t_0) = \gamma \|q_v(t_0)\|^{\alpha-1}$$
$$\dot{k} = \tfrac{1}{2}k(1-\alpha)\beta q_0 \|q_v\|^{\alpha-1} = \tfrac{d\gamma\|q_v\|^{\alpha-1}}{dt} \tag{22}$$

Based on (21) and (22), it could be found that finite-time condition (20) is satisfied, and (19) could be transformed to:

$$\dot{V}_q = 2q_v^T\dot{q}_v = -kq_0\|q_v\|^2 \leq -\beta q_0\|q_v\|^{\alpha+1} = -\beta q_0 V_q^{\alpha+1/2} \tag{23}$$

The system converge time satisfies:

$$t_f \leq \frac{2V^{\frac{1-\alpha}{2}}(t_0)}{\beta q_0(t_0)(1-\alpha)} \tag{24}$$

The next step is to improve on sliding mode (15), $\omega, \dot{\omega}$ are all norm upper bound. It is obvious that angular velocity ω satisfies following the property and is the norm upper bound.

$$\|\omega\| = \|-kq_v\| = \|q_v\|^{\alpha} \tag{25}$$

Calculating the derivative of angular velocity, it could be found that:

$$\begin{aligned}
\dot{\omega} &= -k\dot{q}_v - \dot{k}q_v \\
&= -k(q_0 I_3 + q_v^{\times})(-kq_v) - \tfrac{k}{2}(1-\alpha)\beta q_0\|q_v\|^{\alpha-1}q_v \\
&= q_0 k^2 q_v - \tfrac{k}{2}(1-\alpha)\beta q_0\|q_v\|^{\alpha-1}q_v \\
&= q_0\beta^2\|q_v\|^{2\alpha-1}e - \tfrac{1}{2}(1-\alpha)\beta^2 q_0\|q_v\|^{2\alpha-1}e
\end{aligned} \tag{26}$$

where e is the unit direction vector of attitude quaternion, which is defined as $e = q_v/\|q_v\|$. Noting that $1/2 < \alpha < 1$, hence, $\omega, \dot{\omega}$ are all norm upper bound during the whole maneuver process, and the demand control torque is also the norm upper bound, i.e., the singularity issue is solved. However, it is also worth noticing that on sliding mode (15), the system parameter k tends towards infinite since $\dot{k}$ is not the norm upper bound. In order to avoid this situation, we rewrite the sliding mode as follows.

$$s = \omega + kq_v$$
$$\dot{k} = \begin{cases} 0 & \text{otherwise} \\ k(1-\alpha)\beta q_0\|q_v\|^{\alpha-1} & \|s\| \leq \varepsilon_1 \,\&\, \|q_v\| > \varepsilon_2 \end{cases} \tag{27}$$

where ε_1 and ε_2 are all small positive scalars. The difference between sliding mode (15) and (27) is simple when the system approaches the equilibrium point, the sliding mode parameter stops updating, and the singularity issue of k could be solved, and the system state could converge to the field of $\|q_v\| > \varepsilon_2$ within finite-time, hence, if a small enough parameter ε_2 is selected (in fact, considering the disturbance system could not reach the absolute equilibrium point, hence reaching the field near system equilibrium could be treated as reaching equilibrium point), the systems finite-time stability could be ensured, and this property will be proven in next section.

4.3. Finite-Time Controller

After the finite-time sliding mode surface is derived, the next step is to construct a finite-time controller to ensure the system state could converge to the proposed sliding mode surface within finite time and could converge to its equilibrium point along this sliding mode surface. Furthermore, the proposed controller could resist the disturbance torque caused by flexible appendages.

$$\begin{cases} u = -k_s sig^r(s) + \omega^\times \hat{J}\omega - \frac{1}{2}k\hat{J}F\omega - l_1 sign(s) \\ \qquad otherwise \\ u = -k_s sig^r(s) + \omega^\times \hat{J}\omega - \frac{1}{2}k\hat{J}F\omega - l_2 sign(s) \\ \qquad -k(1-\alpha)\beta q_0 \|q_v\|^{\alpha-1}\hat{J}q_v \\ \qquad if \|s\| \le \varepsilon_1 \,\&\, \|q_v\| > \varepsilon_2 \end{cases} \tag{28}$$

where k_s is a positive scalar, r is a positive scalar which satisfies $0 < r < 1$, $sign(x)$ is the sign function of vector x, vector function $sig^r(x)$ and l_i is defined as follows:

$$sig^r(x) = x/\|x\|^r \tag{29}$$

$$\begin{cases} l_1 = \bar{d} + \lambda\|\omega\|^2 + \frac{k}{2}\lambda\|\omega\| + v \\ l_2 = \bar{d} + \lambda\|\omega\|^2 + \frac{k}{2}\lambda\|\omega\| + k\lambda(1-\alpha)\beta q_0\|q_v\|^\alpha + v \end{cases} \tag{30}$$

where λ is a positive scalar which satisfies $\lambda \ge \lambda_M\left(\tilde{J}\right)$ with $\lambda_M\left(\tilde{J}\right)$ as the maximum eigenvalue value of the error inertia matrix $\tilde{J}$. Furthermore, v in (30) is a positive scalar, which is meant to resist the disturbance torque caused by flexible appendages, and the method to select this parameter will be given in a later text.

The next step is to discuss the system stability governed by the controller (28). We select the Lyapunov function as follows

$$V = \frac{1}{2}s^T Js \tag{31}$$

It is obvious that Function (31) is a semi-positive definite when and only when $s = 0$ Function (31) equals zero. Except for the condition that $\|s\| \le \varepsilon_1 \,\&\, \|q_v\| > \varepsilon_2$, we calculate the derivative of (31) and the substitute controller (28), and it could be found that:

$$\begin{aligned} \dot{V} &= s^T J\dot{s} \\ &= s^T J\left(\dot{\omega} + k\dot{q}_v\right) \\ &= s^T\left(-\delta^T\ddot{\eta} - \omega^\times\left(J\omega + \delta^T\dot{\eta}\right) + u + d\right) + \frac{1}{2}s^T JF\omega \\ &= -\|s\|^{2-r} - s^T\left(-\delta^T\ddot{\eta} - \omega^\times\delta^T\dot{\eta} + v\right) \\ &\quad + s^T\left(\tilde{J}\omega - \frac{1}{2}\tilde{J}F\omega + d\right) - l_1 s^T sign(s) \end{aligned} \tag{32}$$

Similarly, we calculate the derivative of Lyapunov function at the stage of $\|s\| \leq \varepsilon_1$ & $\|q_v\| > \varepsilon_2$, it could be found that:

$$
\begin{aligned}
\dot{V} &= s^T J \dot{s} \\
&= s^T J \left(\dot{\omega} + k\dot{q}_v + k q_v \right) \\
&= s^T \left(-\delta^T \ddot{\eta} - \omega^\times \left(J\omega + \delta^T \dot{\eta} \right) + u + d \right) + \tfrac{1}{2} s^T JF\omega \\
&\quad + \tfrac{k}{2}(1 - \alpha)\beta q_0 \|q_v\|^{\alpha - 1} s^T J q_v \\
&= -\|s\|^{2-r} - s^T \left(-\delta^T \ddot{\eta} - \omega^\times \delta^T \dot{\eta} + v \right) \\
&\quad + s^T \left(\tilde{J}\omega - \tfrac{1}{2}\tilde{J}F\omega + \tfrac{k}{2}(1 - \alpha)\beta q_0 \|q_v\|^{\alpha - 1}\tilde{J}q_v + d \right) \\
&\quad - l_1 s^T \mathbf{sign}(s)
\end{aligned}
\tag{33}
$$

Noting the definition of l_1 and l_2 in (30), it could be easily found that the last two terms in (32) and (33) are negative definites, i.e.,:

$$
\begin{aligned}
& s^T \left(\tilde{J}\omega - \tfrac{1}{2}\tilde{J}F\omega + d \right) - l_1 s^T \mathbf{sign}(s) \leq 0 \\
& s^T \left(\tilde{J}\omega - \tfrac{1}{2}\tilde{J}F\omega + \tfrac{k}{2}(1 - \alpha)\beta q_0 \|q_v\|^{\alpha - 1}\tilde{J}q_v + d \right) \\
& \qquad\qquad - l_1 s^T \mathbf{sign}(s) \leq 0
\end{aligned}
\tag{34}
$$

Hence (32) and (33) could be transformed to:

$$
\dot{V} \leq -\|s\|^{2-r} - s^T \left(-\delta^T \ddot{\eta} - \omega^\times \delta^T \dot{\eta} + v \right)
\tag{35}
$$

It is obvious the first term in (35) satisfies the finite-time stability condition, and if the last term in (35) is a negative definite, the system could converge to the designed sliding mode surface within a finite time. Moreover, it could be found that the term that needs to be suppressed is the flexible vibration; although, its modal state is hard to get by sensors on satellite, its dynamic model is known; hence, the modal state could be estimated by a state observer. We define $\hat{\eta}$ and its update law as follows.

$$
\ddot{\hat{\eta}} + C\dot{\hat{\eta}} + K\hat{\eta} = -\delta\tau
\tag{36}
$$

where:

$$
\tau = \hat{J}^{-1} \left(u - \omega^\times \hat{J}\omega - \omega^\times \delta^T \hat{\eta} - \delta^T \ddot{\hat{\eta}} \right)
\tag{37}
$$

It could easily be found that the estimation variable $\hat{\eta}$ has the same dynamic model as the modal state η. We define the estimation error variable $\tilde{\eta}$ as follows.

$$
\tilde{\eta} = \eta - \hat{\eta}
\tag{38}
$$

Substitute models (1) and (36) into (38), and it could be found that:

$$
\ddot{\tilde{\eta}} + C\dot{\tilde{\eta}} + K\tilde{\eta} = -\delta(\dot{\omega} - \tau)
\tag{39}
$$

Obviously, the only differences are the terms τ and $\dot{\omega}$ caused by model uncertainty and disturbance torque. Noting that C and K are both positive definite matrices, the error variable $\tilde{\eta}$ is typical of the Lagrange system and $\tilde{\eta}$ would track $-\delta(\dot{\omega} - \tau)$ by an exponential rate. Furthermore, noting that $-\delta(\dot{\omega} - \tau)$ is a relatively small term, it could be treated that estimation variable $\hat{\eta}$ could track modal state η, but there exists a small tracking error. Hence, the auxiliary term v could be defined as follows.

$$
v = -\delta^T \ddot{\hat{\eta}} - \omega^\times \delta^T \dot{\hat{\eta}} - v_0 \mathbf{sign}(s)
\tag{40}
$$

The function of the last term, v_0, is to offset the effect caused by the model error $-\delta(\dot{\omega}-\tau)$. Noting the ith modal estimation error could be described as:

$$\ddot{\tilde{\eta}}_i + 2\xi_i\omega_{ni}\dot{\tilde{\eta}}_i + \omega_{ni}^2\tilde{\eta}_i = (-\delta\dot{\omega} + \delta\tau)_i \tag{41}$$

The analytical solution of (41) could be written as follows.

$$\tilde{\eta}_i(t) = (-\delta\dot{\omega} + \delta\tau)_i(t) - e^{-\zeta_i\omega_{ni}t}\sin(\omega_d t + \phi)/\sqrt{1-\zeta_i^2}$$
$$\omega_d = \sqrt{1-\zeta_i^2}\,\omega_{ni}, \phi = \arctan\left(\sqrt{1-\zeta_i^2}/\xi_i\right) \tag{42}$$
$$if \quad 0 < \xi_i < 1$$

$$\tilde{\eta}_i(t) = (-\delta\dot{\omega} + \delta\tau)_i(t) - e^{-\omega_{ni}t}(\omega_{ni}t + 1)$$
$$if \quad \xi_i = 1 \tag{43}$$

$$\tilde{\eta}_i(t) = (-\delta\dot{\omega} + \delta\tau)_i(t) + e^{-(\xi_i+\sqrt{\zeta_i^2-1})\omega_{ni}t}/\left(2\sqrt{\zeta_i^2-1}\left(\zeta_i + \sqrt{\zeta_i^2-1}\right)\right)$$
$$- e^{-(\xi_i-\sqrt{\zeta_i^2-1})\omega_{ni}t}/\left(2\sqrt{\zeta_i^2-1}\left(\zeta_i - \sqrt{\zeta_i^2-1}\right)\right) \tag{44}$$
$$if \quad \xi_i > 1$$

Noting that $(-\delta\dot{\omega} + \delta\tau)_i(t)$ is a relatively small term, the major part is the transient term, i.e., the second term in (42)–(44). The idea to offset the transient term is to design v_0's envelop function, as Figure 1 shows.

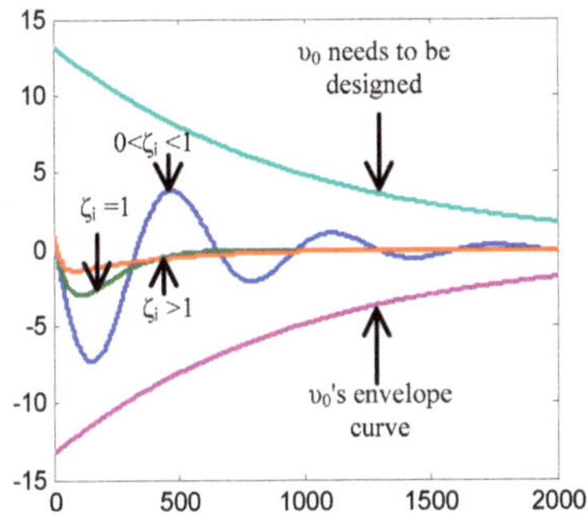

Figure 1. Envelop function.

The basic idea to design v_0 is to ensure its norm is larger than the norm of $\tilde{\eta}_i(t)$, and noting Equations (42)–(44), v_0 could be designed as follows.

$$v_0 = \tilde{\eta}_i(t_0)\exp(-\omega_v t)$$
$$\omega_v < \min\left(\sqrt{\xi_i^2 - 1}\,\omega_{ni}, \omega_{ni}, -\left(\xi_i + \sqrt{\xi_i^2 - 1}\right)\omega_{ni}\right) \tag{45}$$

By selecting the initial value of $\tilde{\eta}_i(t_0)$ (in this paper, $\tilde{\eta}_i(t_0)$ is selected five times larger than the model error and ω_v is selected $1/5$ times less than the minimum value of ω_{ni}), it could be found that:

$$s^T\left(-\delta^T\ddot{\eta} - \omega^{\times}\delta^T\dot{\eta} + v\right)$$
$$= s^T\left(-\delta^T\ddot{\tilde{\eta}} - \omega^{\times}\delta^T\dot{\tilde{\eta}} - \tilde{\eta}_i(t_0)\exp(-\omega_v t)\,\text{sign}(s)\right) \tag{46}$$
$$< 0$$

Hence, the derivative of V in (35) satisfies:

$$\dot{V} \leq -\|s\|^{2-r} = (2V)^{2-r/2} \tag{47}$$

Hence, the sliding mode state s could converge to zero within finite time, and based on the previous discussion, the system state could converge to the field near the equilibrium point within finite timem and the systems finite-time stability has been proven.

Moreover, it could be found that the sign function terms in (40) and (28) (except the disturbance term $\bar{d}$) tend to zero as the system converges to its equilibrium point and would not cause high-frequency vibrations.

5. Simulation

Comparing Group

Set the system parameters as follows.

$$
\begin{aligned}
J &= diag(100,75,50)\text{kg·m}^2, \hat{J} = diag(98,77,49)\text{kg·m}^2 \\
\omega(0) &= \begin{bmatrix} 0.01 & -0.02 & 0.03 \end{bmatrix}\text{rad/s} \\
q(0) &= \begin{bmatrix} 0 & \sqrt{6}/6 & \sqrt{3}/3 & \sqrt{2}/2 \end{bmatrix} \\
\omega_n &= \begin{bmatrix} 0.7 & 1 & 1.8 & 2.5 \end{bmatrix}^T, \\
\xi &= 10^{-2} \times \begin{bmatrix} 5.6 & 8.6 & 12.8 & 25.2 \end{bmatrix}^T \\
\delta &= \begin{bmatrix} 7 & 1.2 & 2.2 \\ -1.2 & 0.9 & -1.7 \\ 1.1 & 2.5 & -0.8 \\ 1.2 & -2.6 & -1.1 \end{bmatrix}
\end{aligned}
\tag{48}
$$

Set the disturbance d as Gauss white noise and its norm upper bound as follows.

$$
d = 5 \times 10^{-3} randn(3,1) Nm, \bar{d} = 5 \times 10^{-3}
\tag{49}
$$

Group A

In order to demonstrate the superiority of the controller in this paper, the standard sliding mode controller (50) is compared.

$$
\begin{aligned}
s &= \omega + kq_v \\
u &= -k_s s + \omega^\times \hat{J}\omega - \tfrac{1}{2}\hat{J}F\omega - l\,\text{sgn}(s) \\
l &= \lambda\left(\|\omega\|^2 + \tfrac{1}{2}k\|\omega\|\right) + \bar{d}
\end{aligned}
\tag{50}
$$

Select the control parameters as follows.

$$
\begin{aligned}
\lambda &= 3, k_s = 2, k = 0.15 \\
T &= 300 \text{ s}, t\_sample = 0.1 \text{ s}
\end{aligned}
\tag{51}
$$

The simulation results of the standard sliding mode controller (50) are given as follows. Based on Figures 2 and 3, it could be found that the system convergence time is more than 300 s, and the steady accuracy of angular velocity and attitude quaternion is about 8×10^{-4}rad/s and 4×10^{-4}. A system governed by a standard sliding mode controller could converge to its equilibrium point, but the convergence rate is relatively slow. Furthermore, based on Figure 4, it could also be found that the modal state also converges to zero as the angular velocity converges to zero, and the maximum vibration is about 0.1 according to numerical simulation.

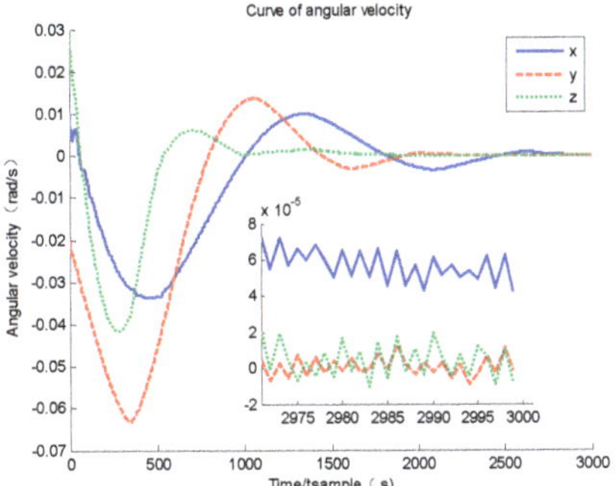

Figure 2. Curve of angular velocity.

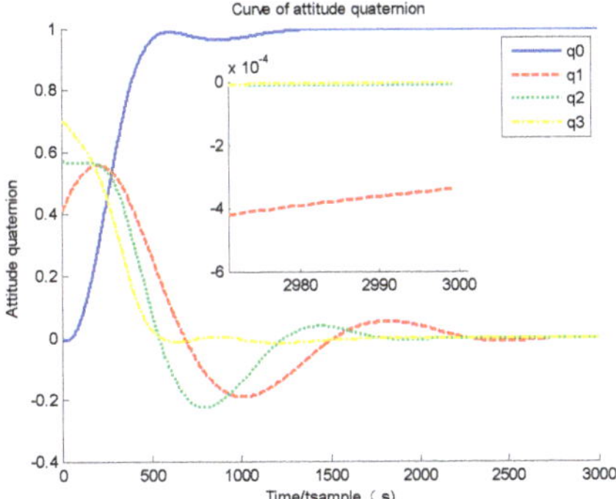

Figure 3. Curve of attitude quaternion.

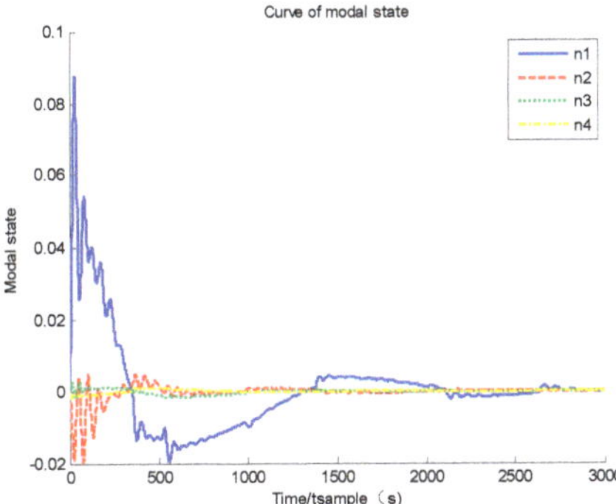

Figure 4. Curve of modal state.

Group B

The next step is to demonstrate the performance of the finite-time controller proposed by Li [14] in 2017. In this paper, the author pointed out that by designing the trajectory of angular velocity properly, the system convergence rate could be largely improved compared to a standard sliding mode controller. Moreover, finite-time stability, as discussed in this paper, is proven in Li's work; hence, this method is compared with the method proposed in this paper. The finite-time controller proposed in Li's work could be written as Equation (52) and select controller parameters as Equation (53) [14].

$$u = \begin{cases} \begin{aligned} &-k\|s\|^p s_e + \varphi_1 \omega - l_1 \mathrm{sgn}(s) & \|q_v\| > \alpha \\ &-k\|s\|^p s_e + \varphi_2 \omega - l_2 \mathrm{sgn}(s) & \beta < \|q_v\| \le \alpha \\ &-k\|s\|^p s_e + \varphi_3 \omega + \frac{k_3^2}{2} r q_0 \|q_v\|^{2r-1} \hat{J} e \\ &\quad - l_3 \mathrm{sgn}(s) & \|q_v\| \le \beta \end{aligned} \end{cases}$$

$$\begin{cases} \varphi_1 = \omega^\times \hat{J} - \frac{k_1}{2} \hat{J} e^\times \left(I_3 + \cot \frac{\varphi}{2} e^\times \right) \\ \varphi_2 = \omega^\times \hat{J} - \frac{k_2}{2} \hat{J} F \\ \varphi_3 = \omega^\times \hat{J} \end{cases}$$

$$\begin{cases} l_1 = \bar{d} + \lambda\|\omega\|^2 + \frac{k_1}{2}\lambda\left(1 + \cot\frac{\varphi}{2}\right)\|\omega\| \\ l_2 = \bar{d} + \lambda\|\omega\|^2 + \frac{k_2}{2}\lambda\|\omega\| \\ l_3 = \bar{d} + \lambda\|\omega\|^2 + \frac{k_3^2}{2} r\lambda|q_0|\|q_v\|^{2r-1} \end{cases} \tag{52}$$

$$\begin{aligned} \lambda &= 3, k_1 = 0.1, k = 5 \\ \alpha &= 1, \beta = 0.2, r = 2/3, p = 0.5 \\ T &= 200\ \mathrm{s}, t\_sample = 0.1\ \mathrm{s} \end{aligned} \tag{53}$$

The simulation results of the finite-time controller (52) are given as follows.

Based on Figures 5 and 6, it could be found that the system converges to its equilibrium point and the convergence time is about 100 s, which is largely improved compared to a standard sliding mode controller. Furthermore, based on Figures 5 and 6, it could be found that the system accuracy at the steady stage is about 1×10^{-4} rad/s and 1×10^{-6} of angular velocity and attitude quaternion, which is also improved compared to Group A. The only drawback of this controller is the flexible deformation. Based on Figure 7, although a modal state could converge to zero along the converge of angular velocity, the maximum modal state is about 0.35, and its frequency is also much higher than that of Group A. The high-frequency vibration causes system state chattering near its equilibrium point (based on Figure 5, three axes of angular velocity all have a chattering issue).

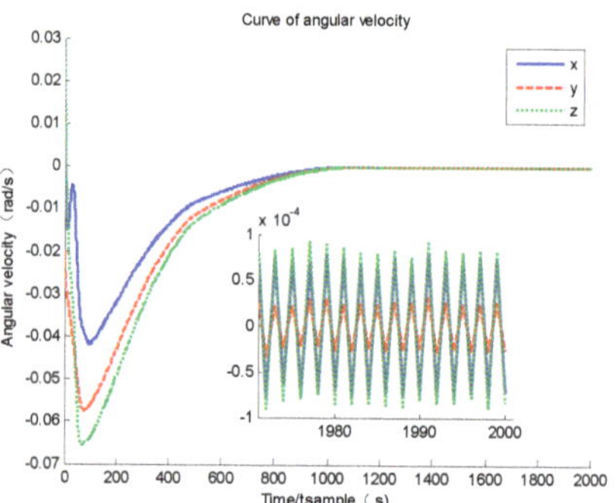

Figure 5. Curve of angular velocity.

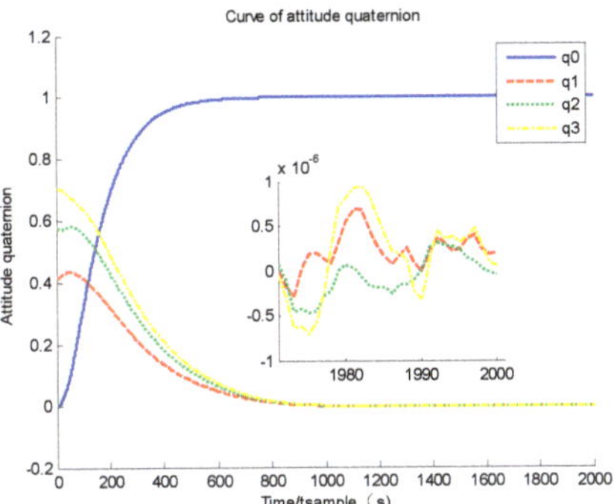

Figure 6. Curve of attitude quaternion.

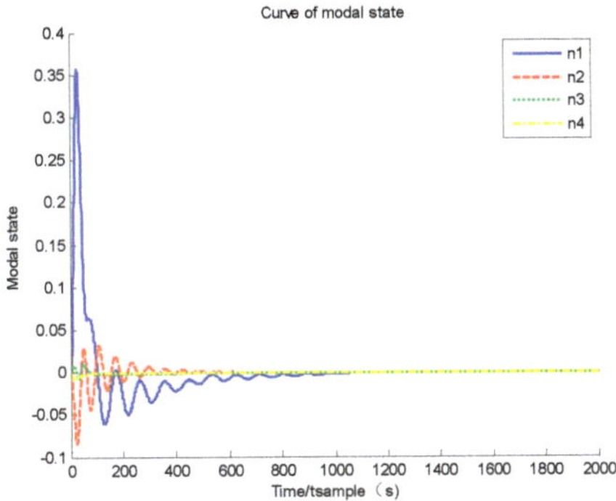

Figure 7. Curve of modal state.

Group C

The PID control algorithm is also a mature and widely used method in satellite attitude control, and the set control parameters are as follows.

$$u = -k_d \boldsymbol{\omega} - k_p \boldsymbol{q}_v - k_I \int \boldsymbol{q}_v dt$$
$$k_d = 10, k_p = 2, k_I = 0.1 \tag{54}$$

The system performance is governed by PID controllers and are shown as follows.

Based on Figures 8 and 9, it could be found that the system governed by a PID controller is stable, and the convergence time is about 200 s, which is slower than the standard sliding mode controller, but the structure is the strongest and most robust. It could be easily found that the shock of the system state near its equilibrium point is much relieved. Moreover, based on Figure 10, it could be found that the PID controller could also suppress the flexible vibration.

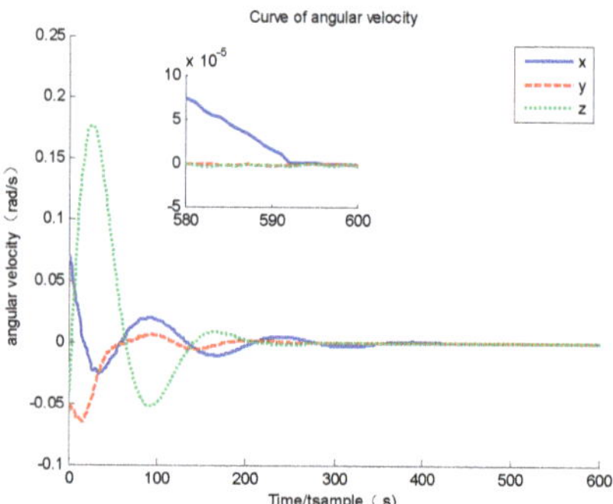

Figure 8. Curve of angular velocity.

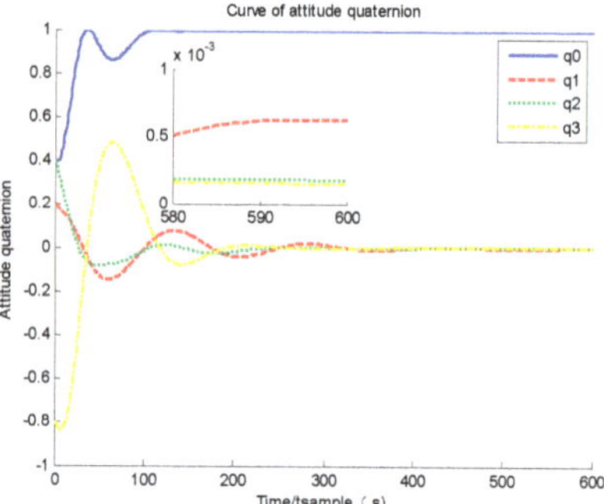

Figure 9. Curve of attitude quaternion.

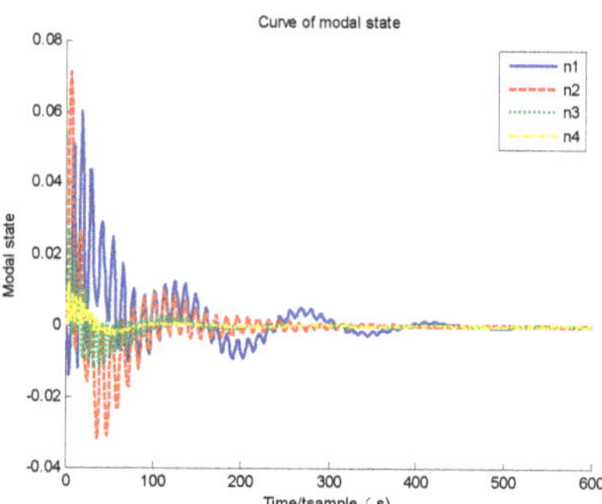

Figure 10. Curve of modal state.

Group D

The final step is to demonstrate the performance of the controller proposed in this paper. We selected the control parameters as follows.

$$\lambda = 3, k(t_0) = 0.05, k_s = 2$$
$$\varepsilon_1 = \varepsilon_2 = 0.001, r = 2/3$$
$$T = 200 \text{ s}, t\_sample = 0.1 \text{ s}$$

(55)

Moreover, we assume the initial modal state estimation error as follows.

$$\widetilde{\eta}(t_0) = \left[\ \begin{matrix} 0.01 & -0.01 & 0.02 & -0.02 \end{matrix}\ \right]^T$$
$$\dot{\widetilde{\eta}}(t_0) = \left[\ \begin{matrix} 0.05 & -0.01 & 0.01 & -0.005 \end{matrix}\ \right]^T \tag{56}$$

We selected the auxiliary term as follows.

$$v_0 = 0.2 \times \exp(-0.01t) \tag{57}$$

Generally, the larger parameter $k(t_0)$ and r could bring a better system convergence rate; however, the system control torque and angular velocity is increased. The parameter k_s determines the rate the system converges to the desired sliding mode. The parameter v_0 determines the suppression for the estimation error; when the initial estimation error, disturbance and model error are relatively large, this parameter should be selected to be larger.

The simulation results are given as follows. Based on Figures 11 and 12, it could be found that the system converges to its equilibrium point and the convergence time is about 45 s, which is the fastest of the three groups, and the system accuracy at a steady stage is 1×10^{-4}rad/s and 4×10^{-8} of angular velocity and attitude quaternion, which is also the best in the three groups. Moreover, it could be found that the initial value of the sliding mode parameter is 1/3 of Group A, but the system convergence time is about 1/8. This proves that by enlarging the sliding mode parameter, the system convergence time could be largely improved. Based on Figure 13, it could be found that the modal state converges to zero, and its maximum value is about 0.2, which is largely improved compared to the finite-time controller in Group B. Comparing the simulation results of Group D with Groups A, B and C, it could be found that the upper bound and frequency of flexible vibration is suppressed. Hence, the chattering issue is largely relieved (seen in Figure 11, the shocking of angular velocity only exists on the Z-axis and its frequency is much lower). Based on Figure 14, it could be found that although there exists la arge initial estimation error, the state observer could also track the real modal state, and the estimation error tends to zero. Based on Figure 15, it could be found that the sliding mode parameter stops updating when the system state nears its equilibrium point; hence, the system does not have a singularity issue.

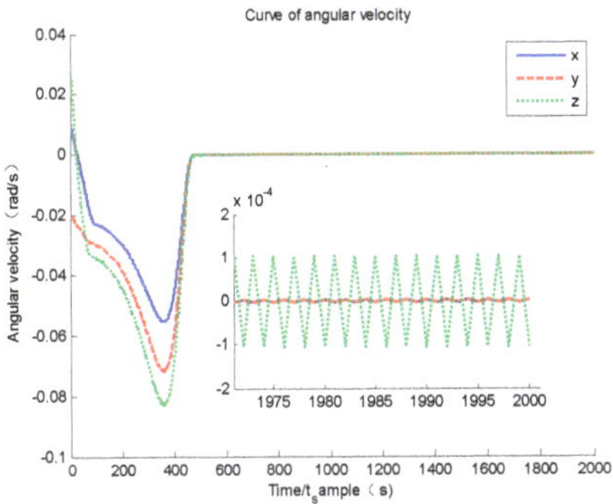

Figure 11. Curve of angular velocity.

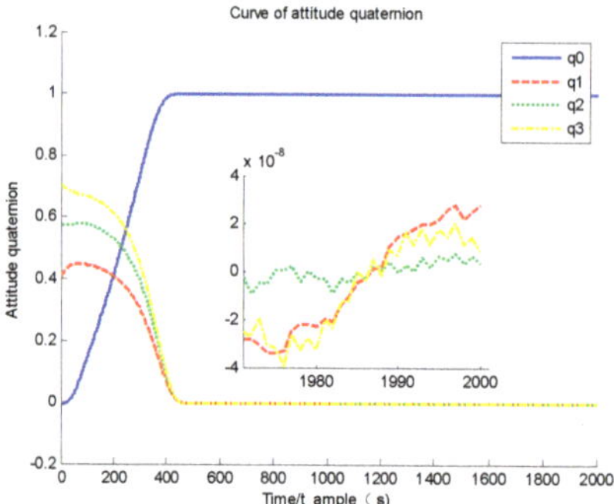

Figure 12. Curve of attitude quaternion.

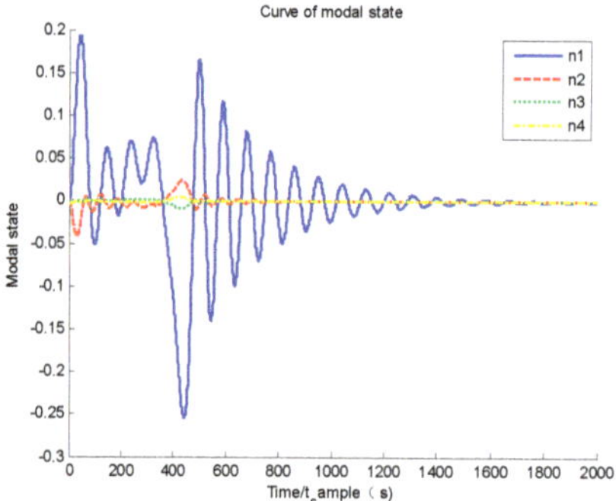

Figure 13. Curve of modal state.

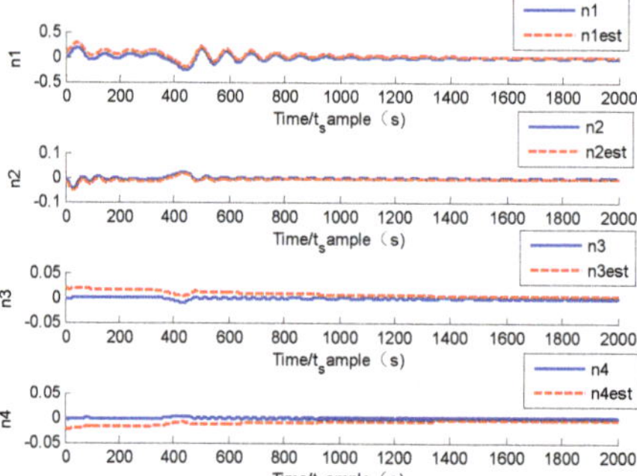

Figure 14. Curve of modal state and its estimation.

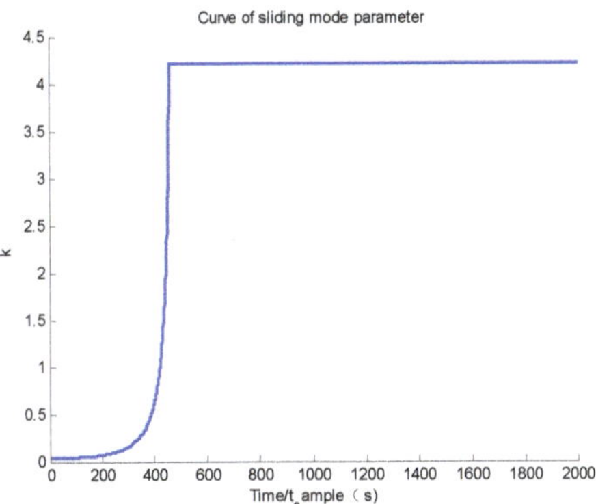

Figure 15. Curve of sliding mode parameter.

The performance of the three groups of controllers is summarized in the following Table 2.

Table 2. Comparison of the three controllers.

	Standard Sliding Mode	**Finite-Time Controller**	**PID Controller**	**Method in This Paper**
Convergence rate	Slow	Medium	Slow	Fast
Steady accuracy	Low	Medium	Low	High
Chattering issue	Weak	Strong	Weak	Medium
Flexible vibration suppression	Small	Large	Small	Medium
Singularity issue	None	Exists	None	None

Based on the comparison in Table 2, it could be found that the major advantages of the method proposed in this paper are its convergence rate, steady accuracy and none singularity issue. Although the updated law of the sliding mode parameter would cause relatively large flexible vibration, the modal state estimation algorithm and auxiliary term could relieve the effect.

6. Conclusions

In this paper: a finite-time controller based on the dynamic sliding mode is proposed, and the system convergence rate is largely improved compared to the standard sliding mode and existing finite-time controller. It is proven that by implementing the updated law of the sliding mode parameter, the system could converge to the field near the equilibrium point within finite time, without causing the singularity issue during the whole control process. Furthermore, it is proven that the key to improving the system performance is to design an angular velocity trajectory properly, and the method proposed in this paper is the updated law of the sliding mode parameter; by implementing this method, the drawback of dropping too fast of angular velocity is avoided. Moreover, a state observer is designed for flexible vibration, and an auxiliary term, which is converging slower than the system state, is designed to suppress the effect of the estimation error. The simulation results prove the effectiveness of the method proposed in this paper, and future work could focus on the modification of the sign function in controllers to avoid high-frequency vibration near the system's equilibrium point.

Author Contributions: Conceptualization, Y.L. and H.L.; methodology, Y.L. and L.X.; software, Y.L. and L.X.; validation, Y.L.; formal analysis, Y.L.; investigation, H.L.; resources, L.X.; data curation, L.X.; writing—original draft preparation, Y.L.; writing—review and editing, Y.L. All authors have read and agreed to the published version of the manuscript.

Funding: This research was funded byNational Natural Science Foundation of China, grant number 61903289, 62003375 and 6210345.

Institutional Review Board Statement: Not applicable.

Informed Consent Statement: Not applicable.

Data Availability Statement: If anyone wants to get the original data of this paper, please contact liyou@xidian.edu.cn.

Acknowledgments: This work was supported partially by the National Natural Science Foundation of China (Project Nos. 61903289, 62003375 and 62103452). The authors greatly appreciate the above financial support. The authors would also like to thank the associate editor and reviewers for their valuable comments and constructive suggestions that helped to improve the paper significantly.

Conflicts of Interest: The authors declare that we have no financial and personal relationships with other people or organizations that can inappropriately influence our work, there is no professional or other personal interest of any nature or kind in any product, service and/or company that could be construed as influencing the position presented in, or the review of, the manuscript entitled.

References

1. Gao, S.; Li, Y.; Yao, S.Y. Robust PD$^+$ Control Algorithm for Satellite Attitude Tracking for Dynamic Targets. Mathematical Problems in Engineering. *Math. Probl. Eng.* **2021**, *2021*, 6680994.
2. Li, Y.; Ye, D. Robust PID controller for flexible satellite attitude control under angular velocity and control torque constraint. *Asian J. Control* **2020**, *22*, 1327–1344. [CrossRef]
3. Li, Y.; Sun, Z.W.; Ye, D. Time Efficient Robust PID Plus Controller for Satellite Attitude Stabilization Control Considering Angular Velocity and Control Torque Constraint. *J. Aerosp. Eng.* **2017**, *30*, 04017030. [CrossRef]
4. Xu, Y.; Lu, Z.; Shan, X.; Jia, W.; Wei, B.; Wang, Y. Study on an Automatic Parking Method Based on the Sliding Mode Variable Structure and Fuzzy Logical Control. *Symmetry* **2018**, *10*, 523. [CrossRef]
5. Zhang, Y.; Nie, Y.; Chen, L. Adaptive Fuzzy Fault-Tolerant Control against Time-Varying Faults via a New Sliding Mode Observer Method. *Symmetry* **2021**, *13*, 1615. [CrossRef]
6. Chakrabarti, D.; Selvaganesan, N. PD and PD beta based sliding mode control algorithms with modified reaching law for satellite attitude maneuver. *Adv. Space Res.* **2020**, *65*, 1279–1295. [CrossRef]
7. Ye, D.; Zhang, H.Z.; Tian, Y.X.; Zhao, Y.; Sun, Z.W. Fuzzy Sliding Mode Control of Nonparallel-ground-track Imaging Satellite with High Precision. *Int. J. Control. Autom. Syst.* **2020**, *18*, 1617–1628. [CrossRef]
8. Gao, S.; Li, Y.; Xue, H.F.; Yao, S.Y. Dynamic Sliding Mode Controller with Variable Structure for Fast Satellite Attitude Maneuver. *Math. Probl. Eng.* **2021**, *2021*, 5539717.
9. Li, Y.; Ye, D.; Sun, Z.W. Time efficient sliding mode controller based on Bang-Bang logic for satellite attitude control. *Aerosp. Sci. Technol.* **2018**, *75*, 342–352. [CrossRef]
10. Xiao, B.; Yin, S.; Wu, L.G. A structure simple controller for satellite attitude tracking maneuver. *IEEE Trans. Ind. Electron.* **2017**, *64*, 1436–1446. [CrossRef]
11. Xiao, B.; Yin, S.; Gao, H.J. Tracking control of robotic manipulators with uncertain kinematics and dynamics. *IEEE Trans. Ind. Electron.* **2016**, *63*, 6439–6449. [CrossRef]
12. Hu, Q.L.; Zhang, Y.M.; Huo, X.; Xiao, B. Adaptive integral-type sliding mode control for spacecraft attitude maneuvering under actuator stuck failures. *Chin. J. Aeronaut.* **2011**, *24*, 32–45. [CrossRef]
13. Ye, D.; Zou, A.M.; Sun, Z.W. Predefined-Time Predefined-Bounded Attitude Tracking Control for Rigid Spacecraft. *IEEE Trans. Aerosp. Electron. Syst.* **2021**. [CrossRef]
14. Li, Y.; Ye, D.; Sun, Z.W. Robust finite time control algorithm for satellite attitude control. *Aerosp. Sci. Technol.* **2017**, *68*, 46–57. [CrossRef]
15. Liang, H.Z.; Wang, J.Y.; Sun, Z.W. Robust decentralized coordinated attitude control of spacecraft formation. *Acta Astronaut.* **2011**, *69*, 280–288. [CrossRef]
16. Liang, H.Z.; Sun, Z.W.; Wang, J.Y. Finite-time attitude synchronization controllers design for spacecraft formations via behavior-based approach. *Part G J. Aerosp. Eng.* **2013**, *227*, 1737–1753. [CrossRef]
17. Wang, J.Y.; Liang, H.Z.; Sun, Z.W. Dual-quaternion-based finite-time control for spacecraft tracking in six degrees of freedom. *J. Aerosp. Eng.* **2013**, *227*, 528–545. [CrossRef]
18. Wang, J.Y.; Liang, H.Z.; Sun, Z.W.; Zhang, S.J.; Liu, M. Finite-time control for spacecraft formation with dual-number-based description. *J. Guid. Control. Dyn.* **2012**, *35*, 950–962. [CrossRef]
19. Wu, S.N.; Wang, R.; Radice, G.; Wu, Z.G. Robust attitude maneuver control of spacecraft with reaction wheel low-speed friction compensation. *Aerosp. Sci. Technol.* **2015**, *43*, 213–218. [CrossRef]
20. Wu, S.N.; Radice, G.; Sun, Z.W. Robust finite-time control for flexible spacecraft attitude maneuver. *J. Aerosp. Eng.* **2014**, *27*, 185–190. [CrossRef]

21. Wie, B.; Barba, P.M. Quaternion Feedback for Spacecraft Large Angle Maneuvers. *J. Guid. Control. Dyn.* **1985**, *8*, 360–365. [CrossRef]
22. Wie, B.; Weiss, H.; Arapostathis, A. Quaternion Feedback Regulator for Spacecraft Eigenaxis Rotations. *J. Guid. Control. Dyn.* **1989**, *12*, 375–380. [CrossRef]
23. Clarke, F.H.; Ledyaev, Y.S.; Sontag, E.D.; Subbotin, A.I. Asymptotic Controllability Inplies Feedback Stabilization. *IEEE Trans. Autom. Control* **1997**, *42*, 1394–1407. [CrossRef]
24. Vadali, S.R.; Junkins, J.L. Closed Loop Slewing of Flexible Spacecraft. *J. Guid. Control. Dyn.* **1990**, *13*, 57–65.

Article

Fault-Diagnosis Sensor Selection for Fuel Cell Stack Systems Combining an Analytic Hierarchy Process with the Technique Order Performance Similarity Ideal Solution Method

Guangying Jin [1,*] and Guangzhe Jin [2]

1   School of Maritime Economics and Management, Dalian Maritime University, Dalian 116026, China
2   College of Engineering Science and Technology, Shanghai Ocean University, Shanghai 201306, China; gzjin@shou.edu.cn
*   Correspondence: guangying.jin@dlmu.edu.cn

Abstract: Multi-Criteria Decision Making (MCDM) methods have rapidly developed and have been applied to many areas for decision making in engineering. Apart from that, the process to select fault-diagnosis sensor for Fuel Cell Stack system in various options is a multi-criteria decision-making (MCDM) issue. However, in light of the choosing of fault diagnosis sensors, there is no MCDM analysis, and Fuel Cell Stack companies also urgently need a solution. Therefore, in this paper, we will use MCDM methods to analysis the fault-diagnosis sensor selection problem for the first time. The main contribution of this paper is to proposed a fault-diagnosis sensor selection methodology, which combines the rank reversal resisted AHP and TOPSIS and supports Fuel Cell Stack companies to select the optimal fault-diagnosis sensors. Apart from that, through the analysis, among all sensor alternatives, the acquisition of the optimal solution can be regarded as solving the symmetric or asymmetric problem of the optimal solution, which just maps to the TOPSIS method. Therefore, after apply the proposed fault-diagnosis sensor selection methodology, the Fuel Cell Stack system fault-diagnosis process will be more efficient, economical, and safe.

Keywords: fault-diagnosis; AHP; TOPSIS; MCDM

Citation: Jin, G.; Jin, G. Fault-Diagnosis Sensor Selection for Fuel Cell Stack Systems Combining an Analytic Hierarchy Process with the Technique Order Performance Similarity Ideal Solution Method. *Symmetry* **2021**, *13*, 2366. https://doi.org/10.3390/sym13122366

Academic Editors: Chengxi Zhang, Jin Wu and Chong Li

Received: 28 October 2021
Accepted: 3 December 2021
Published: 8 December 2021

1. Introduction

A Fuel Cell Stack system (FCS) refers to a power generation system with fuel cell as the core, fuel supply and circulation system, oxidizer supply system, water/heat management system, control system, etc., and able to continuously output electronic power [1]. The main research interests for FCS include using lightweight materials, optimizing design, and improving the specific power of the fuel cell system, improving the fast cold start capability and dynamic response performance of the FCS system, researching fuel processors with load following capabilities, optimizing supercapacitors and hydrogen storage for system design to improve system efficiency and peak shaving capabilities, recover braking energy, etc. [1].

Apart from that, a fault is a state in which the system cannot perform a prescribed function. Generally speaking, a fault refers to an event in which the function of some components in the system fails and the function of the entire system deteriorates [2]. The FCS is mainly composed of a stack, a fuel processor, a power regulator, and an air compressor. There will be many potential faults which can directly influence the FCS system and FCS cause the system to no work properly [3]. Therefore, the effective use of fault-diagnosis sensors to detect FCS systems becomes very important. The fault diagnosis is the process of using various detection and testing methods to find out whether there is a fault in the system and equipment [4].

A proper fault-diagnosis sensor selection is a deeply significant problem for FCS because of the reason that choosing an unsuitable fault-diagnosis sensor can directly influence the main function testing progress of the FCS fault-diagnosis system. Therefore, selecting a suitable fault-diagnosis sensor can be a very important part to obtain the most optimal consequence for FCS.

Therefore, FCS companies urgently need to choose more efficient, safer, and more affordable fault diagnosis sensors. The fault-diagnosis sensor selection problem is a Multi-Criteria Decision-Analysis (MCDM) problem.

Therefore, in this paper, we use the rank aversion resisted AHP and TOPSIS method to approach the FCS fault-diagnosis sensor selection problem. The main objective of this study is to put forward a fault-diagnosis sensor selection methodology which considers difference criteria and sub-criteria, and helps FCS companies to select the optimal fault-diagnosis sensors, and ensure that the FCS fault-diagnosis process becomes more efficient, economical, and safe.

The rest of this article is structured as follows. The "Literature review" section introduces the background and related research. The "Methods" section describes the methods that we used in this research. The "Case study" section characterizes an example to derive experimental information from and to analyze the consequence of imitation. A detailed discussion and ideas for future work are summarized in the "Discussion" section.

2. Literature Review

A fuel cell is a chemical device that directly converts the chemical energy of the fuel into electrical energy, also known as an electrochemical generator, and it is the fourth power generation technology after hydropower, thermal power, and atomic power [1]. Meanwhile, in addition to the fuel cell's main body (power generation system), the FCS also has some peripheral devices, including fuel reforming supply system, gasoline supply system, water management system, thermal management system, dangerous communication system, control system, safety system, etc. [5].

To obtain the optimal FCS fault-diagnosis sensor in all alternatives, we can use the asymmetry of the evaluation result. Symmetry and asymmetry are frequent patterns that are widely studied in a variety of fields. In most cases, symmetries can exist in an arithmetic equation, which has been an important part for approving issues or conduct a more in-depth study. In this research, selecting the fault-diagnosis sensor problem can be regarded as the process of solving the symmetric and asymmetric problem in mathematical formulation. To obtain the optimal fault-diagnosis sensor, we have to evaluate different sensor alternatives depending on various evaluation criteria. The criteria usually include positive criteria and negative criteria. Therefore, in the evaluation formula, we need to consider both positive and negative criteria, and get positive and negative evaluation solutions, which can be seen as the symmetry of formula. Meanwhile, to distinguish and sort all the alternatives and get the final optimal sensors, the positive and negative results should be asymmetrical. In fact, in most cases, the evaluation results are asymmetrical.

Selecting the new fault-diagnosis sensor wastes energy and is a complex process, requiring multi-disciplinary cognition and expert experience. Apart from that, with the rapid development of information and communication technology, a large amount of engineering system information data are being produced. Therefore, to select an economical, efficient, and logically well-performed fault-diagnosis sensor, the decision-maker should consider various sources of information [6,7].

The fault-diagnosis sensor selection problem can be seen as a Multi-Criteria Decision-Making (MCDM) [8,9] issue. Multi-Criteria Decision-Making has always been a well-known part of decision making [10]. It concerns establishing or approving determination and planning issues under multi-standards [11]. MCDM supports managers to make a decision, which quantifies a special standard depending on its significance in relation to other targets [12,13]. MCDM methods provide an opportunity to take into account the multidimensional nature of the considered problem [14]. The MCDM technique can also take into account the economic, community, civilization, and circumstance affairs that can enhance the project [15]. MCDM is considered to be a special decision-making process, and the issue can be emphasized as follows: selection is required, the level needs to be defined, all the alternatives need to be prioritized, and different options behaviors need to be illustrated [16]. MCDM methods are very different depending on the various dimensions, for instance a fuzzy environment, for which interests and assessment standards are represented by different methods, including different methods of value aggregation; whether there is a chance it has certain information can also define the different kinds of MCDM methods [17–20].

There are many MCDM techniques. The traditional classic mainstream MCDM techniques are Analytic Hierarchy Process (AHP) [21,22], Analytic network process (ANP) [23], Technique for Order Performance by Similarity to Ideal Solution (TOPSIS) [24], Classic MAUT [25], ELECTRE [26], PROMETHEE [27], UTA [28], UTADIS [29], etc., while modern classic mainstream MCDM techniques are COMET [30], COPRAS [31], SPOTIS [32], SIMUS [33], and so on.

The AHP technique includes determining the total corresponding standards and compares every two values with each other to estimate the effect of every standard [34]. The AHP can be seen as a measurement based on the comparison between two alternatives, depending on the decision of specialists to obtain the precedence of all the alternatives [35]. AHP is widely applied to MCDM problems in different fields, for instance financial and scheduling, and managers can use this method to establish an MCDM issue under a hierarchy property type [36].

TOPSIS is a generic approach to deal with multi-standard determination issues [37]. Moreover, TOPSIS is also a symmetrical technique used for the ranking of the alternates, and it is the best-known approach for alternative ranking of MCMD problems [38]. The symmetrical TOPSIS technique can estimate critical criteria [39]. The main idea of TOPSIS is that the best alternative among all competitive alternatives should be at the minimum distance from the Positive-Ideal (P-I) solution and have maximum distance from the Negative-Ideal (N-I) solution [38,40], which can be regarded as the process of solving the symmetric problem in mathematical formulation. Furthermore, for determining the best alternative among several others, TOPSIS proves to be a good MCDM method [39]. Apart from that, TOPSIS also has other advantages: (1) The structure of TOPSIS is reasonable; (2) The calculation steps are very easy; (3) It allows to find the optimal options for every standard chosen through an understandable arithmetic from; (4) The significance weights are included in the process of comparison [41].

The COPRAS (Complex Proportional Assessment) technique usually evaluates the maximized/minimized index data, and the influence of these values on the properties in consequence evaluation is taken into account [42]. Moreover, the COPRAS technique supposes a straightforward and different scale, relying on the importance and rank of the effectiveness of the usable alternatives in the appearance of conflicting standards [43,44]. The target of COMET is to approach the issues of MCDM in a fuzzy condition, and this can be seen as a novel way of thinking to approach the issues of MCDM with regards to inconstancy [30]. The method is based on L-R-type GFN, which can obtain the hesitant rank for an option in a definitive standard. SPOTIS is a novel rank reversal-free MCDM method, and it has very low complexity [32]. Moreover, compared with the COMET approach, it requires much less information. SIMUS (Sequential Interactive Model for Urban Systems) is a hybrid model, not only based on Linear Programming but also on traditional approaches, such as 'Weighted sum' and 'Outranking' procedures, and can handle any number of objectives, albeit not reaching an optimal result as in the case of Linear Programming [33].

Faizi et al. proposed the MCGDM (Multi-Criteria Group Decision Making) methodology, which was based on the B-W technique with Hamacher polymerization operations of intuitive binary combination language sets [45]. This method was proposed for modifying efficiency with operations of determining steps. Božanić et al. proposed methodologies for MCDM depending on the D values, the FUCOM technique and RAFSI technique, and it can approach the industrial mechanical equipment choosing problem [46].

COMET (the Characteristic Objects Method) has an attractive function that can avoid rank reversal; however, compared with AHP and TOPSIS, it requires much more information under the situation of the very common value matrix definition in MCDM issues [32]. Moreover, when we consider the stability between the COPRAS and TOPSIS techniques with a value variable, we find that TOPSIS is better and more insensitive when the value does not change much, and compared with other technologies, the priority result is different [47]. Moreover, compared with TOPSIS and AHP, the COMET method also needs much more information when we confront it with traditional MCDM issues [32]. Furthermore, it is very hard to apply the SPOTIS approach as compared with TOPSIS, due to the fact that there exists uncertainty in the option of minimum and maximum bounds of the standard [32]. SIMUS, because it is based on Linear Programming, and LP works with a very different approach in a decision-making scenario when compared with other methods, can be considered a geometric tactic.

Here, to select the most suitable MCDM technique for fault-diagnosis sensor selection, we have to consider some judgment methods. Marco Cinelli et al. proposed a taxonomy depending on the features of MCDM steps [48], which included three important steps: phase 1—issue statement phase, phase 2—determination recommendation establishment phase, and phase 3—characteristics and method assistance phase. Wątróbski et al. also proposed a taxonomy of MCDA methods, and the proposed taxonomy can be selected as a universal scheme to choose a suitable MCDA technique under a certain decision problem domain [17]. Therefore, we can apply these taxonomies with MCDM techniques and choose a similar MCDM technique(s) for fault-diagnosis sensor selection studies.

1. After all the considerations above, we have found that the AHP and TOPSIS method is the most adaptable for the fault-diagnosis sensor selection. We summarized the following reasons: the fault-diagnosis sensor selection problem is a deterministic MCDM problem and, as compared with other methods, AHP and TOPSIS are optimal for deterministic conditions.
2. The weight definition in the fault-diagnosis sensor choosing means subjective judgement steps, and it just maps to the AHP method.
3. Compared with other complex process methods, it is easy to apply and use the AHP and TOPSIS methods.
4. Among all MCDM techniques, the AHP and TOPSIS techniques require less information.
5. Among all MCDM techniques, the TOPSIS technique has good stability in a data variable case.

The main consideration in this study is to select the most suitable fault-diagnosis sensor and also consider the various standards when making a decision. These are multidisciplinary criteria, and there are many intangible or immeasurable factors. Moreover, when choosing engineering methods or equipment, there are many criteria classification methods. Yazdani-Chamzini et al. considered the technical parameters, operational parameters and economical parameters when selecting the most appropriate mining methods [49]. Štirbanović et al. applied the MCDM method for flotation machine selection, which considered the constructional parameters, economical parameters, and technical parameters [50]. Sarrate et al. proposed a methodology depending on fault diagnosis capability optimization for finding a suitable sensors' location for FCS systems, and considered different fault-diagnosis criteria, such as compressor parameters, inlet manifold parameters, air conditioner parameters, humidifier parameters, FCS anode parameters, and so on [51]. When we define the importance weights of the criteria, there are also intangible or immeasurable factors that should be approached.

The AHP is a targeted and practical decision-making method for analyzing qualitative problems, and the characteristic is to structure the various factors in complex issues through inter-connection to make them organized [35]. Moreover, the AHP is the most commonly used technique due to its simplicity, ease of use, and great flexibility [52]. However, there is a rank versal phenomenon in the AHP method. The rank reversal problem means that when ranking the pros and cons of alternatives to an MCDM problem, adding or reducing an alternative and applying the same decision model will cause inconsistent ranking results. Wang and Elhag proposed methodology to avoid rank reversal in AHP [53]; we can use this method to approach the problem of the rank reversal problem in the AHP method.

Therefore, we can use the AHP method to define the weights of fault-diagnosis sensor importance criteria. Additionally, we can use TOPSIS to obtain the result of the feasible fault-diagnosis sensor alternatives. The most important reason why we use TOPSIS is that it can analyze the length to the P-I result and the N-I result, which just conforms to the symmetry of the formula. Additionally, we can get the asymmetry of all the candidate sensor evaluation result through TOPSIS.

Our main objective is to help FCS companies select more efficient, safer, and more affordable fault-diagnosis sensors. The fault-diagnosis sensor selection problem is an MCDM problem. Therefore, this research utilizes the AHP method (the MCDM method) to determine the importance weights of the fault-diagnosis sensor estimation standard, and TOPSIS is to find suitable solutions. It is then possible for FCS companies to apply this methodology and choose the optimal fault-diagnosis sensor, making the FCS fault-detection process more efficient, economical, and safe.

3. Methods

3.1. The AHP Technique

The whole AHP steps are shown as follows [35]:

1. Depending on how thorough the knowledge about the system is, determine the main target and establish the measures and policies involved in the planning and decision-making.
2. Establish a hierarchical framwork, and define the location of all the factors that we use in this framwork according to different goals and different functions.
3. Determine the degree of correlation between neighboring layer factors. Establish pairwise comparison matrices, determine the relative weight of a factor on the previous layer and the significance of the corresponding factors on this layer.
4. Obtain the composite weight of every level factor to the target. Moreover, the sorting needs to be done, and the importance of the main target of the bottom element at the framwork needs to be defined.
5. Establish the weight of each layer element of the system goal, perform the total sorting, and determine the importance of the overall goal of the lowest element in the hierarchical structure.

First of all, we have to decide the main issue and the different intellectual ideas. After that, targets, standards, and options will be sorted depending on the hierarchical composition. In this hierarchical structure, the objectives are in the upper layer, usually related to the determined options in the middle layer and all the options located in lower layer. In the third step, we have to make a comparison between the alternatives. Here, we have to define the importance scale. The importance scale refers to how many more times one element is important than another. The AHP Fundamental Scale is shown in Table 1.

Table 1. The AHP Fundamental Scale [53,54].

The Paired Comparison Reference Rank		
Significance Rank	**Relation Type**	**Interpretation**
1	Same significance	Two options dedicated to the same degree to the target
3	Little significance of an option to the other	Assessment is slightly more inclined to one option over another
5	High significance	Assessment is strongly inclined to one option over another
7	Higher significance	Very high inclination to one option over another
9	Absolute significance	One option is absolutely more significant than another
2,4,6,8	Intermediate values between the two ratios	When there is a need to subdivide

Note. C = Criteria.

After that, the rank of the importance is obtained through a paired comparison matrix among alternative m and alternative n of the form (Equation (1)) [55]:

$$A = \begin{bmatrix} 1 & a_{12} & \cdots & a_{1n} & \cdots & a_{1N} \\ 1/a_{12} & 1 & \cdots & a_{2n} & \cdots & a_{2N} \\ \vdots & \vdots & 1 & \vdots & \cdots & \vdots \\ 1/a_{1n} & 1/a_{2n} & \cdots & 1 & \cdots & a_{mN} \\ \vdots & \vdots & \cdots & \vdots & \cdots & \vdots \\ 1/a_{1N} & 1/a_{2N} & \cdots & 1/a_{mN} & \cdots & 1 \end{bmatrix} \tag{1}$$

In Equation (1), a_{mn} refers to the value appearance of the determination of pairwise comparison values (alternative m, alternative n) for all alternatives (m, n = 1, 2, ... , n). Here, m refers to a row of A and n refers to a column of A. In Equation (1), a_{mn} cannot be equal to 0.

In the next process, we have to calculate the normalized value for every matrix and define the corresponding weights for them. The corresponding weights can be obtained from u and λ_{max} (Equation (2)):

$$A_w = \lambda_{max} u \tag{2}$$

where u refers to the right eigenvector and λ_{max} refers to the largest eigenvector.

To avoid the Rank Reversal problem for the AHP technique, it is possible to calculate the rescaled right eigenvector weights (Equation (3)):

$$\hat{U} = kU = \left(\frac{u_1}{\sum_1^N u_m}, \frac{u_2}{\sum_1^N u_m}, \cdots, \frac{u_N}{\sum_1^N u_m}, \frac{u_{N+1}}{\sum_1^N u_m} \right)^T \tag{3}$$

where $\sum_{m=1}^N u_{nA} = \sum_{m=1}^N k u_n$, $\sum_{m=1}^N u_{nA} = 1$, and $k = 1/\sum_{m=1}^N u_n$.

In Equation (3), $\hat{U}$ can be seen as the normalization with respect to the original N alternatives. We can calculate the rescaled weight $\hat{U}$ to resist the rank reversal phenomenon for the AHP technique when an option is inserted or dropped.

3.2. TOPSIS Technique

The basic process of the TOPSIS technique can be seen as follows:

- First, establish a decision matrix for alternatives (Equation (4)):

$$D = \begin{bmatrix} y_{11} & y_{12} & \cdots & y_{1j} & \cdots & y_{1J} \\ y_{21} & y_{22} & \cdots & y_{2j} & \cdots & y_{2J} \\ \vdots & \vdots & \cdots & \vdots & \cdots & \vdots \\ y_{i1} & y_{j2} & \cdots & y_{ij} & \cdots & y_{iJ} \\ \vdots & \vdots & \cdots & \vdots & \cdots & \vdots \\ y_{I1} & y_{I2} & & y_{Ij} & \cdots & y_{IJ} \end{bmatrix} \tag{4}$$

where D is the decision matrix, y_{ij} is the jth criterion value related to the ith alternative, I is total number of alternatives, and J is total number of criteria.

- Second, obtain the normalized decision matrix $Z(=z_{ij})$ (Equation (5)):

$$z_{ij} = \frac{y_{ij}}{\sqrt{\sum_{i=1}^I y_{ij}^2}} \tag{5}$$

where z_{ij} is the normalized value for the jth criterion value related to the ith alternative and I is the total number of alternatives. The reason why we use the vector normalization technique is that many researchers have analyzed the effects of different normalizations for TOPSIS, and they have found that the vector normalization method is most suitable for TOPSIS [56,57]. Moreover, in this process, they have computed the consistency of the results of all the alternatives, and analyzed the sensitivity of the weight for different normalization methods applied on TOPSIS.

- Third, obtain the weighted normalized decision matrix $X(=x_{ij})$ (Equation (6)):

$$x_{ij} = \omega_j \cdot z_{ij} \tag{6}$$

Here, the normalized value is obtained from the multiplication of the original value and the corresponding weights. In this research, the weight can be defined by the AHP method.

- Fourth, calculate the P-I and N-I results (Equations (7) and (8)):

$$\text{P-I solution}: \ x_j^+ = \begin{cases} \max\limits_i x_{ij}, & i \in l' \\ \min\limits_i x_{ij}, & i \in l'' \end{cases} \tag{7}$$

where l' is the value set associate with benefit criteria and l'' is the value set associate with the cost criteria.

$$\text{N-I solution}: \ x_j^- = \begin{cases} \min\limits_i x_{ij}, & i \in l' \\ \max\limits_i x_{ij}, & i \in l'' \end{cases} \tag{8}$$

where l' is the value set associated with the benefit criteria and l'' is the value set associate with loss standards.

- Fifth, obtain a symmetric n-dimensional Euclidean distance from every result to the P-I result and the N-I result (Equations (9) and (10)):

$$\text{Symmetric distance to P-I solution}: \ d_i^+ = \sqrt{\sum_{j=1}^{J}\left(x_{ij} - x_j^+\right)^2} \tag{9}$$

$$\text{Symmetric distance to N-I solution}: \ d_i^- = \sqrt{\sum_{j=1}^{J}\left(x_{ij} - x_j^-\right)^2} \tag{10}$$

- Sixth, obtain the closeness to the ideal result (Equation (11)):

$$C_i^* = \frac{d_i^-}{\left(d_i^- + d_i^+\right)} \tag{11}$$

- Seventh, determine the order of the C_i^* value to define the performance of the alternatives. The larger the C_i^* value is, the better the performance of the alternatives is.

4. Results

4.1. The Whole Process of Fault-Diagnosis Sensors for FCS

The whole process of fault-diagnosis sensor selection method for FCS is provided in Figure 1.

First of all, we should determine the criteria to be used in the evaluation of fault-diagnosis sensor alternatives. Here, the criteria should be defined by multidisciplinary knowledge (cost, efficiency, impact on the environment, safety, and so on). After that, we can assign criteria weights to the AHP method. Next, we can determine the final alternative evaluation result through the TOPSIS method.

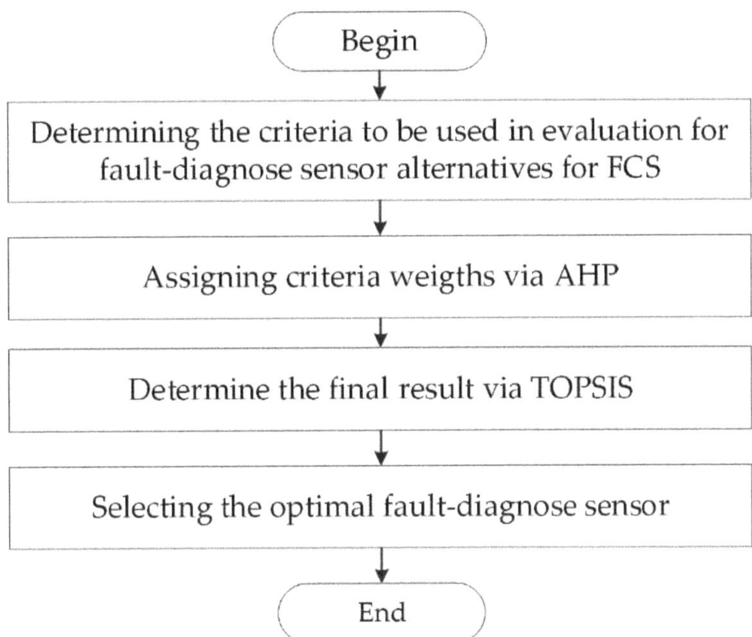

Figure 1. The whole process of the fault-diagnosis sensor selection method for FCS.

4.2. Criteria for Fault-Diagnosis Sensors Selection for FCS

The standard that is used in the chosen fault-diagnosis sensors are decided by the company experts. Here, experts consist of experienced employees from different fault-diagnosis fields and FCS companies. They are very familiar with the size and shape, installation, performance, and expansion of fault-diagnosis sensors. At the same time, they are also very aware of the various safety hazards of sensors, and what a reasonable budget should be for each sensor component.

The proposed criteria and sub-criteria are obtained through related reference research works and from interviews with experts. After that, depending on the initial screening result, four important criteria (Constructional Parameters, Efficient Parameters, Economic Parameters, and Safety Parameters) to be used for fault-diagnosis sensors selection are established. Apart from that, each criterion has several sub-criteria. There are three sub-criteria for Constructional Parameters, six sub-criteria for Economical Parameters, six sub-criteria for Efficient Parameters, and two sub-criteria for Safety Parameters. These criteria are mainly considered to be widely acceptable by the experts. All these criteria are meant to make the fault-diagnosis process for FCS more efficient, more economical, more environmentally friendly, and safer. The four criteria and their descriptions are as follows (see also Table 2).

- Constructional Parameters: Constructional parameters are related to the size and shape of the fault-diagnosis sensors, the installation ease, and expansion ability.
- Efficient Parameters: Efficient parameters are related to the efficiency and performance of fault-diagnosis sensors.
- Economical Parameters: Economic parameters include the cost of all parts of the fault-diagnosis sensor.
- Safety Parameters: Safety parameters include the incidence of sensor breakage or electrical leakage and the safety of the sensor system in an emergency.
- Resilience and tolerance: Resilience and tolerance parameters related to the ability of a fault-diagnosis sensor system to provide the required capability in the face of

adversity and the fault-tolerant design of the sensor. In an FCS system, there are different kinds of adversities, such as electromagnetic and light interference, sudden power cuts, and the best detection distance range, and we have to evaluate the capacity, which is related to the ability to deal with all these interferences, and ensure that the error diagnosis process of the fault-diagnosis sensor system continues. Tolerance is related to the capability of the fault-diagnosis system to continue error-free work in the situation of an unexpected failure (complete unworking, fixed deviation, drift deviation, and accuracy degradation).

Table 2. Criteria and sub-criteria of fault-diagnosis sensors for the Fuel Cell Stack system.

Criteria	Sub-Criteria
Constructional Parameters (C_1)	The size and the shape of the fault-diagnosis sensor (C_{11}) The installation easiness (C_{12}) The expansion ability (C_{13})
Economical Parameters (C_2)	Compressor motor checking sensor cost (C_{21}) Supply manifold checking sensor cost (C_{22}) Air cooler checking sensor cost (C_{23}) Static humidifier checking sensor cost (C_{24}) Outlet manifold checking sensor cost (C_{25}) Stack cathode checking sensor cost (C_{26})
Efficient Parameters (C_3)	Compressor motor checking (C_{31}) Supply manifold checking (C_{32}) Air cooler checking (C_{33}) Static humidifier checking (C_{34}) Outlet manifold checking (C_{35}) Stack cathode and anode checking (C_{36})
Safety Parameters (C_4)	Incidence of sensor breakage or electrical leakage (C_{41}) Safety of the sensor system in an emergency (C_{42})
Resilience and tolerance parameters (C_5)	fault-diagnosis sensor system resilience ability (C_{51}) fault-tolerant design of the sensor (C_{52})

Note. C = Criteria.

4.3. Assigning the Weights of the Criteria via AHP

In this approach, experts use the AHP method to distribute or decide the standard and sub-standard weights, depending on the professional competence of them. A scheme for FCS, which is deeply recognized by the experts in the field of control science, can describe the FCS working process very well [58–60], and the corresponding fault-diagnose sensor system can be seen as Figure 2.

In Figure 2, we can find that the model FCS mainly includes seven elements. Moreover, the FCS can be seen as the machine which can transform fuel energy into electronic energy [3]. Therefore, in order to allow the chemical energy of the fuel to be converted into electrical energy more smoothly, safely, more effectively, and more economical in the FCS system, there is a need to select and set appropriate FCS fault-diagnosis sensor to check different elements in the FCS system. In Figure 2, we can also find that the FCS fault-diagnosis sensor can be divided in six parts (Compressor motor checking, Supply Manifold checking, Air Cooler checking, Static humidifier checking, Outlet manifold checking, and Stack cathode checking), and the detailed functions of these parts can be described as follows:

- Compressor motor checking: the main function of this part is to check the angular speed and motor torque of the compressor, and record the size of the compressor current.
- Supply Manifold checking: the main function of this part it to check the exist air mass flow rate and temperature of the of the inlet manifold.

- Air Cooler checking: the main function of this part is to check the air mass flow rate and temperature of the magic cooler. Moreover, it can also check the humidity of the magic cooler.
- Static humidifier checking: the main function of this part is to check the exit air mass humidity, temperature, and pressure of the stack humidifier, and also check the injected vapor mass flow rate.
- Outlet manifold checking: the main function of this part if to check the outlet manifold exit air mass flow rate, pressure, and humidity.
- Stack cathode checking: this main function of this part is to check the cathode and anode exit hydrogen mass flow rate, hydrogen pressure, relative hydrogen humidity, and exit vapor mass flow rate.

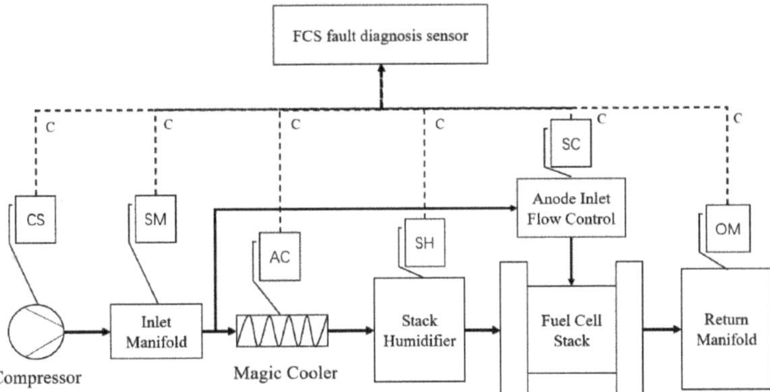

Figure 2. The FCS scheme and the corresponding fault-diagnosis sensor system. Note. C = Collecting data. CS = Compressor motor checking part. SM = Supply Manifold checking part AC = Air Cooler checking part. SH = Static humidifier checking part. OM = Outlet manifold checking part. SC = Stack cathode checking part.

These checking parts can detect all components of FCS in real time, and return the checking data to the fault diagnosis sensor system. Therefore, depending on the information above and the criteria and sub-criteria in Table 2, we can establish a multi-level hierarchical structure as in Figure 3.

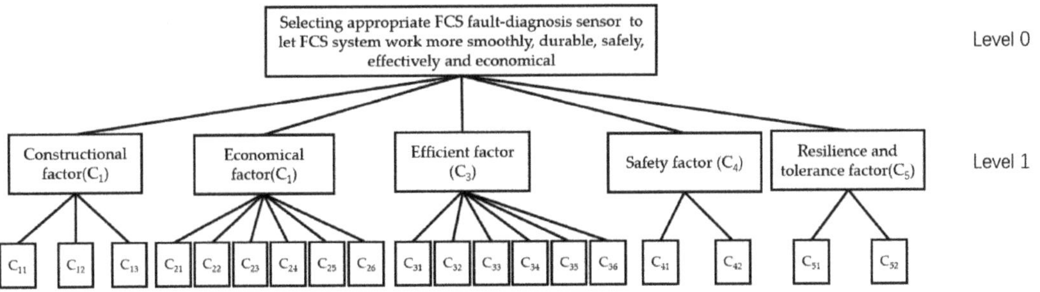

Figure 3. Multi-level hierarchical structure for the FCS fault-diagnosis sensor criteria weight definition.

In Figure 3, we can find that there are two levels of hierarchical structure. Level 0 is the main target of our research, while level 1 shows the multi-standards, and consists of five main factors to select the FCS fault-diagnosis sensor. Moreover, every main factor also consists of several sub-criteria. The reason why there are only two levels of hierarchical structure is that we only need to use the AHP method to define the weight of all the criteria, and the selection of the FCS fault-diagnosis sensor will be defined in the TOPSIS method.

Therefore, depending on Figure 3, experts can give the paired comparison result depending on the comparison rank information in Table 1. The pairwise comparison result of the criteria and sub-criteria for the fault-diagnosis sensor is shown in Tables 3–8.

Table 3. Consequence of comparison of four fault-diagnosis criteria and the rescaled weight of the significance of them.

Criteria	C_1	C_2	C_3	C_4	C_5	u_i	$\hat{u}_i$
Constructional Parameters (C_1)	1	1	2	5	3	0.323	0.36
Economical Parameters (C_2)	1	1	2	7	2	0.329	0.367
Efficient Parameters (C_3)	1/2	1/2	1	3	2	0.178	0.198
Safety Parameters (C_4)	1/5	1/7	1/3	1	1	0.067	0.075
Resilience and tolerance parameters (C_5)	1/3	1/2	1/2	1	1	0.104	0.116

Note. C = Criteria. u_i = Weight for the fault-diagnosis criteria. $\hat{u}_i$ = rescaled weight for the fault-diagnosis criteria.

Table 4. Pairwise comparison result and weight of the Constructional Parameter (C_1) sub-criteria.

Criteria	$\hat{u}_i$	Sub-Criteria	C_{11}	C_{12}	C_{13}	v_k	rv_k	w_k ($\hat{u}_i \times rv_k$)
C_1	0.36	C_{11}	1	4	6	0.710	0.821	0.296
		C_{12}	1/4	1	1	0.155	0.179	0.064
		C_{13}	1/6	1	1	0.135	0.156	0.056

Note. C = Criteria. $\hat{u}_i$ = rescaled weight for the fault-diagnosis criteria. v_k = Weight for the fault-diagnosis sensor sub-criteria. rv_k = rescaled weight for the fault-diagnosis sub-criteria. w_k = Calculated weight for the fault-diagnosis sensor sub-criteria.

Table 5. Pairwise comparison result and weight of the Economical Parameter (C_2) sub-criteria.

C	$\hat{u}_i$	Sub-Criteria	C_{21}	C_{22}	C_{23}	C_{24}	C_{25}	C_{26}	v_k	rv_k	w_k ($\hat{u}_i \times rv_k$)
C_2	0.367	C_{21}	1	3	4	7	4	3	0.395	0.425	0.156
		C_{22}	1/3	1	2	3	5	2	0.211	0.227	0.083
		C_{23}	1/4	1/2	1	2	4	1	0.135	0.145	0.053
		C_{24}	1/7	1/3	1/2	1	2	3	0.096	0.103	0.038
		C_{25}	1/5	1/5	1/4	1/2	1	5	0.093	0.1	0.037
		C_{26}	1/3	1/2	1	1/3	1/5	1	0.069	0.074	0.027

Note. R = Criteria. $\hat{u}_i$ = rescaled weight for the fault-diagnosis criteria. v_k = Weight for the fault-diagnosis sensor sub-criteria. rv_k = rescaled weight for the fault-diagnosis sub-criteria. w_k = Calculated weight for the fault-diagnosis sensor sub-criteria.

Table 6. Pairwise comparison result and the weight of the Efficient Parameter (C_3) sub-criteria.

C	$\hat{u}_i$	Sub-Criteria	CT_{31}	C_{32}	C_{33}	C_{34}	C_{35}	C_{36}	v_k	rv_k	w_k ($\hat{u}_i \times rv_k$)
C_3	0.198	C_{31}	1	2	5	3	4	1	0.329	0.361	0.071
		C_{32}	1/2	1	2	4	6	3	0.247	0.271	0.054
		C_{33}	1/5	1/2	1	2	4	1	0.132	0.145	0.029
		C_{34}	1/3	1/4	1/2	1	2	3	0.105	0.115	0.023
		C_{35}	1/5	1/6	1/4	1/2	1	5	0.098	0.108	0.021
		C_{36}	1	1/3	1	1/3	1/5	1	0.089	0.098	0.019

Note. C = Criteria. $\hat{u}_i$ = rescaled weight for the fault-diagnosis criteria. v_k = Weight for the fault-diagnosis sensor sub-criteria. rv_k = rescaled weight for the fault-diagnosis sub-criteria. w_k = Calculated weight for the fault-diagnosis sensor sub-criteria.

Table 7. Consequence of the comparison of Safety Parameter (C_4) sub-criteria.

Criteria	$\hat{u}_i$	Sub-Criteria	C_{41}	C_{42}	v_k	rv_k	w_k ($\hat{u}_i \times rv_k$)
C_4	0.075	C_{41}	1	3	0.75	1	0.075
		C_{42}	1/3	1	0.25	0.333	0.025

Note. C = Criteria. $\hat{u}_i$ = rescaled weight for the fault-diagnosis criteria. v_k = Weight for the fault-diagnosis sensor sub-criteria. rv_k = rescaled weight for the fault-diagnosis sub-criteria. w_k = Calculated weight for the fault-diagnosis sensor sub-criteria.

Table 8. Consequence of the comparison of Resilience and tolerance parameter (C_5) sub-criteria.

Criteria	$\hat{u}_i$	Sub-Criteria	C_{41}	C_{42}	v_k	rv_k	w_k ($\hat{u}_i \times rv_k$)
C_5	0.116	C_{51}	1	4	0.8	1	0.116
		C_{52}	1/4	1	0.2	0.25	0.029

Note. C = Criteria. $\hat{u}_i$ = rescaled weight for the fault-diagnosis criteria. v_k = Weight for the fault-diagnosis sensor sub-criteria. rv_k = rescaled weight for the fault-diagnosis sub-criteria. w_k = Calculated weight for the fault-diagnosis sensor sub-criteria.

The weight results ($\hat{u}_i$ and rv_k) in Tables 3–8 are released depending on the Equations (2) and (3) (the rescaled weights for the main right eigenvector for the pairwise comparison are shown in Tables 3–8). Here, the calculated weight for the fault-diagnosis sensor sub-criteria (w_k) is obtained by multiplying the criteria of the rescaled weight $\hat{u}_i$ and the sub-criteria of the rescaled weight rv_k.

4.4. Determining the Final Result via TOPSIS

To determine the final fault-diagnosis sensor selection result for FCS, five fault-diagnosis sensors were evaluated using the TOPSIS method. For evaluation, a 1–5 scale, shown in Table 9, was used.

Table 9. Evaluation value used for ranking fault-diagnosis sensors for FCS.

Value	Meaning
5	Excellent
4	Good
3	Normal
2	Bad
1	Terrible

Experts can use ranking in Table 9. to evaluate the five fault-diagnosis sensors. The ranking of the five options, corresponding to the fault-diagnosis sensor evaluation sub-standard, can be seen in Table 10.

Table 10. The ratings of evaluated alternatives with respect to each criterion.

A	CT_1			CT_2						CT_3						CT_4		CT_5	
	CT_{11}	CT_{12}	CT_{13}	CT_{21}	CT_{22}	CT_{23}	CT_{24}	CT_{25}	CT_{26}	CT_{31}	CT_{32}	CT_{33}	CT_{34}	CT_{35}	CT_{36}	CT_{41}	CT_{42}	CT_{51}	CT_{52}
SE_1	2	4	4	1	3	4	4	5	5	4	4	4	1	5	3	5	2	3	3
SE_2	1	1	2	2	4	4	1	2	1	2	2	2	1	2	5	2	1	5	2
SE_3	3	3	3	3	5	2	3	1	2	1	4	2	5	3	4	4	4	3	2
SE_4	4	2	1	5	2	1	2	3	1	4	2	1	1	5	2	2	2	2	1
SE_5	2	4	2	5	2	4	2	3	1	2	1	5	4	3	1	1	2	2	3

Note. A = Alternative. SE = Sensor. CT = Criteria.

The weighted normalized values determined by using Equation (5), Equation (6), and Tables 4–8, are shown in Table 11.

In Table 11, the MI refers to the loss standard (a smaller value means better results). Moreover, MA refers to the gain standard (larger value means better results). The weighted normalized value for alternative Sensor 1 (SE_1) relative to sub-criteria CT_{11} is 0.102. The value is obtained by multiplying the calculated weight of sub-criteria C_{11} (w_k (0.296 in Table 4)) and the normalized decision matrix value for C_{11} ($0.343 = \frac{2}{\sqrt{2^2+1^2+3^2+4^2+2^2}}$). After that, depending on Table 11 and Equations (7) and (8), the P-I and N-I results can be decided. The P-I and N-I results can be seen in Table 12.

Depending on the data from Tables 11 and 12 and Equations (9) and (10), the relative distances of all options from the P-I result and N-I result can be obtained. Finally, we can compute the distance between each option and the P-I result depending on Equation (11).

Symmetry **2021**, 13, 2366

The relative distances between each option and the P-I and N-I solution, and the result of relative closeness to the N-I, can be seen in Table 13.

Table 11. The weighted normalized value of the alternatives with respect to each criterion.

A	CT_1			CT_2						CT_3						CT_4		CT_5	
	CT_{11} MI	CT_{12} MA	CT_{13} MA	CT_{21} MI	CT_{22} MI	CT_{23} MI	CT_{24} MI	CT_{25} MI	CT_{26} MI	CT_{31} MA	CT_{32} MA	CT_{33} MA	CT_{34} MA	CT_{35} MA	CT_{36} MA	CT_{41} MA	CT_{42} MA	CT_{51} MA	CT_{52} MA
SE_1	0.102	0.038	0.038	0.02	0.033	0.029	0.026	0.027	0.024	0.044	0.034	0.016	0.003	0.012	0.008	0.053	0.009	0.049	0.017
SE_2	0.051	0.009	0.019	0.039	0.044	0.029	0.007	0.011	0.005	0.022	0.017	0.008	0.003	0.005	0.013	0.021	0.005	0.081	0.011
SE_3	0.152	0.028	0.029	0.059	0.054	0.015	0.02	0.005	0.01	0.011	0.034	0.008	0.017	0.007	0.01	0.042	0.019	0.049	0.011
SE_4	0.203	0.019	0.010	0.098	0.022	0.007	0.013	0.016	0.005	0.044	0.017	0.004	0.003	0.012	0.005	0.021	0.009	0.032	0.006
SE_5	0.102	0.038	0.019	0.098	0.022	0.029	0.013	0.016	0.005	0.022	0.008	0.021	0.014	0.007	0.003	0.011	0.009	0.032	0.017

Note. A = Alternative. SE = Sensor. CT = Criteria. MI = Min. MA = Max.

Table 12. The P-I and N-I solutions of the considered fault-diagnosis sensor alternatives.

A	CT_1			CT_2						CT_3						CT_4		CT_5	
	CT_{11} MI	CT_{12} MA	CT_{13} MA	CT_{21} MI	CT_{22} MI	CT_{23} MI	CT_{24} MI	CT_{25} MI	CT_{26} MI	CT_{31} MA	CT_{32} MA	CT_{33} MA	CT_{34} MA	CT_{35} MA	CT_{36} MA	CT_{41} MA	CT_{42} MA	CT_{51} MA	CT_{52} MA
SE^+	0.051	0.038	0.038	0.02	0.022	0.007	0.007	0.005	0.005	0.044	0.034	0.021	0.017	0.012	0.013	0.011	0.019	0.081	0.017
SE^-	0.203	0.009	00.01	0.098	0.054	0.029	0.026	0.027	0.024	0.011	0.008	0.004	0.003	0.005	0.003	0.053	0.005	0.032	0.006

Note. A = Alternative. SE = Sensor. CT = Criteria. MI = Min. MA = Max.

Table 13. TOPSIS results.

Alternative	d_i^+	d_i^-	C_i^*
SE_1	0.087	0.144	0.624
SE_2	0.059	0.174	0.747
SE_3	0.128	0.079	0.381
SE_4	0.185	0.033	0.151
SE_5	0.116	0.106	0.476

Note. C_i^* = The relative closeness to the ith ideal solution.

Table 13 shows the evaluation result of the candidate options depending on the calculation of the TOPSIS methodology. From the C_i^* value in Table 13, we can find that the fault-diagnosis sensor SE_2 has the highest value (0.747). In contrast, sensor SE_4 has the smallest value (0.151). Moreover, we also find that companies pay more attention to the Constructional Parameters (C_1) and Economical parameters (C_2) (the rescaled weights ($\hat{u}_i$) for these two parts in Table 3 are 0.36 and 0.367, respectively). Moreover, in the Constructional Parameters (C_1), companies pay more attention to the criteria C_{11} (the rescaled weight ($\hat{u}_i$) for C_{11} is 0.296; the lower the criteria data, the better), and in the Economical parameters (C_2), companies pay more attention to the criteria C_{21} (the rescaled weight ($\hat{u}_i$) for C_{21} is 0.156; the smaller the criteria value, the better). Therefore, when we use these two criteria (C_{11}, C_{21}) to compare all the sensors, we can find that SE_2 has the smallest value (1 for C_{11} and 2 for C_{21}, see Table 10). Therefore, compared with other sensors, SE_2 has the smallest size and shape, while the compressor motor inspection has the lowest cost, and these two parts are also the most important reason why SE_2 is optimal. Additionally, companies should pay more attention to these two parts when considering other new sensors. Therefore, given the result above, companies can select fault-diagnosis sensor SE_2 to ensure that the FCS fault-diagnosis process is more efficient, economical, and safer compared with other alternatives.

5. Discussion

The fault-detection process is one of the most important steps in the effective operation of Fuel Cell Stack Systems, and Fuel Cell Stack Systems are generally complex and need to be systematically organized in order to achieve high levels of efficiency. A fault-detection sensor can ensure the effective operation of Fuel Cell Stack Systems, and companies need to consider several criteria and restrictions for the sensors' selection process. Meanwhile, with the rapid development of manufacturing techniques and manufacturing economy problems [61–63], safety problems and efficiency problems [64] frequently emerge. Therefore, to ensure a safe and efficient fault-diagnosis process, it is very important for companies to select an adaptable fault-diagnosis sensor that meets these conditions.

Therefore, considering the problems above, this article proposed a fault-diagnosis selection method for Fuel Cell Stack Systems, considering the symmetric and asymmetric problem in a mathematical formulation and also considering the different multi-disciplinary criteria (Construction Parameters, Economical Parameters, Efficient Parameters, and Safety Parameters). To define the weight of the fault-diagnosis evaluation criteria for Fuel Cell Stack Systems, we introduced the AHP (Analytic Hierarchy Process) method. We used the AHP technique to define the importance of the fault-diagnosis evaluation standard or sub-standard. Additionally, we used a method [53] to avoid rank reversal in AHP. From Tables 2–8, we can get the weight information for all the four main criteria and the calculated weight information for 17 sub-criteria. The calculated weight information value multiplies the weight of the criteria by the associated weight of the sub-criteria. Next, we used TOPSIS to select and obtain the fault-diagnosis sensor in all the alternatives. Here, we considered both the positive and negative criteria, and obtained positive and negative evaluation solutions (Tables 11 and 12) through the computation of the P-I solution (Equation (7)), NN-I solution (Equation (8)), The symmetric distance from the P-I result (Equation (9)), and the symmetric distance from the N-I result (Equation (10)). From the TOPSIS results in Table 13, companies can obtain and select the most suitable fault-diagnosis sensor (the higher the C_i^* value, the better). The main contribution of this paper is to help Fuel Cell Stack Systems companies to select the appropriate fault-diagnosis sensor to ensure that the Fuel Cell Stack Systems fault-diagnosis process is more economical, efficient, and safer.

Although the methodology was developed for the fault-diagnosis sensor selection problem, it is also adaptable for the selection of other manufacturing process facilities, with slight modifications, such as the cutting or welding facility selection problem in car manufacturing companies. Therefore, further studies will need to focus on other directions. In a future study, we will continue to study the comparison of the Technique for Order Performance by Similarity to Ideal Solution, Višekriterijumsko Kompromisno Rangiranje, Complex Proportional Assessment, and PROMETHEE II (complete ranking) techniques based on the results of using different normalization methods. Additionally, a sensitivity analysis should be performed for the ranking result (Table 13), such as the use of coefficients to measure the similarity of two rankings in decision-making problems [65]. Therefore, this will also be a direction for future research. Additionally, engineering information is sometimes incomplete or in a fuzzy environment, which also needs to be addressed in future research.

Our research started with the illustration of the significance of fault-diagnosis sensor selection for a Fuel Cell Stack system. Next, we introduced the main target and the issues regarding the chosen fault-diagnosis sensor. Then, we used this method (a combination of AHP and TOPSIS) to approach fault-diagnosis sensor selection considering different fault-diagnosis sensor evaluation criteria. Ultimately, depending on the proposed methodology, companies will be able choose the most suitable adaptable fault-diagnosis sensor for their Fuel Cell Stack system. This method can be used to ensure that the Fuel Cell Stack system fault-diagnosis process is more efficient, economical, and safer.

Author Contributions: Ideas, G.J. (Guangying Jin) and G.J. (Guangzhe Jin); Method, G.J. (Guangying Jin); Writing—Original Draft Preparation, G.J. (Guangying Jin); Writing—Review and Editing, G.J. (Guangying Jin); Supervision, G.J. (Guangzhe Jin); Funding Acquisition, G.J. (Guangying Jin). All authors have read and agreed to the published version of the manuscript.

Funding: The study was supported by the Talent Research Start-up Founding of Dalian Maritime University, authorization code: 02502329.

Conflicts of Interest: The authors declare no conflict of interest.

References

1. Aitouche, A.; Olteanu, S.C.; Bouamama, B.O. A survey of diagnostic of fuel cell stack systems. *IFAC Proc. Vol.* **2012**, *45*, 84–89. [CrossRef]
2. Blanke, M.; Frei, W.C.; Kraus, F. What is fault-tolerant control? *IFAC Proc.* **2000**, *33*, 41–52. [CrossRef]
3. Sarrate, R.; Nejjari, F.; Rosich, A. Sensor placement for fault diagnosis performance maximization in distribution networks. In Proceedings of the 20th Mediterranean Conference on Control & Automation (MED), Barcelona, Spain, 3–6 July 2012; IEEE: Barcelona, Spain, 2012; pp. 110–115.
4. Gao, C.; Duan, G. Fault diagnosis and fault tolerant control for nonlinear satellite attitude control systems. *Aerosp. Sci. Technol.* **2014**, *33*, 9–15. [CrossRef]
5. Rosli, R.E.; Sulong, A.B.; Daud, W.R.W.; Zulkifley, M.A.; Husaini, T.; Rosli, M.I.; Majlan, E.H.; Haque, M.A. A review of high-temperature proton exchange membrane fuel cell (HT-PEMFC) system. *Int. J. Hydrogen Energy* **2017**, *42*, 9293–9314. [CrossRef]
6. Ayağ, Z.; Özdemir, R.G. A fuzzy AHP approach to evaluating machine tool alternatives. *J. Intell. Manuf.* **2006**, *17*, 179–190. [CrossRef]
7. Dağdeviren, M.; Yavuz, S.; Kılınç, N. Weapon selection using the AHP and TOPSIS methods under fuzzy environment. *Expert Syst. Appl.* **2009**, *36*, 8143–8151. [CrossRef]
8. Wang, C.N.; Thanh, N.V.; Su, C.C. The Study of a Multicriteria Decision Making Model for Wave Power Plant Location Selection in Vietnam. *Processes* **2019**, *7*, 650. [CrossRef]
9. Yimen, N.; Dagbasi, M. Multi-Attribute Decision-Making: Applying a Modified Brown–Gibson Model and RETScreen Software to the Optimal Location Process of Utility-Scale Photovoltaic Plants. *Processes* **2019**, *7*, 505. [CrossRef]
10. Pohekar, S.D.; Ramachandran, M. Application of multi-criteria decision making to sustainable energy planning—A review. *Renew. Sustain. Energy Rev.* **2004**, *8*, 365–381. [CrossRef]
11. Majumder, M. *Impact of Urbanization on Water Shortage in Face of Climatic Aberrations*; Springer: Berlin/Heidelberg, Germany, 2015.
12. Kumar, A.; Sah, B.; Singh, A.R.; Deng, Y.; He, X.; Kumar, P.; Bansal, R.C. A review of multi criteria decision making (MCDM) towards sustainable renewable energy development. *Renew. Sustain. Energy Rev.* **2017**, *69*, 596–609. [CrossRef]
13. Jin, G.; Sperandio, S.; Girard, P. Selection of design project with the consideration of designers' satisfaction factors and collaboration ability. *Comput. Ind. Eng.* **2019**, *131*, 66–81. [CrossRef]
14. Bączkiewicz, A.; Kizielewicz, B.; Shekhovtsov, A.; Yelmikheiev, M.; Kozlov, V.; Sałabun, W. Comparative Analysis of Solar Panels with Determination of Local Significance Levels of Criteria Using the MCDM Methods Resistant to the Rank Reversal Phenomenon. *Energies* **2021**, *14*, 5727. [CrossRef]
15. Nesticò, A.; Somma, P. Comparative analysis of multi-criteria methods for the enhancement of historical buildings. *Sustainability* **2019**, *11*, 4526. [CrossRef]
16. Nesticò, A.; Elia, C.; Naddeo, V. Sustainability of urban regeneration projects: Novel selection model based on analytic network process and zero-one goal programming. *Land Use Policy* **2020**, *99*, 104831. [CrossRef]
17. Wątróbski, J.; Jankowski, J.; Ziemba, P.; Karczmarczyk, A.; Zioło, M. Generalised framework for multi-criteria method selection. *Omega* **2019**, *86*, 107–124. [CrossRef]
18. Vansnick, J.C. On the problem of weights in multiple criteria decision making (the noncompensatory approach). *Eur. J. Oper. Res.* **1986**, *24*, 288–294. [CrossRef]
19. Saaty, T.L.; Ergu, D. When is a decision-making method trustworthy? Criteria for evaluating multi-criteria decision-making methods. *Int. J. Inf. Technol. Decis. Mak.* **2015**, *14*, 1171–1187. [CrossRef]
20. Roy, B.; Słowiński, R. Questions guiding the choice of a multicriteria decision aiding method. *EURO J. Decis. Process.* **2013**, *1*, 69–97. [CrossRef]
21. Saaty, T.L. Relative measurement and its generalization in decision making why pairwise comparisons are central in mathematics for the measurement of intangible factors the analytic hierarchy/network process. *RACSAM-Rev. Real Acad. Cienc. Exactas Fis. Naturales. Ser. A Mat.* **2008**, *102*, 251–318. [CrossRef]
22. Jin, G.; Sperandio, S.; Girard, P. Management of the design process: Human resource evaluation in factories of the future. *Concurr. Eng.* **2018**, *26*, 313–327. [CrossRef]
23. Saaty, T.L. *Analytic Network Process*; Springer: New York, NY, USA, 2013.
24. Ginting, G.; Fadlina, M.; Siahaan, A.P.U.; Rahim, R. Technical approach of TOPSIS in decision making. *Int. J. Recent Trends Eng. Res.* **2017**, *3*, 58–64.
25. Dyer, J.S. Multiattribute utility theory (MAUT). In *Multiple Criteria Decision Analysis*; Springer: New York, NY, USA, 2016; pp. 285–314.
26. Figueira, J.R.; Mousseau, V.; Roy, B. ELECTRE methods. In *Multiple Criteria Decision Analysis*; Springer: New York, NY, USA, 2016; pp. 155–185.

27. Brans, J.P.; de Smet, Y. PROMETHEE methods. In *Multiple Criteria Decision Analysis*; Springer: New York, NY, USA, 2016; pp. 187–219.
28. Işık, A.T.; Adalı, E.A. UTA Method for the Consulting Firm Selection Problem. *J. Eng. Sci. Technol. Rev.* **2016**, *9*, 56–60. [CrossRef]
29. Manshadi, E.D.; Mehregan MR Safari, H. Supplier classification using UTADIS method based on performance criteria. *Int. J. Acad. Res. Bus. Soc. Sci.* **2015**, *5*, 31. [CrossRef]
30. Faizi, S.; Sałabun, W.; Rashid, T.; Wątróbski, J.; Zafar, S. Group decision-making for hesitant fuzzy sets based on characteristic objects method. *Symmetry* **2017**, *9*, 136. [CrossRef]
31. Goswami, S.S.S.; Behera, D.K.K.; Afzal, A.; Razak Kaladgi, A.; Khan, S.A.A.; Rajendran, P.; Subbiah, R.; Asif, M. Analysis of a Robot Selection Problem Using Two Newly Developed Hybrid MCDM Models of TOPSIS-ARAS and COPRAS-ARAS. *Symmetry* **2021**, *13*, 1331. [CrossRef]
32. Dezert, J.; Tchamova, A.; Han, D.; Tacnet, J.M. The spotis rank reversal free method for multi-criteria decision-making support. In Proceedings of the 2020 IEEE 23rd International Conference on Information Fusion (FUSION), Rustenburg, South Africa, 6–9 July 2020; IEEE: Rustenburg, South Africa, 2020; pp. 1–8.
33. Munier, N. A new approach to the rank reversal phenomenon in MCDM with the SIMUS method. *Mult. Criteria Decis. Mak.* **2016**, *11*, 137–152. [CrossRef]
34. Boudreau-Trudel, B.; Zaras, K. Comparison of analytic hierarchy process and dominance-based rough set approach as multi-criteria decision aid methods for the selection of investment Comparison of analytic hierarchy process and dominance-based rough set approach as multi-criteria decision aid methods for the selection of investment projects. *Am. J. of Ind. Bus. Manag.* **2012**, *2*, 7–12.
35. Saaty, T.L. Decision making with the analytic hierarchy process. *Int. J. Serv. Sci. Indersci.* **2008**, *1*, 83–98. [CrossRef]
36. Expert. The Analytic Hierarchy Process (AHP). DTU Transport Compendium Series Part 2, Technical University of Denmark 2014. Available online: http://www.systemicplanning.dk/AHP.pdf (accessed on 21 November 2021).
37. Liu, D.; Huang, A.; Liu, Y.; Liu, Z. An Extension TOPSIS Method Based on the Decision Maker's Risk Attitude and the Adjusted Probabilistic Fuzzy Set. *Symmetry* **2021**, *13*, 891. [CrossRef]
38. Agrawal, A.; Seh, A.H.; Baz, A.; Alhakami, H.; Alhakami, W.; Baz, M.; Kumar, R.; Khan, R.A. Software security estimation using the hybrid fuzzy ANP-TOPSIS approach: Design tactics perspective. *Symmetry* **2020**, *12*, 598. [CrossRef]
39. Sahu, K.; Alzahrani, F.A.; Srivastava, R.K.; Kumar, R. Hesitant fuzzy sets based symmetrical model of decision-making for estimating the durability of Web application. *Symmetry* **2020**, *12*, 1770. [CrossRef]
40. Lai, Y.J.; Liu, T.Y.; Hwang, C.L. Topsis for MODM. *Eur. J. Oper. Res.* **1994**, *76*, 486–500. [CrossRef]
41. Wang, T.C.; Chang, T.H. Application of TOPSIS in evaluating initial training aircraft under a fuzzy environment. *Expert Syst. Appl.* **2007**, *33*, 870–880. [CrossRef]
42. Alinezhad, A.; Khalili, J. COPRAS Method. In *New Methods and Applications in Multiple Attribute Decision Making (MADM)*; Springer: Cham, Switzerland, 2019; pp. 87–91.
43. Liou, J.J.; Tamošaitienė, J.; Zavadskas, E.K.; Tzeng, G.H. New hybrid COPRAS-G MADM Model for improving and selecting suppliers in green supply chain management. *Int. J. Prod. Res.* **2016**, *54*, 114–134. [CrossRef]
44. Roy, J.; Kumar Sharma, H.; Kar, S.; Kazimieras Zavadskas, E.; Saparauskas, J. An extended COPRAS model for multi-criteria decision-making problems and its application in web-based hotel evaluation and selection. *Econ. Res.-Ekon. Istraživanja* **2019**, *32*, 219–253. [CrossRef]
45. Faizi, S.; Sałabun, W.; Nawaz, S.; ur Rehman, A.; Wątróbski, J. Best-Worst method and Hamacher aggregation operations for intuitionistic 2-tuple linguistic sets. *Expert Syst. Appl.* **2021**, *181*, 115088. [CrossRef]
46. Božanić, D.; Milić, A.; Tešić, D.; Salabun, W.; Pamučar, D. D numbers–fucom–fuzzy rafsi model for selecting the group of construction machines for enabling mobility. *Facta Univ. Ser. Mech. Eng.* **2021**, *19*, 447–471. [CrossRef]
47. Kraujalienė, L. Comparative analysis of multicriteria decision-making methods evaluating the efficiency of technology transfer. *Bus. Manag. Educ.* **2019**, *17*, 72–93.
48. Cinelli, M.; Kadziński, M.; Gonzalez, M.; Słowiński, R. How to support the application of multiple criteria decision analysis? Let us start with a comprehensive taxonomy. *Omega* **2020**, *96*, 102261. [CrossRef]
49. Yazdani-Chamzini, A.; Haji Yakchali, S.; Kazimieras Zavadskas, E. Using a integrated MCDM model for mining method selection in presence of uncertainty. *Econ. Res.-Ekon. Istraživanja* **2012**, *25*, 869–904. [CrossRef]
50. Štirbanović, Z.; Stanujkić, D.; Miljanović, I.; Milanović, D. Application of MCDM methods for flotation machine selection. *Miner. Eng.* **2019**, *137*, 140–146. [CrossRef]
51. Sarrate, R.; Nejjari, F.; Rosich, A. Sensor Placement for Fault Diagnosis Performance Maximization under Budget Constraints. Available online: http://digital.csic.es/bitstream/10261/97613/4/Sensor%20placement.pdf (accessed on 21 November 2021).
52. Guerrero-Baena, M.D.; Gómez-Limón, J.A.; Fruet Cardozo, J.V. Are multi-criteria decision making techniques useful for solving corporate finance problems? A bibliometric analysis. *Rev. Metodos Cuantitativos Para La Econ. La Empresa* **2014**, *17*, 60–79.
53. Wang, Y.M.; Elhag, T.M.S. An approach to avoiding rank reversal in AHP. *Decis. Support Syst.* **2006**, *42*, 1474–1480. [CrossRef]
54. Omar, F.; Bushby, S.T.; Williams, R.D. Assessing the performance of residential energy management control Algorithms: Multi-criteria decision making using the analytical hierarchy process. *Energy Build.* **2019**, *199*, 537–546. [CrossRef]
55. Saaty, T.L. What is the analytic hierarchy process? In *Mathematical Models for Decision Support*; Springer: Berlin/Heidelberg, Germany, 1988; pp. 109–121.

56. Vafaei, N.; Ribeiro, R.A.; Camarinha-Matos, L.M. Normalization Techniques for Multi-Criteria Decision Making: Analytical Hierarchy Process Case Study. In Proceedings of the Doctoral Conference on Computing, Electrical and Industrial Systems, Costa de Caparica, Portugal, 11–13 April 2016; Springer: Cham, Switzerland, 2016; pp. 261–269.
57. Chakraborty, S.; Yeh, C.H. A simulation comparison of normalization procedures for TOPSIS. In Proceedings of the 2009 International Conference on Computers & Industrial Engineering, Troyes, France, 6–9 July 2009; IEEE: Troyes, France, 2009; pp. 1815–1820.
58. Pukrushpan, J.T.; Peng, H.; Stefanopoulou, A.G. Control-oriented modeling and analysis for automotive fuel cell systems. *J. Dyn. Syst. Meas. Control* **2004**, *126*, 14–25. [CrossRef]
59. Pukrushpan, J.T. *Modeling and Control of Fuel Cell Systems and Fuel Processors*; University of Michigan: Ann Arbor, MI, USA, 2003.
60. Rosich Oliva, A. Sensor placement for fault diagnosis based on structural models: Application to a fuel cell stak system. Tesi doctoral, UPC, Institut d'Organització i Control de Sistemes Industrials. 2011. Available online: http://hdl.handle.net/2117/945 10. (accessed on 21 November 2021).
61. Díaz, J.; Fernández, F.J. The impact of hot metal temperature on CO_2 emissions from basic oxygen converter. *Environ. Sci. Pollut. Res.* **2019**, *27*, 33–42. [CrossRef]
62. Fernández, F.J.; Folgueras, M.B.; Suárez, I. Energy study in water loop heat pump systems for office buildings in the Iberian Peninsula. *Energy Procedia* **2017**, *136*, 91–96. [CrossRef]
63. Enguita, J.M.; Alvarez, I.; Bobis, C.F.; Marina, J.; Fernández, Y.; Sirat, G.Y. Conoscopic holography-based long-standoff profilemeter for surface inspection in adverse environment. *Opt. Eng.* **2006**, *45*, 073602. [CrossRef]
64. Han, Z.; Zhang, Q.; Shi, H.; Zhang, J. An improved compact genetic algorithm for scheduling problems in a flexible flow shop with a multi-queue buffer. *Processes* **2019**, *7*, 302. [CrossRef]
65. Sałabun, W.; Urbaniak, K. A new coefficient of rankings similarity in decision-making problems. In Proceedings of the International Conference on Computational Science, Amsterdam, The Netherlands, 3–5 June 2020; Springer: Cham, Switzerland, 2020; pp. 632–645.

Article

Trajectory Tracking Control for Underactuated USV with Prescribed Performance and Input Quantization

Kunyi Jiang, Lei Mao *, Yumin Su and Yuxin Zheng

Science and Technology on Underwater Vehicle Laboratory, Harbin Engineering University, Harbin 150001, China; jiang_kunyi@hrbeu.edu.cn (K.J.); suyumin@hrbeu.edu.cn (Y.S.); zhengyuxin@hrbeu.edu.cn (Y.Z.)
* Correspondence: ml0831@hrbeu.edu.cn; Tel.: +86-17701806828

Abstract: This paper is devoted to the problem of prescribed performance trajectory tracking control for symmetrical underactuated unmanned surface vessels (USVs) in the presence of model uncertainties and input quantization. By combining backstepping filter mechanisms and adaptive algorithms, two robust control architectures are investigated for surge motion and yaw motion. To guarantee the prespecified performance requirements for position tracking control, the constrained error dynamics are transformed to unconstrained ones by virtue of a tangent-type nonlinear mapping function. On the other hand, the inaccurate model can be identified through radial basis neural networks (RBFNNs), where the minimum learning parameter (MLP) algorithm is employed with a low computational complexity. Furthermore, quantization errors can be effectively reduced even when the parameters of the quantizer remain unavailable to designers. Finally, the effectiveness of the proposed controllers is verified via theoretical analyses and numerical simulations.

Keywords: underactuated USV; prescribed performance control; input quantization; model-free control; minimum learning parameter

Citation: Jiang, K.; Mao, L.; Su, Y.; Zheng, Y. Trajectory Tracking Control for Underactuated USV with Prescribed Performance and Input Quantization. *Symmetry* **2021**, *13*, 2208. https://doi.org/10.3390/sym13112208

Academic Editor: Alexander Zaslavski

Received: 5 October 2021
Accepted: 27 October 2021
Published: 19 November 2021

Publisher's Note: MDPI stays neutral with regard to jurisdictional claims in published maps and institutional affiliations.

1. Introduction

At present, USVs are expected to play an increasingly important role in both military and civilian domains, such as reconnaissance and surveillance, marine surveying and mapping, marine resources exploration and development, etc. [1–4]. As one of the most significant components of USVs, trajectory tracking control systems determine the success of various missions and hence have received tremendous interest from the field of ocean engineering. However, the controller design for the trajectory tracking of USVs still poses enormous challenges owing to unexpected marine disturbances and the complex system involved, which features coupling and nonlinearity. On the other hand, the usage scenarios and mission objectives also mean that there are high requirements for the performance of the controllers, the prescribed behavioral metrics, and a constrained communication bandwidth.

For an underactuated vessel, the unique feature is that the control torque provided by actuators only acts in surge and yaw motions and is less than the three degrees of freedom (DOF) used in conventional surface vessel dynamics [5]. To fulfill this practical demand in engineering, numerous control algorithms, including sliding mode control (SMC) [6–9], backstepping control [10–13], model predictive control [14,15], and observer-based control [16,17], enable USVs to accomplish trajectory tracking control. In particular, as SMC is capable of realizing fast responses, is insensitive to interference, and can help to improve robustness, fruitful results have been obtained in many fields (spacecraft rendezvous [7], underwater vehicles [8], and surface vessels [9]) from utilizing newly developed sliding mode methods. In [9], with the aid of a line-of-sight-based integral sliding-mode technique, high-accuracy paths following USVs are achieved even in the presence of unknown dynamics and external disturbances. On the other hand, backstepping strategies always offer superior performances in robust control and adaptive control when a system suffers from uncertain nonlinear dynamics [16]. Though effective, the problem of expansive calculations has caused considerable trouble in traditional backstepping designs. To address this obstruction, the dynamic surface control (DSC) scheme is introduced to facilitate the realization of control for the trajectory tracking problem [1] and leader–follower cooperative formation problem [17] of USVs.

Without a loss of generality, control signals are usually updated through time sampling and are accompanied by ubiquitous redundant data transmission to many practical systems, leading to severe onboard resource occupation. Thus, it is reasonable to consider the issue of realizing trajectory tracking control under the constrained communication bandwidth of the USV. For this purpose, the quantized control method [18–20] and event-triggered algorithm [21–23] have been intensively studied in various agent systems integrated with a set of independent functional modules. In quantization control, the original signal output from the control module is first converted into the discrete sequence by a quantizer and then transmitted to the actuator. In this case, a finite amount of information is directly stored in actuators so that the change in control signals can be executed by transmitting a spot of the code. Therefore, the burden of communication will be significantly reduced by virtue of the quantization mechanisms. In the last decade, quantized control in connection with robust approaches has attracted increasing amounts of attention from researchers and has been studied in the fields of spacecraft formation [24,25], unmanned aerial vehicles [26,27], and underwater vehicles [28]. However, to the best of the author's knowledge, developing controllers for USVs with quantized transmitted information remains an open problem.

In addition to input quantization, another important issue that deserves further investigation is the state constraint control of USVs, which has been ignored in numerous studies [7,8,11,16,17,25–28]. In practical applications, it is realistic and of great significance to consider that the USV position error should be limited strictly by both sides of the feasible channel to ensure the navigation safety of the vessel [6]. With this problem in mind, efforts have been made by researchers to satisfy the output constraint of the system, and there appears to be a variety of control schemes, such as the barrier Lyapunov function (BLF) [29–32], nonlinear mapping (NM) control [13,29], and prescribed performance (PP) control [30–32]. To restrain tracking error variables, the use of logarithmic BLF [33] and tan-type BLF [34], in conjunction with an adaptive algorithm, was proposed for the trajectory control of single and multiple underactuated surface vessels, respectively. However, it must be mentioned that the accompanying problems, such as the complexity and heavy workload of BLF-based procedures, restrict its application. As a superior method, NM-based control, which is dedicated to mapping the constrained output onto the real number set, has been proven to be effective in handling the constraint problem [29]. Different from the above maneuvers, the PP strategy described in [30] is capable of ensuring that the tracking errors of underactuated USVs converge to a predesigned region; more extensions of this method can be found in [31,32].

Inspired by the above observations, this paper mainly concentrates on providing a solution to the problem of the trajectory tracking control of USVs subject to prescribed performance, uncertain dynamics, and communication constraints. The control signal is discretized by a hysteresis logarithmic quantizer (HLQ), which reduces the communication load significantly. A backstepping-based adaptive algorithm combining DSC and RBFNNs is proposed for tackling the negative effects of model uncertainties, unavailable disturbances, and quantization errors on control performance. The main contributions of the proposed controller are as follows:

(i) Compared with the quantized control described in [18–20], the HLQ-based adaptive algorithm is employed in this paper to transform the traditional continuous signal to the discrete one so that a high superiority is ensured in reducing the communication load and improving the control accuracy. Moreover, the application of this technique is further extended to a case where quantizer parameters are unavailable.

(ii) In contrast to the numerous existing control strategies for USV trajectory tracking [11–13], in this paper the state constraint problem of the position tracking error is taken into consideration. For this purpose, a novel error transformation mechanism is developed on the basis of a tangent function, such that the security of marine navigation can be guaranteed.

The remainder of this paper is organized as follows. The preliminaries and mathematic models of the USV are given in Section 2. Subsequently, Section 3 elaborates on the design of the quantized adaptive control strategy. In Section 4, numerical simulations are presented to authenticate the effectiveness of the proposed algorithm. Finally, conclusions are drawn in Section 5.

2. Preliminaries and Problem Formulation

2.1. Mathematical Model of Underactuated USV

With no consideration of heave, roll, and pitch motion, the three degrees of freedom (DOF) kinematics model of underactuated USV is expressed as:

$$\begin{cases} \dot{x} = u\cos(\psi) - v\sin(\psi) \\ \dot{y} = u\sin(\psi) + v\cos(\psi) \\ \dot{\psi} = r \end{cases} \tag{1}$$

where the vector $\eta = [x, y, \psi]^{\mathrm{T}}$ denotes the position and heading angle in the earth-fixed frame (EF) and the vector $v = [u, v, r]^{\mathrm{T}}$ denotes the linear velocity and angular velocity in the body-fixed frame (BF).

The dynamics model of underactuated vessels is formulated as [5]:

$$\begin{cases} \dot{u} = \frac{m_{22}}{m_{11}}vr - f_u(u) + \frac{1}{m_{11}}Q(\tau_u) + \frac{1}{m_{11}}\tau_{wu} \\ \dot{v} = -\frac{m_{11}}{m_{22}}ur - f_v(v) + \frac{1}{m_{22}}\tau_{wv} \\ \dot{r} = \frac{m_{11}-m_{22}}{m_{33}}uv - f_r(r) + \frac{1}{m_{33}}Q(\tau_r) + \frac{1}{m_{33}}\tau_{wr} \end{cases} \tag{2}$$

where $m_{ii}(i = 1, 2, 3)$ represents the added mass and combined inertia of the vessel; τ_u and τ_r stand for the control input of surge propulsion force and yaw moment with $Q(\tau)$ as the quantization conversion function; τ_{wu}, τ_{wv}, and τ_{wr} are defined as the various external disturbances resulting from winds, currents, and waves. In addition, the hydrodynamic damping terms $f_u(u)$, $f_v(v)$, and $f_r(r)$ are described as:

$$\begin{cases} f_u(u) = \frac{d_u}{m_{11}}u + \sum_{i=2}^{3}\frac{d_{ui}}{m_{11}}|u|^{i-1}u \\ f_v(v) = \frac{d_v}{m_{22}}v + \sum_{i=2}^{3}\frac{d_{vi}}{m_{22}}|v|^{i-1}v \\ f_r(r) = \frac{d_r}{m_{33}}r + \sum_{i=2}^{3}\frac{d_{ri}}{m_{133}}|r|^{i-1}r \end{cases} \tag{3}$$

where the hydrodynamic damping coefficients d_u, d_v, d_r, d_{ui}, d_{vi}, and d_{ri} can be obtained by parameter identification using experimental data. Nevertheless, taking into account the complicated and volatile maritime environment, the precise values of the above damping terms are difficult to measure in real time. Therefore, it is assumed that all of the hydrodynamic parameters are bounded and unavailable.

2.2. Formulation of HLQ

In terms of the trajectory tracking control system of USVs, constrained communication bandwidth usually exists between the controller and the actuators. As a novel technology for wireless interaction, the quantifier plays a major role in alleviating the pressure of data transmission; refer to the existing results shown in [21]. The HLQ is introduced here to replace the traditional continuous time signals:

$$
Q(\tau) = \begin{cases}
\tau_i \mathrm{sgn}(\tau), & \begin{aligned} &\frac{\tau_i}{1+\delta} < |\tau| \leq \tau_i, \dot{\tau} < 0 \\ &\tau_i < |\tau| \leq \frac{\tau_i}{1-\delta}, \dot{\tau} > 0 \end{aligned} \\
\tau_i(1+\delta)\mathrm{sgn}(\tau), & \begin{aligned} &\tau_i < |\tau| \leq \frac{\tau_i}{1-\delta}, \dot{\tau} < 0 \\ &\frac{\tau_i}{1-\delta} < |\tau| \leq \frac{\tau_i(1+\delta)}{1-\delta}, \dot{\tau} > 0 \end{aligned} \\
0, & \begin{aligned} &0 \leq |\tau| < \frac{\tau_{\min}}{1+\delta}, \dot{\tau} < 0 \\ &\frac{\tau_{\min}}{1+\delta} < \tau \leq \tau_{\min}, \dot{\tau} > 0 \end{aligned} \\
Q(\tau(t^-)), & \dot{\tau} = 0
\end{cases}
\tag{4}
$$

where $\tau_i = \rho^{1-i}\tau_{\min}\,(i = 1, 2, \cdots)$ with $\tau_{\min} > 0$ denotes the range of the hysteresis zone for $Q(\tau)$. The parameter δ determines the transmitting rate of the communication channel and satisfies $\delta = (1-\rho)/(1+\rho)$ with $0 < \rho < 1$. Obviously, $Q(\tau)$ is in the set $U = \{0, \pm\tau_i, \pm\tau_i(1+\delta)\}$.

Remark 1. *Generally, we consider the USV as symmetric about the longitudinal section of the hull. Introducing this assumption can remarkably simplify the complexity of the mathematical model. Considering HLQ, on the one hand, the HLQ adopted in this paper possesses some common features with conventional quantizers. For instance, a limited number of quantization levels can be directly stored in the actuator, meaning that the output torque can be changed with only small information codes being received. On the other hand, the unique advantage of HLQ lies in the inherent hysteresis property, which has relevance for the reduction in interaction frequency and the anti-chattering performance.*

Inevitably, the introduction of the HLQ results in considerable quantization errors. To eliminate the adverse impact on the control system, the quantized signal can be decomposed into a continuous part and a discontinuous part as follows [35]:

$$
Q(\tau) = \kappa(\tau)\tau - E(\tau)
\tag{5}
$$

where $\kappa(\tau) = \begin{cases} \frac{Q(\tau)}{\tau}, & Q(\tau) \neq 0 \\ 1, & Q(\tau) = 0 \end{cases}$ and $E(\tau) = \begin{cases} 0, & Q(\tau) \neq 0 \\ \tau, & Q(\tau) = 0 \end{cases}$.

Lemma 1. [35]: *Considering Equation (5), it is easy to find that $\kappa(\tau)$ is continuous and $E(\tau)$ is discontinuous; thus, the following inequality holds:*

$$
\begin{cases} 1 - \delta \leq \kappa(\tau) \leq 1 + \delta \\ E(\tau) \leq \tau_{\min} \end{cases}
\tag{6}
$$

where δ and $\tau_{\min}$ are the design parameters of HLQ.

Remark 2. *Observing the output characteristics of the HLQ, there is no doubt that the quantization error will increase as the control input increases. Practically, the selection of δ is capable of affecting the difference between signals before and after the quantizer [35,36]. Nevertheless, as the upper bound of the estimation error is assumed to rely on the size of the control input, it is difficult to obtain the boundness in advance of the control design. To solve this problem, a novel quantization decomposition was proposed in (5), such that the quantization error only depends on the information of the HLQ.*

2.3. Function Approximation Based on RBFNNs

As an effective technology for the nonlinear approximation of uncertain systems, RBFNNs have been extensively used in dynamic analysis and advanced controller design.

In this paper, the model uncertainties caused by unmeasurable hydrodynamic damping will be surmounted by RBFNNs.

Lemma 2. [4]: *Any unknown smooth function $f(x) : \mathbb{R}^n \to \mathbb{R}$ can be expressed by the following formula:*

$$f(x) = W^{\mathrm{T}} h(X) + o \tag{7}$$

where o is the additional approximation error; $W = [w_1, w_2, \cdots, w_m]^{\mathrm{T}}$ is the weight vector; $w_i, i = 1, 2, \ldots, m$ denotes the gain coefficient of the corresponding hidden layer node, with m standing for the node number; $X = [X_1, X_2, \ldots, X_m]^{\mathrm{T}}$ is the input vector; $h(X) = [h_1(X), h_2(X), \ldots, h_m(X)]^{\mathrm{T}}$ denotes the Gaussian function vector formed as shown in (8), with $C_{n \times m} = [c_1, c_2, \ldots, c_m]$ being the center matrix and $\sigma = [\sigma_1, \sigma_2, \ldots, \sigma_m]^{\mathrm{T}}$ being the width vector.

$$h_i(X) = \exp\left(-\frac{\|X - c_i\|^2}{2\sigma_i^2}\right), \quad i = 1, 2, \ldots, m \tag{8}$$

Though effective, it is still costly to identify the network's parameters online with the increasing amount in hidden layer nodes. Consequently, in order to alleviate the problem of the huge computational resources required without deteriorating the performance, the MLP technology is implemented during the backstepping design. In this way, only one scalar instead of the whole weight matrix is estimated adaptively, meaning that the computational burden for nonlinear approximation is significantly decreased.

2.4. Problem Statement

For the purpose of describing the trajectory tracking of USVs, the reference information, including the position and heading angle, is given by the virtual target, whose dynamics are expressed as:

$$\begin{cases} \dot{x}_d = u_d \cos(\psi) - v_d \sin(\psi) \\ \dot{y}_d = u_d \sin(\psi) + v_d \cos(\psi) \\ \dot{\psi}_d = r_d \end{cases} \tag{9}$$

where $\eta_d = [x_d, y_d, \psi_d]^{\mathrm{T}}$ denotes the desired position and heading angle and $v_d = [u_d, v_d, r_d]^{\mathrm{T}}$ represents the desired velocities of the virtual vessel.

Comparing the reference and actual trajectories, the tracking errors are defined as:

$$\begin{aligned} x_e = x_d - x, \quad & y_e = y_d - y, \\ \psi_e = \psi_r - \psi, \quad & R_e = \sqrt{x_e^2 + y_e^2} \end{aligned} \tag{10}$$

where R_e is the relative distance between the pursuer and the target. The azimuth of the vessel is defined as ψ_r, which is determined by the position of the reference trajectory. Thus, ψ_r is calculated as:

$$\psi_r = \frac{1}{2}\pi[1 - \mathrm{sgn}(x_e)]\mathrm{sgn}(y_e) + \arctan\left(\frac{y_e}{x_e}\right) \tag{11}$$

Remark 3. *It can be derived from (11) that $\psi_r \in (-\pi, \pi]$. In addition, in the cases of $x_e = 0$ and $y_e \neq 0$, it can be concluded that $\arctan\left(\frac{y_e}{x_e}\right) \to \pm\frac{\pi}{2}$. Meanwhile, if the position error satisfies $R_e = 0$, $\arctan\left(\frac{y_e}{x_e}\right)$ will be not defined. Therefore, it is specified that $\psi_r = \psi_d$ when $R_e = 0$.*

The structure of the trajectory tracking control system of USV is shown as Figure 1. To promote the controller design, several assumptions and the relevant lemma are given as:

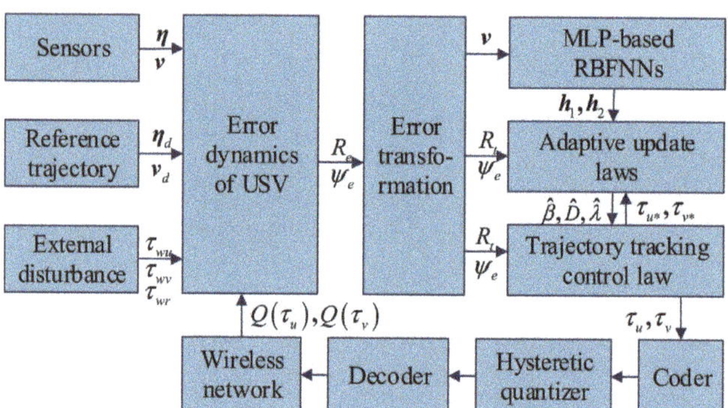

Figure 1. Structure of the trajectory tracking control system of USV.

Assumption 1. *The reference trajectories in (9) should be smooth—that is, x_d, y_d, ψ_d, $\dot{x}_d$, $\dot{y}_d$, and $\dot{\psi}_d$ are all bounded.*

Assumption 2. *The external disturbance, τ_{wu}, τ_{wv}, and τ_{wr}, are unknown but bounded—that is, $\tau_{wu} \leq d_u$, $\tau_{wv} \leq d_v$, and $\tau_{wr} \leq d_r$ hold with d_u, d_v, d_r in being unknown positive constants.*

Lemma 3. [35]: *For any $\varepsilon > 0$ and $x \in \mathbb{R}$, the following inequality holds:*

$$0 \leq |x| - \frac{|x|^2}{\sqrt{x^2 + \varepsilon^2}} \leq |x| - \frac{|x|^2}{|x| + \varepsilon} < \varepsilon \tag{12}$$

The control objective of this study can be summarized as follows. The backstepping filtering algorithm is constructed for the trajectory tracking control of underactuated USVs. On the basis of Assumptions 1 and 2, the constrained communication bandwidth and system uncertainties are taken into consideration. Consequently, the position and attitude tracking errors are stabilized, while the predefined transient performance is guaranteed.

3. Adaptive Backstepping Design Based on Quantized Input

In this section, a quantized prescribed performance control strategy is provided for the trajectory tracking of an underactuated USV. Aiming at stabilizing the position and attitude tracking errors, the backstepping-based dynamic filtering method is developed with compensation for EOC. In particular, the accurate approximation for unmeasured parameters is implemented via MLP algorithm-based RBFNNs, which have a low burden of computational resources. Meanwhile, with the consideration of the constrained communication bandwidth between controllers and actuators, the HLQ is introduced to provide the quantized control signal and the resulting quantization errors are eliminated by resorting to the adaptive estimation.

3.1. Position Controller Design

When reviewing the tracking errors defined in (10), it follows that:

$$x_e = R_e \cos(\psi_r), \quad y_e = R_e \sin(\psi_r) \tag{13}$$

Taking into account (1) and (9), the time derivative of R_e can be obtained as:

$$\begin{aligned}
\dot{R}_e &= (x_e \dot{x}_e + y_e \dot{y}_e)/R_e \\
&= (\dot{x}_d - \dot{x}) \cos(\psi_r) + (\dot{y}_d - \dot{y}) \sin(\psi_r) \\
&= (\dot{x}_d - u \cos(\psi) + v \sin(\psi)) \cos(\psi_r) + (\dot{y}_d - u \sin(\psi) - v \cos(\psi)) \sin(\psi_r) \\
&= \dot{x}_d \cos(\psi_r) + \dot{y}_d \sin(\psi_r) - u \cos(\psi_e) - v \sin(\psi_e)
\end{aligned} \tag{14}$$

To ensure that the tracking error dynamics satisfy the designer-specified behavioral metrics, the error transformation is executed [37] so that an equivalent "state-constrained" model is established for the subsequent controller design. First off, a positive continue function, $b_1(t)$, is introduced as the upper bound of R_e, which means:

$$|R_e| \leq b_1(t) \tag{15}$$

with the performance function (PF) $b_1(t)$ satisfying $b_1(0) > R_e(0)$. Furthermore, the normalized tracking error (NTE) is expressed as:

$$R_n = \frac{R_e}{b_1} \tag{16}$$

By resorting to the tangent function, the error transformation equation is defined as:

$$R_t = \tan\left(\frac{\pi R_n}{2}\right) \tag{17}$$

Remark 4. *Mathematically, if $|R_e| \to b_1(t)$, it has $R_n \to 1$ (or $R_n \to -1$); thus, $R_t \to \infty$ (or $R_t \to -\infty$) holds and vice versa. In other words, the performance constraint (15) can be satisfied through bounding the variable R_t. Consequently, the control objective of position tracking designs an algorithm to guarantee the boundedness and convergence of R_t.*

According to (14), the derivative of R_n and R_t can be obtained as:

$$\dot{R}_n = \frac{\dot{R}_e b_1 - R_e \dot{b}_1}{b_1^2} = \frac{1}{b_1}\left(-R_n \dot{b}_1 + \dot{x}_d \cos(\psi_r) + \dot{y}_d \sin(\psi_r) - u\cos(\psi_e) - v\sin(\psi_e)\right) \tag{18}$$

$$\dot{R}_t = \Gamma_R\left(-R_n \dot{b}_1 + \dot{x}_d \cos(\psi_r) + \dot{y}_d \sin(\psi_r) - u\cos(\psi_e) - v\sin(\psi_e)\right) \tag{19}$$

where $\Gamma_R = \frac{\pi(1+R_t^2)}{2b_1}$ is a positive definite variable.

Step 1. With the definition of the velocity tracking error $u_e = u_d - u$, the Lyapunov function candidate (LFC) is selected as:

$$V_1 = \frac{1}{2}R_t^2 \tag{20}$$

Taking the time derivative of V_1 and substituting (19) yields:

$$
\begin{aligned}
\dot{V}_1 &= R_t \dot{R}_t \\
&= \Gamma_R R_t\left(-R_n \dot{b}_1 + \dot{x}_d \cos(\psi_r) + \dot{y}_d \sin(\psi_r) - u\cos(\psi_e) - v\sin(\psi_e)\right) \\
&= \Gamma_R R_t\left[-R_n \dot{b}_1 + \dot{x}_d \cos(\psi_r) + \dot{y}_d \sin(\psi_r) - u_d\cos(\psi_e) + u_e\cos(\psi_e) - v\sin(\psi_e)\right]
\end{aligned} \tag{21}
$$

To stabilize the transformed position tracking error R_t, the virtual control law u_d is proposed as:

$$u_d = \frac{1}{\cos(\psi_e)}\left(k_1 R_t - R_n \dot{b}_1 + \dot{x}_d \cos(\psi_r) + \dot{y}_d \sin(\psi_r) - v\sin(\psi_e)\right) \tag{22}$$

with the design parameter k_1 being a positive constant.

Substituting it into (21), $\dot{V}_1$ becomes:

$$\dot{V}_1 = \Gamma_R R_t(-k_1 R_t + u_e \cos(\psi_e)) \tag{23}$$

Step 2. To remedy the problem of the EOC, a first-order low-pass filter is introduced to process the virtual command u_d. The output characteristic of the filter is given as:

$$l_1 \dot{u}_c + u_c = u_d, \quad u_c(0) = u_d(0)$$
$$\dot{u}_c = (u_d - u_c)/l_1 \tag{24}$$

where u_c is the output signal and the constant l_1 is the filter parameter. Thus, in the following design $\dot{u}_d$ is replaced by $\dot{u}_c$. The second LFC is chosen as:

$$V_2 = V_1 + \frac{1}{2}u_e^2 \tag{25}$$

According to the dynamic (2) and the result (23), the derivative of V_2 is written as:

$$\begin{aligned}
\dot{V}_2 &= \dot{V}_1 + u_e \dot{u}_e \\
&= \Gamma_R R_t(-k_1 R_t + u_e \cos(\psi_e)) + u_e(\dot{u}_c - \dot{u}) \\
&= \Gamma_R R_t(-k_1 R_t + u_e \cos(\psi_e)) \\
&\quad + u_e\left(\dot{u}_c - \frac{m_{22}}{m_{11}}v\omega - \frac{1}{m_{11}}Q(\tau_u) + f_u(u) - \frac{1}{m_{11}}\tau_{wu}\right)
\end{aligned} \tag{26}$$

In particular, the hydrodynamic damping term $f_u(u)$ is regarded as an unknown continuous function that can be approximated by RBFNN. Learning from Lemma 2, one finds:

$$f_u(u) = W_1^{\mathsf{T}} h_1(v) + o_1 \tag{27}$$

where the approximation error o_1 satisfies $0 < o_1 \le O_1$.

In order to remove the need for the excessive calculations caused by estimating the entire weight matrix, the upper bound β_1 is introduced as $\beta_1 \ge \|W_1\|$ and the parameter h_1 is defined as $h_1 = \|h_1(v)\|$. Hence, $\dot{V}_2$ can be further derived as:

$$\begin{aligned}
\dot{V}_2 &= \Gamma_R R_t(-k_1 R_t + u_e \cos(\psi_e)) + u_e\left(\dot{u}_c - \frac{m_{22}}{m_{11}}v\omega - \frac{1}{m_{11}}Q(\tau_u)\right) \\
&\quad + u_e\left(W_1^{\mathsf{T}} h_1(v) + o_1 - \frac{1}{m_{11}}\tau_{wu}\right) \\
&\le \Gamma_R R_t(-k_1 R_t + u_e \cos(\psi_e)) + u_e\left(\dot{u}_c - \frac{m_{22}}{m_{11}}v\omega - \frac{1}{m_{11}}Q(\tau_u)\right) \\
&\quad + |u_e|\left(\beta_1 h_1 + O_1 + \frac{1}{m_{11}}d_u\right)
\end{aligned} \tag{28}$$

Referring to the discussion in (11, 12), the quantized signal of the control input τ_u can be obtained as:

$$Q(\tau_u) = \kappa_1(\tau_u)\tau_u - E_1(\tau_u) \tag{29}$$

with $1 - \delta_1 \le \kappa_1(\tau_u) \le 1 + \delta_1$ and $E_1(\tau_u) \le \tau_{u\min}$.

To facilitate the adaptive estimation, it is defined that $\lambda_1 = \frac{1}{1-\delta_1}$ and $D_1 = O_1 + \frac{1}{m_{11}}d_u + \frac{1}{m_{11}}\tau_{u\min}$. Moreover, the variables $\hat{\beta}_1$, $\hat{D}_1$, and $\hat{\lambda}_1$ are introduced as estimate values, while the estimate errors are represented as $\tilde{\beta}_1$, $\tilde{D}_1$, and $\tilde{\lambda}_1$.

$$\tilde{\beta}_1 = \beta_1 - \hat{\beta}_1, \tilde{D}_1 = D_1 - \hat{D}_1, \tilde{\lambda}_1 = \lambda_1 - \hat{\lambda}_1 \tag{30}$$

Subsequently, the control signal is constructed as:

$$\tau_u = m_{11}\hat{\lambda}_1 \tau_{u*} \tag{31}$$

$$\tau_{u*} = \frac{\eta_t^2 u_e}{\sqrt{\eta_t^2 u_e^2 + \varepsilon_1}} + (k_2 + k_u)u_e \tag{32}$$

$$k_u = \frac{\hat{\beta}_1^2 h_1^2}{\sqrt{\hat{\beta}_1^2 h_1^2 u_e^2 + \varepsilon_1}} + \frac{\hat{D}_1^2}{\sqrt{\hat{D}_1^2 u_e^2 + \varepsilon_1}} \tag{33}$$

where $\eta_t = \left(\dot{u}_c - \frac{m_{22}}{m_{11}}vr + \Gamma_R R_t \cos(\psi_e)\right)$ and the controller parameter k_2 is a positive constant. Furthermore, the adaptive update laws are given as:

$$\dot{\hat{\beta}}_1 = c_1\left(h_1|u_e| - c_2\hat{\beta}_1\right) \tag{34}$$

$$\dot{\hat{D}}_1 = c_3\left(|u_e| - c_4\hat{D}_1\right) \tag{35}$$

$$\dot{\hat{\lambda}}_1 = c_5\left(u_e\tau_{u*} - c_6\hat{\lambda}_1\right) \tag{36}$$

where $c_i(i = 1,2,3,4,5,6)$ are all positive parameters.

Remark 5. *Considering the ingeniously designed controller (31–34), there exist three highlights that deserve some attention. (i) Different from the conservative quantized control, the algorithm introduced in this paper is developed under the unavailable quantizer parameter δ_1. For this reason, the constant λ_1 is introduced for adaptive estimation and procedure (31) is constructed to compensate the resulting concussion. (ii) The unknown constant D_1 is defined as the lumped additive uncertainties, which is constituted by the approximation error of NN, the environmental disturbance, and the hysteresis zone of the quantizer. On the other hand, the scalar β_1 is utilized for estimating the weight matrix of RBFNN, which emphatically reflects the superiority of MLP in terms of the computation burden. (iii) The first term in τ_{u*} stands for system dynamics compensation, and the remainder suppresses the phenomenon of chattering, while the convergence of tracking errors and estimate errors can be ensured.*

For the sake of convenience, $\kappa_1(\tau_u)$ and $E_1(\tau_u)$ are abbreviated as κ_1 and E_1, respectively. Thus, with the substitution of (29), (28) can be further written as:

$$\begin{aligned}
\dot{V}_2 &\leq \Gamma_R R_t(-k_1 R_t + u_e \cos(\psi_e)) + u_e\left[\dot{u}_c - \frac{m_{22}}{m_{11}}vr - \frac{1}{m_{11}}(\kappa_1\tau_u - E_1)\right] \\
&\quad + |u_e|\left(\beta_1 h_1 + O_1 + \frac{1}{m_{11}}d_u\right)
\end{aligned} \tag{37}$$

According to the property of κ_1, λ_1 and substituting (31), the term $-\frac{1}{m_{11}}(\kappa_1\tau_u - E_1)$ can be calculated as:

$$\begin{aligned}
-\frac{1}{m_{11}}(\kappa_1\tau_u - E_1)u_e &= -\frac{1}{m_{11}}\left(\kappa_1 m_{11}\hat{\lambda}\tau_{u*} - E_1\right)u_e \\
&= -\kappa_1\left(\lambda_1 - \tilde{\lambda}_1\right)u_e\tau_{u*} + \frac{1}{m_{11}}E_1 u_e \\
&\leq -(1-\delta_1)\left(\lambda_1 - \tilde{\lambda}_1\right)u_e\tau_{u*} + \frac{1}{m_{11}}E_1 u_e \\
&= -u_e\tau_{u*} + (1-\delta_1)\tilde{\lambda}_1 u_e\tau_{u*} + \frac{1}{m_{11}}\tau_{u\min}|u_e|
\end{aligned} \tag{38}$$

By recalling the control signal in (32) and utilizing Lemma 3, one finds:

$$\begin{aligned}
-\frac{1}{m_{11}}(\kappa_1\tau_u - E_1)u_e &\leq (1-\delta_1)\tilde{\lambda}_1 u_e\tau_{u*} + \frac{1}{m_{11}}E_1 u_e - u_e\left(\frac{\eta_t^2 u_e}{\sqrt{\eta_t^2 u_e^2 + \varepsilon_1}} + (k_2 + k_u)u_e\right) \\
&\leq -|\eta_t u_e| - (k_2 + k_u)u_e^2 + (1-\delta_1)\tilde{\lambda}_1 u_e\tau_{u*} + \frac{1}{m_{11}}\tau_{u\min}|u_e| + \varepsilon_1
\end{aligned} \tag{39}$$

Substituting the result into (37), $\dot{V}_2$ can be scaled as:

$$\begin{aligned}
\dot{V}_2 &\leq \Gamma_R R_t(-k_1 R_t + u_e \cos(\psi_e)) + u_e\left(\dot{u}_c - \frac{m_{22}}{m_{11}}v\omega\right) \\
&\quad - \left|\left(\dot{u}_c - \frac{m_{22}}{m_{11}}vr + \Gamma_R R_t \cos(\psi_e)\right)u_e\right| - (k_2 + k_u)u_e^2 \\
&\quad + (1-\delta_1)\tilde{\lambda}_1 u_e\tau_{u*} + |u_e|\beta_1 h_1 + |u_e|\left(O_1 + \frac{1}{m_{11}}d_u + \frac{1}{m_{11}}\tau_{u\min}\right) + \varepsilon_1 \\
&\leq -k_1\Gamma_R R_t^2 - (k_2 + k_u)u_e^2 + (1-\delta_1)\tilde{\lambda}_1 u_e\tau_{u*} + |u_e|\beta_1 h_1 + |u_e|D_1 + \varepsilon_1
\end{aligned} \tag{40}$$

Consequently, taking into account (33), a concise result can be obtained:

$$
\begin{aligned}
\dot{V}_2 \;\leq\; & -k_1 \Gamma_R R_t^2 - k_2 u_e^2 - u_e^2 \left(\frac{\hat{\beta}_1^2 h_1^2}{\sqrt{\hat{\beta}_1^2 h_1^2 u_e^2 + \varepsilon_1}} + \frac{\hat{D}_1^2}{\sqrt{\hat{D}_1^2 u_e^2 + \varepsilon_1}} \right) \\
& + (1 - \delta_1)\tilde{\lambda}_1 u_e \tau_{u*} + |u_e|\beta_1 h_1 + |u_e| D_1 + \varepsilon_1 \\
\leq\; & -k_1 \Gamma_R R_t^2 - k_2 u_e^2 - |u_e|\hat{\beta}_1 h_1 - |u_e|\hat{D}_1 + (1 - \delta_1)\tilde{\lambda}_1 u_e \tau_{u*} + |u_e|\beta_1 h_1 + |u_e| D_1 + 2\varepsilon_1 \\
=\; & -k_1 \Gamma_R R_t^2 - k_2 u_e^2 + |u_e|\widetilde{\beta}_1 h_1 + |u_e|\widetilde{D}_1 + (1 - \delta_1)\tilde{\lambda}_1 u_e \tau_{u*} + 2\varepsilon_1
\end{aligned}
\tag{41}
$$

3.2. Attitude Controller Design

According to (10), the derivative of ψ_e is expressed as:

$$
\dot{\psi}_e = \dot{\psi}_r - r
\tag{42}
$$

Step 1. To stabilize the attitude of the pursuer, the LFC is selected as:

$$
V_3 = \frac{1}{2}\psi_e^2
\tag{43}
$$

With the definition of the angular velocity tracking error being $r_e = r_d - r$, the time derivative of (44) is deduced as:

$$
\begin{aligned}
\dot{V}_3 \;&= \psi_e \dot{\psi}_e \\
&= \psi_e \left(\dot{\psi}_r - r \right) \\
&= \psi_e \left(\dot{\psi}_r + r_e - r_d \right)
\end{aligned}
\tag{44}
$$

Specially, the virtual control law r_d is designed as:

$$
r_d = k_3 \psi_e + \dot{\psi}_r
\tag{45}
$$

with k_3 being a positive constant. Hence, $\dot{V}_3$ becomes:

$$
\dot{V}_3 = \psi_e (r_e - k_3 \psi_e)
\tag{46}
$$

Step 2. Referring to (24), let the virtual control law r_d pass through a first-order filter:

$$
\begin{aligned}
&\iota_2 \dot{r}_c + r_c = r_d, \quad r_c(0) = r_d(0) \\
&\dot{r}_c = (r_d - r_c)/\iota_2
\end{aligned}
\tag{47}
$$

where r_c is the filtered signal and ι_2 is the filter parameter. In this step, $\dot{r}_d$ is substituted by $\dot{r}_c$.

$$
V_4 = V_3 + \frac{1}{2} r_e^2
\tag{48}
$$

Differentiating V_4 with respect to time and substituting (10), one obtains:

$$
\begin{aligned}
\dot{V}_4 \;&= \dot{V}_3 + r_e \dot{r}_e \\
&= \psi_e(r_e - k_3\psi_e) + r_e\left(\dot{r}_c - \dot{r}\right) \\
&= \psi_e(r_e - k_3\psi_e) + r_e\left(\dot{r}_c - \frac{m_{11} - m_{22}}{m_{33}} uv - \frac{1}{m_{33}} Q(\tau_r) + f_r(r) - \frac{1}{m_{33}} \tau_{wr}\right)
\end{aligned}
\tag{49}
$$

One caveat here is that the nonlinear function $f_r(r)$ cannot be observed accurately. Therefore, the MLP-based RBFNN is applied to approximate the time-varying dynamics. According to Lemma 2, the following equation is valid.

$$
f_r(r) = W_2^{\mathrm{T}} h_2(v) + o_2
\tag{50}
$$

where the approximation error satisfies $0 < o_2 \leq O_2$. Moreover, two parameters are defined as $\beta_2 \geq \|W_2\|$ and $h_2 = \|h_2(v)\|$, meaning that only one scalar needs to be adaptively estimated, which reduces the computational burden. In this way, (50) can be derived as:

$$
\begin{aligned}
\dot{V}_4 &= \psi_e(r_e - k_3\psi_e) + r_e\left(\dot{r}_c - \frac{m_{11}-m_{22}}{m_{33}}uv - \frac{1}{m_{33}}Q(\tau_R) + W_2^{\mathsf{T}}h_2(v) + o_2 - \frac{1}{m_{33}}\tau_{wr}\right) \\
&\leq \psi_e(r_e - k_3\psi_e) + r_e\left(\dot{r}_c - \frac{m_{11}-m_{22}}{m_{33}}uv - \frac{1}{m_{33}}Q(\tau_R)\right) + |r_e|\left(\beta_2 h_2 + O_2 + \frac{1}{m_{33}}d_r\right)
\end{aligned}
\tag{51}
$$

Learning from Lemma 1, the quantized signal for control input τ_r can be written as:

$$
Q(\tau_r) = \kappa_2(\tau_r)\tau_r - E_2(\tau_r)
\tag{52}
$$

with $1 - \delta_2 \leq \kappa_2(\tau_r) \leq 1 + \delta_2$ and $E_2(\tau_r) \leq \tau_{r\min}$.

Before giving the control algorithm, it is necessary to define the parameter $\lambda_2 = \frac{1}{1-\delta_2}$ and the lumped unknown term $D_2 = O_2 + \frac{1}{m_{33}}d_r + \frac{1}{m_{33}}\tau_{r\min}$. Meanwhile, the adaptive estimate values are introduced as $\hat{D}_2$, $\hat{\beta}_2$, and $\hat{\lambda}_2$, meaning that the estimate errors can be expressed as:

$$
\tilde{D}_2 = D_2 - \hat{D}_2,\; \tilde{\beta}_2 = \beta_2 - \hat{\beta}_2,\; \tilde{\lambda}_2 = \lambda_2 - \hat{\lambda}_2
\tag{53}
$$

Consequently, the control input is elaborated as:

$$
\tau_r = m_{33}\hat{\lambda}_2\tau_{r*}
\tag{54}
$$

$$
\tau_{r*} = \frac{\eta_r^2 r_e}{\sqrt{\eta_r^2 r_e^2 + \varepsilon_2}} + (k_4 + k_r)r_e
\tag{55}
$$

$$
k_r = \frac{\hat{\beta}_2^2 h_2^2}{\sqrt{\hat{\beta}_2^2 h_2^2 r_e^2 + \varepsilon_2}} + \frac{\hat{D}_2^2}{\sqrt{\hat{D}_2^2 r_e^2 + \varepsilon_2}}
\tag{56}
$$

where $\eta_r = \left(\dot{r}_c - \frac{m_{11}-m_{22}}{m_{33}}uv + \psi_e\right)$; k_4 is a positive controller parameter. The adaptive learning laws are given as:

$$
\dot{\hat{\beta}}_2 = c_7\left(h_2|r_e| - c_8\hat{\beta}_2\right)
\tag{57}
$$

$$
\dot{\hat{D}}_2 = c_9\left(|r_e| - c_{10}\hat{D}_2\right)
\tag{58}
$$

$$
\dot{\hat{\lambda}}_2 = c_{11}\left(r_e\tau_{r*} - c_{12}\hat{\lambda}_2\right)
\tag{59}
$$

where $c_i(i = 7,8,9,10,11,12)$ are all positive constants.

According to the definitions of κ_2 and λ_2, the term $-\frac{1}{m_{33}}(\kappa_2\varepsilon_w - E_2)$ can be calculated as:

$$
\begin{aligned}
-\frac{1}{m_{33}}(\kappa_2\tau_r - E_2)r_e &= -\frac{1}{m_{33}}\left(\kappa_2 m_{33}\hat{\lambda}_2\tau_r - E_2\right)r_e \\
&= -\kappa_2\hat{\lambda}_2 r_e\tau_r + \frac{1}{m_{33}}E_2 r_e \\
&\leq -(1 - \delta_2)\left(\lambda_2 - \tilde{\lambda}_2\right)r_e\tau_r + \frac{1}{m_{33}}E_2 r_e \\
&= -\varepsilon_r r_e + (1 - \delta_2)\tilde{\lambda}_2 r_e\tau_r + \frac{1}{m_{33}}\tau_{r\min}r_e
\end{aligned}
\tag{60}
$$

Substituting the control signal in (55) leads to:

$$
\begin{aligned}
-\frac{1}{m_{33}}(\kappa_2\tau_r - E_2)r_e &\leq -\frac{\eta_r^2 r_e^2}{\sqrt{\eta_r^2 r_e^2 + \varepsilon_2}} - (k_4 + k_w)r_e^2 + (1 - \delta_2)\tilde{\lambda}_2 r_e\tau_r + \frac{1}{m_{33}}\tau_{r\min}r_e \\
&\leq -|\eta_r r_e| - (k_4 + k_r)r_e^2 + (1 - \delta_2)\tilde{\lambda}_2 r_e\tau_r + \frac{1}{m_{33}}\tau_{r\min}r_e + \varepsilon_2
\end{aligned}
\tag{61}
$$

Taking the results into consideration, (52) follows:

$$
\begin{aligned}
\dot{V}_4 &\leq \psi_e(r_e - k_3\psi_e) + r_e\left(\dot{r}_c - \frac{m_{11}-m_{22}}{m_{33}}uv\right) - \left|\left(\dot{r}_c - \frac{m_{11}-m_{22}}{m_{33}}uv + \psi_e\right)r_e\right| \\
&\quad -(k_4 + k_r)r_e^2 + (1 - \delta_2)\tilde{\lambda}_2 r_e\tau_r + |r_e|\beta_2 h_2 + |r_e|\left(O_2 + \frac{1}{m_{33}}d_r + \frac{1}{m_{33}}\tau_{r\min}\right) \\
&\leq -k_3\psi_e^2 - (k_4 + k_r)r_e^2 + (1 - \delta_2)\tilde{\lambda}_2 r_e\tau_r + |r_e|\beta_2 h_2 + |r_e|D_2
\end{aligned}
\tag{62}
$$

Finally, with the substitution of (57), a concise result is obtained:

$$\dot{V}_4 \leq -k_3\psi_e^2 - k_4 r_e^2 - r_e^2\left(\frac{\hat{\beta}_2^2 h_2^2}{\sqrt{\hat{\beta}_2^2 h_2^2 r_e^2 + \varepsilon_2}} + \frac{\hat{D}_2^2}{\sqrt{\hat{D}_2^2 r_e^2 + \varepsilon_2}}\right)$$
$$+(1-\delta_2)\tilde{\lambda}_2 r_e \tau_r + |r_e|\beta_2 h_2 + |r_e|D_2$$
$$\leq -k_3\psi_e^2 - k_4 r_e^2 - |r_e|\hat{\beta}_2 h_2 - |r_e|\hat{D}_2 + (1-\delta_2)\tilde{\lambda}_2 r_e \tau_r + |r_e|\beta_2 h_2 + |r_e|D_2 + 2\varepsilon_2$$
$$= -k_3\psi_e^2 - k_4 r_e^2 + |r_e|\tilde{\beta}_2 h_2 + |r_e|\tilde{D}_2 + (1-\delta_2)\tilde{\lambda}_2 r_e \tau_r + 2\varepsilon_2 \tag{63}$$

3.3. Stability Analysis

In this section, the stability analysis proceeds using the Lyapunov theory. First off, the theorem is given as follows.

Theorem 1. *Consider the underactuated AUV model described in (1) and (2) and the tracking error dynamics (14) under Assumptions 1 and 2. On basis of the quantizer (4), the error transformation (21), the filters (24) and (48), and the controllers (31)–(36) and (55)–(60) are capable of ensuring that all signals in the closed-loop system are bounded and that the position tracking error R_e satisfies the predefined behavioral metrics.*

Proof. In order to prevent the adverse effects caused by the introduction of filters, the filter errors are defined and the validation will be provided to guarantee their convergence.

$$\tilde{\alpha}_u = u_c - u_d, \tilde{\alpha}_r = r_c - r_d \tag{64}$$

Associated with the descriptions of (24) and (48), their time derivatives satisfy the relation:

$$\dot{\tilde{\alpha}}_u = \dot{u}_c - \dot{u}_d = -\frac{\tilde{\alpha}_u}{\iota_1} + \dot{u}_d \leq -\frac{\tilde{\alpha}_u}{\iota_1} + B_1,$$
$$\dot{\tilde{\alpha}}_r = \dot{r}_c - \dot{r}_d = -\frac{\tilde{\alpha}_r}{\iota_2} + \dot{r}_d \leq -\frac{\tilde{\alpha}_r}{\iota_2} + B_2. \tag{65}$$

where the continuous function $B_i (i = 1, 2)$ possesses an unknown certain maximum—i.e., $|B_i| \leq \vartheta_i$.

Subsequently, the overall LFC is defined as:

$$V = V_2 + \frac{1}{2c_1}\tilde{\beta}_1^2 + \frac{1}{2c_3}\tilde{D}_1^2 + \frac{1-\delta_1}{2c_5}\tilde{\lambda}_1^2 + \frac{1}{2}\tilde{\alpha}_u^2$$
$$+ V_4 + \frac{1}{2c_7}\tilde{\beta}_2^2 + \frac{1}{2c_9}\tilde{D}_2^2 + \frac{1-\delta_2}{2\gamma_{11}}\tilde{\lambda}_2^2 + \frac{1}{2}\tilde{\alpha}_r^2 \tag{66}$$

With the aid of (41) and (63), differentiating V with respect to time yields:

$$\dot{V} = \dot{V}_1 + \frac{1}{c_1}\tilde{\beta}_1\dot{\tilde{\beta}}_1 + \frac{1}{c_3}\tilde{D}_1\dot{\tilde{D}}_1 + \frac{1-\delta_1}{\gamma_2}\tilde{\lambda}_1\dot{\tilde{\lambda}}_1 + \tilde{\alpha}_u\dot{\tilde{\alpha}}_u$$
$$+ \dot{V}_2 + \frac{1}{c_5}\tilde{\beta}_2\dot{\tilde{\beta}}_2 + \frac{1}{c_7}\tilde{D}_2\dot{\tilde{D}}_2 + \frac{1-\delta_2}{\gamma_6}\tilde{\lambda}_2\dot{\tilde{\lambda}}_2 + \tilde{\alpha}_r\dot{\tilde{\alpha}}_r$$
$$= -k_1\Gamma_R R_t^2 - k_2 u_e^2 + |u_e|\tilde{\beta}_1 h_1 + |u_e|\tilde{D}_1 + (1-\delta_1)\tilde{\lambda}_1 u_e \tau_{u*} + 2\varepsilon_1$$
$$- \tilde{\beta}_1(h_1|u_e| - c_2\hat{\beta}_1) - \tilde{D}_1(|u_e| - c_4\hat{D}_1) - (1-\delta_1)\tilde{\lambda}_1(u_e\tau_{u*} - c_6\hat{\lambda}_1) + \tilde{\alpha}_u\dot{\tilde{\alpha}}_u$$
$$- k_3\psi_e^2 - k_4 r_e^2 + |r_e|\tilde{\beta}_2 h_2 + |r_e|\tilde{D}_2 + (1-\delta_2)\tilde{\lambda}_2 r_e \tau_r + 2\varepsilon_2$$
$$- \tilde{\beta}_2(h_2|r_e| - c_8\hat{\beta}_2) - \tilde{D}_2(|r_e| - c_{10}\hat{D}_2) - (1-\delta_2)\tilde{\lambda}_2(r_e\tau_{r*} - c_{12}\hat{\lambda}_2) + \tilde{\alpha}_r\dot{\tilde{\alpha}}_r$$
$$= -k_1\Gamma_R R_t^2 - k_2 u_e^2 + c_2\hat{\beta}_1\tilde{\beta}_1 + c_4\hat{D}_1\tilde{D}_1 + c_6(1-\delta_1)\hat{\lambda}_1\tilde{\lambda}_1 + \tilde{\alpha}_u\dot{\tilde{\alpha}}_u + 2\varepsilon_1$$
$$- k_3\psi_e^2 - k_4 r_e^2 + c_8\hat{\beta}_2\tilde{\beta}_2 + c_{10}\hat{D}_2\tilde{D}_2 + c_{12}(1-\delta_2)\hat{\lambda}_2\tilde{\lambda}_2 + \tilde{\alpha}_r\dot{\tilde{\alpha}}_r + 2\varepsilon_2 \tag{67}$$

By resorting to Young's inequality, one of the above terms can be scaled as:

$$c_2\hat{\beta}_1\tilde{\beta}_1 = c_2\left(-\tilde{\beta}_1^2 + \beta_1\tilde{\beta}_1\right)$$
$$\leq c_2\left(-\tilde{\beta}_1^2 + \frac{1}{2\sigma_1}\tilde{\beta}_1^2 + \frac{\sigma_1}{2}\beta_1^2\right)$$
$$\leq -\frac{c_2(2\sigma_1 - 1)}{2\sigma_1}\tilde{\beta}_1^2 + \frac{c_2\sigma_1}{2}\beta_1^2 \tag{68}$$

where $\sigma_1 > 0.5$. Then, a similar calculation can be implemented for $c_4\hat{D}_1\tilde{D}_1$, $c_6(1-\delta_1)\hat{\lambda}_1\tilde{\lambda}_1$, $c_8\hat{\beta}_2\tilde{\beta}_2$, $c_{10}\hat{D}_2\tilde{D}_2$, and $c_{12}(1-\delta_2)\hat{\lambda}_2\tilde{\lambda}_2$, whose results are shown as:

$$
\begin{aligned}
c_4\hat{D}_1\tilde{D}_1 &\leq -\tfrac{c_4(2\sigma_2-1)}{2\sigma_2}\tilde{D}_1^2 + \tfrac{c_4\sigma_2}{2}D_1^2 \\
c_6(1-\delta_1)\hat{\lambda}_1\tilde{\lambda}_1 &\leq -\tfrac{c_6(1-\delta_1)(2\sigma_3-1)}{2\sigma_3}\tilde{\lambda}_1^2 + \tfrac{c_6(1-\delta_1)\sigma_3}{2}\lambda_1^2 \\
c_8\hat{\beta}_2\tilde{\beta}_2 &\leq -\tfrac{c_8(2\sigma_4-1)}{2\sigma_4}\tilde{\beta}_2^2 + \tfrac{c_8\sigma_4}{2}\beta_2^2 \\
c_{10}\hat{D}_2\tilde{D}_2 &\leq -\tfrac{c_{10}(2\sigma_5-1)}{2\sigma_5}\tilde{D}_2^2 + \tfrac{c_{10}\sigma_5}{2}D_2^2 \\
c_{12}(1-\delta_2)\hat{\lambda}_2\tilde{\lambda}_2 &\leq -\tfrac{c_{12}(1-\delta_2)(2\sigma_6-1)}{2\sigma_6}\tilde{\lambda}_2^2 + \tfrac{c_{12}(1-\delta_2)\sigma_6}{2}\lambda_2^2
\end{aligned}
\tag{69}
$$

with $\sigma_i > 0.5 \, (i = 2,3,4,5,6)$.

On the other hand, with the consideration of (65), the terms for the filter error lead to:

$$
\begin{aligned}
\tilde{\alpha}_u\dot{\tilde{\alpha}}_u &= -\tfrac{\tilde{\alpha}_u^2}{l_1} + \tilde{\alpha}_u u_d \leq -\tfrac{\tilde{\alpha}_u^2}{l_1} + |\tilde{\alpha}_u|\vartheta_1 \leq -\tfrac{\tilde{\alpha}_u^2}{l_1} + \tilde{\alpha}_u^2 + \tfrac{\vartheta_1^2}{4}, \\
\tilde{\alpha}_r\dot{\tilde{\alpha}}_r &= -\tfrac{\tilde{\alpha}_r^2}{l_2} + \tilde{\alpha}_r r_d \leq -\tfrac{\tilde{\alpha}_r^2}{l_2} + |\tilde{\alpha}_r|\vartheta_2 \leq -\tfrac{\tilde{\alpha}_r^2}{l_2} + \tilde{\alpha}_r^2 + \tfrac{\vartheta_2^2}{4}.
\end{aligned}
\tag{70}
$$

Substituting the results of (68)–(70), $\dot{V}$ is finally obtained as:

$$
\begin{aligned}
\dot{V} \;\leq\; & -k_1\Gamma_R R_t^2 - k_2 u_e^2 - \tfrac{c_2(2\sigma_1-1)}{2\sigma_1}\tilde{\beta}_1^2 - \tfrac{c_4(2\sigma_2-1)}{2\sigma_2}\tilde{D}_1^2 - \tfrac{c_6(1-\delta_1)(2\sigma_3-1)}{2\sigma_3}\tilde{\lambda}_1^2 - \left(\tfrac{1}{\gamma_1}-1\right)\tilde{\alpha}_u^2 \\
& -k_3\psi_e^2 - k_4 r_e^2 - \tfrac{c_8(2\sigma_4-1)}{2\sigma_4}\tilde{\beta}_2^2 - \tfrac{c_{10}(2\sigma_5-1)}{2\sigma_5}\tilde{D}_2^2 - \tfrac{c_{12}(1-\delta_2)(2\sigma_6-1)}{2\sigma_6}\tilde{\lambda}_2^2 - \left(\tfrac{1}{\gamma_2}-1\right)\tilde{\alpha}_r^2 \\
& +\tfrac{c_2\sigma_1}{2}\beta_1^2 + \tfrac{c_4\sigma_2}{2}D_1^2 + \tfrac{c_6(1-\delta_1)\sigma_3}{2}\lambda_1^2 + \tfrac{\vartheta_1^2}{4} + 2\varepsilon_1 \\
& +\tfrac{c_8\sigma_4}{2}\beta_2^2 + \tfrac{c_{10}\sigma_5}{2}D_2^2 + \tfrac{c_{12}(1-\delta_2)\sigma_6}{2}\lambda_2^2 + \tfrac{\vartheta_2^2}{4} + 2\varepsilon_2 \\
\leq\; & -\rho V + \Delta
\end{aligned}
\tag{71}
$$

with

$$
\rho = \min \left\{
\begin{array}{l}
2k_1\Gamma_R,\, 2k_2,\, \tfrac{c_1 c_2(2\sigma_1-1)}{\sigma_1},\, \tfrac{c_3 c_4(2\sigma_2-1)}{\sigma_2},\, \tfrac{c_5 c_6(2\sigma_3-1)}{\sigma_3},\, 2\left(\tfrac{1}{\gamma_1}-1\right) \\[2mm]
2k_3,\, 2k_4,\, \tfrac{c_7 c_8(2\sigma_4-1)}{\sigma_4},\, \tfrac{c_9 c_{10}(2\sigma_5-1)}{\sigma_5},\, \tfrac{c_{11} c_{12}(2\sigma_6-1)}{\sigma_6},\, 2\left(\tfrac{1}{\gamma_2}-1\right)
\end{array}
\right\}
\tag{72}
$$

$$
\begin{aligned}
\Delta = \;& \tfrac{c_2\sigma_1}{2}\beta_1^2 + \tfrac{c_4\sigma_2}{2}D_1^2 + \tfrac{c_6(1-\delta_1)\sigma_3}{2}\lambda_1^2 + \tfrac{\vartheta_1^2}{4} + 2\varepsilon_1 \\
& +\tfrac{c_8\sigma_4}{2}\beta_2^2 + \tfrac{c_{10}\sigma_5}{2}D_2^2 + \tfrac{c_{12}(1-\delta_2)\sigma_6}{2}\lambda_2^2 + \tfrac{\vartheta_2^2}{4} + 2\varepsilon_2
\end{aligned}
$$

Thus far, this indicates that the transformed tracking errors, the estimate errors, and the filter errors tend to converge exponentially into a tiny neighborhood around zero. Hence, as illustrated in Remark 4, the predefined performance for the position tracking errors is guaranteed all the while.

Ultimately, the validity of Theorem 1 has been illustrated. $\square$

4. Simulation

In this section, four simulation examples are conducted to validate the stability and exhibit the superiority of the proposed controller. First, a digital model is built on the basis of the USV's motion dynamics and the specific parameters are set reasonably. Subsequently, the simulation results will be displayed in the form of figures, which are capable of demonstrating the detailed performance of the developed algorithm. Furthermore, the experiments under different environments and the comparison with a conservative controller provide to highlight the advantages of this work.

4.1. Parameter Setting

The vessel model parameters from [5] are adopted for numerical simulation, which are presented in Table 1. For the reference trajectory, we have $u_d = 0.5$ m/s, $v_d = 0$ m/s,

$r_d = 0.5 * \pi/180$ rad/s, and the initial states are chosen as $\eta_d(0) = [0.5 \text{ m}, 0 \text{ m}, 3 * \pi/180 \text{ rad}]^{\text{T}}$ and $v_d(0) = [0 \text{ m/s}, 0 \text{ m/s}, 0 \text{ rad/s}]^{\text{T}}$. The external disturbances are given as (73).

$$\begin{bmatrix} \tau_{wu}(t) \\ \tau_{wv}(t) \\ \tau_{wr}(t) \end{bmatrix} = k_d \begin{bmatrix} \sin(0.01 * t) \\ \cos(0.01 * t) \\ \sin(0.01 * t) \end{bmatrix} \tag{73}$$

where k_d is the amplitude parameter of the disturbances.

Table 1. Main parameters.

Parameter	Value	Unit	Parameter	Value	Unit
m_{11}	1.1274	kg	d_v	0.1183	kg/s
m_{22}	1.8902	kg	d_{v2}	0.05915	kg/m
m_{23}	0.1278	kg/m^2	d_{v3}	0.029575	kg/m^2
d_u	0.0358	kg/s	d_r	0.0308	kg/s
d_{u2}	0.0179	kg/s	d_{r2}	0.0154	kg/m
d_{u3}	0.00895	kg/m^2	d_{r3}	0.0077	kg/m^2

Simulations are conducted in four sets, which are distinguished by different initial states and disturbances. The configurations of the Scenarios are given as follows:

Scenario I: $\eta(0) = [-5 \text{ m}, -5 \text{ m}, 0.45 \text{ rad}]^{\text{T}}$, $v(0) = [0 \text{ m/s}, 0 \text{ m/s}, 0 \text{ rad/s}]^{\text{T}}$, $k_d = 0.01$

Scenario II: $\eta(0) = [-5 \text{ m}, -5 \text{ m}, 0.45 \text{ rad}]^{\text{T}}$, $v(0) = [0 \text{ m/s}, 0 \text{ m/s}, 0 \text{ rad/s}]^{\text{T}}$, $k_d = 0.04$

Scenario III: $\eta(0) = [-10 \text{ m}, -10 \text{ m}, 0.45 \text{ rad}]^{\text{T}}$, $v(0) = [0 \text{ m/s}, 0 \text{ m/s}, 0 \text{ rad/s}]^{\text{T}}$, $k_d = 0.01$

Scenario IV: $\eta(0) = [-10 \text{ m}, -10 \text{ m}, 0.45 \text{ rad}]^{\text{T}}$, $v(0) = [0 \text{ m/s}, 0 \text{ m/s}, 0 \text{ rad/s}]^{\text{T}}$, $k_d = 0.04$

In this case, the performance function for R_e is selected as $b_1(t) = (30 - 0.05) * \exp(-0.1 * t) + 0.05$. The parameters of the quantizer are chosen as: $\delta_1 = 0.01$, $\tau_{u\min} = 0.005$, $\delta_2 = 0.01$, $\tau_{r\min} = 0.005$. The parameters of the position tracking controller (31)–(36) are set as: $k_1 = 0.01$, $k_2 = 20$, $c_1 = 1$, $c_2 = 10$, $c_3 = 1$, $c_4 = 10$, $c_5 = 0.1$, $c_6 = 5$, $\varepsilon_1 = 0.001$. On the other hand, the parameters of the attitude tracking controller (55)–(60) are set as: $k_3 = 5$, $k_4 = 20$, $c_7 = 1$, $c_8 = 5$, $c_9 = 1$, $c_{10} = 5$, $c_{11} = 0.001$, $c_{12} = 5$, $\varepsilon_2 = 0.001$. The time constant of the filter is chosen as $\iota_1 = 0.01$, $\iota_2 = 0.01$.

4.2. Robustness Test under Different Intensity of Disturbance

Trajectory tracking in the $x - y$ plane is depicted in Figure 2. Specifically speaking, the position tracking error and the heading tracking error are plotted in Figure 3, respectively. From these results, it is observed that the virtual USV can follow the desired trajectory with a fast convergence speed and satisfactory accuracy. Figure 3 shows that position tracking with a prescribed performance and high stability can be achieved by the proposed controllers. The output constraint of the position is guaranteed with the help of the tangent-type error transformations, thereby meeting the requirements of the science objectives for the given mission.

Figure 4 gives the velocity tracking errors of the surge and yaw motions, which can converge to a compact set around the origin quickly and maintain a stable state thereafter. The estimations of the adaptive parameters in surge motion and yaw motion are plotted in Figures 5–7. This indicates that all of the adaptive parameters are bounded, which conforms to Theorem 1. Quantized control signals that take values in a finite set are presented in Figure 8. With the consideration of input quantization constraints, it is clear that the control inputs are discrete and keep a constant value within a time interval, meaning that the data transmission will be considerably improved.

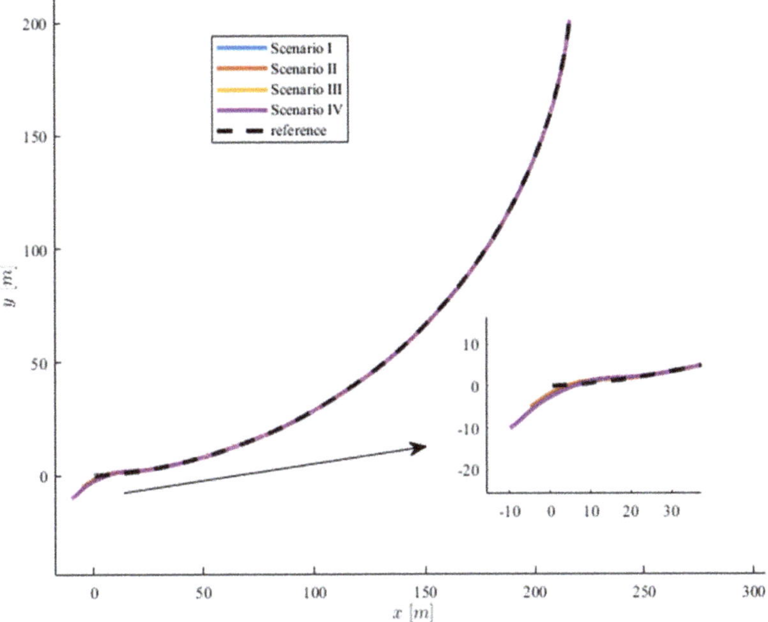

Figure 2. Trajectories of the USV in the 2D plane.

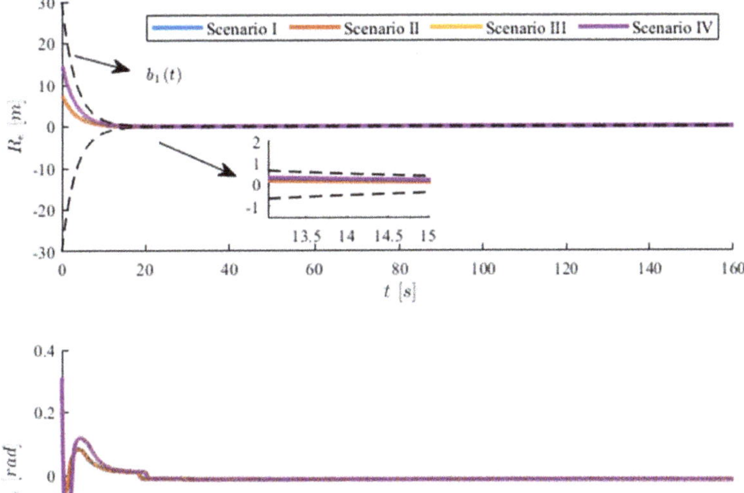

Figure 3. Time responses of position tracking errors and heading tracking errors.

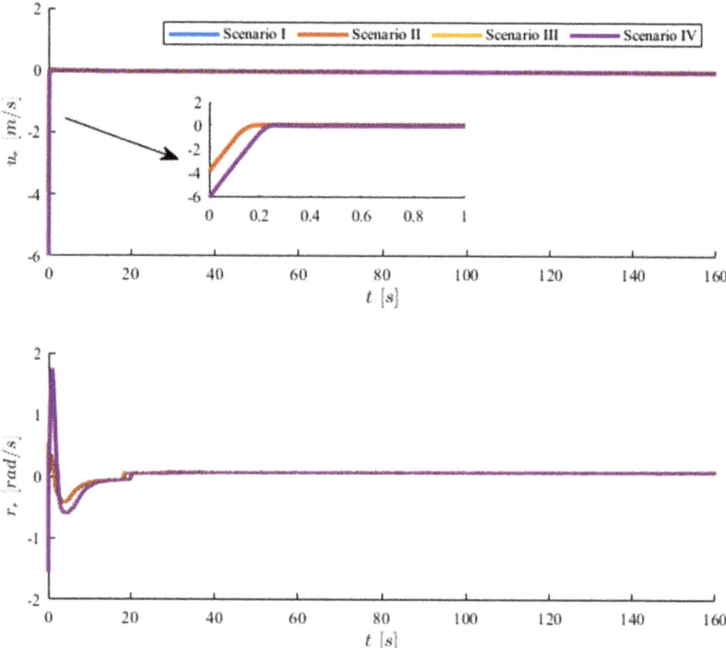

Figure 4. Time responses of velocity tracking errors.

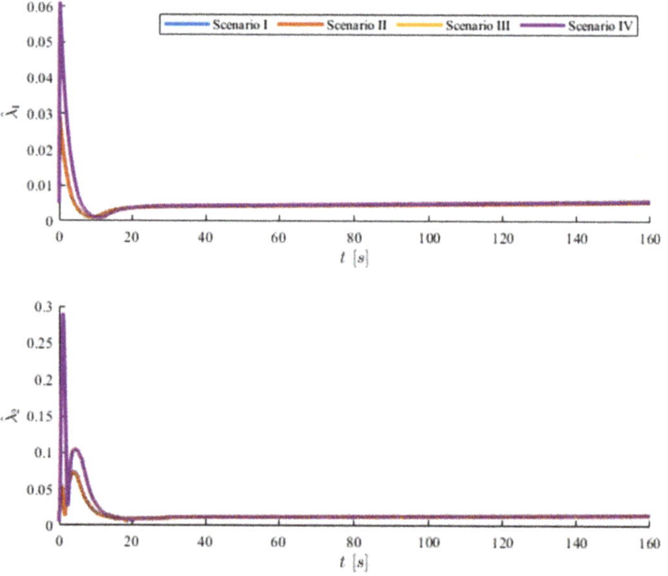

Figure 5. Time responses of adaptive parameter estimations $\hat{\lambda}_1$ and $\hat{\lambda}_2$.

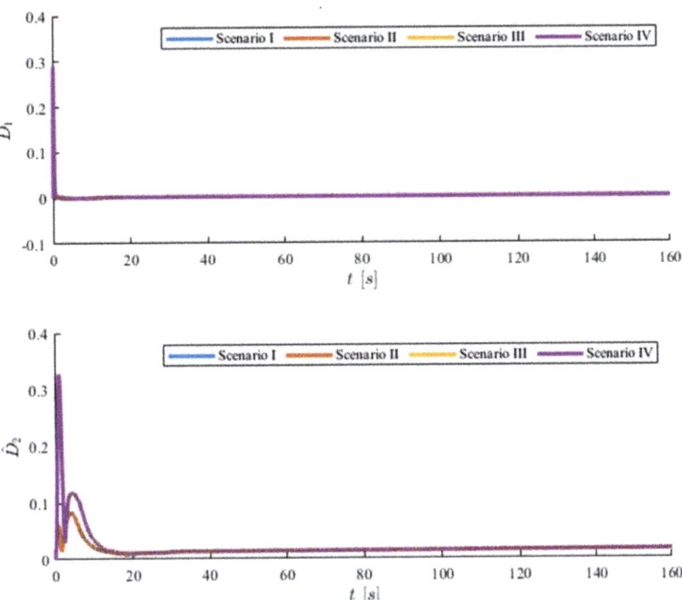

Figure 6. Time responses of adaptive parameter estimations $\hat{D}_1$ and $\hat{D}_2$.

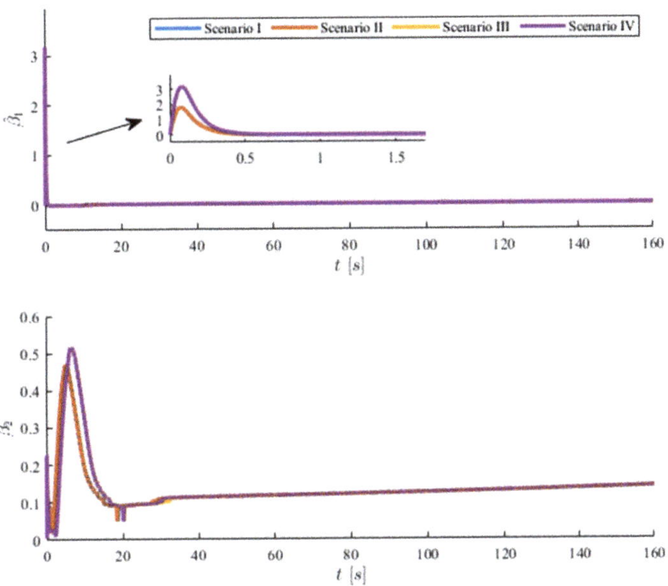

Figure 7. Time responses of adaptive parameter estimations $\hat{\beta}_1$ and $\hat{\beta}_2$.

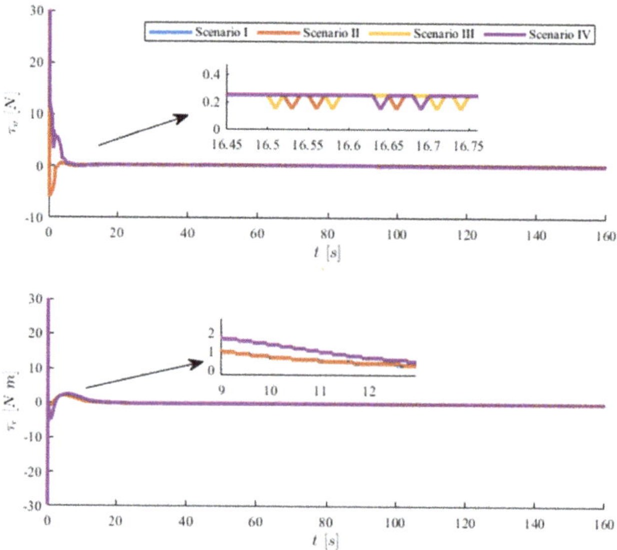

Figure 8. Time responses of quantized control forces.

From these figures, the boundedness of all of the closed loop signals can be easily verified. That is to say that accurate trajectory tracking and attitude tracking are achieved in this simulation, even in the presence of the quantized inter-vessel transmitted state information, model uncertainties, and external disturbances. Furthermore, by comparing all of the results under different scenarios, the proposed scheme enjoys a high robustness and control performance under different initial tracking errors and disturbances. The above-mentioned simulation analysis is in accordance with Theorem 1.

5. Conclusions

In this paper, a quantized prescribed performance control mechanism that considers uncertain model dynamics and quantizer parameters is proposed for the trajectory tracking of USVs. By combining the adaptive algorithm and the MLP-based NN technique, the global boundedness and asymptotical stability of the closed-loop system are guaranteed. Furthermore, the position tracking error has been constrained to the predefined region with the aid of a tangent-type error transformation. Compared with the exiting quantized framework, the significant advantage of the presented controller is that the information of the HLQ does not need to be accurately known by the designer, while the data transmission burden can be effectively alleviated. Numerical simulations have exhibited the effectiveness and advantages of the control strategy developed.

Author Contributions: Conceptualization, Y.Z., K.J. and L.M.; methodology, Y.Z., K.J. and L.M.; software, Y.Z.; validation, Y.Z., K.J. and L.M.; formal analysis, Y.Z. and K.J.; investigation, L.M.; resources, Y.Z.; data curation, L.M.; writing—original draft preparation, K.J.; writing—review and editing, L.M.; visualization, Y.Z.; supervision, Y.S.; project administration, Y.S.; funding acquisition, Y.S. All authors have read and agreed to the published version of the manuscript.

Funding: This research received no external funding.

Institutional Review Board Statement: Not applicable.

Informed Consent Statement: Not applicable.

Data Availability Statement: Not applicable.

Conflicts of Interest: The authors declare that they have no conflict of interest concerning the publication of this manuscript.

References

1. Zheng, Z.; Ruan, L.; Zhu, M. Output-constrained tracking control of an underactuated autonomous underwater vehicle with uncertainties. *Ocean. Eng.* **2019**, *175*, 241–250. [CrossRef]
2. Zheng, Z.; Huang, Y.; Xie, L.; Zhu, B. Adaptive Trajectory Tracking Control of a Fully Actuated Surface Vessel with Asymmetrically Constrained Input and Output. *IEEE Trans. Control. Syst. Technol.* **2018**, *26*, 1851–1859. [CrossRef]
3. Qin, H.; Li, C.; Sun, Y.; Li, X.; Du, Y.; Deng, Z. Finite-time trajectory tracking control of unmanned surface vessel with error constraints and input saturations. *J. Frankl. Inst.* **2019**, *357*, 11472–11495. [CrossRef]
4. Shen, Z.; Bi, Y.; Wang, Y.; Guo, C. MLP Neural Network-based Recursive Sliding Mode Dynamic Surface Control for Trajectory Tracking of Fully Actuated Surface Vessel Subject to Unknown Dynamics and Input Saturation. *Neurocomputing* **2020**, *377*, 103–112. [CrossRef]
5. Fu, M.; Wang, T.; Wang, C. Adaptive Neural-Based Finite-Time Trajectory Tracking Control for Underactuated Marine Surface Vessels with Position Error Constraint. *IEEE Access* **2019**, *7*, 16309–16322. [CrossRef]
6. Guo, Y.; Huang, B.; Li, A.; Wang, C. Integral sliding mode control for Euler-Lagrange systems with input saturation. *Int. J. Robust Nonlinear Control.* **2018**, *29*, 1088–1100. [CrossRef]
7. Shi, Z.; Deng, C.; Zhang, S.; Xie, Y.; Cui, H.; Hao, Y. Hyperbolic Tangent Function-Based Finite-Time Sliding Mode Control for Spacecraft Rendezvous Maneuver without Chattering. *IEEE Access* **2020**, *8*, 60838–60849. [CrossRef]
8. Zhang, M.; Liu, X.; Yin, B.; Liu, W. Adaptive terminal sliding mode based thruster fault tolerant control for underwater vehicle in time-varying ocean currents. *J. Frankl. Inst.* **2015**, *352*, 4935–4961. [CrossRef]
9. Yu, Y.; Guo, C.; Yu, H. Finite-Time PLOS-Based Integral Sliding-Mode Adaptive Neural Path Following for Unmanned Surface Vessels with Unknown Dynamics and Disturbances. *IEEE Trans. Autom. Sci. Eng.* **2019**, *16*, 1500–1511. [CrossRef]
10. Liao, Y.-L.; Wan, L.; Zhuang, J.-Y. Backstepping Dynamical Sliding Mode Control Method for the Path Following of the Underactuated Surface Vessel. *Procedia Eng.* **2011**, *15*, 256–263. [CrossRef]
11. Wen, G.; Ge, S.S.; Chen, C.L.P.; Tu, F.; Wang, S. Adaptive Tracking Control of Surface Vessel Using Optimized Backstepping Technique. *IEEE Trans. Syst. Man Cybern.* **2019**, *49*, 3420–3431. [CrossRef]
12. Xie, T.; Li, Y.; Jiang, Y.; An, L.; Wu, H. Backstepping active disturbance rejection control for trajectory tracking of underactuated autonomous underwater vehicles with position error constraint. *Int. J. Adv. Robot. Syst.* **2020**, *17*, 1729881420909633. [CrossRef]
13. Zhang, T.; Xia, M.; Zhu, J. Adaptive backstepping neural control of state-delayed nonlinear systems with full-state constraints and unmodeled dynamics. *Int. J. Adapt. Control. Signal Process.* **2017**, *31*, 1704–1722. [CrossRef]
14. Zheng, H.; Wu, J.; Wu, W.; Zhang, Y. Robust dynamic positioning of autonomous surface vessels with tube-based model predictive control. *Ocean. Eng.* **2020**, *199*, 106820. [CrossRef]
15. Zhang, J.; Sun, T.; Liu, Z. Robust model predictive control for path-following of underactuated surface vessels with roll constraints. *Ocean. Eng.* **2017**, *143*, 125–132. [CrossRef]
16. Yu, J.; Shi, P.; Zhao, L. Finite-time command filtered backstepping control for a class of nonlinear systems. *Automatica* **2018**, *92*, 173–180. [CrossRef]
17. Lu, Y.; Zhang, G.; Sun, Z.; Zhang, W. Adaptive cooperative formation control of autonomous surface vessels with uncertain dynamics and external disturbances. *Ocean. Eng.* **2018**, *167*, 36–44. [CrossRef]
18. Xing, L.; Wen, C.; Zhu, Y.; Su, H.; Liu, Z. Output feedback control for uncertain nonlinear systems with input quantization. *Automatica* **2016**, *65*, 191–202. [CrossRef]
19. Xing, L.; Wen, C.; Wang, L.; Liu, Z.; Su, H. Adaptive output feedback regulation for a class of nonlinear systems subject to input and output quantization. *J. Frankl. Inst.* **2017**, *354*, 6536–6549. [CrossRef]
20. Zhou, J.; Wen, C.; Wang, W. Adaptive control of uncertain nonlinear systems with quantized input signal. *Automatica* **2018**, *95*, 152–162. [CrossRef]
21. Zhang, C.H.; Yang, G.H. Event-Triggered Adaptive Output Feedback Control for a Class of Uncertain Nonlinear Systems with Actuator Failures. *IEEE Trans. Syst. Man Cybern.* **2020**, *50*, 201–210. [CrossRef] [PubMed]
22. Wang, A.; Liu, L.; Qiu, J.; Feng, G. Event-Triggered Robust Adaptive Fuzzy Control for a Class of Nonlinear Systems. *IEEE Trans. Fuzzy Syst.* **2019**, *27*, 1648–1658. [CrossRef]
23. Li, T.; Wen, C.; Yang, J.; Li, S.; Guo, L. Event-triggered tracking control for nonlinear systems subject to time-varying external disturbances. *Automatica* **2020**, *119*, 109070. [CrossRef]
24. Wu, B.; Xu, C.; Zhang, Y. Decentralized adaptive control for attitude synchronization of multiple spacecraft via quantized information exchange. *Acta Astronaut.* **2020**, *175*, 57–65. [CrossRef]
25. Liu, R.; Cao, X.; Liu, M.; Zhu, Y. 6-DOF fixed-time adaptive tracking control for spacecraft formation flying with input quantization. *Inf. Sci.* **2019**, *475*, 82–99. [CrossRef]
26. Hu, J.; Sun, X.; Liu, S.; He, L. Adaptive finite-time formation tracking control for multiple nonholonomic UAV system with uncertainties and quantized input. *Int. J. Adapt. Control. Signal Process.* **2019**, *33*, 114–129. [CrossRef]
27. Wang, Y.; He, L.; Huang, C. Adaptive time-varying formation tracking control of unmanned aerial vehicles with quantized input. *ISA Trans.* **2019**, *85*, 76–83. [CrossRef] [PubMed]

28. Yan, Y.; Yu, S. Sliding mode tracking control of autonomous underwater vehicles with the effect of quantization. *Ocean. Eng.* **2018**, *151*, 322–328. [CrossRef]
29. Guo, T.; Wu, X. Backstepping control for output-constrained nonlinear systems based on nonlinear mapping. *Neural Comput. Appl.* **2014**, *25*, 1665–1674. [CrossRef]
30. Jia, Z.; Hu, Z.; Zhang, W. Adaptive output-feedback control with prescribed performance for trajectory tracking of underactuated surface vessels. *ISA Trans.* **2019**, *95*, 18–26. [CrossRef] [PubMed]
31. Dai, S.-L.; He, S.; Wang, M.; Yuan, C. Adaptive Neural Control of Underactuated Surface Vessels with Prescribed Performance Guarantees. *IEEE Trans. Neural Netw.* **2019**, *30*, 3686–3698. [CrossRef]
32. Park, B.S.; Yoo, S.J. Robust fault–tolerant tracking with predefined performance for underactuated surface vessels. *Ocean. Eng.* **2016**, *115*, 159–167. [CrossRef]
33. He, W.; Yin, Z.; Sun, C. Adaptive Neural Network Control of a Marine Vessel with Constraints Using the Asymmetric Barrier Lyapunov Function. *IEEE Trans Cybern.* **2017**, *47*, 1641–1651. [CrossRef] [PubMed]
34. Jin, X. Fault tolerant finite-time leader–follower formation control for autonomous surface vessels with LOS range and angle constraints. *Automatica* **2016**, *68*, 228–236. [CrossRef]
35. Shao, X.; Shi, Y. Neural adaptive control for MEMS gyroscope with full-state constraints and quantized input. *IEEE Trans. Ind. Inform.* **2020**, *16*, 6444–6454. [CrossRef]
36. Chen, Z.; Zhang, Y.; Zhang, Y.; Nie, Y.; Tang, J.; Zhu, S. Disturbance-Observer-Based Sliding Mode Control Design for Nonlinear Unmanned Surface Vessel with Uncertainties. *IEEE Access* **2019**, *7*, 148522–148530. [CrossRef]
37. An, H.; Wu, Q.; Wang, G.; Guo, Z.; Wang, C. Simplified longitudinal control of air-breathing hypersonic vehicles with hybrid actuators. *Aerosp. Sci. Technol.* **2020**, *104*, 105936. [CrossRef]

 *symmetry*

 MDPI

Article

Energy Efficiency Enhanced Landing Strategy for Manned eVTOLs Using L_1 Adaptive Control

Zian Wang [1,*,†], Shengchen Mao [1,†], Zheng Gong [1,*], Chi Zhang [2] and Jun He [3]

[1] School of Aerospace Engineering, Nanjing University of Aeronautics and Astronautics, Nanjing 210016, China; maoshengchen@nuaa.edu.cn

[2] COMAC Beijing Aircraft Technology Research Institute, Beijing 100083, China; zhang-c20@mails.tsinghua.edu.cn

[3] School of Aerospace Engineering, Shenyang Aerospace University, Shenyang 110136, China; 13840342646@139.com

* Correspondence: wangzian@nuaa.edu.cn (Z.W.); matthewzhenggong@nuaa.edu.cn (Z.G.)

† These authors contributed equally to this work and should be considered co-first authors.

Abstract: A new landing strategy is presented for manned electric vertical takeoff and landing (eVTOL) vehicles, using a roll maneuver to obtain a trajectory in the horizontal plane. This strategy rejects the altitude surging in the landing process, which is the fatal drawback of the conventional jumping strategy. The strategy leads to a smoother transition from the wing-borne mode to the thrust-borne mode, and has a higher energy efficiency, meaning a better flight experience and higher economic performance. To employ the strategy, a five-stage maneuver is designed, using the lateral maneuver instead of longitudinal climbing. Additionally, a control system based on L_1 adaptive control theory is designed to assist manned driving or execute flight missions independently, consisting of the guidance logic, stability augmentation system and flight management unit. The strategy is verified with the ET120 platform, by Monte Carlo simulation for robustness and safety performance, and an experiment was performed to compare the benefits with conventional landing strategies. The results show that the performance of the control system is robust enough to reduce perturbation by at least 20% in all modeling parameters, and ensures consistent dynamic characteristics between different flight modes. Additionally, the strategy successfully avoids climbing during the landing process with a smooth trajectory, and reduces the energy consumed for landing by 64%.

Keywords: eVTOL; flight dynamics modeling; L_1 adaptive control; guidance; deceleration and landing strategy; energy efficiency; Monte Carlo simulation

Citation: Wang, Z.; Mao, S.; Gong, Z.; Zhang, C.; He, J. Energy Efficiency Enhanced Landing Strategy for Manned eVTOLs Using L_1 Adaptive Control. *Symmetry* **2021**, *13*, 2125. https://doi.org/10.3390/sym13112125

Academic Editors: Chengxi Zhang, Jin Wu and Chong Li

Received: 3 October 2021
Accepted: 21 October 2021
Published: 8 November 2021

Publisher's Note: MDPI stays neutral with regard to jurisdictional claims in published maps and institutional affiliations.

1. Introduction

Electric vertical takeoff and landing (eVTOL) vehicles are an emerging class of aircraft configurations. This configuration highlights the distributed propulsion (DEP) system, which enables vehicles to cruise economically and take off and land vertically by switching between the fixed-wing mode and a multi-rotor mode. According to Ref. [1], DEP gives the vehicle the potential to be more efficient, flexible and reliable than conventional VTOLs, by changing the flight mode from thrust-borne flight to wing-borne flight. This in turn changes the dynamic characteristics of the flight significantly, and poses challenges with respect to the design of strategies and control laws to ensure uniform flight in the entire envelope [2]. The existing studies pay much attention to the takeoff and acceleration phase and cruise trajectory scheduling, but little attention has been paid to deceleration and landing strategies [3].

The existing landing strategies are insufficient for manned eVTOL vehicles [4–6]. The conventional landing strategy, which causes the vehicle to jump steeply, results in a poor comfort level, due to the high degree of normal overload and the rapid change in pressure with altitude, as well as low efficiency, because most of the kinetic energy is not dissipated, but rather converted into gravity potential energy, resulting in considerable time and rotor energy wasted in the subsequent vertical landing phase. The post-stall maneuver strategy is highly efficient for unmanned aerial vehicles (UAVs), because the angle of attack is pulled up into the stall region, meaning there is no excessive lift generated in order to rise, and greater drag coefficients for a produced for the energy to dissipate. However, a fatal defect of this strategy is that it is highly possible for the vehicles to go out of control in the stall region [7], which is unacceptable for manned air service.

The fact that the eVTOL aircraft possesses symmetry in the lateral channels prompted us to design a landing strategy making the most of these lateral symmetry characteristics. The landing procedure is conducted in a three-dimensional (3D) environment, while the aforementioned strategies all plan a two-dimensional (2D) trajectory [8]. Ref. [9] proposes a rolling-horizon landing arrival scheduling method for eVTOLs from the perspective of the management of limited vertiports in peak hours. This prompted us to design a deceleration and landing strategy that can take advantage of the vast 3D airspace around the vertiports, avoiding this abrupt, altitude-increasing maneuver. The symmetrical characteristics in the roll control channels makes the strategy more feasible in real-world application [10].

Usually, a strategy is performed with the assistance of control systems, especially for vehicles with complicated dynamic characteristics [11–15]. The conventional Proportional-Integral-Derivative control systems are widely used in normal fixed-wing aircrafts, but are seldom used for advanced configured flights, due to their limited robustness performance and the vast flight envelope of the advanced vehicles [16]. Artificial neural network systems with model predictive control are able to adapt to the variations in the dynamics of the plants [17,18]. However, the long iteration cycles and computational time are fatal, as change in flight mode is a rapid process. The model-based control theories, namely, the nonlinear dynamic inversion (NDI) and the quantitative feedback theory, are rapid enough and can ensure precise tracking, but they strongly rely on model precision, which is unrealistic for new configurations [19,20]. Aircrafts with vertical takeoff and landing abilities usually require customized control systems that can adapt to changes in control strategies and dynamic characteristics, and can provide sufficient robustness margins and fault tolerance [21–25]. Targeted at the multiple flight modes and the modeling uncertainties of eVTOLs, a control system implemented with L_1 controllers is designed.

The L_1 adaptive control algorithm is sufficiently fast and robust to be equipped on a flight control system for eVTOLs. This algorithm has been applied in various plants successfully, including in aerospace, nuclear technics, marine, etc., contexts [26,27]. The basic principle of the L_1 adaptive theory is the use of a low-pass filter to purify the model error of fast-varying external disturbances, thus decoupling the adaptive performance from the robustness performance [28,29]. Its successful application in multiple flight tests has convinced aerospace engineers of its ability to reject rapidly varying uncertainties, significantly changing the plant dynamics [12,30]. Additionally, controllers based on L_1 theory avoid complicated and time-consuming gain-scheduling, making the control system more customer-friendly [12,31]. With the assistance of the L_1 control system, pilots can perform the strategy at a low level of work load, because the autopilot is able to plan the landing trajectory and make the vehicle track it using the guidance logic, while the stability augmentation system ensures consistent flight performance in all flight modes. The details of the strategy and assistance control system are presented in this paper. The main contents of this work are as follows:

1. Flight dynamics modeling of a large-scale 120 kg electric-vertical takeoff and landing vehicle (ET120), and an analysis of the dynamic characteristics of the system.
2. The design of a four-layer control system based on the L_1 adaptive controller for baseline angular rate control, and a stability analysis of the controller.
3. The design of a roll-horizon deceleration and vertical landing strategy that avoids altitude surging using a smooth transition from fixed-wing mode to multi-rotor mode, along with the maneuver and guidance logic.

4. Monte Carlo simulations for controller performance and strategy performance verification and parameter setting. Comparison simulations with the conventional strategy are performed for validation.

2. Platform Modeling and Dynamics Analysis

2.1. Platform Introduction

The study object of this work was the ET120 platform (as shown in Figure 1), a laterally symmetric eVTOL configuration developed for future urban air mobility. The main body of the platform is a lifting body consistent with normal fixed-wing configurations, consisting of the fuselage, wings and T-tails. The DEP system distinguishes the ET120 from conventional aircraft, where four pairs of vertically mounted rotors provide the hovering power, and one horizontally mounted rotor provides the propulsion power. The geometric parameters of the protype ET120 are given in Table 1.

Table 1. The geometric parameters of ET120.

Parameters	Values
Reference area (m^2)	3.0103
Wing span (m)	5.8
Mean aerodynamic chord (m)	0.6
Mass (kg)	120

The DEP system, together with the aerodynamic surfaces, engages in the flight control in different ways, according to the flight mode. In fixed-wing mode, the attitude control is practiced by the deflection aerodynamic control surfaces. Namely, the aileron deflection δ_a, elevator deflection δ_e and the rudder deflection δ_r. The velocity is controlled by the speed of the propulsion rotor δ_t. In multi-rotor mode, the attitudes are controlled by the speed difference of the hovering rotors, n_ϕ, n_θ and n_φ. The altitude controllable variable is the total speed of the hovering rotors n_h. The flight speed is controlled by the attitude, given the total hovering rotor speed, which makes the vehicle an under-actuated system in multi-rotor mode. The physical control principles are illustrated in Table 2.

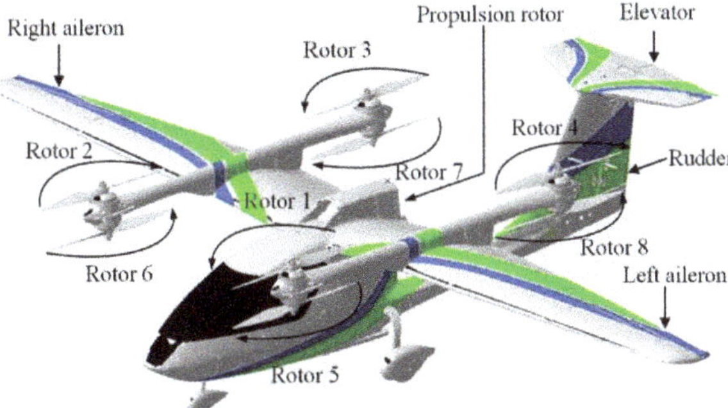

Figure 1. The actuators of the ET120 platform.

Table 2. The control methods of the ET120 in different modes.

Mode	Multi-Rotors	Transitional	Fixed-Wing
Roll	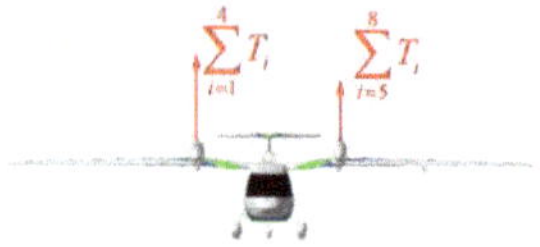 $\sum_{i=1}^{4} T_i \quad \sum_{i=5}^{8} T_i$	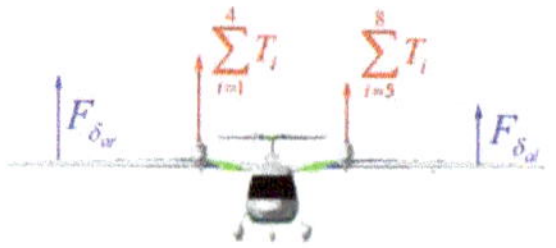 $F_{\delta_a} \quad \sum_{i=1}^{4} T_i \quad \sum_{i=5}^{8} T_i \quad F_{\delta_a}$	 $F_{\delta_a} \quad F_{\delta_a}$
Pitch	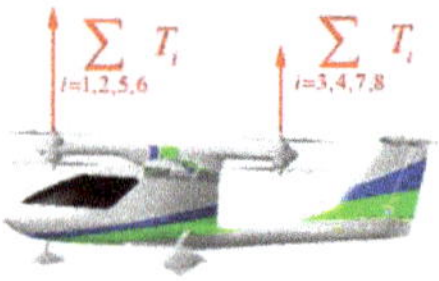 $\sum_{i=1,2,5,6} T_i \quad \sum_{i=3,4,7,8} T_i$	 F_{δ_e}	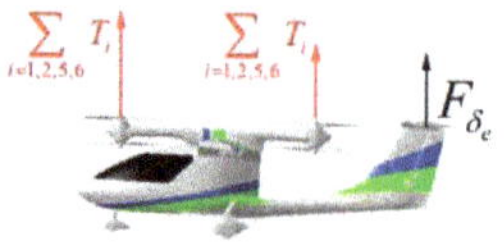 $\sum_{i=1,2,5,6} T_i \quad \sum_{i=1,2,5,6} T_i \quad F_{\delta_e}$
Yaw	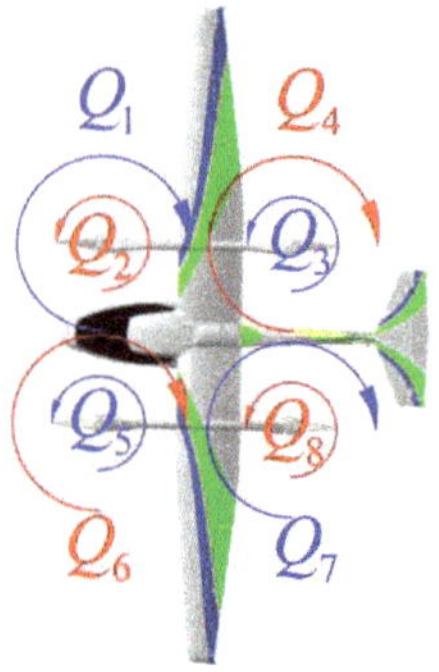 $Q_1 \quad Q_4 \quad Q_2 \quad Q_3 \quad Q_5 \quad Q_8 \quad Q_6 \quad Q_7$	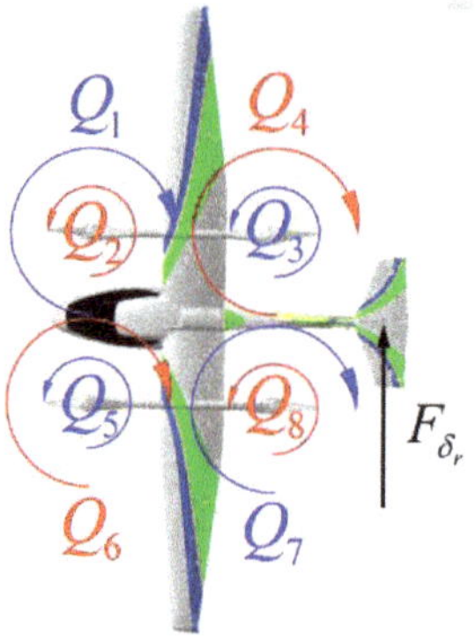 $Q_1 \quad Q_4 \quad Q_2 \quad Q_3 \quad Q_5 \quad Q_8 \quad Q_6 \quad Q_7 \quad F_{\delta_r}$	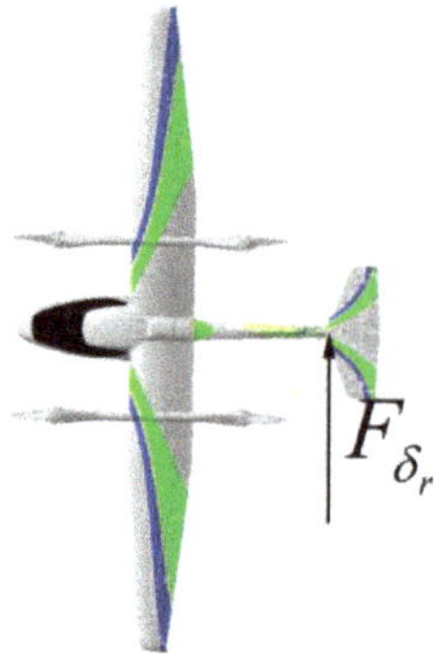 F_{δ_r}
Vertical		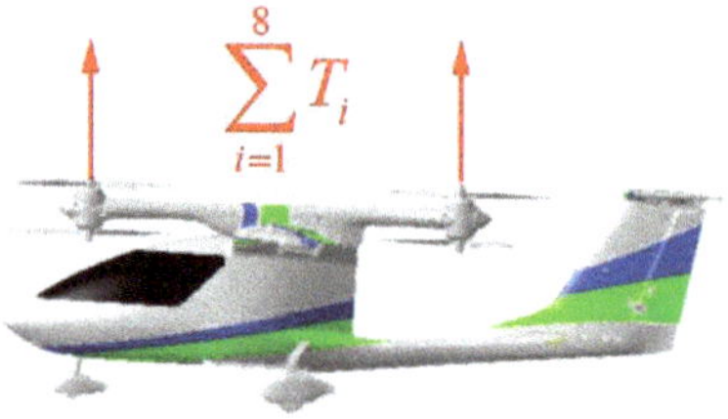 $\sum_{i=1}^{8} T_i$	

2.2. Flight Dynamics Model

The dynamics model of ET120 is established based on the actual aircraft parameters and wind tunnel tests and are separated into 3 main parts:

The mass balance part. This part gives the gravity and the position of its center.

The aerodynamics part. This part concerns the pure aerodynamic forces and moments provided by the fixed-wing part, including the fuselage, the wings, the tails and the aero-surfaces. High-quality aerodynamics data are obtained from wind tunnel tests.

The propulsion system part. This part provides the forces and moments produced by each rotor. The thrust and torque and tilt moments at different inflow velocity and inflow angles are obtained using the blade element momentum theory (BEMT) mentioned in Ref. [7].

The forces and moments given by the above three parts are added up and projected to the body frame, outputting the overall force vector $F = \begin{bmatrix} F_x & F_y & F_z \end{bmatrix}^T$ and moment vector $M = \begin{bmatrix} L & M & N \end{bmatrix}^T$. Then the 6 DOF function of the ET120 can be modeled as:

$$\dot{V} = \frac{F}{m} - \Omega \times V \tag{1}$$

$$\dot{\Omega} = J^{-1}[M - \Omega \times (J \cdot \Omega)] \tag{2}$$

$$\dot{\theta} = E\Omega \tag{3}$$

$$\dot{I} = R_b^e V \tag{4}$$

where $V = [u, v, w]^T$ is the velocity vector in the body frame; m is the mass property; $\theta = [\phi, \theta, \varphi]^T$ are Euler angles, with ϕ, θ and φ being the roll, pitch and yaw, respectively; $\Omega = [p, q, r]^T$ are the angular rate in the body frame, with p, q and r being the roll, pitch, yaw angular rate, respectively; I is the position vector in earth frame, J is the inertia matrix, and E and R_b^e are the transform matrix from the angular vector to the Euler angular vector, and the rotation matrix from the body frame to the inertia frame, respectively.

2.3. Dynamics Anaylsis

The dynamics analysis is based on the trimming point of the ET120, using the linearized model of the flight dynamics model Equations (1)–(4). Due to the discrepancy in control actuators in different flight modes, the flight dynamic characteristics of the ET120 differ significantly. This consequently influences the control methods and control system design. The dynamic characteristics are reflected by the eigenvalue of the system matrix of the linearized model.

- Longitudinal

The longitudinal eigenvalues in different flight modes are depicted in Figure 2. The trimming states of these modes are set at:

1. Fixed-wing mode: airspeed (20 m/s~50 m/s), level flight.
2. Transitional mode: airspeed (2 m/s~20 m/s), level flight.
3. Multi-rotor mode: pitch angle (0~10°), hovering (airspeed zero).

In fixed-wing mode, the longitudinal eigenvalues are all complex numbers distributed on the left side of the coordinate planes, indicating oscillatory convergence with static stability. In transitional and hovering mode, the eigenvalues are two complex values with a negative real part and two complex values with a positive real part. This indicates static instability in longitudinal modes.

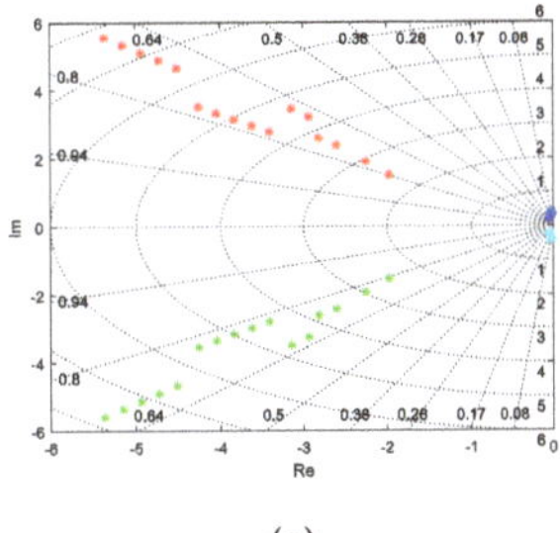

(a)

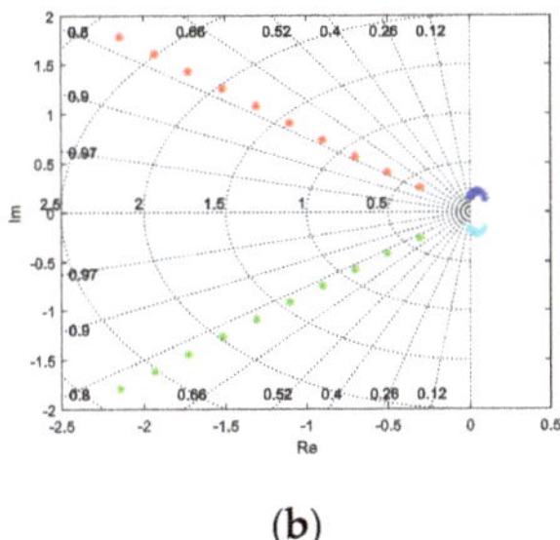

(b)

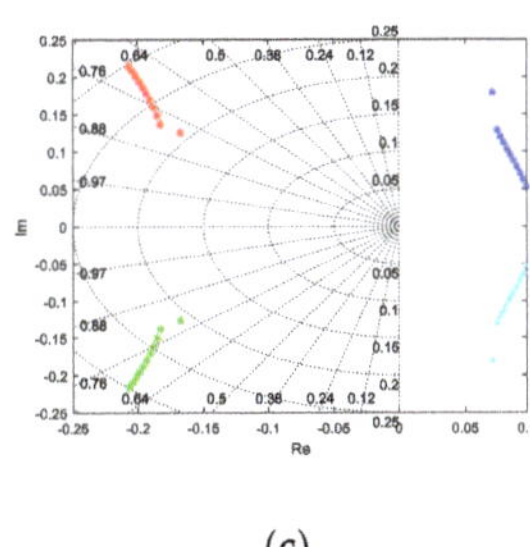

(c)

Figure 2. The longitudinal eigenvalues of the ET120 in different flight modes. (**a**) fixed-wing mode; (**b**) transitional mode; (**c**) multi-rotor mode.

- Lateral

The lateral eigenvalues are depicted in Figure 3. The trimming states of different flight modes are set as in the longitudinal analysis.

In fixed-wing mode and transitional mode, the eigenvalues are two conjugate complex numbers with negative real parts and a negative number and a positive number. This is consistent with normal fixed-wing vehicles, presenting an oscillatory converged Dutch roll mode, a converged roll mode and a slowly diverged spiral mode. However, in multi-rotor mode at low airspeed, the Dutch roll eigenvalues cross the imaginary axis. This indicates an abrupt change in the lateral dynamic characteristics in hovering mode.

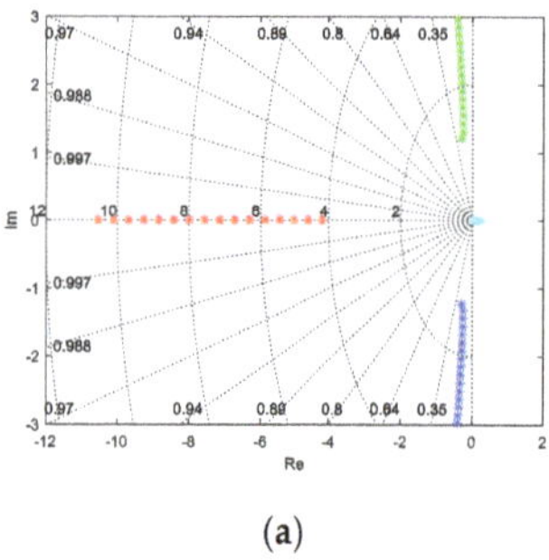

(a)

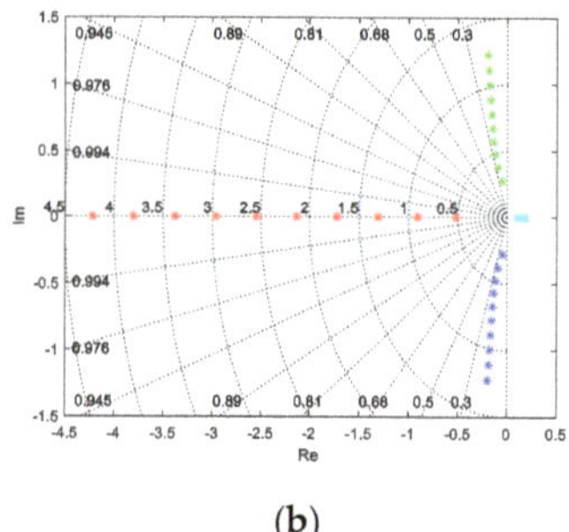

(b)

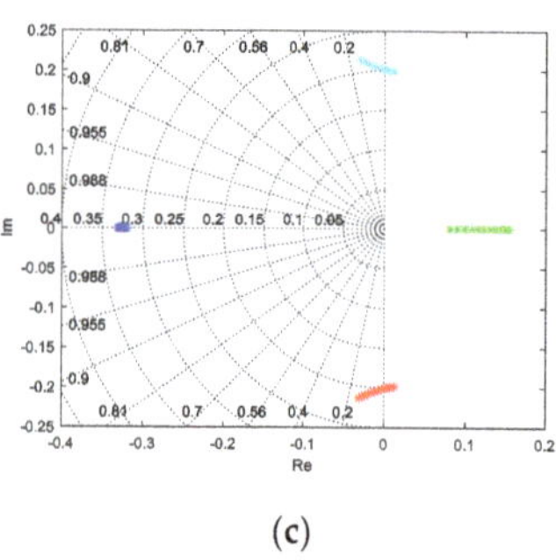

(c)

Figure 3. The lateral eigenvalues of the ET120 in different flight modes. (**a**) fixed-wing mode; (**b**) transitional mode; (**c**) multi-rotor mode.

- Numerical results

The numerical results for the dynamics analysis are given in Table 3. Clearly, the dynamic characteristics varies significantly both in longitudinal and lateral channels between the three flight modes. This feature of the ET120 prompted us to employ the L_1 adaptive control theory to address the model uncertainties and the dynamic characteristic changes during an entire envelope flight.

Table 3. The dynamic characteristics of the ET120.

	Channels		Fixed-Wing	Transition	Multi-Rotor
Longitudinal	Long-term	Frequency	0.39~0.41	diverged	diverged
		Damping ratio	0.042~0.306	diverged	diverged
	Short-term	Frequency	2.4~7.73	0.3982~2.78	0.21~0.298
		Damping ratio	0.672~0.79	0.766~0.768	0.6936~0.8
Lateral	Roll	Time constant	0.095~0.24	0.23~1.92	3.08~3.13
	Spiral	Time constant	$-40{\sim}-130$	$-5.063{\sim}-11.23$	$-6.49{\sim}-11.56$
	Dutch roll	Frequency	1.25~3.0	0.2816~1.23	cross imaginary axis
		Damping ratio	0.133~0.234	0.1619~0.1794	cross imaginary axis

3. Control System Design

As shown in Figure 4, a four-layer flight control system is designed for the ET120, including: a trajectory planner, a flight management unit, a guidance layer, and a control stability augmentation layer. The trajectory planner generates the flight path with navigation information. The flight management unit decides the navigation and control modes of the vehicle, e.g., the waypoint mode for navigation, the fixed-wing mode for flight control. The guidance layer drives the ET120 to follow the path at the desired airspeed, according to the chosen flight mode management unit. The control stability augmentation layer is used to enhance control stability in both rotor and fixed-wing modes. By using the single-input–single-output (SISO) structure, the guidance and control algorithms in the longitudinal, lateral, and directional channels are designed independently. As this work focuses on strategy design, this section only discusses the inner loop control stability augmentation, and the guidance logic is discussed with respect to the maneuver strategy design, while the other layers are omitted.

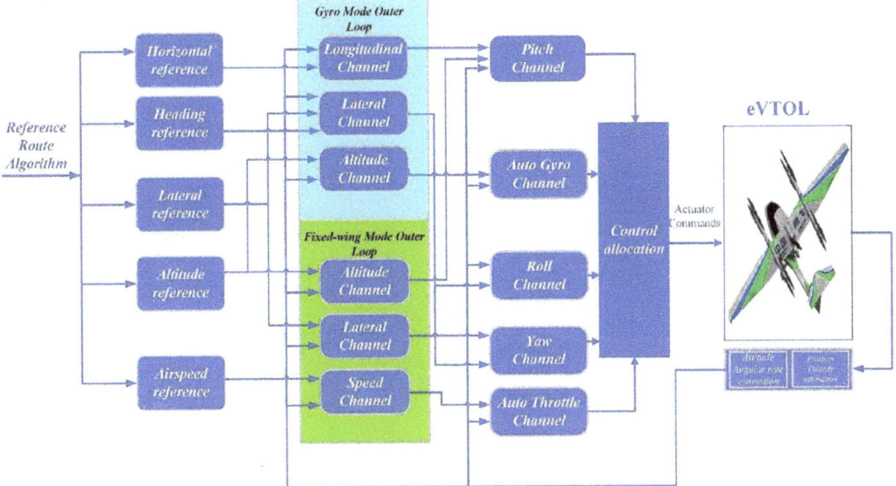

Figure 4. The control architecture of ET120.

3.1. Control Law

A three-axis stability augmentation system (CSAS) is designed to generate virtual angular acceleration commands and moment instructions to the control allocator. The baseline angular rate control, which is the basis of the whole control strategy, is practiced by this system. The CSAS is designed based on L_1 adaptive theory, which stands out in the following respects, compared to the conventional PID controller: (1) The performance of the L_1 controller relies less on model precision. (2) The L_1 controller avoids complicated gain scheduling. (3) It is easier for the L_1 controller to ensure level 1 flight quality, even though the plant is not ascertained. (4) In-time adjustment during flight is accessible to the L_1 controller. For slow loop control, namely, the pitch angle and the bank angle loop, an NDI controller is enough to provide the desired dynamic characteristics. As shown in Figure 5, the pitch and roll cascade channel naturally decouple the rapid angular rate control and slow attitude angle control according to the time-separation principle, where q_c, p_c and r_c are the pitch, roll and yaw rate command, respectively, and θ_c and ϕ_c are the pitch and roll commands, respectively. Owing to the similar control structures of the roll, pitch, and yaw channels, only the pitch channel is discussed here.

3.1.1. The Attitude NDI Controller

Based on the 6DOF function of the ET120, (2), (3), the control plant can be described as:

$$\dot{\theta}(t) = f_1(\theta(t), t) + g_1(\theta(t), t)q(t) \tag{5}$$

where $f_1(\theta(t), t)$, $g_1(\theta(t), t)$ are affine functions with $f_1(\theta(0), 0) = 0$ and $g_1(\theta(0), 0) = 1$.

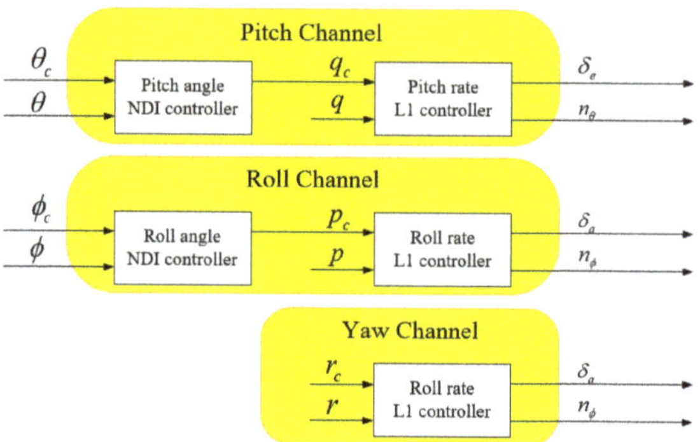

Figure 5. Three-axis decoupled control channels.

The tracking error $\Delta\theta(t)$ is the difference between the pitch angle command $\theta_c(t)$ and the actual pitch angle response $\theta(t)$, that is

$$\Delta\theta(t) = \theta_c(t) - \theta(t) \tag{6}$$

The NDI controller is designed as:

$$q_c(t) = g_1^{-1}(K_\theta \Delta\theta(t) - f_1(\theta(t), t)) \tag{7}$$

where the K_θ is the gain of the pitch angle.

3.1.2. The Angular Rate L_1 Controller

The L_1 controller is composed of four parts: the control plant, the control law, the state predictor and the adaptive law, which will be explained in this section.

- Control plant

The pitching rate function is derived from the 6DOF function (2); with the perturbation assumption, the system can be constructed by first-order coefficients M_α, M_q and $M_{\dot{q}_{vc}}$:

$$\dot{q}(t) = M_\alpha \alpha(t) + M_q q(t) + M_{\dot{q}_{vc}} \dot{q}_{vc}(t) \tag{8}$$

where $\dot{q}_{vc}(t)$ is the total virtual pitch angular acceleration. By moving the $\dot{q}_{vc}(t)$ term to the left and all the other terms to the right, we get:

$$\dot{q}_{vc}(t) = \frac{1}{M_{\dot{q}_{vc}}} [\dot{q}_c(t) - M_\alpha \alpha(t) - M_q q(t)] \tag{9}$$

Physically, the above function means the pitch acceleration is provided by the moments, and the right terms are generated by the aero-surfaces and the rotor system. If we define $\dot{q}_{ac}(t)$ to represent the angular acceleration provided by the aero-surfaces and $\dot{q}_{rc}(t)$ the rotor system, Equation (9) can be rewritten as:

$$\dot{q}_{vc}(t) = \dot{q}_{rc}(t) + \dot{q}_{ac}(t) \tag{10}$$

In hovering mode, the efficiency of the aero-surfaces is limited due to the low airspeed, and the majority of the control energy is generated by the hovering rotor system. It is reasonable that $\dot{q}_{vc}(t) \triangleq \dot{q}_{rc}(t)$, $\dot{q}_{ac}(t) \triangleq 0$. In fixed-wing mode, the hovering rotor is inactivated, and the situation is reversed, that is, $\dot{q}_{rc}(t) \triangleq 0$, $\dot{q}_{vc}(t) \triangleq \dot{q}_{ac}(t)$. During transition flight, both the aero-surfaces and hovering rotors achieve the commanded pitch angular acceleration.

Considering the modeling uncertainty, Equation (8) is rewritten as:

$$\dot{q}(t) = (M_\alpha + \hat{M}_\alpha)\alpha(t) + (M_q + \hat{M}_q)q(t) + (M_{\dot{q}_{vc}} + \hat{M}_{\dot{q}_{vc}})\dot{q}_{vc}(t) + \sigma_1 \tag{11}$$

where $\hat{M}_\alpha$, $\hat{M}_q$ and $\hat{M}_{\dot{q}_{vc}}$ are the coefficient uncertainties, and σ_1 is the disturbance factor. Substituting Equation (9) into Equation (10), we get:

$$\dot{q}(t) = \hat{M}_\alpha\alpha(t) + \hat{M}_q q(t) + \frac{\hat{M}_{\dot{q}_{vc}}}{M_{\dot{q}_{vc}}}(\dot{q}_c(t) - M_\alpha\alpha(t) - M_q q(t)) + \dot{q}_c(t) + \sigma_1 \tag{12}$$

With Equation (12), the first-order reference model can be constructed as follows:

$$\dot{q}(t) = -K_q q(t) + K_q q_c(t) \tag{13}$$

By combining Equations (12) and (13), we can rewrite the control plant of the pitch channel as:

$$\begin{cases} \dot{q}(t) = -K_q q(t) + K_q \eta(t) \\ \eta(t) = \omega_q \dot{q}_c(t) + f_2(t, q(t)) \\ f_2(t, q(t)) = \theta_q q(t) + \sigma_q \end{cases} \tag{14}$$

where $\omega_q = 1 + \frac{\hat{M}_{\dot{q}_{uc}}}{M_{\dot{q}_{uc}}}$ is the virtual control coefficient, $\theta_q = K_q - \frac{\hat{M}_{\dot{q}_{uc}}}{M_{\dot{q}_{uc}}}M_q$ is aerodynamic coefficient, and $\sigma_q = -\frac{\hat{M}_{\dot{q}_{uc}}}{M_{\dot{q}_{uc}}}M_\alpha\alpha + \sigma_1$ is the aerodynamic disturbance.

- The L_1 controller

The L_1 controller is designed based on the following assumptions, which can be measured in practical application.

Assumption 1. *The plant unknown coefficient ω_q is uniformly bounded in $[\omega_{ql}, \omega_{qu}]$, where ω_{ql} and ω_{qu} are the lower and upper bounds of ω_q.*

Assumption 2. *$f_2(t,0)$ in Equation (14) is uniformly bounded, that is $\|f_2(t,0)\|_\infty \leq b$, with $b > 0$, where $\|\bullet\|_\infty$ is the ∞-norm.*

Assumption 3. *The partial derivative of f_2 is semi-globally uniformly bounded: for each $\delta > 0$, there exists $d_{fq}(\delta) > 0$ and $d_{ft}(\delta) > 0$ that ensures the partial derivative of $f_2(t, q(t))$ is piecewise continuous regardless of time, which is written as:*

$$\begin{cases} \|\frac{\partial f_2(t,q(t))}{\partial q}\|_\infty \leq d_{fq}(\delta), \\ \|\frac{\partial f_2(t,q(t))}{\partial t}\|_\infty \leq d_{ft}(\delta). \end{cases} \tag{15}$$

For the inner loop control system, the uncertainty possesses a certain magnitude and limit, and these can be realized using engineering measures. These assumptions are easily satisfied in practical applications. Given these assumptions, the L_1 controller, which includes a state predictor, the adaptive law and the control law, can be designed with lemma 1:

Lemma 1. *For each $\tau \geq 0$, if $\|q_\tau\|_{L_\infty} \leq \rho$ and $\|\dot{q}_\tau\|_{L_\infty} \leq d$, where ρ and d are positive constants, and $\theta_q(t)$ and $\sigma_q(t)$ are continuous [29]. In addition, their derivatives for $t \in [0, \tau]$ are*

$$f(t, q(t)) = \theta_q(t)\|q_t\|_{L_\infty} + \sigma_q(t) \tag{16}$$

$$|\theta_q(t)| < d_{fq}(\rho), |\dot{\theta}_q(t)| \leq d_\theta \tag{17}$$

$$|\sigma_q(t)| < b, |\dot{\sigma}_q(t)| \leq d_\sigma \tag{18}$$

where d_θ and d_σ are calculable limits; $\|\bullet\|_{L_\infty}$ is the L_∞-norm.

- State predictor

The state predictor is a system that reflects the control plants, and which has similar dynamic characteristics to the control plant. According to Equation (14), the state predictor can be designed as:

$$\begin{cases} \dot{\hat{q}}(t) = -K_q\hat{q}(t) + K_q\hat{\eta}(t) \\ \hat{\eta}(t) = \hat{\omega}_q(t)\dot{q}_c(t) + \hat{\theta}_q(t)q(t) + \hat{\sigma}_q(t) \\ \hat{y}(t) = \hat{q}(t) \end{cases} \tag{19}$$

where $\hat{\omega}_q(t)$ is the estimated uncertainty of the control factor, $\hat{\theta}_q(t)$ is the estimated uncertainty of the aerodynamic factor, and $\hat{\sigma}_q(t)$ is the estimated uncertainty of aerodynamic disturbance.

- Adaptive law

The adaptive gains are produced by:

$$\begin{cases} \dot{\hat{\theta}}_q(t) = \Gamma K_{proj}(\hat{\theta}_q(t), -\tilde{q}(t)PK_q\|q(t)\|_\infty) \\ \dot{\hat{\sigma}}_q(t) = \Gamma K_{proj}(\hat{\sigma}_q(t), -\tilde{q}(t)PK_q) \\ \dot{\hat{\omega}}_q(t) = \Gamma K_{proj}(\hat{\omega}_q(t), -\tilde{q}(t)PK_q\dot{q}_c(t)) \end{cases} \tag{20}$$

where Γ is adaptive gain, $\tilde{q}(t) = \hat{q}(t) - q(t)$ is tracking error, P is the solution of the Lyapunov equation $-K_q^T P - PK_q = -Q, Q > 0$. K_{proj} is the projection operator that can guarantee the boundedness of the adaptive parameters.

- Control law

The control law is given as:

$$\begin{cases} \dot{q}_c = K_d D[K_g q(t) - \hat{\omega}_q(t)\dot{q}_c(t) - \hat{\theta}_q(t)q(t) - \hat{\sigma}_q(t)] \\ \dot{q}_{vc} = \frac{1}{M_{\dot{q}_{vc}}}[\dot{q}_c - M_\alpha\alpha(t) - M_q q(t)] \end{cases} \tag{21}$$

where K_g is adaptive feedback gain, D is a low-pass filter and K_d is the adaptive feed forward gain, as depicted in Figure 6. The filter is expected to have dynamic characteristics that satisfy the following transfer function:

$$C(s) = \omega_q K_d D(s)(I + \omega_q K_d D(s))^{-1}, C(0) = I \tag{22}$$

where I is the identity matrix.

Additionally, the values of K_d and D should ensure that for a given ρ_0, there exists $\rho_r > \rho_{in}$ to maintain the L_1 norm condition:

$$\|G(s)\|_{L_1} < \frac{\rho_r - \|H(s)C(s)K_g\|_{L_1}\|q_c\|_{L_\infty} - \rho_{in}}{L_{\rho_r}\rho_r + b} \tag{23}$$

where $\|\bullet\|_{L_1}$ is the L_1-norm, and

$$\begin{cases} \rho_{in} := \|s(sI + K_q)^{-1}\|_{L_1}\rho_0 \\ H(s) = (sI + K_q)^{-1}K_q \\ G(s) = H(s)[I - C(s)] \\ L_{\rho_r} = \frac{\rho_r + \bar{\gamma}_1}{\rho_r}d_{fq}[\rho_r + \bar{\gamma}_1] \end{cases} \tag{24}$$

where $\bar{\gamma}_1$ is an arbitrary positive constant and d_{fq} is defined in Equation (15).

The L_1 adaptive controller is constructed on the basis of Equations (19) and (21), with the L_1-norm condition satisfied.

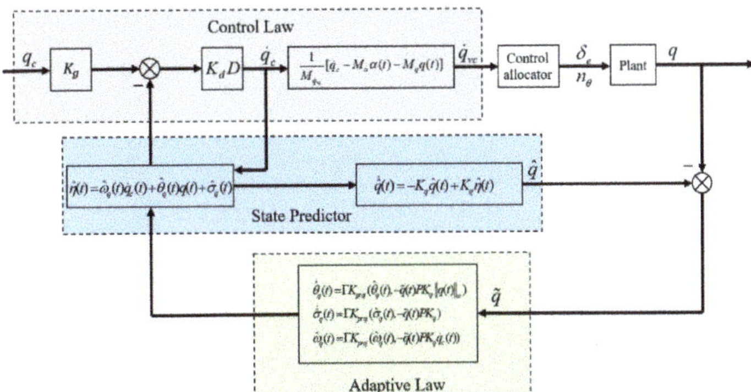

Figure 6. The L_1 adaptive controller architecture.

- Stability analysis

The stability analysis was performed using Lyapunov's second method, as described in Ref. [21].

Firstly, the tracking error the difference between the response of the state predictor (Equation (19)) and the control plant (Equation (14)), that is:

$$\begin{cases} \tilde{q}(t) = \hat{q}(t) - q(t) \\ \dot{\tilde{q}}(t) = -K_q \tilde{q}(t) + K_q \tilde{\eta}(t) \end{cases} \tag{25}$$

where $\tilde{*}, * \in \{q, \dot{q}, \eta\}$ represents the tracking errors, $\tilde{\eta}(t) = \hat{\eta}(t) - \eta(t)$. Given Equation (14), we have:

$$\tilde{\eta}(t) = \hat{\eta}(t) - \eta(t) = \dot{q}_c(t)\tilde{\omega}_q(t) + q(t)\tilde{\theta}_q(t) + \tilde{\sigma}_q(t) \tag{26}$$

with $\tilde{\omega}_q(t) = \hat{\omega}_q(t) - \omega_q(t)$, $\tilde{\theta}_q(t) = \hat{\theta}_q(t) - \theta_q(t)$ and $\tilde{\sigma}_q(t) = \hat{\sigma}_q(t) - \sigma_q(t)$. The angular acceleration error is derived by substituting Equation (26) into Equation (25), that is:

$$\dot{\tilde{q}}(t) = -K_q \tilde{q}(t) + K_q [\dot{q}_c(t)\tilde{\omega}_q(t) + q(t)\tilde{\theta}_q(t) + \tilde{\sigma}_q(t)] \tag{27}$$

The goal of adaptive laws is to drive the $\tilde{\omega}(t)$, $\tilde{\theta}(t)$ and $\tilde{\sigma}(t)$ tend to zero to achieve stable error dynamics $\dot{\tilde{q}}(t) = -K_q \tilde{q}(t)$.

The candidate Lyapunov function is formulated as:

$$V(\tilde{q}, \tilde{\omega}_q, \tilde{\theta}_q, \tilde{\sigma}_q) = \tilde{q}^T(t)P\tilde{q}(t) + \frac{1}{\Gamma}(\tilde{\omega}_q{}^2(t) + \tilde{\theta}_q{}^2(t) + \tilde{\sigma}_q{}^2(t)) \tag{28}$$

The time derivatives of Equation (28) is:

$$\dot{V}(\tilde{q}, \tilde{\omega}_q, \tilde{\theta}_q, \tilde{\sigma}_q) = \dot{\tilde{q}}^T(t)P\tilde{q}(t) + \tilde{q}^T(t)P\dot{\tilde{q}}(t) + \frac{2}{\Gamma}(\tilde{\omega}_q(t)\dot{\tilde{\omega}}_q(t) + \tilde{\theta}_q(t)\dot{\tilde{\theta}}_q(t) + \tilde{\sigma}_q(t)\dot{\tilde{\sigma}}_q(t)) \tag{29}$$

It is assumed that ω_q, θ_q and σ_q vary slowly enough to satisfy $\dot{\tilde{\omega}}_q \approx \dot{\hat{\omega}}_q$, $\dot{\tilde{\theta}}_q \approx \dot{\hat{\theta}}_q$, $\dot{\tilde{\sigma}}_q \approx \dot{\hat{\sigma}}_q$. Substituting Equation (27) into Equation (29), we get

$$\begin{aligned} \dot{V}(\tilde{q}, \tilde{\omega}_q, \tilde{\theta}_q, \tilde{\sigma}_q) = \ & \left\{ -\tilde{q}^T(t)K_q^T + [\tilde{\omega}_q{}^T(t)\dot{q}_c{}^T(t) + \tilde{\theta}_q{}^T(t)q^T(t) + \tilde{\sigma}_q{}^T(t)]K_q^T \right\}P\tilde{q}(t) \\ & + \tilde{q}^T(t)P\left\{ -K_q\tilde{q}(t) + K_q[\dot{q}_c(t)\tilde{\omega}_q(t) + q(t)\tilde{\theta}_q(t) + \tilde{\sigma}_q(t)] \right\} \\ & + \frac{2}{\Gamma}(\tilde{\omega}_q{}^T(t)\dot{\hat{\omega}}_q(t) + \tilde{\theta}_q{}^T(t)\dot{\hat{\theta}}_q(t) + \tilde{\sigma}_q{}^T(t)\dot{\hat{\sigma}}_q(t)) \end{aligned} \tag{30}$$

With adaptive law defined in Equations (20) and (24), Equation (30) is re-constructed as

$$\begin{aligned}
\dot{V}(\tilde{q},\tilde{\omega}_q,\tilde{\theta}_q,\tilde{\sigma}_q) = \ & -\tilde{q}^T(t)Q\tilde{q}(t) + 2\tilde{\omega}_q(t)(\tilde{q}^T(t)PK_q\dot{q}_c{}^T(t) \\
& +K_{proj}(\hat{\omega}_q(t), -\tilde{q}^T(t)PK_q\dot{q}_c(t))) \\
& +2\hat{\theta}_q{}^T(t)(q(t)\tilde{q}^T(t)PK_q + K_{proj}(\hat{\theta}_q(t), -q(t) - \tilde{q}^T(t)PK_q)) \\
& +2\tilde{\sigma}_q{}^T(t)(\tilde{q}^T(t)PK_q + K_{proj}(\hat{\sigma}_q(t), -\tilde{q}^T(t)PK_q)) \\
& -\tfrac{2}{\Gamma}(\tilde{\omega}_q{}^T(t)\dot{\omega}_q(t) + \tilde{\theta}_q{}^T(t)\dot{\theta}_q(t) + \tilde{\sigma}_q{}^T(t)\dot{\sigma}_q(t))
\end{aligned} \tag{31}$$

The projection operator in the adaptive laws ensures that the adaptive parameters are limited to a known compact set Λ. The projection operator is written as $\dot{\theta}_q(t) = K_{proj}(\theta_q, \Gamma z)$, and the properties of the projection function guarantee that for any point $\theta_q(\tau_1) \in \Lambda$, where $\tau_1 \in [0, t)$ and z is a parameter. Then, we have:

$$(\theta_q - \theta_q(\tau_1))^T(\Gamma^{-1}K_{proj}(\theta_q, \Gamma z) - z) \leq 0 \tag{32}$$

With (32), Equation (31) can be simplified to an inequation:

$$\dot{V}(\tilde{q},\tilde{\omega}_q,\tilde{\theta}_q,\tilde{\sigma}_q) \leq -\tilde{q}^T(t)Q\tilde{q}(t) + \frac{2}{\Gamma}\left(\left|\tilde{\omega}_q{}^T(t)\dot{\omega}_q(t) + \tilde{\theta}_q{}^T(t)\dot{\theta}_q(t) + \tilde{\sigma}_q{}^T(t)\dot{\sigma}_q(t)\right|\right) \tag{33}$$

As ω_q is a constant, $\dot{\omega}_q = 0$. Then, (33) can be expressed as:

$$\dot{V}(\tilde{q},\tilde{\omega}_q,\tilde{\theta}_q,\tilde{\sigma}_q) \leq -\tilde{q}^T(t)Q\tilde{q}(t) + \frac{2}{\Gamma}\left(\left|\tilde{\theta}_q{}^T(t)\dot{\theta}_q(t) + \tilde{\sigma}_q{}^T(t)\dot{\sigma}_q(t)\right|\right) \tag{34}$$

According to the bounds Equations (17) and (18) defined in Lemma 1, function (34) is simplified as:

$$\dot{V}(\tilde{q},\tilde{\omega}_q,\tilde{\theta}_q,\tilde{\sigma}_q) \leq -\tilde{q}^T(t)Q\tilde{q}(t) + \frac{4}{\Gamma}(d_{fq}(\rho)d_\theta + bd_\sigma) \tag{35}$$

Using the properties of projection operator again, Equation (28) is reduced to an inequation as:

$$\tilde{\omega}_q{}^2(t) + \tilde{\theta}_q{}^2(t) + \tilde{\sigma}_q{}^2(t) \leq (\omega_{qu} - \omega_{ql})^2 + 4d_{fq}{}^2(\rho) + 4b^2 \tag{36}$$

The condition $\tilde{q}(0) = 0$ derives

$$V(0) \leq \frac{1}{\Gamma}((\omega_{qu} - \omega_{ql})^2 + 4d_{fq}(\rho)d_\theta + 4bd_\sigma) \tag{37}$$

Assume that:

$$V(t) > \frac{\lambda_m(\rho_r)}{\Gamma} \tag{38}$$

where $\lambda_m(\rho_r) \triangleq (\omega_{qu} - \omega_{ql})^2 + 4d_{fq}{}^2(\rho) + 4b^2 + 4\frac{\lambda_{\max}(P)}{\lambda_{\min}(Q)}(d_{fq}(\rho)d_\theta + bd_\sigma)$, $\lambda_{\max}(P)$ is the max eigenvalue of matrix P, and $\lambda_{\min}(P)$ is the min eigenvalue of matrix Q.

Substitute (36) and (38) into Equation (28) and you get:

$$\tilde{q}^T(t)Q\tilde{q}(t) \geq \frac{\lambda_{\max}(Q)}{\lambda_{\min}(P)}\tilde{q}^T(t)P\tilde{q}(t) \geq \frac{4}{\Gamma}(d_{fq}(\rho)d_\theta + bd_\sigma) \tag{39}$$

Using (39) and (35) yields $\dot{V}(\tilde{q},\tilde{\omega}_q,\tilde{\theta}_q,\tilde{\sigma}_q) < 0$
Therefore, we have:

$$V(t) \leq V(0) \leq \frac{1}{\Gamma}((\omega_{qu} - \omega_{ql})^2 + 4d_{fq}(\rho)d_\theta + 4bd_\sigma) \leq \frac{\lambda_m(\rho_r)}{\Gamma} \tag{40}$$

Since the result of Equation (40) contradicts the assumption of Equation (32), the actual assumption of Equation (38) should be rewritten as:

$$V(t) \leq \frac{\lambda_m(\rho_r)}{\Gamma} \tag{41}$$

The Lyapunov second method indicates that the system satisfying (41) is stable.

3.2. Control Allocation

A control allocator is designed to determine the command of the actuators based on the angular rate command generated by the CSAS. For the abundant control actuators in all channels, the allocation criterion uses the efficiency of the actuators, which is related to the airspeed, to determine the actuator commands.

Since the body frame of the ET120 is almost symmetrical, the products of inertia J_{xy}, J_{yz} and J_{zx} can be ignored, and the virtual commands to the actuators in the roll, pitch and yaw channels can be given according to the outputs of the controllers in corresponding channels $\dot{p}_c$, $\dot{q}_c$ and $\dot{r}_c$:

$$\begin{cases} \delta_a = \frac{\dot{p}_c}{L_{\max}}\delta_a^{\min} \\ \delta_e = \frac{\dot{q}_c}{M_{\max}}\delta_e^{\min} \\ \delta_r = \frac{\dot{r}_c}{N_{\max}}\delta_r^{\max} \end{cases} , \quad \begin{cases} n_\phi = \frac{\dot{p}_c}{L_{\max}}n_\phi^{\max} \\ n_\theta = \frac{\dot{q}_c}{M_{\max}}n_\theta^{\max} \\ n_\varphi = \frac{\dot{r}_c}{N_{\max}}n_\phi^{\max} \end{cases} \tag{42}$$

where the superscript max and min represent the maximum or minimum outputs of the corresponding controllable variables.$L_{\max}$, $M_{\max}$ and $N_{\max}$ represent the maximum control effectiveness in the roll, pitch, and yaw channels, respectively, and are given by the following equations:

$$\begin{cases} L_{\max} = \eta_{\delta_a}\overline{L}_{\delta_a}\delta_a^{\min} + \overline{L}_{n_\phi}n_\phi^{\max} \\ M_{\max} = \eta_{\delta_e}\overline{M}_{\delta_e}\delta_e^{\min} + \overline{M}_{n_\theta}n_\theta^{\max} \\ N_{\max} = \overline{N}_{\delta_r}\delta_r^{\max} + \overline{N}_{n_\varphi}n_\varphi^{\max} \end{cases} \tag{43}$$

where $\overline{L}_{\delta_a}$, $\overline{M}_{\delta_e}$, $\overline{N}_{\delta_r}$, are the moments provided by each unit deflection of the corresponding control surface in roll, pitch, and yaw channels. $\overline{L}_{n_\phi}$, $\overline{M}_{n_\theta}$, $\overline{N}_{n_\varphi}$ are the moments produced by each unit's virtual control inputs for the rotors in the roll, pitch, and yaw channels. η_{δ_a} and η_{δ_e} are gain-scheduled coefficients that are relevant to the airspeed V_t.

4. Deceleration Transition Process Maneuver Design

4.1. Process Analysis and Maneuver Design

The currently used deceleration and landing process for the ET120 vehicle is depicted in Figure 7. The process begins from level flight in fixed-wing cruising mode. When the autonomous landing logic of the autopilot is activated, the ET120 vehicle turns off the propulsion rotor and performs the pitch-up maneuver. In the early stage of deceleration, a large pitch-up input increases the angle of attack at the cruising speed (usually at a high airspeed), which produces a large aerodynamic lift (more than the aircraft weight), resulting in climbing behavior. This is done for two reasons. One is to transform the kinetic energy into potential energy by increasing altitude. Another is to cut off the forward thrust force to reduce the kinetic energy input, so the decrease in airspeed is accompanied by an increase in altitude. When the airspeed is reduced to near the hover decision speed, the ET120 vehicle follows the multi-rotor hover mode in preparation for vertical descent and final landing. This strategy has the following shortcomings: 1. The deceleration corridor is a straight and long line, which imposes restrictions on high-density traffic in terminal airspace. 2. The increase in altitude goes against the common sense of decreasing altitude during the landing process. 3. The increase in altitude also consumes more battery energy and increases the workload of the hovering rotors, resulting in lower flight performance.

4. It is also a very poor driving experience for passengers. This ragged deceleration strategy is obviously not suitable for urban transportation.

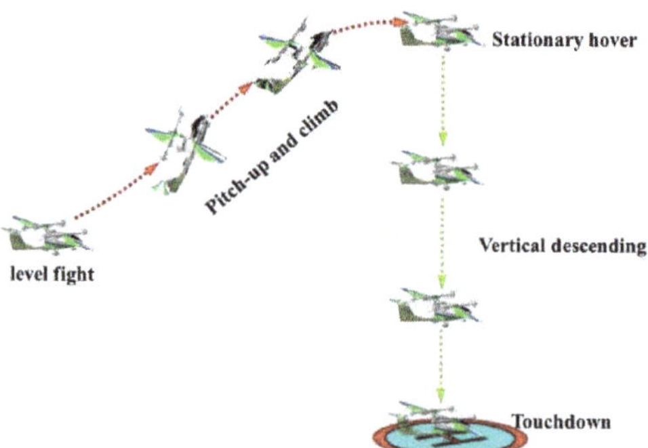

Figure 7. Conventional jumping deceleration and landing process.

A major improvement would be to avoid the climbing behavior caused by the large lift. One idea is to perform a fast pull-up maneuver like a fighter jet to cause the ET120 vehicle to enter post-stall flight, as shown in Figure 8. Instead of the pitch-up motion, the large normal overload input quickly produces a large AOA, and thus a large drag that leads to a continuous drop in lift and airspeed. When the AOA exceeds the stall AOA, the ET120 vehicle falls into a stall flight maneuver and re-engages to hover stationary in preparation for vertical descent to the landing pad (vertiport). During this deceleration process, the altitude can be slightly perturbed. However, this extreme deceleration approach comes at the expense of safety and reliability, and is not suitable for urban transportation.

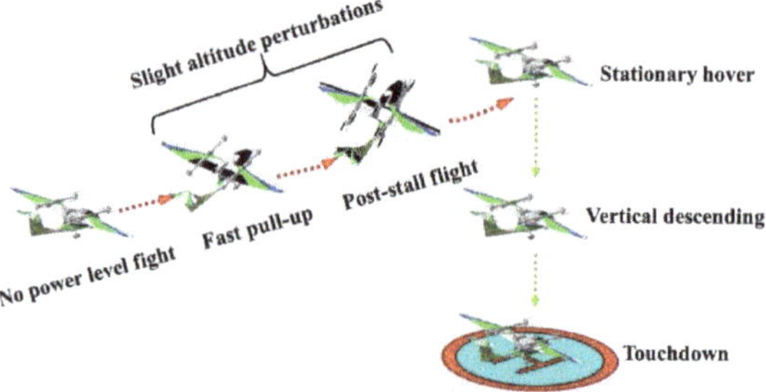

Figure 8. High AOA post-stall maneuver deceleration and landing process.

Another idea is to bank the ET120 vehicle to cleverly guide the large lift caused by pitch-up motion sideways, just like the bank-to-turn technology for missile autopilot. This allows the ET120 vehicle to turn the maximum lift plane and project the partial lift (used to balance the gravity) to the longitudinal (vertical) plane to suppress climbing tendency.

Motivated by the above analysis, a five-stage spiral control strategy for a comfortable deceleration transition and landing process is designed for the ET120 vehicle, as shown in Figure 9. The logic begins from a fixed-wing cruise flight at Point A, after thoroughly performing five stages, the ET120 vehicle finally vertically touches down on the landing pad, at Point F. The main five-stage spiral control strategy is as follows:

- Stage I: (red dotted lines)

Stage I is used to get close to the landing pad from far away. At Point A, the ET120 vehicle begins to bank and enter a turn to track the loiter circle centered on the landing pad in fixed-wing flight mode. The turn ends at Point B, when the cross-track error from the current position to the loiter circle is less than 3 m. At this stage, altitude and airspeed maintain constant.

- Stage II: (green dotted lines)

In stage II, the aircraft begins to enter the deceleration transition phase. At point B, the ET120 vehicle turns off the propulsion rotor and enters no-power flight in fixed-wing mode. It also maintains a fixed altitude to increase the AOA and thus increase the drag to reduce airspeed. Meanwhile, a fixed bank angle ϕ_0 is set to spirally turn toward the landing pad. This stage ends at Point C, where the pitch angle reaches the set value θ_0.

- Stage III: (blue dotted lines)

During this stage, the ET120 vehicle is still in mid-airspeed flight, and the efficiency of aero-surface is far greater than the multi-rotors. The multi-rotors are activated, and their throttle is set to 30% of the maximum throttle. Meanwhile, the bank angle is fixed at ϕ_1 to continue turning, and it continues to maintain altitude to reduce airspeed. This stage ends at Point D, when the pitch angle reaches the set value θ_1.

- Stage IV: (yellow dotted lines)

At this stage, the ET120 vehicle is in mid- to low-airspeed flight, and the efficiency of the aero-surface is equal to the multi-rotors. Then, the throttle of the multi-rotors is set to 60% of the maximum throttle. It enters rotor mode, and the altitude is controlled by the multi-rotors. Meanwhile, the aircraft begins to align its nose to the route via attitude motion. This stage ends at Point E, when the airspeed is less than 5 m/s.

- Stage V: (purple dotted lines)

At this stage, the ET120 vehicle engages a stationary hover state in preparation for vertical descent and touchdown. Once the vehicle receives the landing instruction, it commences vertical descent to the landing pad at Point F.

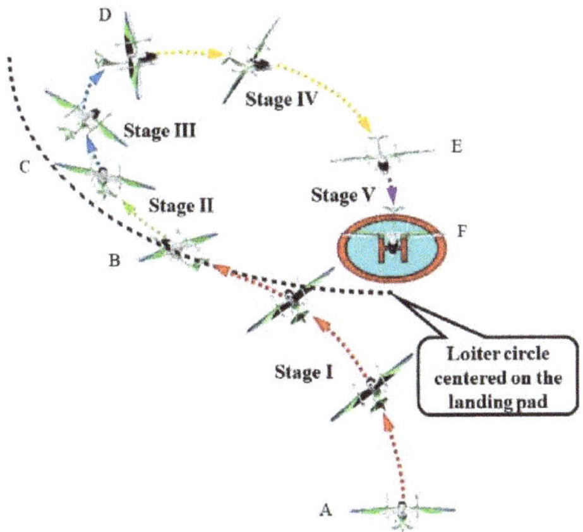

Figure 9. Five stages of the roll-horizon deceleration and landing strategy.

4.2. Control Module Design

- Stage I:

In stage I, it is necessary to design the lateral guidance law to guide the ET120 to the loiter circle in fixed-wing flight mode, as shown in Figure 10. The typical implementation of lateral guidance converts the cross-track error and the track angle error to the acceleration reference. The controller for lateral channel is designed via a cascaded-loop form. The cross-track error, Δy, and the desired track angle reference, ψ_d, are inputs of the lateral channel, and the desired lateral acceleration command, $a_{y,c}$, is the output command.

The desired lateral speed, $\dot{y}_d$, is first obtained from the cross-track error Δy:

$$\dot{y}_d = k_{Py}\Delta y \tag{44}$$

where k_{Py} is the proportional gain, and Equation (44) expects the cross-track error to converge linearly according to the time constant $1/k_{py}$.

Then, the desired track angle caused by lateral motion, $\Delta\psi_c$, can be converted as:

$$\Delta\psi_c = \sin^{-1}(\dot{y}_d/V_g) \tag{45}$$

where V_g is groundspeed. $\Delta\psi_c$ is usually considered to be small, and can be approximated as $\Delta\psi_c = \dot{y}_d/V_g$, which is also limited to $[-\pi/2, \pi/2]$. ψ_c is combined with the desired track angle reference, ψ_d, to form a track angle command ψ_c.

$$\psi_c = \psi_d + \Delta\psi_c \approx \psi_d + \dot{y}_d/V_g \tag{46}$$

Then, the lateral acceleration command a_{yc} can be constructed as:

$$a_{yc} = k_{P\psi}(\psi_c - \psi)V_g \tag{47}$$

where $k_{P\psi}$ is proportional gain and $(\psi_c - \psi)$ should be limited to $[-\pi, \pi]$.

With a coordinated turn assumption, the lateral acceleration command a_{yc} can be converted to a roll angle command ϕ_c.

$$\phi_c = \tan^{-1}(a_{yc}/g) \tag{48}$$

where g is acceleration of gravity.

In the roll angle controller, ϕ is controlled by a proportional control:

$$\dot{\phi}_c = k_{P\phi}(\phi_c - \phi) \tag{49}$$

where $k_{P\phi}$ is proportional gain.

Then, the relationship of Euler angular rates to body angular rates are constructed as:

$$\begin{cases} \dot{\phi} = p + \tan\theta(q\sin\phi + r\cos\phi) \\ \dot{\theta} = q\cos\phi - r\sin\phi \\ \dot{\psi} = (q\sin\phi + r\cos\phi)/\cos\theta \end{cases} \tag{50}$$

From Equation (50), the desired roll angular rate p_c can be calculated as

$$p_c = \dot{\phi}_c - \tan\theta(q\sin\phi + r\cos\phi) \tag{51}$$

For the yaw channel, a yaw damper is designed to improve the damping features of the Dutch roll. In addition, a high-pass filter is also added to weaken the steady yaw angular rate signal by the stable loiter. The controller is constructed as:

$$\delta_r = k_{Pr}\frac{\tau s}{\tau s + 1} \tag{52}$$

where k_{Pr} is proportional gain, and τ is time constant of high pass filter.

In the altitude channel, the fixed-wing implementations of the altitude motion are achieved via the path angle change. The inputs for the altitude control channel are the altitude h, the climbing rate $\dot{h}$, and the desired path climbing rate reference $\dot{h}_d$. The output is the path angle command γ_c.

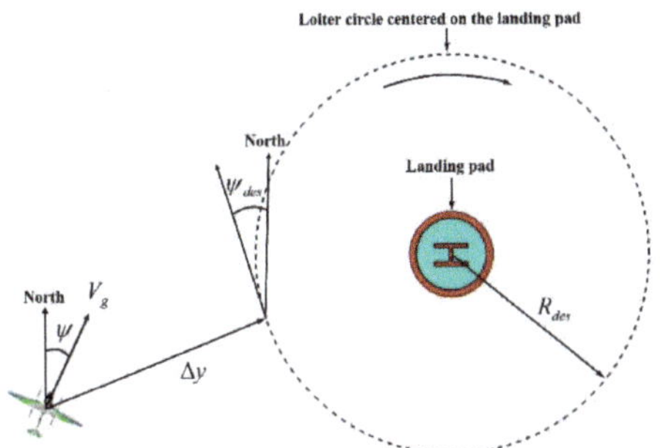

Figure 10. The lateral guidance principle in fixed-wing mode.

As in the design of the lateral channel, the desired climbing rate command is constructed as:

$$\dot{h}_c = k_{Ph}(h_d - h) + \dot{h}_d \tag{53}$$

where k_{Ph} is proportional gain. It is the converted to path angle command γ_c:

$$\gamma_c = \tan^{-1}\theta\left(\dot{h}_c / V_g\right) \tag{54}$$

Then, pitch angle control is achieved via a proportional-integral controller, organized as:

$$\theta_c = k_{P\gamma}(\gamma_c - \gamma) + k_{I\gamma}\int_0^t(\gamma_c - \gamma)d\tau \tag{55}$$

where $k_{P\gamma}$ is proportional gain; $k_{I\gamma}$ is integral gain.

Similarly to the roll channel, θ is also controlled via a proportional controller:

$$\dot{\theta}_c = k_{P\theta}(\theta_c - \theta) \tag{56}$$

where $k_{P\theta}$ is proportional gain; $\dot{\theta}_c$ is pitch angular rate command.

From Equation (50), the desired pitch angular rate q_c can be calculated as:

$$q_c = \frac{\dot{\theta}_c}{\cos\phi} + r\tan\phi \tag{57}$$

For the airspeed channel, the desired airspeed reference V_{td} is controlled via a proportional-integral controller. The input is airspeed V_t and the output is throttle command δ_t.

$$\delta_t = k_{Pv}(V_{td} - V_t) + k_{Iv}\int(V_{td} - V_t)dt \tag{58}$$

where k_{Pv} is proportional gain and k_{Iv} is integral gain.

In addition, the hover rotor speed command n_h is set to zero. The above commands together form $U = [p_c, q_c, \delta_r, \delta_t, n_h = 0]$ for the reference input of the angular rate loop.

- Stage II:

In stage II, the ET120 enters fixed-wing gliding flight mode. The ET120 turns off the propulsion rotor and maintains a fixed bank angle, ϕ_0, at the end of Stage I. In general, the altitude channel is similar as that in Stage I. In addition, the hovering and propulsion rotors are set zero. By combining Equations (51), (52) and (57), p_c, q_c and δ_r are derived in the output U. Then, the new U is organized as $U = [p_c, q_c, \delta_r, \delta_t = 0, n_h = 0]$ for the reference input of the angular rate loop.

- Stage III:

In Stage III, the ET120 reaches the set pitch angle θ_0 at the end of Stage II. Correspondingly, the airspeed is reduced to mid speed. At this stage, the control module is similar to that in Stage II; in addition, an open-loop hovering rotor system is applied.

The hovering rotor command is set to $\delta_t = 30\%$ and the roll angle is fixed at $\phi = \phi_0$. By combining Equations (51), (52) and (57), p_c, q_c and δ_r are derived in the output U. Then, the new U is organized as $U = [p_c, q_c, \delta_r, \delta_t = 0, n_h = 30\%]$ for the reference input of the angular rate loop.

- Stage IV:

In Stage IV, the ET120 reaches the set pitch angle θ_1 at the end of Stage III. In addition, it is in mid- to low-airspeed flight. The hovering rotor system gradually occupies the dominant position of the control capability. The autopilot puts ET120 into rotor mode.

The control of forward speed is achieved by body pitch, and control of altitude is performed via total hovering rotor thrust. In addition, the roll angle is fixed at $\phi = \phi_1$. For the design of altitude channel, the desired climbing rate command, $\dot{h}_c$, is first obtained from Equation (53). It is then used to derive the vertical acceleration command a_{hc} via a proportional controller.

$$a_{hc} = k_{Pa_h}\left(\dot{h}_c - \dot{h}\right) \tag{59}$$

where k_{Pa_h} is a proportional gain.

The inputs for forward speed channel are groundspeed V_g and its command V_{gc}, and the output is θ_c via a Proportional-Integral controller. The forward acceleration is first calculated from forward speed error:

$$a_{xc} = k_{PV_g}\left(V_{gc} - V_g\right) + k_{IV_g}\int_0^t\left(V_{gc} - V_g\right)d\tau \tag{60}$$

where k_{PV_g} is a proportional gain, and k_{IV_g} is an integral gain.

The a_{xc} is usually small and the approximate relationship between a_{xc} and θ_c can be calculated as:

$$\theta_c \approx -\frac{a_{xc}}{g} \tag{61}$$

The control of θ_c is mentioned in Stage 1.

In addition, the roll angle is fixed at $\phi = \phi_1$. To control yaw channel, a proportional controller is applied. The heading angle can be derived as:

$$\dot{\psi}_c = k_{P\varphi}\left(\tan^{-1}(V_e/V_n) - \varphi\right) \tag{62}$$

where $k_{P\varphi}$ is a proportional controller.

From Equation (50), the yaw rate command can be calculated as:

$$r_c = \frac{\dot{\psi}_c\cos\theta - q\sin\phi}{\cos\phi} \tag{63}$$

where r_c is then inserted into $U = [p_c, q_c, r_c, \delta_t, n_h]$ for output to the reference input of the angular rate loop.

The hovering rotors are set 60%. By combining Equations (51) and (57), p_c and q_c are derived in the output U. Then, the new U is organized as $U = [p_c, q_c, r_r, \delta_t = 0, n_h]$ for the reference input of the angular rate loop.

- Stage V:

In stage V, the ET120 enters stationary hover mode. The lateral position in this stage should be controlled to within the landing window. Implementation of the guidance algorithm is depicted in Figure 11. Control of forward position is driven by body pitch, and control of lateral position is achieved via body roll. In addition, its heading aligns with the reference route via body yaw.

In the control module, the desired lateral speed, $\dot{y}_d$, and forward speed, $\dot{x}_d$ are first acquired from cross-track error, Δy and forward distance error Δx:

$$\begin{cases} \dot{y}_d = k_{Py}\Delta y \\ \dot{x}_d = k_{Px}\Delta x \end{cases} \tag{64}$$

where k_{Px} is proportional gain.

Then, we project the $\dot{y}_d$, $\dot{x}_d$ and ground speed into the ET120's direction of movement to acquire forward speed error, $\Delta \dot{x}$ and lateral speed error, $\Delta \dot{y}$:

$$\begin{cases} \Delta \dot{x} = \dot{x}_d \cos(\psi - \psi_d) + \dot{y}_d \sin(\psi - \psi_d) - V_n \cos(\psi) - V_e \sin(\psi) \\ \Delta \dot{y} = (\dot{y}_d)\sin(\psi - \psi_d) + (\dot{y}_d)\cos(\psi - \psi_d) - V_e \cos(\psi) + V_n \sin(\psi) \end{cases} \tag{65}$$

These are then used to derive forward and lateral acceleration commands, a_{xc} and a_{yc}, respectively:

$$\begin{cases} a_{xc} = k_{P\dot{x}}(\Delta \dot{x}) + k_{I\dot{x}} \int_0^t (\Delta \dot{x})d\tau \\ a_{yc} = k_{P\dot{y}}(\Delta \dot{y}) + k_{I\dot{y}} \int_0^t (\Delta \dot{y})d\tau \end{cases} \tag{66}$$

where $k_{P\dot{x}}$ and $k_{P\dot{y}}$ are proportional gains. $k_{I\dot{x}}$ and $k_{I\dot{y}}$ are integral gains.

The a_{xc} and a_{yc} are usually limited to be numerically small, and the approximate relationship between $a_{xc}(a_{yc})$ and $\theta_c(\phi_c)$ can be calculated as:

$$\begin{cases} \theta_c \approx -\dfrac{a_{xc}}{g} \\ \phi_c \approx \dfrac{a_{yc}}{g} \end{cases} \tag{67}$$

The control of θ_c and ϕ_c are mentioned in Stage 1.

To control yaw channel, a proportional controller is applied. The heading angle can be derived as:

$$\dot{\psi}_c = k_{P\psi}(\psi_c - \psi) \tag{68}$$

From Equation (50), the yaw rate command can be calculated as:

$$r_c = \frac{\dot{\psi}_c \cos\theta - q\sin\phi}{\cos\phi} \tag{69}$$

r_c is then inserted into $U = [p_c, q_c, r_c, \delta_t, n_h]$ for output to the reference input of the baseline L_1 adaptive controllers.

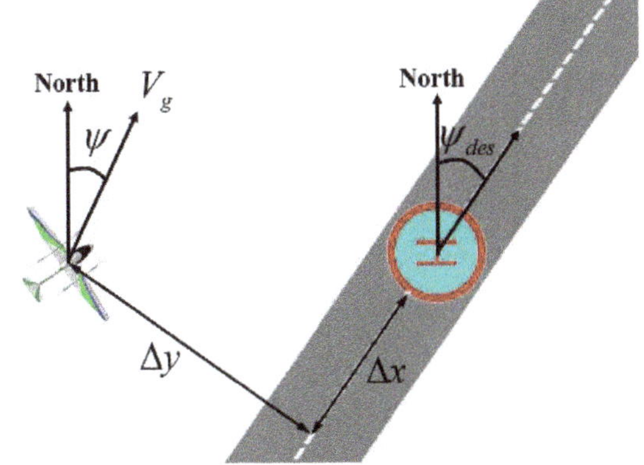

Figure 11. The lateral guidance principle in multi-rotor mode.

The control logic of the roll-horizon deceleration process is illustrated in Figure 12, mainly describing the guidance, control mode and path plans.

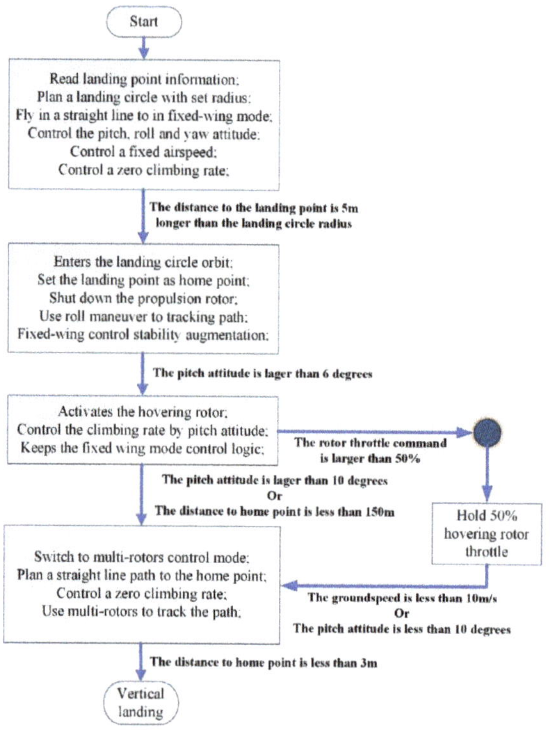

Figure 12. The roll-horizon deceleration logic.

The complete control scheme for deceleration transition and landing process is illustratively depicted in Figure 13. It includes the L_1 angular rate adaptive controller and the corresponding control modules designed for the five stages of the deceleration landing process, which will be discussed in detail in the following sections. In Figure 13, each control module is marked with content-related sections and equations of the control algorithm.

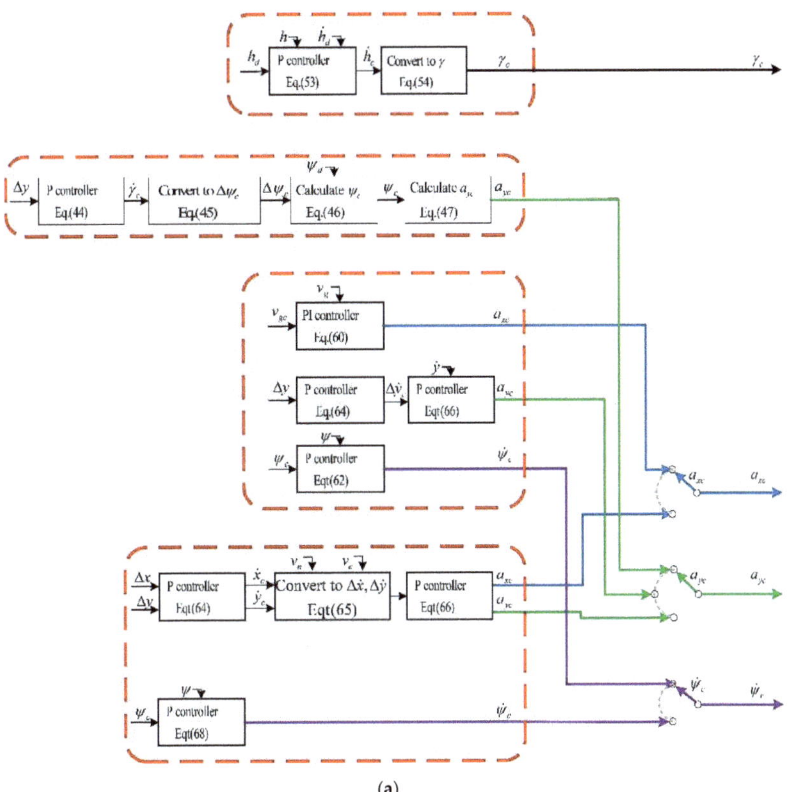

(a)

Figure 13. *Cont.*

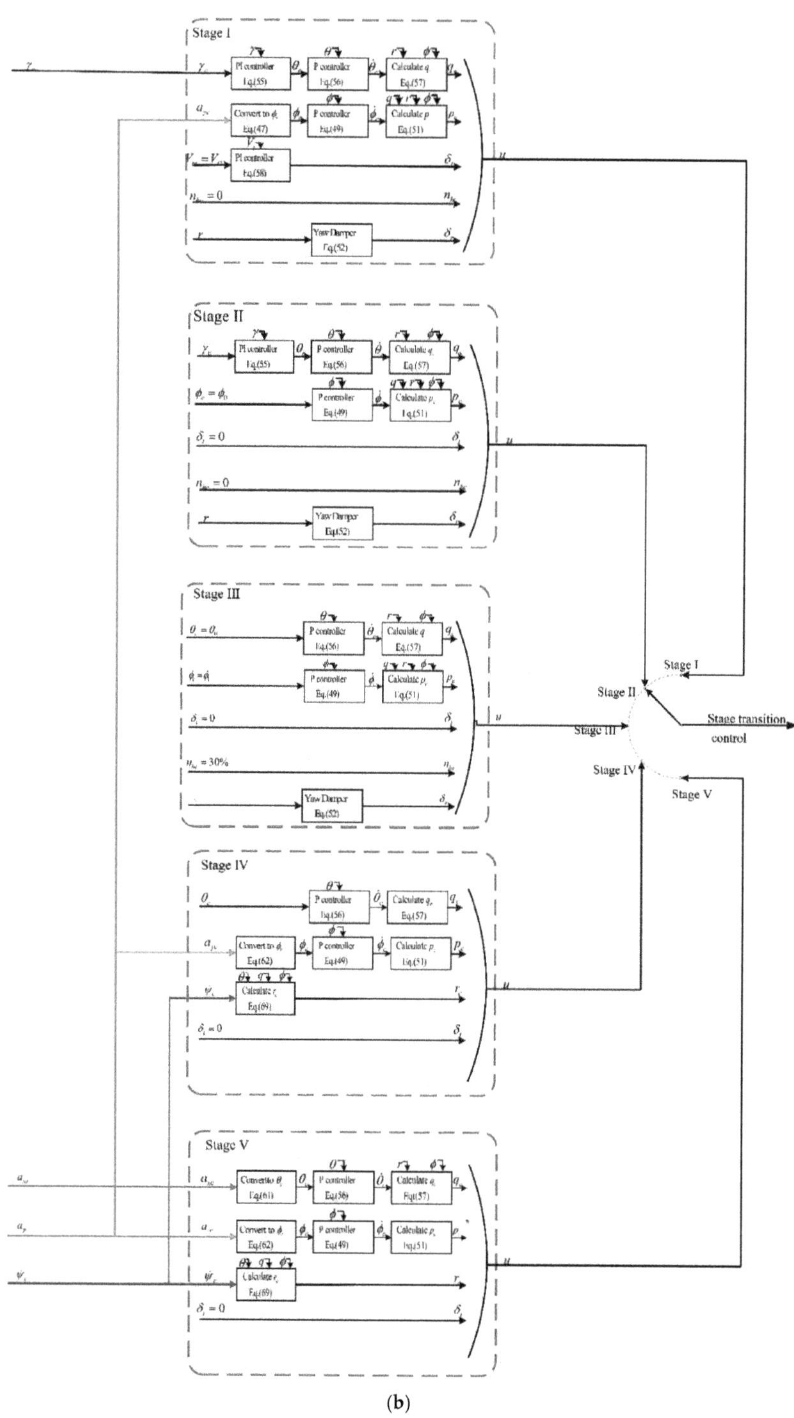

(b)

Figure 13. The control logic of the five-stage maneuver. (**a**) Guidance layer; (**b**) Strategy maneuver logic.

5. Simulation and Verification

5.1. Monte Carlo Simulations of Angular Rate L_1 Controller

The virtual control coefficient and virtual state coefficient contain the estimation of the model coefficients, influencing the tracking performance of the controller. Usually, the values of these factors are set empirically, which requires verification. This section takes the roll channel as an example to verify the parameter settings. The parameters to verify are the roll rate virtual control coefficient ω_p and virtual state coefficient K_p.

A 2 rad/s frequent square roll rate command was given as the input to an L_1 controller with $\Gamma = 2500$ and $D = 1/s$. The results for different combinations of ω_p and K_p are given in Figures 14 and 15. The numerical performance results are given in Table 4.

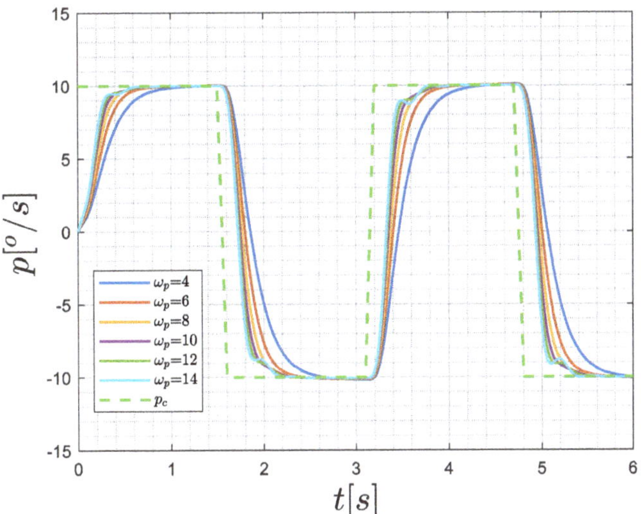

Figure 14. Time histories of roll rate with $K_p = 8$.

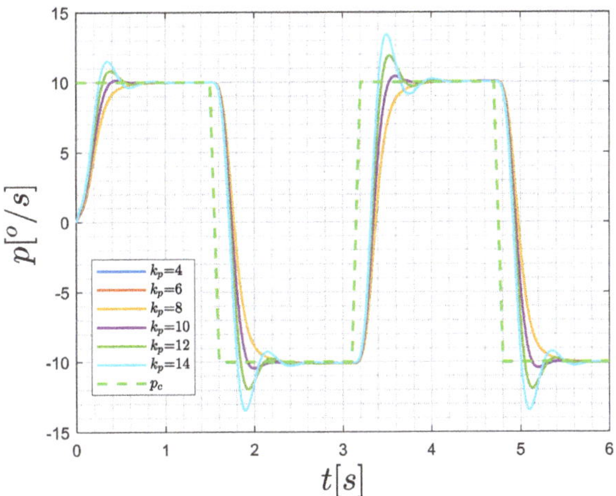

Figure 15. Time histories of roll rate with $\omega_p = 8$.

Table 4. The performance of different L_1 parameters.

$[K_p, \omega_p]$	Risetime	Overshoot
[4,8]	1.17	-
[6,8]	0.72	-
[8,8]	0.65	-
[10,8]	0.63	-
[12,8]	0.62	-
[14,8]	0.6	-
[8,4]	1.14	-
[8,6]	0.78	-
[8,10]	0.63	1.5%
[8,12]	0.76	8%
[8,14]	0.85	15%

We chose $K_p = 8$, $\omega_p = 8$ as the values for the virtual control coefficient and virtual state coefficient, as these values make a tradeoff between rapidity and stability, with a small risetime and zero overshoots.

5.2. Monte Carlo Simulations of Angular Rate L_1 Controller

To verify the performance of the L_1 controller, a Monte Carlo simulation was executed. The modeling parameters that influence the control efficiency and trimming states were subjected to perturbation, because these factors are the most sensitive to the performance of control systems. The perturbation parameters are given in Table 5.

Table 5. The perturbation parameters.

Parameters	Perturbations
$C_{m_{\delta_e}}$	$\pm 20\%$
C_{m_α}	$\pm 20\%$
$C_{L_{\delta_e}}$	$\pm 20\%$
C_{S_β}	$\pm 20\%$
$C_{S_{\delta_r}}$	$\pm 20\%$
C_{l_β}	$\pm 20\%$
$C_{l_{\delta_a}}$	$\pm 20\%$
C_{l_p}	$\pm 50\%$
C_{l_r}	$\pm 50\%$
C_{n_β}	$\pm 20\%$
$C_{n_{\delta_r}}$	$\pm 20\%$
C_{n_p}	$\pm 50\%$
C_{n_r}	$\pm 50\%$
CG_x	± 0.3 m
J_{xx}	$\pm 20\%$
J_{yy}	$\pm 20\%$
J_{zz}	$\pm 20\%$

1. percentages mean multiplication gain; 2. decimals mean addition value.

In Table 5, $C_{L_{\delta_e}}$, $C_{m_{\delta_e}}$, $C_{S_{\delta_r}}$, $C_{n_{\delta_r}}$, $C_{l_{\delta_a}}$ are the control derivatives, C_{m_α}, C_{S_β}, C_{l_β}, C_{n_β} are the stability derivatives, C_{l_p}, C_{n_r} are the damping derivatives, C_{l_r}, C_{n_p} are the cross damping derivatives, CG_x is the center of gravity position in body x-axis, and J_{xx}, J_{yy}, J_{zz} are the inertia moments.

The simulation states of the three flight modes are given as:

- Fixed-wing: airspeed 35 m/s.
- Transition: airspeed 10 m/s.
- Multi-rotor: airspeed 0 m/s.

The tracking performances are tested with continuous reversed step commands:

- Roll angle command: $0°$ at 0 s, $25°$ at 5 s, $-25°$ at 12 s, $0°$ at 20 s.

- Pitch angle command: $0°$ at 0 s, $10°$ at 5 s, $-10°$ at 12 s, $0°$ at 20 s.

The simulation results of the three flight modes are depicted in Figures 16–18.

The results show that (1) the distribution of the attitude angles is narrow, which indicates a strong robustness of the L_1 controller, which is able to reject the perturbation listed above; (2) the tracking performance of the L_1 controller is excellent, with a steady state error of zero and a trivial time latency.

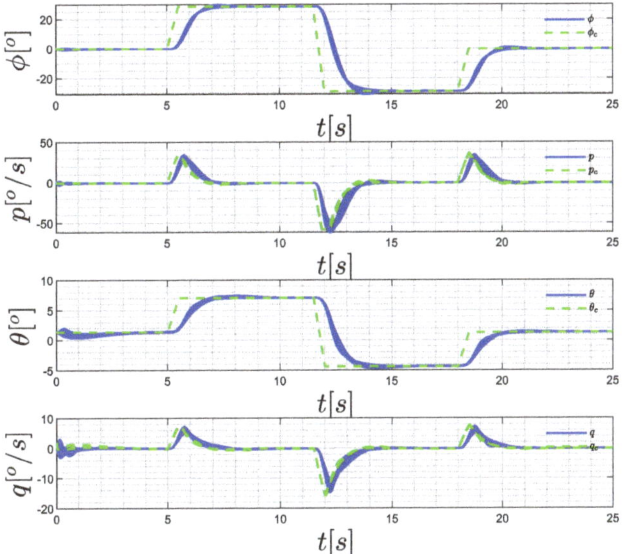

Figure 16. Time histories of longitudinal and lateral flight parameters in fixed–wing mode.

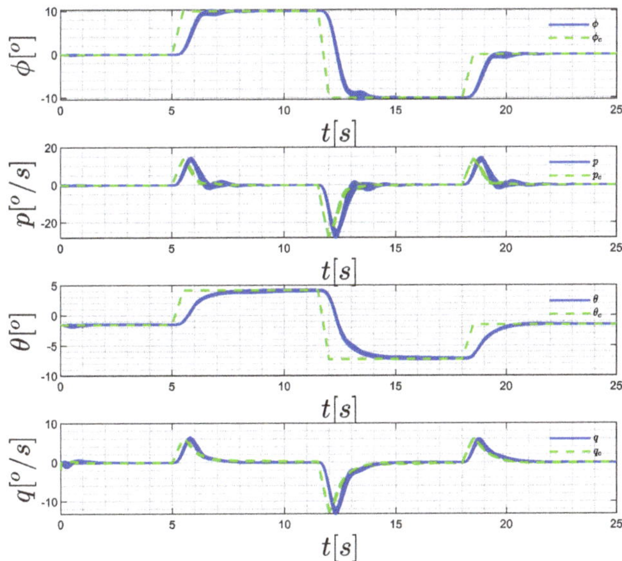

Figure 17. Time histories of longitudinal and lateral flight parameters in transitional mode.

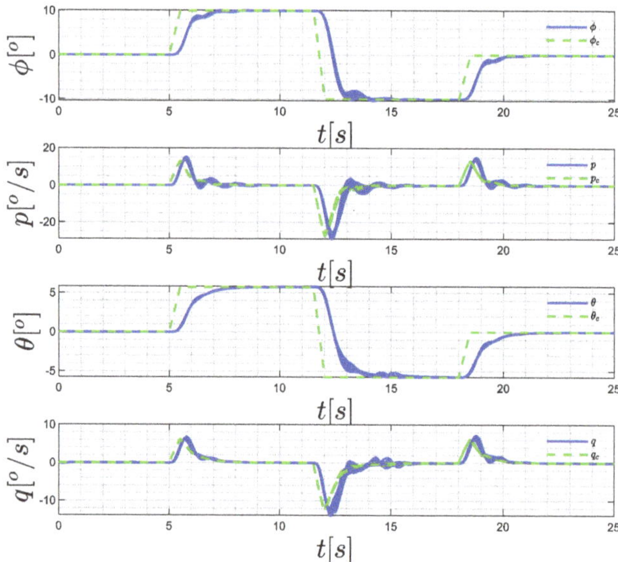

Figure 18. Time histories of longitudinal and lateral flight parameters in multi–rotor mode.

5.3. Simulations for Flight Path Verification

In practical applications, an eVTOL may enter vertiports in different initial states, namely with respect to cruise speed and landing circle radius. A feasible deceleration strategy should constrain the endings of the deceleration phase in the landing window for vertical landing in any possible initial states. To verify this, possible ranges of cruise speed and landing circle radius were set for the simulation as:

- Cruise airspeed: 30 m/s~50 m/s.
- Landing circle radius: 240 m~300 m.

The horizontal flight path results are given in Figure 19. It can be seen that the endings of all curves are distributed in a limited area, located in the landing window. These excellent results ensure a precise landing for vertiport management.

5.4. Monte Carlo Simulations of Rolling-Horizon Deceleration and Landing Strategy

Monte Carlo simulations are a common practice for robustness performance verification. The stability derivatives, control derivatives and damping derivatives are sensitive to the baseline angular rate control, but count for little in trajectory planning, while the basic values of the aerodynamic forces influence the deceleration efficiency and climbing rate control, which are sensitive to the trajectory results of the strategy; only the basic values of the aerodynamic forces C_{L_0}, C_{D_0}, C_{S_0} are perturbed in this simulation. The perturbation parameters are given in Table 6.

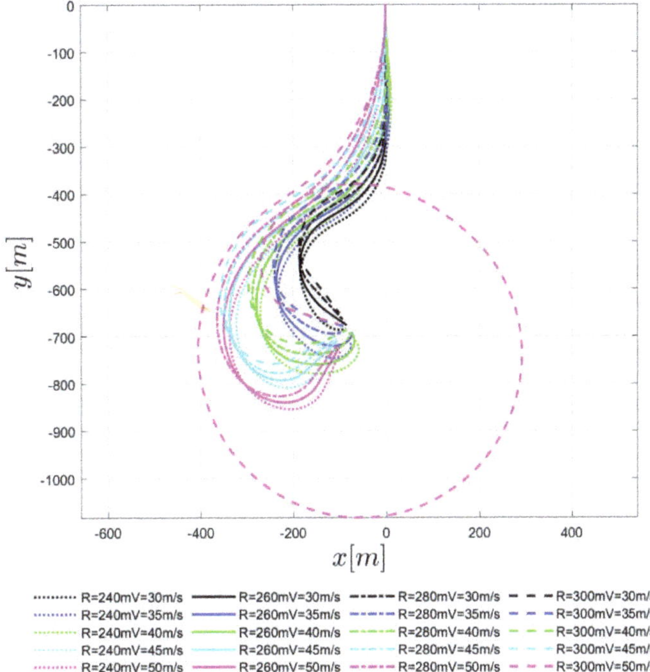

Figure 19. Trajectory in horizontal planes in different initial states.

Table 6. Perturbation parameters.

Parameters	Perturbations
C_{L_0}	±20%
C_{D_0}	±20%
C_{S_0}	±0.6

1. percentages mean multiplication gain; 2. decimals mean addition value.

The initial state of the Monte Carlo simulation is set as:

- Altitude: 50 m.
- Airspeed: 30 m/s.
- Pitch angle: 0°.
- Roll angle: 0°.
- Radius of landing circle: 268 m.
- Distance to landing point: 680 m.

The simulation results are given in Figures 20–22, where the black lines, blue lines, canyon lines and pinkish red lines mark the fixed-wing flight phase, transitional flight phase, multi-rotors flight phase, and vertical landing phase, respectively, and the red line marks the nominal state result.

Figure 20 shows the three-dimensional trajectory results, where the green circle is the landing circle orbit, the green point is the landing point and canyon point is the landing window. The results show:

- The deceleration and landing process have no altitude surging, and the trajectory is controlled to be narrowly distributed inside the landing circle. The intention to avoid climbing is achieved, and the robust performance of the strategy is excellent.

- The end of the deceleration phase (at the end of the canyon lines) is controlled inside the landing window. Therefore, the strategy ensures a precise landing point, enabling its practical use in vertiport management.

  Figures 21 and 22 give the time history of the flight parameters. The results show:

- In the Stage I maneuver (black lines before 8 s), the strategy firstly adjusts the attitudes of the flight to enter a straight-line level flight in fixed-wing mode. This phase has narrowly distributed endings, owing to the fact that the airspeed is fixed until the flight approaches the landing circle.
- In the Stage II maneuver (black lines after 8 s), an abrupt change in the bank angle and pitching angle is observed, owing to the fact that the flight uses the bank maneuver to enter the landing circle orbit and tries to maintain a stable altitude using the pitching angle.
- In the stage III maneuver (blue lines), the attitudes of the aircraft exhibit another abrupt change. During this phase, the guidance logic takes the landing point as the home point to plan a new trajectory, and the flight has to bank to the other side while varying the pitching angle to maintain a stable altitude. Additionally, the hovering rotors are activated in this phase, as shown in Figure 22. The airspeed begins to decline to 20 m/s.
- In the stage IV maneuver (canyon lines), the bank angle settles at $-25°$ and the pitch angle increases monotonically with the decrease in airspeed. During this phase, the lateral attitude is stable, and the partial lift to maintain longitudinal balance is fixed. At mid airspeed, the flight uses the aero-surfaces to provide lift (before 28 s), and then at low airspeed, the hovering rotors are used.
- In the stage V maneuver (pinkish red lines), the altitude decline is obvious, with a low airspeed and wide-ranging angle of attack. During this phase, the flight is landing vertically, and the angle of attack is insignificant, as the flight is thrust-driven.

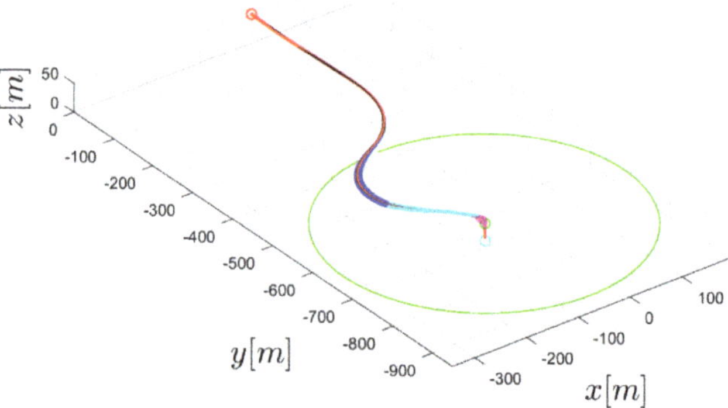

Figure 20. Monte Carlo trajectory results.

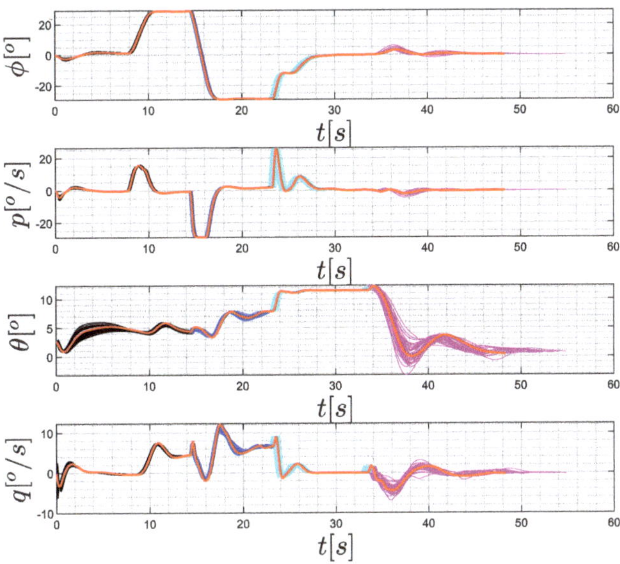

Figure 21. Time histories of longitudinal and lateral flight parameters of the roll-horizon strategy.

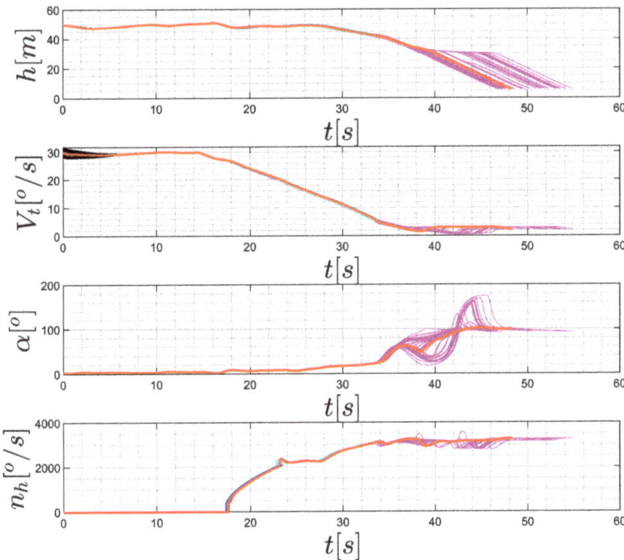

Figure 22. Time histories of altitude, airspeed, angle of attack and rotor speed of the roll-horizon strategy.

5.5. Comparison with Jumping Deceleration and Landing Strategy

The deceleration and landing strategy is proposed to achieve higher efficiency and driving comfort. A comparison with the conventional jumping deceleration strategy, whose trajectory is given in Figure 23, is carried out to verify the energy efficiency.

Figure 24 presents the altitude and velocity channel comparison results. Clearly, the roll-horizon landing strategy has a slight altitude variation, which controls the altitude to within 40 m to 50 m, while the conventional strategy has a large altitude range that covers

40 m to 70 m. In the velocity channel, we found that the time taken to decelerate to 5 m/s (which is the hovering speed) is almost the same (that is, around 19 s), which means the time efficiency for deceleration of the two strategies is equally matched. However, when looking at the slopes of the curves, it can be seen that the slope of the roll-horizon curve is basically unchanged, which indicates a smooth deceleration phase, while the jumping strategy curve has a larger deceleration rate before the altitude inflection point (at about 9 s), which makes the deceleration process less comfortable. In the angle of attack diagram, the conventional jumping strategy has an angle ranging from 0° to 40°, while the roll-horizon strategy angle range is from 0° to 20°, the large angle to 40° is close to the control inability region, which is unacceptable for manned flight.

Figure 25 presents the work and power consumed by the hovering rotors, where P means power and W means work. Obviously, the hovering rotor power of the roll-horizon strategy is lower than that of the jumping strategy throughout the deceleration process. This means a lower workload for the hovering rotors, which is friendly to the rotor life span. Additionally, the lower power means less hovering rotor work, which is depicted more directly in the work diagram. The total work consumed by the roll-horizon strategy is almost half that of the jumping strategy, which is a considerable advantage in terms of energy and economical efficiency.

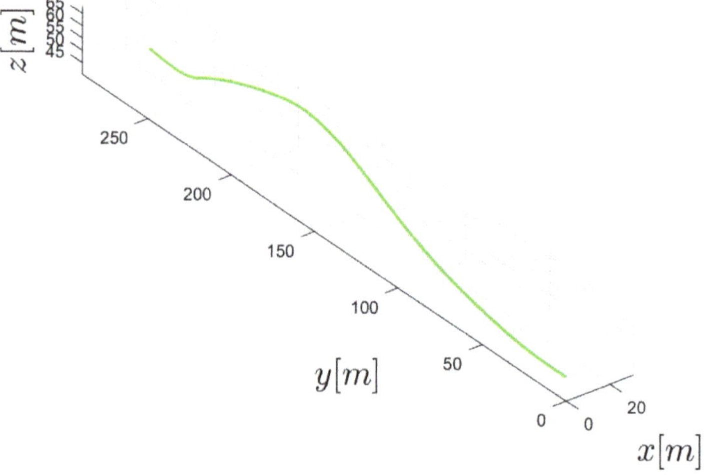

Figure 23. The trajectory of the jumping strategy.

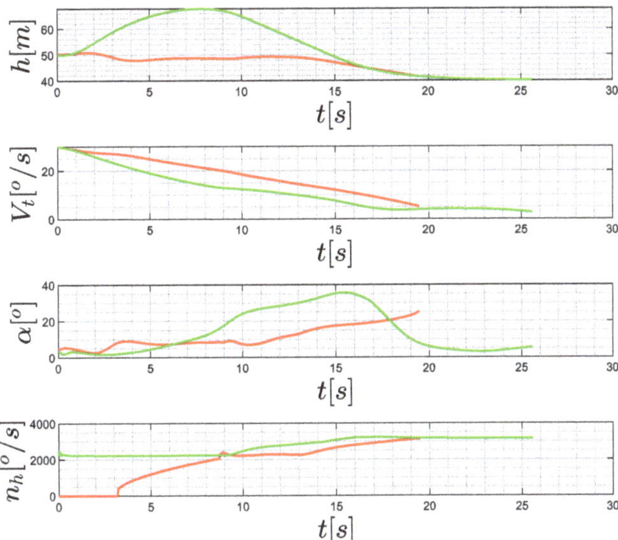

Figure 24. Altitude and velocity channel comparison.

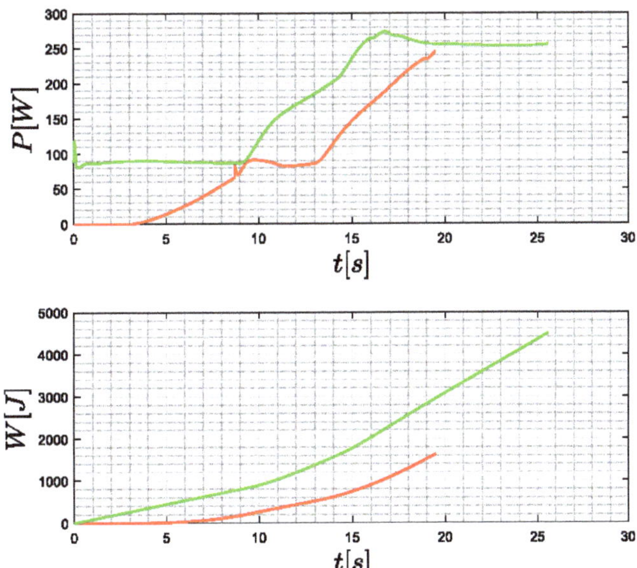

Figure 25. Altitude and velocity channel comparison.

A quantitative comparison is given in Table 7.

Table 7. The comparison results.

Indexes	Roll-Horizon	Jumping
Time to deceleration to 5 m/s	19.4 s	16.5 s
Time to descend to 40 m	19.8 s	21 s
The maximum Altitude	50.67 m	67.8 m
The maximum rotor power	244.7 W	273.9 W
The total rotor work	1613 J	4484 J
The maximum AOA	25.6°	35.65°

It can be found that:

- The time efficiency of the two strategies are similar, where roll-horizon is more efficient at descending while jumping is more efficient at decelerating.
- The roll-horizon strategy successfully avoids climbing during the landing process.
- The power requirement for the hovering rotors of the roll-horizon strategy is less than that of the jumping strategy.
- The total energy consumption of the roll-horizon strategy is less than half that of the jumping strategy.
- The maximum angle of attack of the roll-horizon strategy is lower and appears in the multi-rotor mode.

6. Conclusions

In this work, an L_1 adaptive controller was designed to control the ET120, an eVTOL with complicated flight dynamic characteristics. Furthermore, a roll-horizon deceleration and vertical landing strategy was presented for an improved driving experience and to promote energy efficiency during manned flight. Monte Carlo simulations and comparison simulations were carried out to verify the performance of the control system and the efficiency of the roll-horizon deceleration strategy. The results show that the L_1 adaptive controller-based control system is robust enough to reject at least 20% of perturbation on all modeling parameters. The guidance logic is reliable for completing the maneuvers designed in this strategy and guarantee a safe and bounded deceleration and landing path. The promoted strategy has a smoothly varying airspeed curve, resulting in a more comfortable manned flight, and has a superior energy efficiency, which is able to reduce the hovering rotor work by 64%. Additionally, the strategy avoids dangerous attitudes that may cause the flight to go out of control.

Author Contributions: Conceptualization, Z.W. and S.M.; methodology, Z.W. and Z.G.; software, S.M. and Z.G.; validation, Z.W., S.M. and J.H.; formal analysis, C.Z.; investigation, S.M.; resources, C.Z., J.H. and Z.G.; data curation, S.M.; writing—original draft preparation, S.M.; writing—review and editing, Z.W. and S.M.; visualization, S.M.; supervision, Z.G.; project administration, Z.G.; funding acquisition, Z.G. All authors have read and agreed to the published version of the manuscript.

Funding: This research received no external funding.

Conflicts of Interest: The authors declare no conflict of interest.

References

1. Kim, H.D.; Perry, A.T.; Ansell, P.J. A Review of Distributed Electric Propulsion Concepts for Air Vehicle Technology. In Proceedings of the AIAA/IEEE Electric Aircraft Technologies Symposium, Cincinnati, OH, USA, 9–11 July 2018.
2. Wang, Z.; Zheng., G.; Cheng, Y.; Sun, M.; Xu, J. Practical control implementation of tri-tilt rotor flying wing unmanned aerial vehicles based upon active disturbance rejection control. *Proc. IMechE Part G J. Aerosp. Eng.* **2020**, *234*, 943–960. [CrossRef]
3. Chauhan, S.S.; Martins, J.R.R.A. Tilt-Wing eVTOL Takeoff Trajectory Optimization. *J. Aircr.* **2019**, *57*, 1–20. [CrossRef]
4. Zian, W.; Zheng, G. Longitudinal Flight Characteristics Analysis and Flight Control Design for Hybrid Vtol UAV in Accelerative Transition. *J. Aerosp. Power* **2019**, *34*, 2177–2190.
5. Liu, N.; Cai, Z.; Wang, Y.; Zhao, J. Fast Level-flight to Hover Mode Transition and Altitude Control in Tiltrotor's Landing Operation. *CJA* **2021**, *34*, 181–193. [CrossRef]

6. Daskilewicz, M.; German, B.; Warren, M.; Garrow, L.A.; Boddupalli, S.S.; Douthat, T.H. Progress in Vertiport Placement and Estimating Aircraft Range Requirements for eVTOL Daily Commuting. In Proceedings of the 2018 Aviation Technology, Integration, and Operations Conference, Atlanta, GA, USA, 25–29 June 2018.
7. Yi, L.U.; Zhang, S.; Zhang, Z.; Zhang, X.; Peng, T.A.N.G.; Shan, F.U. Multiple hierarchy risk assessment with hybrid model for safety enhancing of unmanned subscale BWB demonstrator flight test. *Chin. J. Aeronaut.* **2019**, *32*, 2612–2626.
8. Jiang, G.; Liu, G.; Yu, H. A Model Free Adaptive Scheme for Integrated Control of Civil Aircraft Trajectory and Attitude. *Symmetry* **2021**, *13*, 347. [CrossRef]
9. Kleinbekman, I.C.; Mitici, M.; Wei, P. Rolling-horizon Electric Vertical Takeoff and Landing Arrival Scheduling for On-demand Urban Air Mobility. *J. Aerosp. Inf. Syst.* **2020**, *17*, 150–159. [CrossRef]
10. Ayala, V.; Torreblanca, M.; Valdivia, W. Toward Applications of Linear Control Systems on the Real World and Theoretical Challenges. *Symmetry* **2021**, *13*, 167. [CrossRef]
11. Saeed, N.A.; Mahrous Awwad, E.; Abdelhamid, T.; El-Meligy, M.A.; Sharaf, M. Adaptive Versus Conventional Positive Position Feedback Controller to Suppress a Nonlinear System Vibrations. *Symmetry* **2021**, *13*, 255. [CrossRef]
12. Leman, T.; Xargay, E.; Dullerud, G.; Hovakimyan, N.; Wendel, T. L_1 Adaptive Control Augmentation System for the X-48b AircrafAIA Guidance. In Proceedings of the Navigation, and Control Conference, Chicago, IL, USA, 10–13 August 2009; AIAA, Urbana-Champaign: Champaign, IL, USA, 2009.
13. Swarnkar, S.; Parwana, H.; Kothari, M.; Abhishek, A. Biplane-quadrotor Tail-sitter Uav: Flight Dynamics and Control. *J. Guid. Control Dyn.* **2018**, *41*, 1049–1067. [CrossRef]
14. Nigam, N.; Bieniawski, S.; Kroo, I.; Vian, J. Control of Multiple UAVs for Persistent Surveillance: Algorithmand Flight Test Results. *IEEE Trans. Control Syst. Technol.* **2012**, *20*, 1236–1251. [CrossRef]
15. Bordignon, K.; Bessolo, J. Control Allocation for the X-35b. In Proceedings of the 2002 Biennial International Powered Lift Conference and Exhibit, Williamsburg, VA, USA, 5–7 November 2002.
16. Chang, J.; Zhu, J.; Liu, R.; Dong, W. Lateral Control for UItra-low Altitude Airdrop Based on the Ci Adaptive ControlAugmentation. *Int. J. Control Autom. Syst.* **2018**, *16*, 461–477. [CrossRef]
17. Calise, A.J.; Rysdyk, R.T. Nonlinear adaptive flight control using neural networks. *Control Syst. Mag. IEEE* **1998**, *18*, 14–25.
18. Kamel, M.; Burri, M.; Siegwart, R. Linear vs nonlinear MPC for trajectory tracking applied to rotary wing micro aerial vehicles. *IFAC-PapersOnLine* **2017**, *50*, 3463–3469. [CrossRef]
19. Gu, Y.; Seanor, B.; Campa, G.; Napolitano, M.R.; Rowe, L.; Gururajan, S.; Wan, S. Design and Flight Testing Evaluation of Formation Control Laws. *IEEE Trans. Control Syst. Technol.* **2006**, *14*, 1105–1112. [CrossRef]
20. Woodbury, T.; Valasek, J. Synthesis and Flight Test of Automatic Landing Controller Using Quantitative Feedback Theory. *J. Guid. Control Dyn.* **2016**, *39*, 1994–2010. [CrossRef]
21. Xia, X.; Yang, M.; Chen, G.; Zhang, L.; Hou, J. Transition Flight Control and Simulation of a Novel Tail-Sitter UAV With Varying Fuselage Shape. *IEEE Access* **2021**, *99*, 1. [CrossRef]
22. Zou, Y.; Xia, K. Robust Fault-Tolerant Control for Underactuated Takeoff and Landing UAVs. *IEEE Trans. Aerosp. Electron. Syst.* **2020**, *56*, 3545–3555. [CrossRef]
23. Wang, X.; Zhu, B.; Zhu, J.; Cheng, Z. Thrust vectoring control of vertical/short takeoff and landing aircraft. *Sci. China Inf. Sci.* **2020**, *63*, 19–26. [CrossRef]
24. Zou, Y.; Zhang, H.; He, W. Adaptive Coordinated Formation Control of Heterogeneous Vertical Takeoff and Landing UAVs Subject to Parametric Uncertainties. *IEEE Trans. Cybern.* **2020**, *99*, 1–12. [CrossRef]
25. Betancourt, J.; Castillo, P.; Lozano, R. Stabilization and Tracking Control Algorithms for VTOL Aircraft: Theoretical and Practical Overview. *J. Intell. Robot. Syst.* **2020**, *100*, 1249–1263. [CrossRef]
26. Vajpayee, V.; Becerra, V.; Bausch, N.; Deng, J.; Shimjith, S.R.; Arul, A.J. L1-Adaptive Robust Control Design for a Pressurized Water-type Nuclear Power Plant. *IEEE Trans. Nucl. Sci.* **2021**, *68*, 1381–1398. [CrossRef]
27. Zhao, J.; He, X.; Li, H.; Lu, L. An adaptive optimization algorithm based on clustering analysis for return multi-flight-phase of VTVL reusable launch vehicle. *Acta Astronaut.* **2021**, *183*, 112–125. [CrossRef]
28. Cao, C.; Hovakimyan, N. Design and Analysis of a Novel L1 Adaptive Controller, Part I: Control Signal and Asymptotic Stability. In Proceedings of the 2006 American Control Conference, Minneapolis, MN, USA, 14–16 June 2006; IEEE: Los Alamitos, CA, USA, 2006.
29. Cao, C.; Hovakimyan, N. Design and Analysis of a Novel L1 Adaptive Controller, Part II: Guaranteed. In Proceedings of the Transient Performance. 2006 American Control Conference, Minneapolis, MN, USA, 14–16 June 2006; IEEE: Los Alamitos, CA, USA, 2006.
30. Xargay, E.; Hovakimyan, N.; Dobrokhodov, V.; Statnikov, R.; Kaminer, I.; Cao, C.; Gregory, I. L_1 Adaptive Flight Control System: Systematic Design and Verification and Validation of Control Metrics. In Proceedings of the AIAA Guidance, Navigation, and Control Conference, Toronto, ON, Canada, 2–5 August 2010.
31. Wise, K.; Lavretsky, E.; Hovakimyan, N.; Cao, C.; Wang, J. Verifiable Adaptive Flight Control. In Proceedings of the AIAA Guidance, Navigation, and Control Conference, San Francisco, CA, USA, 15–18 August 2005.

MDPI
St. Alban-Anlage 66
4052 Basel
Switzerland
www.mdpi.com

Symmetry Editorial Office
E-mail: symmetry@mdpi.com
www.mdpi.com/journal/symmetry